Fluid Power with Applications

Seventh Edition

Anthony Esposito

Professor Emeritus
Department of Manufacturing Engineering
Miami University
Oxford, Ohio

PEARSON

Prentice
Hall

Upper Saddle River, New Jersey
Columbus, Ohio

Library of Congress Cataloging-in-Publication Data

Esposito, Anthony
    Fluid power with applications/Anthony Esposito.—7th ed.
        p. cm.
    Includes index.
    ISBN-13: 978-0-13-513690-4
    ISBN-10:   0-13-513690-3
    1. Fluid power technology. I. Title.
TJ843. E86 2009
620.1'06—dc22   2008010859

Editor in Chief: Vernon Anthony
Acquisitions Editor: Eric Krassow
Managing Editor, Development: Christine Buckendahl
Editorial Assistant: Sonya Kottcamp
Production Coordination: S4Carlisle Publishing Services
Project Manager: Alexandrina Benedicto Wolf
Operations Specialist: Laura Weaver
Art Director: Diane Ernsberger
Cover Designer: Jeff Vanik
Director of Marketing: David Gesell
Marketing Manager: Derill Trakalo
Marketing Assistant: Les Roberts

This book was set in Times Roman by S4Carlisle Publishing Services and was printed and bound by Courier Westford, Inc. The cover was printed by Phoenix Color Corp.

Pearson Education Ltd., London
Pearson Education Singapore Pte. Ltd.
Pearson Education Canada, Inc.
Pearson Education—Japan

Pearson Education Australia Pty. Limited
Pearson Education North Asia Ltd. , Hong Kong
Pearson Educación de Mexico, S.A. de C.V.
Pearson Education Malaysia Pte. Ltd.

9 8 7 6 5 4 3 2 1
ISBN-13: 978-0-13-513690-4
ISBN-10:   0-13-513690-3

To my wife, Mary,
and to my grandchildren:
Carly, Chelsea, Kevin, Bryan, Drew,
Chloe, Sophia, Aversa, Logan, and Nicol

Contents

PREFACE   **xv**

1 INTRODUCTION TO FLUID POWER   **1**

Learning Objectives   1
1.1   What Is Fluid Power?   1
1.2   History of Fluid Power   4
1.3   Advantages of Fluid Power   8
1.4   Applications of Fluid Power   11
1.5   Components of a Fluid Power System   15
1.6   The Fluid Power Industry   20
Exercises   21

2 PHYSICAL PROPERTIES OF HYDRAULIC FLUIDS   **23**

Learning Objectives   23
2.1   Introduction   23
2.2   Fluids: Liquids and Gases   25
2.3   Specific Weight, Density, and Specific Gravity   27
2.4   Force, Pressure, and Head   32
2.5   The SI Metric System   38
2.6   Bulk Modulus   42
2.7   Viscosity   42

2.8   Viscosity Index   48
2.9   Automotive Shock Absorbers: Viscosity at Work   51
2.10   Illustrative Examples Using the SI Metric System   53
2.11   Key Equations   54
        Exercises   55

3 ENERGY AND POWER IN HYDRAULIC SYSTEMS   59

        Learning Objectives   59
3.1   Introduction   59
3.2   Review of Mechanics   61
3.3   Multiplication of Force (Pascal's Law)   67
3.4   Applications of Pascal's Law   72
3.5   Hydroforming of Metal Components   78
3.6   Conservation of Energy   82
3.7   The Continuity Equation   84
3.8   Hydraulic Power   86
3.9   Bernoulli's Equation   91
3.10   Torricelli's Theorem   97
3.11   The Siphon   99
3.12   Energy, Power, and Flow Rate in the SI Metric System   101
3.13   Illustrative Examples Using the SI Metric System   103
3.14   Key Equations   106
        Exercises   108

4 FRICTIONAL LOSSES IN HYDRAULIC PIPELINES   117

        Learning Objectives   117
4.1   Introduction   117
4.2   Laminar and Turbulent Flow   119
4.3   Reynolds Number   120
4.4   Darcy's Equation   123
4.5   Frictional Losses in Laminar Flow   124
4.6   Frictional Losses in Turbulent Flow   125
4.7   Losses in Valves and Fittings   129
4.8   Equivalent-Length Technique   132

4.9   Hydraulic Circuit Analysis   136
4.10   Circuit Analysis Using the SI Metric System   138
4.11   Key Equations   142
        Exercises   143

5 HYDRAULIC PUMPS   148

        Learning Objectives   148
5.1   Introduction   149
5.2   Waterjet Cutting   150
5.3   Pumping Theory   151
5.4   Pump Classification   153
5.5   Gear Pumps   156
5.6   Vane Pumps   164
5.7   Piston Pumps   170
5.8   Pump Performance   176
5.9   Pump Noise   184
5.10   Pump Selection   189
5.11   Pump Performance Ratings in Metric Units   190
5.12   Key Equations   193
        Exercises   195

6 HYDRAULIC CYLINDERS AND CUSHIONING DEVICES   199

        Learning Objectives   199
6.1   Introduction   199
6.2   Hydraulic Cylinder Operating Features   201
6.3   Cylinder Mountings and Mechanical Linkages   203
6.4   Cylinder Force, Velocity, and Power   205
6.5   Cylinder Loads Due to Moving of Weights   208
6.6   Special Cylinder Designs   210
6.7   Cylinder Loadings Through Mechanical Linkages   211
6.8   Hydraulic Cylinder Cushions   217
6.9   Hydraulic Shock Absorbers   220
6.10   Key Equations   224
        Exercises   225

7 HYDRAULIC MOTORS   230

Learning Objectives   230
7.1   Introduction   230
7.2   Limited Rotation Hydraulic Motors   231
7.3   Gear Motors   234
7.4   Vane Motors   237
7.5   Piston Motors   238
7.6   Hydraulic Motor Theoretical Torque, Power, and Flow Rate   242
7.7   Hydraulic Motor Performance   246
7.8   Hydrostatic Transmissions   250
7.9   Hydraulic Motor Performance in Metric Units   253
7.10   Key Equations   255
       Exercises   257

8 HYDRAULIC VALVES   261

Learning Objectives   261
8.1   Introduction   262
8.2   Directional Control Valves   262
8.3   Pressure Control Valves   275
8.4   Flow Control Valves   284
8.5   Servo Valves   292
8.6   Proportional Control Valves   295
8.7   Cartridge Valves   295
8.8   Hydraulic Fuses   300
8.9   Key Equations   303
       Exercises   303

9 HYDRAULIC CIRCUIT DESIGN AND ANALYSIS   307

Learning Objectives   307
9.1   Introduction   307
9.2   Control of a Single-Acting Hydraulic Cylinder   310
9.3   Control of a Double-Acting Hydraulic Cylinder   311
9.4   Regenerative Cylinder Circuit   312
9.5   Pump-Unloading Circuit   316

9.6   Double-Pump Hydraulic System   317
9.7   Counterbalance Valve Application   320
9.8   Hydraulic Cylinder Sequencing Circuits   320
9.9   Automatic Cylinder Reciprocating System   322
9.10   Locked Cylinder Using Pilot Check Valves   322
9.11   Cylinder Synchronizing Circuits   323
9.12   Fail-Safe Circuits   325
9.13   Speed Control of a Hydraulic Cylinder   328
9.14   Speed Control of a Hydraulic Motor   333
9.15   Hydraulic Motor Braking System   334
9.16   Hydrostatic Transmission System   334
9.17   Air-Over-Oil Circuit   337
9.18   Analysis of Hydraulic System with Frictional Losses Considered   338
9.19   Mechanical-Hydraulic Servo System   342
9.20   Key Equations   343
        Exercises   344

10 HYDRAULIC CONDUCTORS AND FITTINGS   353

        Learning Objectives   353
10.1   Introduction   353
10.2   Conductor Sizing for Flow-Rate Requirements   354
10.3   Pressure Rating of Conductors   356
10.4   Steel Pipes   360
10.5   Steel Tubing   362
10.6   Plastic Tubing   368
10.7   Flexible Hoses   369
10.8   Quick Disconnect Couplings   374
10.9   Metric Steel Tubing   374
10.10   Key Equations   377
        Exercises   377

11 ANCILLARY HYDRAULIC DEVICES   380

        Learning Objectives   380
11.1   Introduction   380
11.2   Reservoirs   381
11.3   Accumulators   383

11.4   Applications of Accumulators   390
11.5   Pressure Intensifiers   394
11.6   Sealing Devices   400
11.7   Heat Exchangers   408
11.8   Pressure Gages   413
11.9   Flowmeters   415
11.10   Key Equations   418
        Exercises   419

12 MAINTENANCE OF HYDRAULIC SYSTEMS   422

        Learning Objectives   422
12.1   Introduction   423
12.2   Oxidation and Corrosion of Hydraulic Fluids   425
12.3   Fire-Resistant Fluids   426
12.4   Foam-Resistant Fluids   428
12.5   Fluid Lubricating Ability   428
12.6   Fluid Neutralization Number   429
12.7   Petroleum-Base Versus Fire-Resistant Fluids   430
12.8   Maintaining and Disposing of Fluids   430
12.9   Filters and Strainers   431
12.10   Beta Ratio of Filters   438
12.11   Fluid Cleanliness Levels   439
12.12   Wear of Moving Parts Due to Solid-Particle Contamination of the Fluid   441
12.13   Problems Caused by Gases in Hydraulic Fluids   442
12.14   Troubleshooting Hydraulic Systems   445
12.15   Safety Considerations   449
12.16   Environmental Issues   449
12.17   Water Hydraulics   450
12.18   Key Equations   452
        Exercises   452

13 PNEUMATICS: AIR PREPARATION AND COMPONENTS   455

        Learning Objectives   455
13.1   Introduction   455
13.2   Properties of Air   457

13.3   The Perfect Gas Laws   460

13.4   Compressors   465

13.5   Fluid Conditioners   474

13.6   Analysis of Moisture Removal from Air   482

13.7   Air Flow-Rate Control with Orifices   485

13.8   Air Control Valves   487

13.9   Pneumatic Actuators   495

13.10   Key Equations   502

         Exercises   503

14 PNEUMATICS: CIRCUITS AND APPLICATIONS   508

         Learning Objectives   508

14.1   Introduction   508

14.2   Pneumatic Circuit Design Considerations   510

14.3   Air Pressure Losses in Pipelines   512

14.4   Economic Cost of Energy Losses in Pneumatic Systems   514

14.5   Basic Pneumatic Circuits   517

14.6   Pneumatic Vacuum Systems   522

14.7   Sizing of Gas-Loaded Accumulators   527

14.8   Pneumatic Circuit Analysis Using Metric Units   530

14.9   Key Equations   533

         Exercises   533

15 BASIC ELECTRICAL CONTROLS FOR FLUID POWER CIRCUITS   540

         Learning Objectives   540

15.1   Introduction   540

15.2   Electrical Components   544

15.3   Control of a Cylinder Using a Single Limit Switch   547

15.4   Reciprocation of a Cylinder Using Pressure or
         Limit Switches   548

15.5   Dual-Cylinder Sequence Circuits   548

15.6   Box-Sorting System   551

15.7   Electrical Control of Regenerative Circuit   551

15.8   Counting, Timing, and Reciprocation of Hydraulic Cylinder   555

         Exercises   556

16 FLUID LOGIC CONTROL SYSTEMS   561

Learning Objectives   561

16.1   Introduction   561

16.2   Moving-Part Logic (MPL) Control Systems   563

16.3   MPL Control of Fluid Power Circuits   567

16.4   Introduction to Boolean Algebra   570

16.5   Illustrative Examples Using Boolean Algebra   576

16.6   Key Equations   580

Exercises   582

17 ADVANCED ELECTRICAL CONTROLS FOR FLUID POWER SYSTEMS   585

Learning Objectives   585

17.1   Introduction   585

17.2   Components of an Electrohydraulic Servo System   590

17.3   Analysis of Electrohydraulic Servo Systems   594

17.4   Programmable Logic Controllers (PLCs)   602

17.5   Key Equations   610

Exercises   611

18 AUTOMATION STUDIO™ COMPUTER SOFTWARE   614

Learning Objectives   614

18.1   Introduction   614

18.2   Design/Creation of Circuits   616

18.3   Simulation of Circuits   617

18.4   Animation of Components   618

18.5   Interfaces to PLCs and Equipment   618

18.6   Virtual Systems   619

APPENDIXES   621

A   Sizes of Steel Pipe (English Units)   622
B   Sizes of Steel Pipe (Metric Units)   624
C   Sizes of Steel Tubing (English Units)   626
D   Sizes of Steel Tubing (Metric Units)   628
E   Unit Conversion Factors   629
F   Nomenclature   631
G   Fluid Power Symbols   634
H   Derivation of Key Equations   637

ANSWERS TO SELECTED ODD-NUMBERED EXERCISES   643   642

INDEX   647

Preface

INTRODUCTION

The primary purpose of the seventh edition of *Fluid Power with Applications* remains the same as that of the previous editions: to provide the student with in-depth background in the vast field of fluid power. As such, this book covers all subjects essential to understanding the design, analysis, operation, maintenance, and application of fluid power systems. It is written for associate and bachelor's degree candidates studying technology at community colleges, technical colleges, and four-year colleges and universities.

In keeping with the previous editions, theory is presented where necessary, but the emphasis is on understanding how fluid power systems operate and their practical applications. The student then learns not only the "why" but also the "how" of fluid power systems operation.

NEW IN THE SEVENTH EDITION

Based on input from users of the sixth edition and from my colleagues, the following changes have been incorporated into this book.

- Included with this textbook is a CD (produced by Famic Technologies, Inc.) that uses Automation Studio™ computer software to simulate 16 of the fluid power circuit presented throughout the book. By playing the CD on a personal computer, the student obtains a dynamic and visual presentation of the creation, simulation, analysis, and animation of many of the fluid power circuits studied in class or assigned as homework exercises.

- New Chapter 18 provides coverage on the salient features and capabilities of Automation Studio computer software. Automation Studio allows the user to design/create virtual circuits, simulate circuit behavior, animate component operation, and perform a mathematical analysis to optimize system performance.

- New Section 12.17 provides coverage on the applications of water hydraulics, including its advantages and disadvantages relative to oil hydraulics. One key advantage is that water hydraulics systems are friendlier to the environment. Because of the ever-increasing concern about environment issues, the number of promising new applications of water hydraulics will continue to grow. Examples where water hydraulics will see increased utilization are in the food and paper processing industries as well as in low-pressure hydraulic systems.

- A number of photographs and illustrations have been updated to reflect current fluid power technology. This allows the student to become more aware of the latest advances in fluid power systems and applications.

- Many of the example problems and end-of-chapter exercises have been changed to better reflect current industrial applications. The example problems and exercises encompass a wide range of subject matter and levels of difficulty to allow students to progress in an orderly manner.

- New material on fluid power systems that enhance the use of renewable energy reflect the increasing emphasis on lowering air pollution levels, and reducing greenhouse gases to minimize global warming. For example, fluid power systems are being applied to agricultural machines to more efficiently harvest crops such as corn and soybeans, which are not only an important food supply but also are used to make ethanol and biodiesel to replace fossil fuels for transportation vehicles. Similarly, fluid power systems are being applied to control the pitch and yaw of wind turbines to more efficiently produce electricity. Using the clean, renewable energy of wind instead of fossil fuels to produce electricity results in cleaner air and a reduction of greenhouse gases that contribute to global warming.

- New Section 3.5 provides coverage of a manufacturing application of hydraulics called *hydroforming*. Hydroforming utilizes a hydraulic liquid to form three-dimensional sheet-metal panels (sheet hydroforming) or stiff and strong structural parts (tube hydroforming). Hydroforming allows for the production of more-complex shaped panels and lighter structural parts than are possible using strictly mechanical methods.

- New Section 5.2 provides coverage of a manufacturing application of hydraulics called *waterjet cutting*. A waterjet cutting system pressurizes water up to about 60,000 psi and pushes it through an orifice to produce a precise jet of water traveling at about three times the speed of sound. This jet is used to cut through a variety of materials to economically and accurately produce complex shaped parts.

INSTRUCTOR RESOURCES

To access supplementary materials online, instructors need to request an instructor access code. Go to *www.pearsonhighered.com/irc,* where you can register for an instructor access code. Within 48 hours after registering, you will receive a confirming e-mail, including an instructor access code. Once you have received your code, go to the site and log on for full instructions on downloading the materials you wish to use.

ACKNOWLEDGMENTS

As in the case of the previous editions, I am indebted to the numerous fluid power equipment-manufacturing companies for permitting the inclusion of their photographs and other illustrations in this textbook.

I thank Jorge L. Alvarado, Texas A&M University; Maurice Bluestein, Indiana University-Purdue University Indianapolis (IUPUI); Mark Meyer, College of DuPage; Terry Nicoletti, University of Central Missouri; and Scott B. Wilson, Ph.D., University of Central Missouri for reviewing the sixth edition and providing many helpful suggestions and comments for improving this seventh edition. Thanks to Nancy Kesterson at Prentice Hall for handling the production of this edition.

I also wish to thank the users of the previous editions for their constructive suggestions, which have also been incorporated.

Anthony Esposito

Introduction to Fluid Power

Learning Objectives

Upon completing this chapter, you should be able to:

1. Explain what fluid power is.
2. Differentiate between the terms *hydraulics* and *pneumatics*.
3. Understand the difference between fluid power systems and fluid transport systems.
4. Appreciate the history of the fluid power industry.
5. Discuss the advantages and disadvantages of fluid power.
6. Describe key applications of fluid power.
7. Specify the basic components of fluid power systems.
8. Appreciate the size and scope of the fluid power industry.
9. Identify the categories of personnel who are employed in the fluid power industry.

1.1 WHAT IS FLUID POWER?

Definition and Terminology

Fluid power is the technology that deals with the generation, control, and transmission of power, using pressurized fluids. It can be said that fluid power is the muscle that moves industry. This is because fluid power is used to push, pull, regulate, or drive virtually all the machines of modern industry. For example, fluid power steers and brakes automobiles, launches spacecraft, moves earth, harvests crops, mines coal, drives machine tools, controls airplanes, processes food, and even drills teeth. In fact, it is almost impossible to find a manufactured product that hasn't been "fluid-powered" in some way at some stage of its production or distribution.

Fluid power is called *hydraulics* when the fluid is a liquid and is called *pneumatics* when the fluid is a gas. Thus fluid power is the general term used for both *hydraulics* and *pneumatics*. Hydraulic systems use liquids such as petroleum oils, synthetic oils, and water. The first hydraulic fluid to be used was water because it is readily available. However, water has many deficiencies in comparison to hydraulic oils. For example water freezes more readily, is not as good a lubricant, and tends to rust metal components. In spite of these deficiencies, there is a renewed effort underway to return to water in certain applications because of water's abundance, nonflammability, and environmental cleanliness. When water hydraulics is used, the water contains additives to improve lubricity and rust protection and prevent freezing where necessary. Hydraulic oils are currently much more widely used than water, but as environmental concerns continue to become more serious, water hydraulics is expected to become more prevalent. Section 12.17 discusses the applications where water hydraulics should be used rather than oil hydraulics and the positive impact this would have on the environment. Pneumatic systems use air as the gas medium because air is very abundant and can be readily exhausted into the atmosphere after completing its assigned task.

There are actually two different types of fluid systems: *fluid transport* and *fluid power*.

Fluid transport systems have as their sole objective the delivery of a fluid from one location to another to accomplish some useful purpose. Examples include pumping stations for pumping water to homes, cross-country gas lines, and systems where chemical processing takes place as various fluids are brought together.

Fluid power systems are designed specifically to perform work. The work is accomplished by a pressurized fluid bearing directly on an operating fluid cylinder or fluid motor. A fluid cylinder produces a force resulting in linear motion, whereas a fluid motor produces a torque resulting in rotary motion. Thus in a fluid power system, cylinders and motors (which are also called *actuators*), provide the muscle to do the desired work. Of course, control components such as valves are needed to ensure that the work is done smoothly, accurately, efficiently, and safely.

Hydraulic Chain Saw

Liquids provide a very rigid medium for transmitting power and thus can operate under high pressures to provide huge forces and torques to drive loads with utmost accuracy and precision. Figure 1-1 shows a hydraulic chain saw that is ideal for large tree trimming applications from an aerial bucket as well as for cut-up removal jobs. These chain saws are commonly used by electric power line crews because they are lightweight, dependable, quiet, and safer than gasoline-powered saws. The chain saw, which uses a hydraulic gear motor, has a total weight of 6.7 lb. It operates with a flow rate range of 4 to 8 gpm and a pressure range of 1000 to 2000 psi.

Pneumatic Chain Hoist

On the other hand, pneumatics systems exhibit spongy characteristics due to the compressibility of air. However, pneumatic systems are less expensive to build and operate. In addition, provisions can be made to control the operation of the pneumatic actuators that drive the loads. Thus pneumatic systems can be used effectively in applications where low pressures can be used because the loads to be driven do not require large forces.

Figure 1-2 shows a pneumatically powered link chain hoist that has a 4400-lb capacity. The hoist motor receives air at a pressure of 90 psi and flow rates up to 70 cubic ft per min. Loads can be lifted and lowered at variable speeds up to a maximum of 12 and 24 ft per min respectively. The power trolley traverses along the length of the support beam at a speed of 70 ft per min.

Figure 1-1. Hydraulic chain saw. *(Courtesy of Greenlee Textron, Inc., Rockford, Illinois.)*

Figure 1-2. Pneumatically powered link chain hoist. *(Courtesy of Ingersoll-Rand Corp., Southern Pines, North Carolina.)*

1.2 HISTORY OF FLUID POWER

Initial Development

Fluid power is probably as old as civilization itself. Ancient historical accounts show that water was used for centuries to produce power by means of water wheels, and air was used to turn windmills and propel ships. However, these early uses of fluid power required the movement of huge quantities of fluid because of the relatively low pressures provided by nature.

Fluid power technology actually began in 1650 with the discovery of Pascal's law: *Pressure is transmitted undiminished in a confined body of fluid.*

Pascal found that when he rammed a cork down into a jug completely full of wine, the bottom of the jug broke and fell out. Pascal's law indicated that the pressures were equal at the top and bottom of the jug. However, the jug has a small opening area at the top and a large area at the bottom. Thus, the bottom absorbs a greater force due to its larger area.

In 1750, Bernoulli developed his law of conservation of energy for a fluid flowing in a pipeline. Pascal's law and Bernoulli's law operate at the very heart of all fluid power applications and are used for analysis purposes. However, it was not until the Industrial Revolution of 1850 in Great Britain that these laws would actually be applied to industry. Up to this time, electrical energy had not been developed to power the machines of industry. Instead, it was fluid power that, by 1870, was being used to drive hydraulic equipment such as cranes, presses, winches, extruding machines, hydraulic jacks, shearing machines, and riveting machines. In these systems, steam engines drove hydraulic water pumps, which delivered water at moderate pressures through pipes to industrial plants for powering the various machines. These early hydraulic systems had a number of deficiencies such as sealing problems because the designs had evolved more as an art than a science.

Then, late in the nineteenth century, electricity emerged as a dominant technology. This resulted in a shift of development effort away from fluid power. Electrical power was soon found to be superior to hydraulics for transmitting power over great distances. There was very little development in fluid power technology during the last 10 years of the nineteenth century.

Beginning of Modern Era

The modern era of fluid power is considered to have begun in 1906 when a hydraulic system was developed to replace electrical systems for elevating and controlling guns on the battleship *USS Virginia*. For this application, the hydraulic system developed used oil instead of water. This change in hydraulic fluid and the subsequent solution of sealing problems were significant milestones in the rebirth of fluid power.

In 1926 the United States developed the first unitized, packaged hydraulic system consisting of a pump, controls, and actuator. The military requirements leading up to World War II kept fluid power applications and developments going at a good pace. The naval industry had used fluid power for cargo handling, winches, propeller pitch control, submarine control systems, operation of shipboard aircraft elevators, and drive systems for radar and sonar.

During World War II the aviation and aerospace industry provided the impetus for many advances in fluid power technology. Examples include hydraulic-actuated landing gears, cargo doors, gun drives, and flight control devices such as rudders, ailerons, and elevons for aircraft.

Today's Fluid Power

The expanding economy that followed World War II led to the present situation where there are virtually a limitless number of fluid power applications. Today fluid power is used extensively in practically every branch of industry. Some typical applications are in automobiles, tractors, airplanes, missiles, boats, robots, and machine tools. In the automobile alone, fluid power is used in hydraulic brakes, automotive transmissions, power steering, power brakes, air conditioning, lubrication, water coolant, and gasoline pumping systems. The innovative use of modern technology such as electrohydraulic closed-loop systems, microprocessors, and improved materials for component construction will continue to advance the performance of fluid power systems.

Figure 1-3(a) is a photograph of a robotized panel bender system that bends metal sheets into parts called panels. The panels are produced by taking incoming flat sheets and bending them one or more times along one or more sides. The bending forces required for the operation of this machine (also called a press-brake) are provided by a hydraulic press cylinder with a 150-ton capacity. The piston of the hydraulic cylinder has a stroke of 14 in, a rapid traverse speed of 450 in per min, and a maximum bending speed of 47 in per min. The system illustrated is computer controlled and utilizes a robot whose movements are coordinated with the movements of the press-brake. This allows the robot to automatically feed the press-brake with the metal sheets to be formed into panels. The robot also unloads and stacks the processed panels so that the entire system can be operated unattended without interruption. The machine station where the bending operations occur is located just to the right of the computer console shown in Figure 1-3(a).

Figure 1-3(b) shows examples of several finished panels formed by this system. This press-brake can handle steel sheets having a thickness in the range of 0.020 in to 0.120 in. The maximum length and width of incoming sheets that this press-brake can handle are 100 in and 60 in respectively. The robot gripping device uses pneumatic suction cups, which allow for the handling of sheets and panels weighing up to 175 lb.

Figure 1-3(c) provides five views of the action of the press-brake tools as they manipulate the sheets and perform the various bending operations. These tools can bend sheets through an angle up to 270°. Computer programming and bending process simulations are generated from a 3D model of the desired panel using CAD/CAM software. An online graphics interface allows the entire bending sequence to be displayed on the computer screen. The software used for simulation and system operation monitoring are run on the press-brake computer.

Figure 1-4(a) shows the U.S. Air Force B-2 Stealth Bomber, which relies on an advanced technology hydraulic flight control actuation system for its excellent handling

(a) Robotized panel bender system.

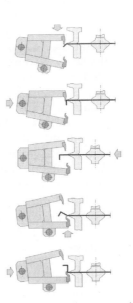

(b) Several finished panels.   (c) Five views of bending operations.

Figure 1-3.   Robotized panel bender system with hydraulic press and robot gripper with pneumatic suction cups. *(Courtesy of Salvagnini America, Hamilton, Ohio.)*

qualities. The B-2 is a low-observable, or stealth, long-range, heavy bomber capable of penetrating sophisticated and dense air-defense shields. It is capable of all-altitude attack missions up to 50,000 ft. It has a range of more than 6000 miles unrefueled, a capacity to carry up to 40,000 lb of weapons and a maximum speed of 475 mph.

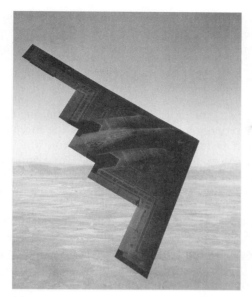

(a) B-2 in flight.

(b) B-2 hydraulic flight control servoactuator.

Figure 1-4. The B-2 Stealth Bomber. *(Courtesy of Moog, Inc., East Aurora, New York.)*

The B-2's distinctive profile, which comes from a unique flying-wing construction, and a special radar-absorbing coating, provide the stealth characteristics. Its unconventional shape presented a technical challenge to meet the additional requirement of possessing excellent handling characteristics. This was accomplished by a sophisticated computerized flight control system that uses a hydraulic flight control servoactuation system. Figure 1-4(b) shows the flight control servoactuator, which is a major component of the flight control system. This servoactuator includes hydraulic actuators with direct drive servovalves for controlling aerodynamic surfaces of the all-wing aircraft.

1.3 ADVANTAGES OF FLUID POWER

There are three basic methods of transmitting power: electrical, mechanical, and fluid power. Most applications actually use a combination of the three methods to obtain the most efficient overall system. To properly determine which method to use, it is important to know the salient features of each type. For example, fluid systems can transmit power more economically over greater distances than can mechanical types. However, fluid systems are restricted to shorter distances than are electrical systems.

The secret of fluid power's success and widespread use is its versatility and manageability. Fluid power is not hindered by the geometry of the machine, as is the case in mechanical systems. Also, power can be transmitted in almost limitless quantities because fluid systems are not so limited by the physical limitations of materials as are electrical systems. For example, the performance of an electromagnet is limited by the saturation limit of steel. On the other hand, the power capacity of fluid systems is limited only by the physical strength of the material (such as steel) used for each component.

Industry is going to depend more and more on automation in order to increase productivity. This includes remote and direct control of production operations, manufacturing processes, and materials handling. Fluid power is well suited for these automation applications because of advantages in the following four major categories.

1. Ease and accuracy of control. By the use of simple levers and push buttons, the operator of a fluid power system can readily start, stop, speed up or slow down, and position forces that provide any desired horsepower with tolerances as precise as one ten-thousandth of an inch. Figure 1-5 shows a fluid power system that allows an aircraft pilot to raise and lower his landing gear. When the pilot moves the lever of a small control valve in one direction, oil under pressure flows to one end of the cylinder to lower the landing gear. To retract the landing gear, the pilot moves the valve lever in the opposite direction, allowing oil to flow into the other end of the cylinder.

2. Multiplication of force. A fluid power system (without using cumbersome gears, pulleys, and levers) can multiply forces simply and efficiently from a fraction of an ounce to several hundred tons of output. Figure 1-6 shows an application where a rugged, powerful drive is required for handling huge logs. In this case, a turntable, which is driven by a hydraulic motor, can carry a 20,000-lb load at a 10-ft radius (a torque of 200,000 ft · lb) under rough operating conditions.

3. Constant force or torque. Only fluid power systems are capable of providing constant force or torque regardless of speed changes. This is accomplished whether the work output moves a few inches per hour, several hundred inches per minute, a few revolutions per hour, or thousands of revolutions per minute. Figure 1-7 shows a commercial lawn mower that uses a hydrostatic transmission in lieu of gears or pulleys to change ground speed. The transmission consists of a hydraulic pump that

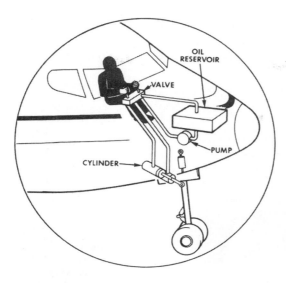

Figure 1-5. Hydraulic operation of aircraft landing gear. *(Courtesy of National Fluid Power Association, Milwaukee, Wisconsin.)*

Figure 1-6. Hydraulically driven turntable for handling huge logs. *(Courtesy of Eaton Corp., Fluid Power Division, Eden Prairie, Minnesota.)*

provides oil under pressure to drive a hydraulic motor at the desired rotational speed. The two-lever hydrostatic drive controls provide a smooth ride when changing ground speeds up to a maximum value of 11 mph. This lawn mower, with its zero turning radius capability, 29-hp gas engine, 72-in deck, and smooth variable ground speed control, makes quick work of large mowing jobs.

4. Simplicity, safety, economy. In general, fluid power systems use fewer moving parts than comparable mechanical or electrical systems. Thus, they are simpler to maintain

Figure 1-7. Commercial lawn mower with hydrostatic transmission. *(Copyright © 1996–2007. All rights reserved. Courtesy of Deere & Company, Cary, North Carolina.)*

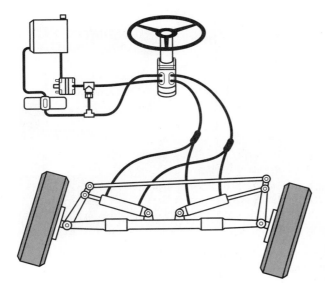

Figure 1-8. Fluid power steering control system for transportation vehicles. *(Courtesy of Eaton Corp., Fluid Power Division, Eden Prairie, Minnesota.)*

and operate. This, in turn, maximizes safety, compactness, and reliability. Figure 1-8 shows a fluid power steering control system designed for transportation vehicles. The steering unit (shown attached to the steering wheel column in Figure 1-8) consists of a manually operated directional control valve and meter in a single body. Because the steering unit is fully fluid-linked, mechanical linkages, universal joints, bearings, reduction gears, and so forth, are eliminated. This provides a simple, compact system. In addition, very little input torque is required to produce the steering control needed.

Additional benefits of fluid power systems include instantly reversible motion, automatic protection against overloads, and infinitely variable speed control. Fluid power systems also have the highest power-per-weight ratio of any known power source.

Drawbacks of Fluid Power

In spite of all the previously mentioned advantages of fluid power, it is not a panacea for all power transmission applications. Fluid power systems also have some drawbacks. For example, hydraulic components must be properly designed and installed to prevent oil leakage from the hydraulic system into the surroundings. Hydraulic pipeline can burst due to excessive oil pressure if proper system design is not implemented. In pneumatic systems, components such as compressed air tanks and accumulators must be properly selected to handle the system maximum air pressure. In addition, proper measures must be taken to control the level of noise in the vicinity of fluid power systems. Noise emanates from components such as pumps, compressors, and pipelines. The underlining theme here is that fluid power systems must be properly designed, installed, and maintained so that they operate in a safe, reliable, efficient, and cost-effective manner.

1.4 APPLICATIONS OF FLUID POWER

Although a number of fluid power applications have already been presented, the following additional examples show more fully the widespread use of fluid in today's world.

1. Fluid power drives high-wire overhead tram. Most overhead trams require a tow cable to travel up or down steep inclines. However, the 22-passenger, hydraulically powered and controlled Sky-tram shown in Figure 1-9 is unique. It is self-propelled and travels on a stationary cable. Because the tram moves instead of the cable, the operator can stop, start, and reverse any one car completely independently of any other car in the tram system. Integral to the design of the Sky-tram drive is a pump (driven by a standard eight-cylinder gasoline engine), which supplies pressurized fluid to four hydraulic motors. Each motor drives two friction drive wheels.

Eight drive wheels on top of the cables support and propel the tram car. On steep inclines, high driving torque is required for ascent and high braking torque for descent.

2. Fluid power is applied to harvesting soybeans. Figure 1-10 shows an agricultural application of fluid power in which a combine is harvesting a field of soybeans. This combine uses a hydraulically controlled cutting platform that increases harvesting capacity by 30% over rigid platforms that use mechanical linkage systems. The fluid power system uses hydraulic cylinders that allow the cutter bar of the platform to float over uneven ground terrain with optimum operator control. The cutter bar can float through a 6-in range to eliminate crop loss while harvesting down or tangled crops. Soybeans are not only an important food supply but also a source for making the renewable fuel called biodiesel in a fashion similar to corn being used

Figure 1-9. Hydraulically powered Sky-tram. *(Courtesy of Sky-tram Systems, Inc., Scottsdale, Arizona.)*

Figure 1-10. Combine with hydraulically controlled cutting platform in process of harvesting a field of soybeans. *(Copyright © 1996–2007. All rights reserved. Courtesy of Deere & Company, Cary, North Carolina.)*

to make ethanol. The use of biodiesel and ethanol for transportation vehicles, in lieu of fossil fuels, reduces greenhouse gases and thus global warming. This agricultural application of hydraulics is of vital importance to mankind due to the increasing worldwide demand for food as well as transportation fuels.

Figure 1-11. Industrial hydraulic lift truck. *(Courtesy of Mitsubishi Caterpillar Forklift America, Houston, Texas.)*

3. Fluid power is the muscle in industrial lift trucks. Figure 1-11 shows a 6500-lb capacity industrial hydraulic lift truck in the process of lifting a large stack of lumber in a warehouse. The hydraulic system includes dual-action tilt cylinders and a hoist cylinder. Lift truck performance modifications are quickly made with a switch. Using this switch the operator can adjust performance parameters such as travel speed, torque, lift and tilt speed, acceleration rate, and braking.

4. Fluid power drives excavators. Figure 1-12 shows an excavator whose hydraulically actuated bucket digs soil from the ground and drops the soil into a dump truck at a construction site. A total of four hydraulic cylinders are used to drive the three pin-connected members called the *boom, stick,* and *bucket.* The boom is the member that is pinned at one end to the cab frame. The stick is the central member that is pinned at one end to the boom and pinned at its other end to the bucket. Two of the cylinders connect the cab frame to the boom. A third cylinder connects the boom to the stick and the fourth cylinder connects the stick to the bucket. For the excavator shown in Figure 1-12, the volume capacity of the largest size bucket is 4.2 cu yd and the maximum lifting capacity at ground level is 41,000 lb.

5. Hydraulics power robot to rescue humans. Figure 1-13 shows a 6-ft tall, 210-lb hydraulically powered robot (in kneeling and standing positions) designed to safely lift humans in a dangerous environment, carry them to a safe location, and put them down. This robot was initially developed for use by the U.S. Armed Forces to rescue soldiers and other casualties off of battlefields. It can travel most places a human can travel and is ideally suited for disaster rescues such as from buildings rendered unsafe due to fire, mudslide, and explosion as well as areas contaminated with biological toxins or radiation. Figure 1-13(a) shows the high-power hydraulic upper body, which allows for the lifting of a human or other object weighing as much as 400 lb. This robot is currently being adapted for use in health care and

Figure 1-12. Hydraulic-powered excavator. *(Courtesy of John Deere Co., Moline, Illinois.)*

elder care because lifting patients is a major cause of serious injury among health-care workers.

6. Hydraulics control the pitch and yaw of wind turbines. Providing sustainable power generation is one of the greatest challenges facing mankind. This is because there is a dire need to reduce greenhouse gases caused by the burning of fossil fuels, which contributes to global warming. It is also vitally important to reduce the world's dependence on fossil fuels for producing energy. The wind is one of the promising renewable energy sources being harnessed to meet this challenge. Wind turbines are being installed at a rapidly increasing rate to drive generators that produce electrical power. Figure 1-14(a) shows two such wind turbines with the sky in the background. Current technology has made it possible to produce wind turbines with up to a 375-ft rotor diameter, a 400-ft hub height above ground level, and a 5-megawatt (MW) electrical power output. A 5-MW wind turbine generates enough electricity for about 4000 households and replaces the burning of 15,000 tons of coal per year.

A new innovation in the field of wind turbine technology is the use of hydraulics to control the pitch and yaw of wind turbines. In order to efficiently produce the electrical power, it is necessary to accurately control of the pitch (angle) of the rotor

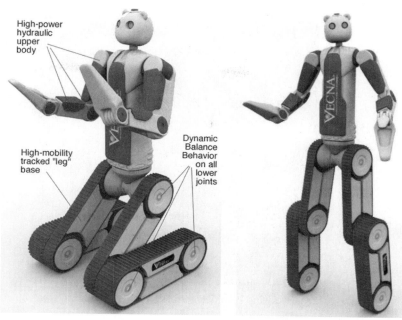

High-power
hydraulic
upper
body

Dynamic
Balance
Behavior
on all
lower
joints

High-mobility
tracked "leg"
base

(a) Kneeling position.

(b) Standing position.

Figure 1-13. The BEAR: Battlefield Extraction-Assist Robot. *(Courtesy of Vecna Technologies, Inc., Cambridge, Massachusetts.)*

blades as the speed of the wind changes. During normal operation while the rotor is revolving, the hydraulic control system will continuously change the pitch a fraction of a degree at a time. In addition, in order to maximize the power and efficiency, it is necessary for the axis of the rotor, which contains the blades, to be parallel to the direction of the wind. This is accomplished by what is called yaw control.

Figure 1-14(b) shows a portion of a wind turbine rotor containing three blades. In this wind turbine the angle of each blade is accurately controlled by a hydraulic cylinder. Similarly, Figure 1-14(c) shows a hydraulic yaw system that continuously points the head of the wind turbine rotor into the wind so that it is always facing the wind. These compact pitch and yaw hydraulic systems not only minimize space requirements and materials but also provide excellent protection from environment degradation and physical damage.

1.5 COMPONENTS OF A FLUID POWER SYSTEM

Hydraulic System

There are six basic components required in a hydraulic system (refer to Figure 1-15):

1. A tank (reservoir) to hold the hydraulic oil

2. A pump to force the oil through the system

(a) Two wind turbines with sky in background.

(b) Hydraulic pitch system.

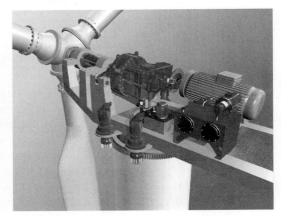

(c) Hydraulic yaw system.

Figure 1-14. Hydraulics control pitch and yaw of wind turbines. *(Courtesy of Bosch Rexroth Corporation, Hoffman Estates, Illinois.)*

    3. An electric motor or other power source to drive the pump

    4. Valves to control oil direction, pressure, and flow rate

    5. An actuator to convert the pressure of the oil into mechanical force or torque to do useful work. Actuators can either be cylinders to provide linear motion, as shown in Figure 1-15, or motors (hydraulic) to provide rotary motion, as shown in Figure 1-16.

    6. Piping, which carries the oil from one location to another

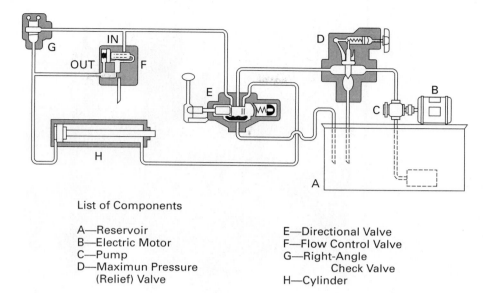

List of Components

A—Reservoir   E—Directional Valve
B—Electric Motor   F—Flow Control Valve
C—Pump   G—Right-Angle
D—Maximun Pressure   Check Valve
    (Relief) Valve   H—Cylinder

Figure 1-15.   Basic hydraulic system with linear hydraulic actuator (cylinder). *(Courtesy of Sperry Vickers, Sperry Rand Corp., Troy, Michigan.)*

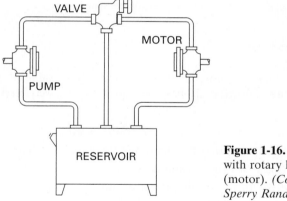

Figure 1-16.   Basic hydraulic system with rotary hydraulic actuator (motor). *(Courtesy of Sperry Vickers, Sperry Rand Corp., Troy, Michigan.)*

Of course, the sophistication and complexity of hydraulic systems will vary depending on the specific applications. This is also true of the individual components that comprise the hydraulic system. As an example, refer to Figure 1-17, which shows two different-sized, complete hydraulic power units designed for two uniquely different applications. Each unit is a complete, packaged power system containing its own electric motor, pump, shaft coupling, reservoir and miscellaneous piping, pressure gages, valves, and other components as required for proper operation. These hydraulic components and systems are studied in detail in subsequent chapters.

Figure 1-17. Two different-sized, complete hydraulic power units. *(Courtesy of Continental Hydraulics, Division of Continental Machines, Inc., Savage, Minnesota.)*

Pneumatic System

Pneumatic systems have components that are similar to those used in hydraulic systems. Essentially the following six basic components are required for pneumatic systems:

1. An air tank to store a given volume of compressed air
2. A compressor to compress the air that comes directly from the atmosphere
3. An electric motor or other prime mover to drive the compressor
4. Valves to control air direction, pressure, and flow rate
5. Actuators, which are similar in operation to hydraulic actuators
6. Piping to carry the pressurized air from one location to another

Figure 1-18 shows a portable pneumatic power unit with an air compressor that is pulley driven by a 13-hp gas engine. The compressor operates at pressures up to 175 psi and has a flow rate capacity of 19 cubic ft per min. This system, which has a 30-gal compressed air tank, features truck-bed mounting to meet the needs of field service applications.

Figure 1-19(a) displays a pneumatic impact wrench designed to tighten and loosen bolts for maintenance and automotive applications. This impact wrench, which

Figure 1-18. Portable pneumatic power unit with a gas engine-driven air compressor. *(Courtesy of Ingersoll-Rand Corp., Davidson, North Carolina.)*

(a) Impact wrench.   (b) Impact wrench in action.

Figure 1-19. Pneumatic impact wrench. *(IR logo is a registered trademark of Ingersoll-Rand Company. Courtesy of Ingersoll-Rand Company, Montvale, New Jersey.)*

weighs 2.4 lb, has an average air consumption rate of 4 cubic ft per min and a maximum torque capacity of 280 ft · lb. In Figure 1-19(b) we see this impact wrench in action as it is being used by a person to torque bolts in an automotive application. An example of the source of air supply for this impact wrench is the air compressor system of Figure 1-18.

In pneumatic systems, after the pressurized air is spent driving actuators, it is then exhausted back into the atmosphere. On the other hand, in hydraulic systems the spent oil drains back to the reservoir and is repeatedly reused after being repressurized by the pump as needed by the system.

1.6 THE FLUID POWER INDUSTRY

Size and Scope

The fluid power industry is huge and is truly a global industry. Statistics from the National Fluid Power Association show the year 2006 sales figure for fluid power products to be $12.7 billion for U.S. companies. This large annual sales figure is reflected in the fact that nearly all U.S. manufacturing plants rely on fluid power in the production of goods. Over half of all U.S. industrial products have fluid power systems or components as part of their basic design. About 75% of all fluid power sales are hydraulic and 25% are pneumatic.

Personnel

Technical personnel who work in the fluid power field can generally be placed into three categories:

1. Fluid power mechanics. Workers in this category are responsible for repair and maintenance of fluid power equipment. They generally are high school graduates who have undertaken an apprenticeship training program. Such a program usually consists of three or four years of paid, on-the-job training plus corresponding classroom instruction.

2. Fluid power technicians. These people usually assist engineers in areas such as design, troubleshooting, testing, maintenance, and installation of fluid power systems. They generally are graduates of two-year technical and community colleges, which award associate degrees in technology. The technician can advance into supervisory positions in sales, manufacturing, or service management.

3. Fluid power engineers. This category consists of people who perform design, development, and testing of new fluid power components or systems. The fluid power engineer typically is a graduate of a four-year college program. Most engineers who work on fluid power systems are manufacturing, sales, or mechanical design oriented. They can advance into management positions in design, manufacturing, or sales.

Future Outlook

The future of the fluid power industry is very promising, especially when one considers that the vast majority of all manufactured products have been processed in some way by fluid power systems. As a result, career opportunities are very bright.

Figure 1-20. Technical Engineer Jessica Reed and Robotics Product Manager Jamie Nichol, developing the next generation of the Battlefield Extraction-Assist Robot, or BEAR, at Vecna's Cambridge Research Laboratory in Cambridge, Massachusetts. Software Engineer Benjamin Bau works on related systems in background. *(Photo by Jonathan Klein. Courtesy of Vecna Technologies, Inc., Cambridge, Massachusetts.)*

The fantastic growth of the fluid power industry has opened many new opportunities in all areas, including supervisors, engineers, technicians, mechanics, sales personnel, and operators.

Figure 1-20 shows several engineers using computer technology to develop the next generation of a hydraulically powered robot.

In addition, a shortage of trained, qualified instructors currently exists. This shortage exists at universities and four-year colleges as well as at two-year technical and community colleges. It is hoped that this book will help in some way in the education of these fluid power–inspired people.

EXERCISES

    1-1. Define the term *fluid power.*

    1-2. Why is hydraulic power especially useful when performing heavy work?

    1-3. What is the difference between the terms *fluid power* and *hydraulics* and *pneumatics?*

    1-4. Compare the use of fluid power to a mechanical system by listing the advantages and disadvantages of each.

1-5. Differentiate between *fluid transport* and *fluid power* systems.

1-6. Comment on the difference between using pneumatic fluid power and hydraulic fluid power.

1-7. What hydraulic device creates a force that can push or pull a load?

1-8. What hydraulic device creates a torque that can rotate a shaft?

1-9. What two factors are responsible for the high responsiveness of hydraulic devices?

1-10. Why can't air be used for all fluid power applications?

1-11. What is the prime mover?

1-12. Name the six basic components required in a hydraulic circuit.

1-13. Name the six basic components required in a pneumatic circuit.

1-14. Take a plant tour of a company that manufactures fluid power components such as pumps, cylinders, valves, or motors. Write a report stating how at least one component is manufactured. List the specifications and include potential customer applications.

1-15. List five applications of fluid power in the automotive industry.

1-16. Give one reason why automotive hydraulic brakes might exhibit a spongy feeling when a driver pushes on the brake pedal.

1-17. Name five hydraulic applications and five pneumatic applications.

1-18. Approximately what percentage of all fluid power sales are for hydraulic and for pneumatic components?

1-19. What three types of personnel work in the fluid power industry?

1-20. Discuss the phrase "the size and scope of fluid power." Cite two facts that show the size of the fluid power industry.

1-21. Obtain from the Fluid Power Society (*www.ifps.org*) the requirements to become a certified fluid power technician.

1-22. Contact the National Fluid Power Association (*www.nfpa.com*) to determine the requirements for becoming a fluid power engineer.

2 Physical Properties of Hydraulic Fluids

Learning Objectives

Upon completing this chapter, you should be able to:

1. Explain the primary functions of a hydraulic fluid.
2. Define the term *fluid*.
3. Distinguish between a liquid and a gas.
4. Appreciate the properties desired of a hydraulic fluid.
5. Define the terms *specific weight, density,* and *specific gravity.*
6. Understand the terms *pressure, head,* and *force.*
7. Differentiate between gage pressures and absolute pressures.
8. Calculate the force created by a pressure.
9. Understand the terms *kinematic viscosity* and *absolute viscosity.*
10. Convert viscosity from one set of units to another set of units.
11. Explain the difference between viscosity and viscosity index.

2.1 INTRODUCTION

The single most important material in a hydraulic system is the working fluid itself. Hydraulic fluid characteristics have a crucial effect on equipment performance and life. It is important to use a clean, high-quality fluid in order to achieve efficient hydraulic system operation.

Most modern hydraulic fluids are complex compounds that have been carefully prepared to meet their demanding tasks. In addition to having a base fluid, hydraulic fluids contain special additives to provide desired characteristics.

A hydraulic fluid has the following four primary functions:

1. Transmit power
2. Lubricate moving parts
3. Seal clearances between mating parts
4. Dissipate heat

In addition a hydraulic fluid must be inexpensive and readily available. To accomplish properly the four primary functions and be practical from a safety and cost point of view, a hydraulic fluid should have the following properties:

1. Good lubricity
2. Ideal viscosity
3. Chemical stability
4. Compatibility with system materials
5. High degree of incompressibility
6. Fire resistance
7. Good heat-transfer capability
8. Low density
9. Foam resistance
10. Nontoxicity
11. Low volatility

This is a challenging list, and no single hydraulic fluid possesses all of these desirable characteristics. The fluid power designer must select the fluid that comes the closest to being ideal overall for a particular application.

Hydraulic fluids must also be changed periodically, the frequency depending not only on the fluid but also on the operating conditions. Laboratory analysis is the best method for determining when a fluid should be changed. Generally speaking, a fluid should be changed when its viscosity and acidity increase due to fluid breakdown or contamination. Preferably, the fluid should be changed while the system is at operating temperature. In this way, most of the impurities are in suspension and will be drained off.

Much hydraulic fluid has been discarded in the past due to the possibility that contamination existed—it costs more to test the fluid than to replace it. This situation has changed as the need to conserve hydraulic fluids has developed. Figure 2-1 shows a hydraulic fluid test kit that provides a quick, easy method to test hydraulic system contamination. Even small hydraulic systems may be checked. The test kit may be used on the spot to determine whether fluid quality permits continued use. Three key quality indicators can be evaluated: viscosity, water content, and foreign particle contamination level.

In this chapter we examine the physical properties of fluids dealing with the transmission of power. These properties include density, pressure, compressibility, viscosity, and viscosity index. In Chapter 12 we discuss the types of fluids used in

Figure 2-1. Hydraulic fluid test kit. *(Courtesy of Gulf Oil Corp., Houston, Texas.)*

hydraulic systems and the chemical-related properties dealing with maintenance of the quality of these hydraulic fluids. These properties include rate of oxidation, fire resistance, foam resistance, lubricating ability, and acidity.

2.2 FLUIDS: LIQUIDS AND GASES

Liquids

The term *fluid* refers to both gases and liquids. A liquid is a fluid that, for a given mass, will have a definite volume independent of the shape of its container. This means that even though a liquid will assume the shape of the container, it will fill only that part of the container whose volume equals the volume of the quantity of the liquid. For example, if water is poured into a vessel and the volume of water is not sufficient to fill the vessel, a free surface will be formed, as shown in Figure 2-2(a). A free surface is also formed in the case of a body of water, such as a lake, exposed to the atmosphere [see Figure 2-2(b)].

Liquids are considered to be incompressible so that their volume does not change with pressure changes. This is not exactly true, but the change in volume due to pressure changes is so small that it is ignored for most engineering applications. Variations from this assumption of incompressibility are discussed in Section 2.6, where the parameter *bulk modulus* is defined.

Gases

Gases, on the other hand, are fluids that are readily compressible. In addition, their volume will vary to fill the vessel containing them. This is illustrated in Figure 2-3,

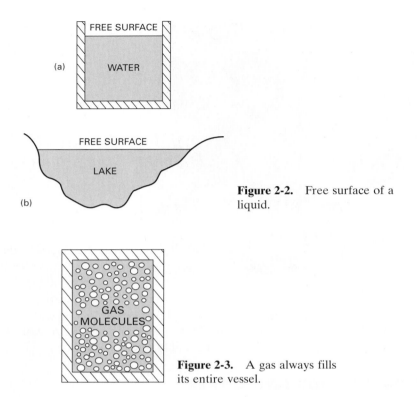

Figure 2-2. Free surface of a liquid.

Figure 2-3. A gas always fills its entire vessel.

Parameter	Liquid	Gas
Volume	Has its own volume	Volume is determined by container
Shape	Takes shape of container but only to its volume	Expands to completely fill and take the shape of the container
Compressibility	Incompressible for most engineering applications	Readily compressible

Figure 2-4. Physical differences between liquids and gases.

where a gas is allowed to enter an empty, closed vessel. As shown, the gas molecules always fill the entire vessel. Therefore, unlike liquids, which have a definite volume for a given mass, the volume of a given mass of a gas will increase to fill the vessel that contains it. Gases are greatly influenced by the pressure to which they are subjected. An increase in pressure causes the volume of the gas to decrease, and vice versa. Figure 2-4 summarizes the key physical differences between liquids and gases for a given mass.

Air is the only gas commonly used in fluid power systems because it is inexpensive and readily available. Air also has the following desirable features as a power fluid:

1. It is fire resistant.
2. It is not messy.
3. It can be exhausted back into the atmosphere.

The disadvantages of using air versus using hydraulic oil are:

1. Due to its compressibility, air cannot be used in an application where accurate positioning or rigid holding is required.
2. Because air is compressible, it tends to be sluggish.
3. Air can be corrosive, since it contains oxygen and water.
4. A lubricant must be added to air to lubricate valves and actuators.
5. Air pressures of greater than 250 psi are typically not used due to the explosion dangers involved if components such as air tanks should rupture. This is because air (due to its compressibility) can store a large amount of energy as it is compressed in a manner similar to that of a mechanical spring.

2.3 SPECIFIC WEIGHT, DENSITY, AND SPECIFIC GRAVITY

Weight Versus Mass

All objects, whether solids or fluids, are pulled toward the center of the earth by a force of attraction. This force is called the weight of the object and is proportional to the object's mass, as defined by

$$F = W = mg \qquad \qquad \textbf{(2-1)}$$

where, in the English system of units (also called U.S. customary units and used extensively in the United States) we have

F = force in units of lb,
W = weight in units of lb,
m = mass of object in units of slugs,
g = proportionality constant called the acceleration of gravity, which equals 32.2 ft/s² at sea level.

A mass of 1 slug is defined as the mass of a platinum-iridium bar at the National Institute of Standards and Technology near Washington, DC.

From Eq. (2-1), W equals 32.2 lb if m is 1 slug. Therefore, 1 slug is the amount of mass that weighs 32.2 lb. We can also conclude from Eq. (2-1) that 1 lb is defined as the force that will give a mass of 1 slug an acceleration of 1 ft/s².

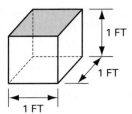

Figure 2-5. Cubic container full of water.

EXAMPLE 2-1

Find the weight of a body having a mass of 4 slugs.

Solution Substituting into Eq. (2-1) yields

$$W = mg = 4 \text{ slugs} \times 32.2 \text{ ft/s}^2 = 129 \text{ lb}$$

Specific Weight

Figure 2-5 shows a cubic container full of water as an example for discussing the fluid property called *specific weight*. Since the container has the shape of a rectangular solid, its volume can be calculated using Eq. (2-2).

$$\text{volume} = (\text{area of base}) \times (\text{height}) \qquad \textbf{(2-2)}$$

Substituting values yields

$$\text{volume} = (1 \text{ ft} \times 1 \text{ ft}) \times (1 \text{ ft}) = 1 \text{ ft}^3$$

It has been found by measurement that 1 ft³ of water weighs 62.4 lb. Specific weight is defined as weight per unit volume. Stated mathematically, we have

$$\text{specific weight} = \frac{\text{weight}}{\text{volume}}$$

or

$$\gamma = \frac{W}{V} \qquad \textbf{(2-3)}$$

where γ = Greek symbol gamma = specific weight (lb/ft³),
   W = weight (lb),
   V = volume (ft³).

Knowing that 1 ft^3 of water weighs 62.4 lb, we can now calculate the specific weight of water using Eq. (2-3):

$$\gamma_{water} = \frac{W}{V} = \frac{62.4 \text{ lb}}{1 \text{ ft}^3} = 62.4 \text{ lb/ft}^3$$

If we want to calculate the specific weight of water in units of lb/in^3, we can perform the following units manipulation:

$$\gamma\left(\frac{\text{lb}}{\text{in}^3}\right) = \gamma\left(\frac{\text{lb}}{\text{ft}^3}\right) \times \frac{1 \text{ ft}^3}{1728 \text{ in}^3}$$

$$\uparrow \qquad\qquad \uparrow \qquad\qquad \uparrow$$

$$\text{units wanted} \quad \text{units have} \quad \text{conversion factor}$$

The conversion factor of 1/1728 is valid since 1 ft^3 = 1728 in^3. This provides a consistent set of units of lb/in^3 on both sides of the equal sign since the units of ft^3 cancel out in the numerator and denominator. The resulting units conversion equation is

$$\gamma(\text{lb/in}^3) = \frac{\gamma(\text{lb/ft}^3)}{1728}$$

Substituting the known value for the specific weight of water in units of lb/ft^3, we have

$$\gamma_{water}(\text{lb/in}^3) = \frac{62.4}{1728} = 0.0361 \text{ lb/in}^3$$

Most oils have a specific weight of about 56 lb/ft^3, or 0.0325 lb/in^3. However, depending on the type of oil, the specific weight can vary from a low of 55 lb/ft^3 to a high of 58 lb/ft^3.

EXAMPLE 2-2

If the body of Example 2-1 has a volume of 1.8 ft^3, find its specific weight.

Solution　Using Eq. (2-3) we have

$$\gamma = \frac{W}{V} = \frac{129 \text{ lb}}{1.8 \text{ ft}^3} = 71.6 \text{ lb/ft}^3$$

Specific Gravity

The specific gravity (SG) of a given fluid is defined as the specific weight of the fluid divided by the specific weight of water. Therefore, the specific gravity of water is unity by definition. The specific gravity of oil can be found using

$$(SG)_{\text{oil}} = \frac{\gamma_{\text{oil}}}{\gamma_{\text{water}}} \qquad (2\text{-}4)$$

Substituting the most typical value of specific weight for oil we have

$$(SG)_{\text{oil}} = \frac{56 \text{ lb/ft}^3}{62.4 \text{ lb/ft}^3} = 0.899$$

Note that specific gravity is a dimensionless parameter (has no units).

EXAMPLE 2-3

Air at 68°F and under atmospheric pressure has a specific weight of 0.0752 lb/ft³. Find its specific gravity.

Solution

$$(SG)_{\text{air}} = \frac{\gamma_{\text{air}}}{\gamma_{\text{water}}} = \frac{0.0752 \text{ lb/ft}^3}{62.4 \text{ lb/ft}^3} = 0.00121$$

Thus, water is 1/0.00121 times, or about 830 times, as heavy as air at 68°F and under atmospheric pressure. It should be noted that since air is highly compressible, the value of 0.00121 for SG is valid only at 68°F and under atmospheric pressure.

Density

In addition to specific weight, we can also talk about the fluid property called *density,* which is defined as mass per unit volume:

$$\rho = \frac{m}{V} \qquad (2\text{-}5)$$

where ρ = Greek symbol rho = density (slugs/ft³),
       m = mass (slugs),
       V = volume (ft³).

Since weight is proportional to mass (from the equation $W = mg$), it follows that specific gravity can also be defined as the density of the given fluid divided by the density of water. This is shown as follows:

$$W = mg$$

or

$$\gamma V = \rho V g$$

Solving for the density we have

$$\rho = \frac{\gamma}{g} \tag{2-6}$$

where γ has units of lb/ft³,
g has units of ft/s²,
ρ has units of slugs/ft³

Hence density equals specific weight divided by the acceleration of gravity. This allows us to obtain the desired result.

$$SG = \frac{\gamma}{\gamma_{\text{water}}} = \frac{g\rho}{g\rho_{\text{water}}} = \frac{\rho}{\rho_{\text{water}}} \tag{2-7}$$

The density of oil having a specific weight of 56 lb/ft³ can be found from Eq. (2-6).

$$\rho_{\text{oil}} = \frac{56 \, \frac{\text{lb}}{\text{ft}^3}}{32.2 \, \frac{\text{ft}}{\text{s}^2}} = 1.74 \, \frac{\text{lb}}{\text{ft}^3} \times \frac{\text{s}^2}{\text{ft}} = 1.74 \, \text{lb} \cdot \text{s}^2/\text{ft}^4$$

$$= 1.74 \, \text{slugs/ft}^3$$

Note that from Eq. (2-1), $W = mg$. Thus, we have the following equality of units between weight and mass: lb = slugs · ft/s².

The density and specific weight of a given fluid changes with pressure and temperature. For most practical engineering applications, changes in the density and specific weight of liquids with pressure and temperature are negligibly small; however, the changes in density and specific weight of gases with pressure and temperature are significant and must be taken into account.

EXAMPLE 2-4

Find the density of the body of Examples 2-1 and 2-2.

Solution   Using Eq. (2-5) yields

$$\rho = \frac{m}{V} = \frac{4 \text{ slugs}}{1.8 \text{ ft}^3} = 2.22 \text{ slugs/ft}^3$$

Also, noting that $\gamma = 71.6$ lb/ft^3 from Example 2-2, we can solve for ρ using Eq. (2-6).

$$\rho = \frac{\gamma}{g} = \frac{71.6 \text{ lb/ft}^3}{32.2 \text{ ft/s}^2} = 2.22 \frac{\text{lb} \cdot \text{s}^2}{\text{ft}^4} = 2.22 \text{ slugs/ft}^3$$

2.4 FORCE, PRESSURE, AND HEAD

Force and Pressure

Pressure is defined as force per unit area. Hence, pressure is the amount of force acting over a unit area, as indicated by

$$p = \frac{F}{A} \qquad\qquad (2\text{-}8)$$

where p = pressure,
        F = force,
        A = area.

Note that p will have units of lb/ft^2 if F and A have units of lb and ft^2, respectively. Similarly, by changing the units of A from ft^2 to in^2, the units for p will become lb/in^2. Let's go back to our 1-ft^3 container of Figure 2-5. The pressure acting on the bottom of the container can be calculated using Eq. (2-8), knowing that the total force acting at the bottom equals the 62.4-lb weight of the water:

$$p = \frac{62.4 \text{ lb}}{1 \text{ ft}^2} = 62.4 \text{ lb/ft}^2 = 62.4 \text{ psf}$$

Units of lb/ft^2 are commonly written as psf. Also, since 1 ft^2 = 144 in^2, the pressure at the bottom of the container can be found in units of lb/in^2 as follows using Eq. (2-8):

$$p = \frac{62.4 \text{ lb}}{144 \text{ in}^2} = 0.433 \text{ lb/in}^2 = 0.433 \text{ psi}$$

Units of lb/in^2 are commonly written as psi.

Head

We can now conclude that, due to its weight, a 1-ft column of water develops at its base a pressure of 0.433 psi. The 1-ft height of water is commonly called a *pressure head.*

Let's now refer to Figure 2-6, which shows a 10-ft high column of water that has a cross-sectional area of 1 ft². Since there are 10 ft³ of water and each cubic foot weighs 62.4 lb, the total weight of water is 624 lb. The pressure at the base is

$$p = \frac{F}{A} = \frac{624 \text{ lb}}{144 \text{ in}^2} = 4.33 \text{ psi}$$

Thus, each foot of the 10-ft head develops a pressure increase of 0.433 psi from its top to bottom.

What happens to the pressure if the fluid is not water? Figure 2-7 shows a 1-ft³ volume of oil. Assuming a weight density of 57 lb/ft³, the pressure at the base is

$$p = \frac{F}{A} = \frac{57 \text{ lb}}{144 \text{ in}^2} = 0.40 \text{ psi}$$

Therefore, as depicted in Figure 2-7, a 2-ft column of oil develops a pressure at its bottom of 0.80 psi. These values for oil are slightly less than for water because the

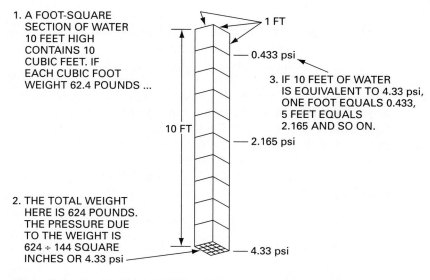

1. A FOOT-SQUARE SECTION OF WATER 10 FEET HIGH CONTAINS 10 CUBIC FEET. IF EACH CUBIC FOOT WEIGHT 62.4 POUNDS ...

1 FT

0.433 psi

3. IF 10 FEET OF WATER IS EQUIVALENT TO 4.33 psi, ONE FOOT EQUALS 0.433, 5 FEET EQUALS 2.165 AND SO ON.

10 FT

2.165 psi

2. THE TOTAL WEIGHT HERE IS 624 POUNDS. THE PRESSURE DUE TO THE WEIGHT IS 624 ÷ 144 SQUARE INCHES OR 4.33 psi

4.33 psi

Figure 2-6. Pressure developed by a 10-ft column of water. *(Courtesy of Sperry Vickers, Sperry Rand Corp., Troy, Michigan.)*

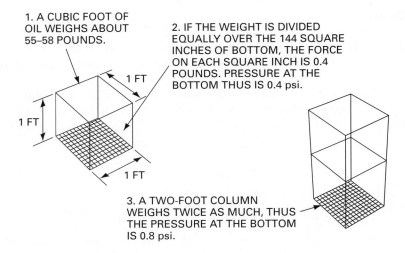

1. A CUBIC FOOT OF OIL WEIGHS ABOUT 55–58 POUNDS.

2. IF THE WEIGHT IS DIVIDED EQUALLY OVER THE 144 SQUARE INCHES OF BOTTOM, THE FORCE ON EACH SQUARE INCH IS 0.4 POUNDS. PRESSURE AT THE BOTTOM THUS IS 0.4 psi.

1 FT

1 FT

1 FT

3. A TWO-FOOT COLUMN WEIGHS TWICE AS MUCH, THUS THE PRESSURE AT THE BOTTOM IS 0.8 psi.

Figure 2-7.   Pressures developed by 1- and 2-ft columns of oil. *(Courtesy of Sperry Vickers, Sperry Rand Corp., Troy, Michigan.)*

specific weight of oil is somewhat less than that for water. Equation (2-9) allows calculation of the pressure developed at the bottom of a column of any liquid.

$$p = \gamma H \qquad \textbf{(2-9)}$$

where p = pressure at bottom of liquid column,
             γ = specific weight of liquid,
             H = liquid column height or head.

Observe per Eq. (2-9) that the pressure does not depend on the cross-sectional area of the liquid column but only on the column height and the specific weight of the liquid. The reason is simple: Changing the cross-sectional area of the liquid column changes its weight (and thus the force at its bottom) by a proportional amount. Hence F/A (which equals pressure) remains constant.

Substituting the correct units for γ and H into Eq. (2-9) produces the proper units for pressure:

$$p(\text{lb/in}^2) = p(\text{psi}) = \gamma(\text{lb/in}^3) \times H(\text{in})$$

or

$$p(\text{lb/ft}^2) = p(\text{psf}) = \gamma(\text{lb/ft}^3) \times H(\text{ft})$$

A derivation of Eq. (2-9) is provided in Appendix H.

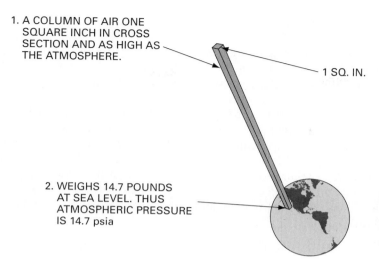

1. A COLUMN OF AIR ONE SQUARE INCH IN CROSS SECTION AND AS HIGH AS THE ATMOSPHERE.

1 SQ. IN.

2. WEIGHS 14.7 POUNDS AT SEA LEVEL. THUS ATMOSPHERIC PRESSURE IS 14.7 psia

Figure 2-8. The atmosphere as a pressure head. (*Courtesy of Sperry Vickers, Sperry Rand Corp., Troy, Michigan.*)

EXAMPLE 2-5

Find the pressure on a skin diver who has descended to a depth of 60 ft in fresh water.

Solution Using Eq. (2-9) we have

$$p(\text{lb/in}^2) = \gamma(\text{lb/in}^3) \times H(\text{in}) = 0.0361 \times (60 \times 12) = 26.0 \text{ psi}$$

Atmospheric Pressure

What about the pressure developed on Earth's surface due to the force of attraction between the atmosphere and Earth? For all practical purposes we live at the bottom of a huge sea of air, which extends hundreds of miles above us. Equation (2-9) cannot be used to find this pressure because of the compressibility of air. As a result the density of the air is not constant throughout the atmosphere. The density is greatest at Earth's surface and diminishes as the distance from Earth increases.

Let's refer to Figure 2-8, which shows a column of air with a cross-sectional area of 1 in² and as high as the atmosphere extends above Earth's surface. This entire column of air weighs about 14.7 lb and thus produces a pressure of about 14.7 lb/in² on Earth's surface at sea level. This pressure is called *atmospheric pressure* and the value of 14.7 lb/in² is called *standard atmosphere pressure* because atmospheric pressure varies a small amount depending on the weather conditions which affect the density of the air. Unless otherwise specified, the actual atmospheric pressure will be assumed to equal the standard atmosphere pressure.

Gage and Absolute Pressure

Why, then, does a deflated automobile tire read zero pressure instead of 14.7 psi when using a pressure gage? The answer lies in the fact that a pressure gage reads gage pressure and not absolute pressure. Gage pressures are measured relative to the atmosphere, whereas absolute pressures are measured relative to a perfect vacuum such as that existing in outer space. To distinguish between them, gage pressures are labeled psig, or simply psi, whereas absolute pressures are labeled psi (abs), or simply psia.

This means that atmospheric pressure equals 14.7 psia or 0 psig. Atmospheric pressure is measured with special devices called *barometers*. Figure 2-9 shows how a mercury barometer works. The atmospheric pressure to be measured can support a column of mercury equal to 30.0 in because this head produces a pressure of 14.7 psi.

This can be checked by using Eq. (2-9) and noting that the specific weight of mercury is 0.490 lb/in³:

$$p = \gamma H$$

$$14.7 \text{ lb/in}^2 = 0.490 \text{ lb/in}^3 \times H(\text{in})$$

$$H = 30.0 \text{ in of mercury}$$

Figure 2-10 has a chart showing the difference between gage and absolute pressures. Let's examine two pressure levels: p_1 and p_2. Relative to a perfect vacuum they are

$$p_1 = 4.7 \text{ psia (a pressure less than atmospheric pressure)}$$

$$p_2 = 24.7 \text{ psia (a pressure greater than atmospheric pressure)}$$

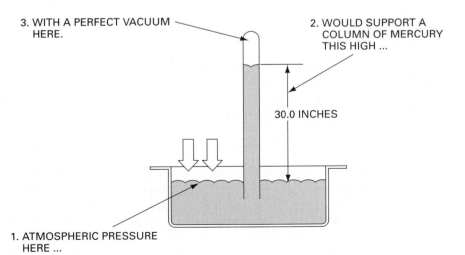

Figure 2-9. Operation of a mercury barometer. *(Courtesy of Sperry Vickers, Sperry Rand Corp., Troy, Michigan.)*

EXAMPLE 2-6

How high would the tube of a barometer have to be if water were used instead of mercury?

Solution Per Eq. (2-9) we have

$$p = \gamma H$$

$$14.7 \text{ lb/in}^2 = 0.0361 \text{ lb/in}^3 \times H(\text{in})$$

$$H = 407 \text{ in} = 34.0 \text{ ft}$$

Thus a water barometer would be impractical because it takes a 34.0-ft column of water to produce a pressure of 14.7 psi at its base.

Relative to the atmosphere, they are

$$p_1 = 10 \text{ psig suction (or vacuum)} = -10 \text{ psig}$$

$$p_2 = 10 \text{ psig}$$

The use of the terms *suction* or *vacuum* and the use of the minus sign mean that pressure p_1 is 10 psi below atmospheric pressure. Also note that the terms *psi* and *psig* are used interchangeably. Hence, p_1 also equals -10 psi and p_2 equals 10 psi.

As can be seen from Figure 2-10, the following equation can be used in converting gage pressures to absolute pressures, and vice versa.

$$p_{\text{abs}} = p_{\text{gage}} + p_{\text{atm}} \tag{2-10}$$

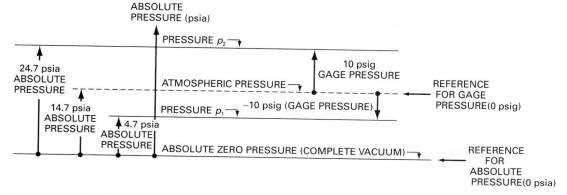

Figure 2-10. Difference between absolute and gage pressures.

EXAMPLE 2-7

Convert a −5-psi pressure to an absolute pressure.

Solution   Using Eq. (2-10) we have

$$\text{absolute pressure} = -5 + 14.7 = 9.7 \text{ psia}$$

EXAMPLE 2-8

Find the absolute pressure on the skin diver of Example 2-5.

Solution   Using Eq. (2-10) yields

$$\text{absolute pressure} = 26.0 + 14.7 = 40.7 \text{ psia}$$

Vacuum or suction pressures exist in certain locations of fluid power systems (for example, in the inlet or suction lines of pumps). Therefore, it is important to understand the meaning of pressures below atmospheric pressure. One way to generate a suction pressure is to remove some of the fluid from a closed vessel initially containing fluid at atmospheric pressure.

2.5 THE SI METRIC SYSTEM

Introduction

The SI metric system was standardized in June 1960 when the International Organization for Standardization approved a metric system called *Le Système International d'Unités*. This system, which is abbreviated SI, has supplanted the old CGS (centimeter-gram-second) metric system, and U.S. adoption of the SI metric system is considered to be imminent.

In the SI metric system, the units of measurement are as follows:

Length is the meter (m).
Mass is the kilogram (kg).
Force is the newton (N).
Time is the second (s).
Temperature is the degree Celsius (°C).

A mass of 1 kilogram is defined as the mass of a platinum-iridium bar at the International Bureau of Weights and Measures near Paris, France.

Length, Mass, and Force Comparisons with English System

The relative sizes of length, mass, and force units between the metric and English systems are given as follows:

One meter equals 39.4 in = 3.28 ft.
One kilogram equals 0.0685 slugs.
One newton equals 0.225 lb.

Per Eq. (2-1) a newton is defined as the force that will give a mass of 1 kg an acceleration of 1 m/s². Stated mathematically, we have

$$1 \text{ N} = 1 \text{ kg} \times 1 \text{ m/s}^2$$

Since the acceleration of gravity at sea level equals 9.80 m/s², a mass of 1 kg weighs 9.80 N. Also, since 1 N = 0.225 lb, a mass of 1 kg also weighs 2.20 lb.

Pressure Comparisons

The SI metric system uses units of pascals (Pa) to represent pressure. A pressure of 1 Pa is equal to a force of 1 N applied over an area of 1 m² and thus is a very small unit of pressure.

$$1 \text{ Pa} = 1 \text{ N/m}^2$$

The conversion between pascals and psi is as follows:

$$1 \text{ Pa} = 0.000145 \text{ psi}$$

Atmospheric pressure in units of pascals is found as follows, by converting 14.7 psi into its equivalent pressure in pascals:

$$p_{atm}(\text{Pa}) = 14.7 \text{ psi (abs)} \times \frac{1 \text{ Pa}}{0.000145 \text{ psi}} = 101,000 \text{ Pa (abs)}$$

Thus, atmospheric pressure equals 101,000 Pa (abs) as well as 14.7 psia. Since the pascal is a very small unit, the bar is commonly used:

$$1 \text{ bar} = 10^5 \text{ N/m}^2 = 10^5 \text{ Pa} = 14.5 \text{ psi}$$

Thus, atmospheric pressure equals 14.7/14.5 bars (abs), or 1.01 bars (abs).

Temperature Comparisons

The temperature (T) in the metric system is measured in units of degrees Celsius (°C), whereas temperature in the English system is measured in units of degrees Fahrenheit (°F). Figure 2-11 shows a graphical representation of these two temperature scales using a mercury thermometer reading a room temperature of 68°F (20°C).

   Relative to Figure 2-11 the following should be noted: The Fahrenheit temperature scale is determined by dividing the temperature range between the freezing point of water (set at 32°F) and the boiling point of water (set at 212°F) at atmospheric pressure into 180 equal increments. The Celsius temperature scale is determined by dividing the temperature range between the freezing point of water (set at 0°C) and the boiling point of water (set at 100°C) at atmospheric pressure into 100 equal increments.

   The mathematical relationship between the Fahrenheit and Celsius scales is

$$T(°F) = 1.8 T(°C) + 32 \qquad\qquad \textbf{(2-11)}$$

Thus, to find the equivalent Celsius temperature corresponding to room temperature (68°F), we have:

$$T(°C) = \frac{T(°F) - 32}{1.8} = \frac{68 - 32}{1.8} = 20°C$$

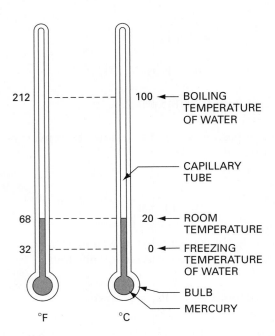

Figure 2-11. Comparison of the Fahrenheit and Celsius temperature scales.

Absolute temperatures (Rankine units in English system and Kelvin units in metric system) are presented in Chapter 13 for use in the gas law equations.

SI System Prefixes

Figure 2-12 provides the prefixes used in the metric system to represent powers of 10. Thus, for example:

$$1 \text{ kPa} = 10^3 \text{ Pa} = 1000 \text{ Pa}$$

This means that atmospheric pressure equals 101 kPa (abs) as well as 1010 milli-bars (abs).

EXAMPLE 2-9

An oil has a specific weight of 56 lb/ft³. Determine its specific weight in units of N/m³.

Solution   units wanted = units given × conversion factors

$$\gamma\left(\frac{N}{m^3}\right) = \gamma\left(\frac{lb}{ft^3}\right) \times \left(\frac{1 \text{ N}}{0.225 \text{ lb}}\right) \times \left(\frac{3.28 \text{ ft}}{1 \text{ m}}\right)^3 = 157\gamma\left(\frac{lb}{ft^3}\right)$$

$$= 157 \times 56 = 8790 \text{ N/m}^3$$

Prefix Name	SI Symbol	Multiplication Factor
tera	T	10^{12}
giga	G	10^9
mega	M	10^6
kilo	k	10^3
centi	c	10^{-2}
milli	m	10^{-3}
micro	μ	10^{-6}
nano	n	10^{-9}
pico	p	10^{-12}

Figure 2-12.   Prefixes used in metric system to represent powers of 10.

EXAMPLE 2-10

At what temperature are the Fahrenheit and Celsius values equal?

Solution   Per the problem statement we have

$$T(°F) = T(°C)$$

Substituting from Eq. (2-11) yields

$$1.8T(°C) + 32 = T(°C) \text{ or } T(°C) = \frac{-32}{0.8} = -40°$$

Thus, $-40°C = -40°F$.

2.6 BULK MODULUS

The highly favorable power-to-weight ratio and the stiffness of hydraulic systems make them the frequent choice for most high-power applications. The stiffness of a hydraulic system is directly related to the incompressibility of the oil. Bulk modulus is a measure of this incompressibility. The higher the bulk modulus, the less compressible or stiffer the fluid.

Mathematically the bulk modulus is defined by Eq. (2-12), where the minus sign indicates that as the pressure increases on a given amount of oil, the oil's volume decreases, and vice versa:

$$\beta = \frac{-\Delta p}{\Delta V/V} \qquad (2\text{-}12)$$

where β = bulk modulus (psi, kPa),
Δp = change in pressure (psi, kPa),
ΔV = change in volume (in³, m³),
V = original volume (in³, m³).

The bulk modulus of an oil changes somewhat with changes in pressure and temperature. However, for the pressure and temperature variations that occur in most fluid power systems, this factor can be neglected. A typical value for oil is 250,000 psi $(1.72 \times 10^6 \text{ kPa})$.

2.7 VISCOSITY

Introduction

Viscosity is probably the single most important property of a hydraulic fluid. It is a measure of a fluid's resistance to flow. When the viscosity is low, the fluid flows

---

EXAMPLE 2-11

A 10-in³ sample of oil is compressed in a cylinder until its pressure is increased from 100 to 2000 psi. If the bulk modulus equals 250,000 psi, find the change in volume of the oil.

Solution   Rewriting Eq. (2-12) to solve for ΔV, we have

$$\Delta V = -V\left(\frac{\Delta p}{\beta}\right) = -10\left(\frac{1900}{250,000}\right) = -0.076 \text{ in}^3$$

This represents only a 0.76% decrease in volume, which shows that oil is highly incompressible.

---

easily and is thin in appearance. A fluid that flows with difficulty has a high viscosity and is thick in appearance.

In reality, the ideal viscosity for a given hydraulic system is a compromise. Too high a viscosity results in

1. High resistance to flow, which causes sluggish operation.
2. Increased power consumption due to frictional losses.
3. Increased pressure drop through valves and lines.
4. High temperatures caused by friction.

On the other hand, if the viscosity is too low, the result is

1. Increased oil leakage past seals.
2. Excessive wear due to breakdown of the oil film between mating moving parts. These moving parts may be internal components of a pump (such as pistons reciprocating in cylinder bores of a piston pump) or a sliding spool inside the body of a valve, as shown in Figure 2-13.

Absolute Viscosity

The concept of viscosity can be understood by examining two parallel plates separated by an oil film of thickness y, as illustrated in Figure 2-14. The lower plate is stationary, whereas the upper plate moves with a velocity v as it is being pushed by a force F as shown. Because of viscosity, the oil adheres to both surfaces. Thus, the velocity of the layer of fluid in contact with the lower plate is zero, and the velocity of the layer in contact with the top plate is v. The consequence is a linearly varying velocity profile whose slope is v/y. The absolute viscosity of the oil can be represented mathematically as follows:

$$\mu = \frac{\tau}{v/y} = \frac{F/A}{v/y} = \frac{\text{shear stress in oil}}{\text{slope of velocity profile}} \tag{2-13}$$

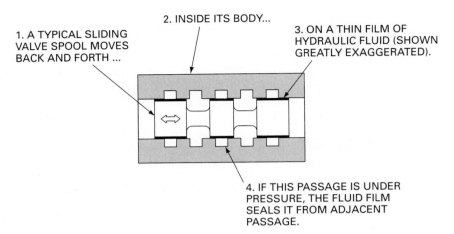

Figure 2-13. Fluid film lubricates and seals moving parts. *(Courtesy of Sperry Vickers, Sperry Rand Corp., Troy, Michigan.)*

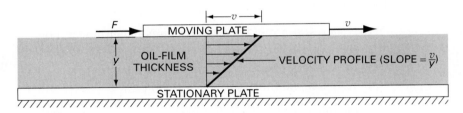

Figure 2-14. Fluid velocity profile between parallel plates due to viscosity.

where τ = Greek symbol tau = the shear stress in the fluid in units of force per unit area (lb/ft^2, N/m^2); the shear stress (which is produced by the force F) causes the sliding of adjacent layers of oil;

v = velocity of the moving plate (ft/s, m/s);

y = oil film thickness (ft, m);

μ = Greek symbol mu = the absolute viscosity of the oil;

F = force applied to the moving upper plate (lb, N);

A = area of the moving plate surface in contact with the oil (ft^2, m^2).

Checking units for μ in the English system using Eq. (2-13), we have

$$\mu = \frac{\text{lb/ft}^2}{(\text{ft/s})/\text{ft}} = \text{lb} \cdot \text{s/ft}^2$$

Similarly, μ has units of N $\cdot$ s/m^2 in the SI metric system.

If the moving plate has unit surface area in contact with the oil, and the upper plate velocity and oil film thickness are given unit values, Eq. (2-13) becomes

$$\mu = \frac{F/A}{v/y} = \frac{F/1}{1/1} = F$$

We can, therefore, define the absolute viscosity of a fluid as the force required to move a flat plate (of unit area at unit distance from a fixed plate) with a unit velocity when the space between the plates is filled with the fluid. Thus, using a fluid of higher viscosity requires a larger force, and vice versa. This shows that viscosity is a measure of a fluid's resistance to flow.

Viscosity is often expressed in the CGS (centimeter-gram-second) metric system. In the CGS metric system, the units per Eq. (2-13) are

$$\mu = \frac{dyn/cm^2}{(cm/s)/cm} = dyn \cdot s/cm^2$$

where a dyne is the force that will accelerate a 1-g mass at a rate of 1 cm/s². The conversions between dynes and newtons is as follows:

$$1\,N = 10^5\,dyn$$

A viscosity of 1 dyn · s/cm² is called a *poise*. The poise is a large unit of viscosity. A more convenient unit is the centipoise, abbreviated cP.

Kinematic Viscosity

Calculations in hydraulic systems often involve the use of kinematic viscosity rather than absolute viscosity. Kinematic viscosity equals absolute viscosity divided by density:

$$\nu = \frac{\mu}{\rho} \qquad\qquad \textbf{(2-14)}$$

where ν = Greek symbol nu = kinematic viscosity.

Units for kinematic viscosity are given as follows: English: ft²/s, SI metric: m²/s, and CGS metric: cm²/s.

A viscosity of 1 cm²/s is called a *stoke*. Because the stoke is a large unit, viscosity is typically reported in centistokes (cS).

Saybolt Viscometer

The viscosity of a fluid is usually measured by a Saybolt viscometer, which is shown schematically in Figure 2-15. Basically, this device consists of an inner chamber containing the sample of oil to be tested. A separate outer compartment, which

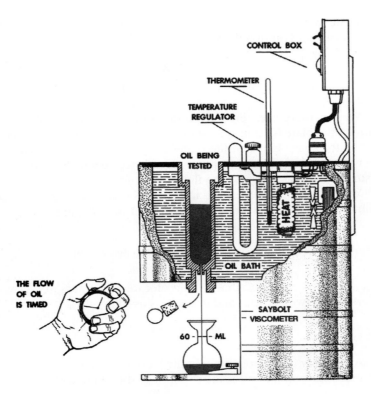

Figure 2-15. Saybolt viscometer.

completely surrounds the inner chamber, contains a quantity of oil whose temperature is controlled by an electrical thermostat and heater. A standard orifice is located at the bottom of the center oil chamber. When the oil sample is at the desired temperature, the time it takes to fill a 60-cm³ container through the metering orifice is then recorded. The time, (t), measured in seconds, is the viscosity of the oil in official units called Saybolt Universal Seconds (SUS). Since a thick liquid flows slowly, its SUS viscosity value will be higher than that for a thin liquid.

A relationship exists between the viscosity in SUS and cS. This relationship is provided by the following empirical equations:

$$\nu(\text{cS}) = 0.226t - \frac{195}{t}, \quad t \leq 100 \text{ SUS} \tag{2-15}$$

$$\nu(\text{cS}) = 0.220t - \frac{135}{t}, \quad t > 100 \text{ SUS} \tag{2-16}$$

where the symbol ν represents the viscosity in cS and t is the viscosity as measured in SUS, or simply seconds.

Kinematic viscosity is defined as absolute viscosity divided by density. Since, in the CGS metric system, density equals specific gravity (because $\rho_{H_2O} = 1$ g/cm^3 and thus has a value of one), the following equation can be used to find the kinematic viscosity in cS if the absolute viscosity in cP is known, and vice versa.

$$\nu(cS) = \frac{\mu(cP)}{SG} \qquad (2\text{-}17)$$

As previously noted, in the SI metric system, N · s/m^2 and m^2/s are the units used for absolute and kinematic viscosity, respectively. Although these are good units for calculation purposes, it is common practice in the fluid power industry to use viscosity expressed in units of SUS or cS.

EXAMPLE 2-12

An oil has a viscosity of 230 SUS at 150°F. Find the corresponding viscosity in units of centistokes and centipoise. The specific gravity of the oil is 0.9.

Solution

$$\nu(cS) = 0.220t - \frac{135}{t} = (0.220)(230) - \frac{135}{230} = 50 \text{ cS}$$

$$\mu(cP) = SG \times \nu(cS) = 0.9\nu(cS) = (0.9)(50) = 45 \text{ cP}$$

Capillary Tube Viscometer

A quick method for determining the kinematic viscosity of fluids in cS and absolute viscosity in cP is shown in Figure 2-16. This test measures the time it takes for a given amount of fluid to flow through a capillary tube under the force of gravity. The time in seconds is then multiplied by the calibration constant for the viscometer to obtain the kinematic viscosity of the sample fluid in centistokes. The absolute viscosity in centipoise is then calculated using Eq. (2-17).

EXAMPLE 2-13

An oil having a density of 0.89 g/cm^3 is tested using a capillary tube viscometer. The given amount of oil flowed through the capillary tube in 250 s. The calibration constant is 0.100. Find the kinematic and absolute viscosities in units of cS and cP, respectively.

Solution   Kinematic viscosity equals the time in seconds multiplied by the calibration constant:

$$\nu(cS) = (250)(0.100) = 25 \text{ cS}$$

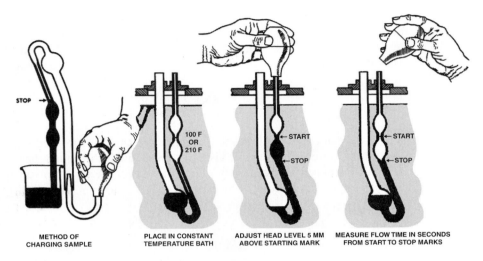

Figure 2-16. Capillary tube viscometer.

In the CGS metric system density numerically equals specific gravity. Therefore, $SG = 0.89$. Solving Eq. (2-17) for μ, we have

$$\mu(cP) = SG \times \nu(cS) = 0.89 \times 25 = 22.3 \text{ cP}$$

2.8 VISCOSITY INDEX

Oil becomes thicker as the temperature decreases and thins when heated. Hence, the viscosity of a given oil must be expressed at a specified temperature. For most hydraulic applications, the viscosity normally equals about 150 SUS at 100°F. It is a general rule of thumb that the viscosity should never fall below 45 SUS or rise above 4000 SUS regardless of the temperature. Figure 2-17 shows how viscosity changes with temperature for several liquid petroleum products. Note that the change in viscosity of a hydraulic oil as a function of temperature is represented by a straight line when using American Society for Testing and Materials (ASTM) standard viscosity temperature charts such as the one used in Figure 2-17. Displayed in the chart of Figure 2-17 is the preferred range of viscosities and temperatures for optimum operation of most hydraulic systems. Where extreme temperature changes are encountered, the fluid should have a high viscosity index.

Viscosity index (VI) is a relative measure of an oil's viscosity change with respect to temperature change. An oil having a low VI is one that exhibits a large change in viscosity with temperature change. A high-VI oil is one that has a relatively stable viscosity, which does not change appreciably with temperature change. The original VI scale ranged from 0 to 100, representing the poorest to best VI characteristics known at that time. Today, with improved refining techniques and chemical additives, oils exist with VI values well above 100. A high-VI oil is a good all-weather-type oil

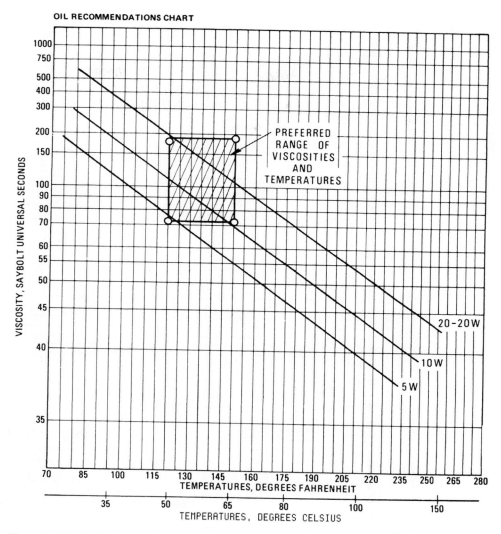

Figure 2-17. Preferred range of oil viscosities and temperatures. *(Courtesy of Sperry Vickers, Sperry Rand Corp., Troy, Michigan.)*

for use with outdoor machines operating in extreme temperature swings. This is where viscosity index is especially significant. For a hydraulic system where the oil temperature does not change appreciably, the viscosity index of the fluid is not as critical.

The VI of any hydraulic oil can be found by using

$$VI = \frac{L - U}{L - H} \times 100 \qquad\qquad (2\text{-}18)$$

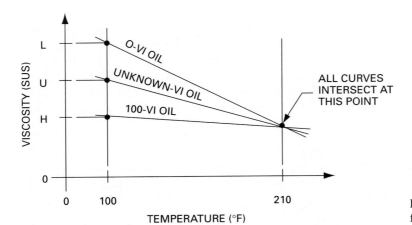

Figure 2-18. Typical curves for a viscosity index test.

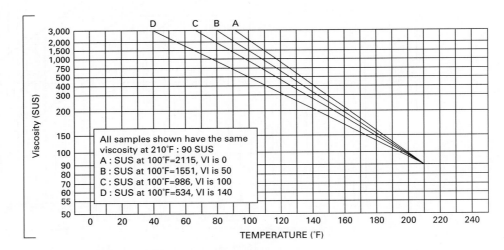

Figure 2-19. ASTM Standard Viscosity-Temperature Chart for Liquid Petroleum Products.

where L = viscosity in SUS of 0-VI oil at 100°F,

          U = viscosity in SUS of unknown-VI oil at 100°F,

          H = viscosity in SUS of 100-VI oil at 100°F.

The VI of an unknown-VI oil is determined from tests. A reference oil of 0 VI and a reference oil of 100 VI are selected, each of which has uniquely the same viscosity at 210°F as the unknown-VI oil. The viscosities of the three oils are then measured at 100°F to give values for L, U, and H. Figure 2-18 provides a pictorial representation of the H, L, and U terms in Eq. (2-18).

Figure 2-19 shows the results of a viscosity index test where oil A is the 0-VI oil and oil C is the 100-VI oil. The two unknown-VI oils B and D were found to have VI values of 50 and 140, respectively. For example, the VI for oil B is calculated as follows using the viscosity values at 100°F for oils A (*L* value), B (*U* value), and C (*H* value) from Figure 2-19:

$$\text{VI(oil B)} = \frac{L - U}{L - H} \times 100 = \frac{2115 - 1551}{2115 - 986} \times 100 = 50$$

EXAMPLE 2-14

A sample of oil with a VI of 80 is tested with a 0-VI oil and a 100-VI oil whose viscosity values at 100°F are 400 and 150 SUS, respectively. What is the viscosity of the sample oil at 100°F in units of SUS?

Solution Substitute directly into Eq. (2-18):

$$\text{VI} = \frac{L - U}{L - H} \times 100$$

$$80 = \frac{400 - U}{400 - 150} \times 100$$

$$U = 200 \text{ SUS}$$

Another characteristic relating to viscosity is called the *pour point*, which is the lowest temperature at which a fluid will flow. It is a very important parameter to specify for hydraulic systems that will be exposed to extremely low temperatures. As a rule of thumb, the pour point should be at least 20°F below the lowest temperature to be experienced by the hydraulic system.

2.9 AUTOMOTIVE SHOCK ABSORBERS: VISCOSITY AT WORK

Figure 2-20(a) provides a cutaway view of an automotive shock absorber that utilizes a pressurized gas and viscous oil for its operation. The operation of this shock absorber involves the flow of oil through orifices and the compression/expansion of a column of gas inside a cylinder. Shock absorbers reduce the bouncing effect of traveling over rough ground leading to improved ride quality and safety. Shock absorbers are designed to help automobiles ride more smoothly with shorter braking distances, handle better, and corner with more authority. Without shock absorbers the automobile would have a bouncing ride because of the energy that is stored in the suspension springs and then released back to the automobile. This bouncing action can cause motion that exceeds the allowable range of the suspension system.

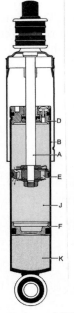

(a) Cutaway view of KONI
gas-charged shock absorber.

(b) Dodge viper with KONI shock absorbers.

Figure 2-20.   Automotive shock absorber application. *(Courtesy of KONI North America, Hebron, Kentucky.)*

The shock absorber shown in Figure 2-20(a) consists of an inner cylinder (B) containing a piston (E) connected to piston rod (A), a piston rod guide (D), and a floating piston (F). The gas (K) which is precharged to a pressure of about 350 psi, is located in the space below the floating piston. The oil (J) is located between the two pistons and in the space around the piston below the piston guide. The upper end of the piston rod extends outside the top of the shock absorber so that the rod can be connected to the suspension system of the automobile.

The rod-connected piston contains small orifices through which the oil can flow in either direction. When an automobile tire hits a road irregularity such as a bump or pothole, the piston rod is driven downward into the cylinder. This is called the bump stroke. This action forces the oil to flow upward through the piston orifices and the floating piston to move downward to further compress the gas below it. The volume of gas displaced by the floating piston equals the volume of oil displaced by the piston rod entering the cylinder during the bump stroke. The resistance of the orifices to the flow of the viscous oil provides the damping force that dissipates energy into heat. During the rebound stroke the opposite direction of motions takes place as the energy stored in the suspension springs is dissipated into heat.

Figure 2-20(b) is a photograph of the Dodge Viper ACR (American Club Racer) that uses the shock absorber design shown in Figure 2-20(a). The combination of the oil being forced to flow through the orifices and the compression/expansion of the gas (as both pistons move through the bump and rebound stroke) provides an optimum shock absorber stiffness and means of absorbing and dissipating vibration caused by the suspension springs as they absorb and release energy. Shock absorbers use a high-VI oil so that the viscosity of the oil doesn't change appreciably during changes in weather conditions from hot summers to cold winters and vice versa. Thus shock absorbers maintain essentially the same damping characteristics in all weather conditions.

2.10 ILLUSTRATIVE EXAMPLES USING THE SI METRIC SYSTEM

In this section we use the metric system of units to solve several example problems.

EXAMPLE 2-15

Find the pressure on a skin diver who has descended to a depth of 18.3 m in fresh water ($\gamma = 9800$ N/m³).

Solution   Per Eq. (2-9) we have

$$p = \gamma h$$

Substituting a consistent set of units yields:

$$p = (9800 \text{ N/m}^3)(18.3 \text{ m})$$
$$= 179,000 \text{ Pa} = 179 \text{ kPa}$$

EXAMPLE 2-16

Convert a −34,000 Pa pressure to an absolute pressure.

Solution

absolute pressure = gage pressure + atmospheric pressure

absolute pressure = −34,000 + 101,000 = 67,000 Pa abs

The "abs" is used to indicate that the 67,000 Pa value is an absolute pressure (measured relative to a perfect vacuum).

EXAMPLE 2-17

A 164-cm³ sample of oil is compressed in a cylinder until its pressure is increased from 687 kPa to 13,740 kPa. If the bulk modulus equals 1718 MPa, find the percent change in volume of the oil.

Solution

$$\frac{\Delta V}{V} = \frac{-\Delta p}{\beta} = \frac{-(13{,}740 - 687)}{1{,}718{,}000} = -0.0076 = -0.76\%$$

EXAMPLE 2-18

In Figure 2-14 the moving plate is 1 m on a side (in contact with the oil) and the oil film is 5 mm thick. A 10-N force is required to move the plate at velocity 1 m/s. Find the absolute viscosity of the oil in units of N · s/m² and cP.

Solution

Using Eq. (2-13) with SI metric units yields

$$\mu = \frac{F/A}{v/y} = \frac{Fy}{vA} = \frac{10 \text{ N} \times 0.005 \text{ m}}{1 \text{ m/s} \times 1 \text{ m}^2} = 0.05 \text{ N} \cdot \text{s/m}^2$$

Using Eq. (2-13) with CGS metric units, and noting that 1 N = 10⁵ dyn, we have

$$\mu = \frac{Fy}{vA} = \frac{10 \times 10^5 \text{ dyn} \times 0.5 \text{ cm}}{100 \text{ cm/s} \times (100 \text{ cm} \times 100 \text{ cm})}$$

$$= 0.5 \text{ dyn} \cdot \text{s/cm}^2 = 0.5 \text{ poise} = 50 \text{ cP}$$

The solution to this example shows that 0.05 N · s/m² = 50 cP or 1 N · s/m² = 1000 cP.

2.11 KEY EQUATIONS

Weight-mass relationship:	$F = W = mg$	**(2-1)**
Specific weight:	$\gamma = \dfrac{W}{V}$	**(2-3)**
Density:	$\rho = \dfrac{m}{V}$	**(2-5)**
Density-specific weight relationship:	$\rho = \dfrac{\gamma}{g}$	**(2-6)**
Specific gravity:	$SG = \dfrac{\gamma}{\gamma_{\text{water}}} = \dfrac{\rho}{\rho_{\text{water}}}$	**(2-7)**

Pressure-force relationship:

$$p = \frac{F}{A}$$

(2-8)

Pressure-head relationship:

$$p = \gamma H$$

(2-9)

Absolute pressure-gage pressure relationship:

$p_{abs} = p_{gage} + p_{atm}$

(2-10)

Fahrenheit–Celsius temperature relationship:

$T(°F) = 1.8T(°C) + 32$

(2-11)

Bulk modulus:

$$\beta = \frac{-\Delta p}{\Delta V/V}$$

(2-12)

Absolute viscosity:

$$\mu = \frac{\tau}{v/y}$$

(2-13)

Kinematic viscosity:

$$\nu = \frac{\mu}{\rho}$$

(2-14)

Conversion from SUS to cS

For $t \leq 100$ SUS:

$$\nu = 0.226t - \frac{195}{t}$$

(2-15)

For $t \geq 100$ SUS:

$$\nu = 0.220t - \frac{135}{t}$$

(2-16)

Viscosity index:

$$VI = \frac{L - U}{L - H} \times 100$$

(2-18)

EXERCISES

Questions, Concepts, and Definitions

2-1. What are the four primary functions of a hydraulic fluid?

2-2. Name 10 properties that a hydraulic fluid should possess.

2-3. Generally speaking, when should a hydraulic fluid be changed?

2-4. What are the differences between a liquid and a gas?

2-5. Name two advantages and two disadvantages that air has in comparison to oil when used in a fluid power system.

2-6. Define the terms *specific weight, density,* and *specific gravity.*

2-7. What is the difference between pressure and force?

2-8. Differentiate between gage and absolute pressures.

2-9. What is meant by the term *bulk modulus?*

2-10. Differentiate between the terms *viscosity* and *viscosity index.*

2-11. Name two undesirable results when using an oil with a viscosity that is too high.

2-12. Name two undesirable results when using an oil with a viscosity that is too low.

2-13. Relative to viscosity measurement, what is a Saybolt Universal Second (SUS)?

2-14. Define the term *pour point*.

2-15. What is meant by the term *pressure head?*

2-16. How is the mercury column supported in a barometer?

2-17. How is the viscosity of hydraulic oil affected by temperature change?

2-18. Under what condition is viscosity index important?

2-19. As a fluid becomes more compressible, does the bulk modulus decrease or increase?

2-20. It is desired to select an oil of a viscosity index suitable for a hydraulic powered front end loader that will operate year-round. What VI characteristic should be specified? Why?

Problems

Note: The letter *E* following an exercise number means that English units are used. Similarly, the letter *M* indicates metric units.

Weight, Density, and Specific Gravity

2-21E. A hydraulic fluid has a specific weight of 55 lb/ft³. What is its specific gravity? What is its density?

2-22E. Fifty gallons of hydraulic oil weigh 372 lb. What is the specific weight in units of lb/ft³?

2-23M. Calculate the density of a hydraulic oil in units of kg/m³ knowing that the density is 1.74 slugs/ft³.

2-24E. A container weighs 3 lb when empty, 53 lb when filled with water, and 66 lb when filled with glycerin. Find the specific gravity of the glycerin.

2-25M. Air at 20°C and atmospheric pressure has a density of 1.23 kg/m³.
   a. Find its specific gravity.
   b. What is the ratio of the specific gravity of water divided by the specific gravity of air at 20°C and atmospheric pressure?
   c. What is the significance of the ratio in part b?

2-26M. A cylindrical container has a diameter of 0.5 m and a height of 1 m. If it is to be filled with a liquid having a specific weight of 2000 N/m³, how many kg of this liquid must be added?

2-27M. A liter of SAE 30 oil weighs 8.70 N. Calculate the oil's
   a. Specific weight
   b. Density
   c. Specific gravity

2-28M. A tank truck contains 125,000 liters of a hydraulic fluid having a specific gravity of 0.9. Determine the fluid's specific weight, density, and weight.

Pressure, Head, and Force

2-29E. What is the pressure at the bottom of a 30-ft column of the hydraulic fluid of Exercise 2-21?

2-30M. Find the absolute pressure on the skin diver of Example 2-5 in units of Pa.

2-31M. Convert a −2-kPa pressure to an absolute pressure in kPa.

2-32E. A 100-gal reservoir is to be mounted within a 2-ft-by-2-ft square. What is the minimum height of the reservoir?

2-33E. A 100-ft-long pipe is inclined at a 30° angle with the horizontal. It is filled with oil of specific gravity 0.90. What is the pressure at the base of the pipe if the top of the pipe is vented to the atmosphere?

2-34M. For the fluid power automotive lift system of Figure 2-21, the hydraulic piston has a 250-mm diameter. How much oil pressure (kPa) is required to lift a 13,300-N automobile?

Bulk Modulus

2-35E. A 20-in³ sample of oil is compressed in a cylinder until its pressure is increased from 50 to 1000 psi. If the bulk modulus equals 300,000 psi, find the change in volume of the oil.

2-36M. A 500-cm³ sample of oil is compressed in a cylinder until its pressure is increased from 1 atm to 50 atm. If the bulk modulus equals 1750 MPa, find the percent change in volume of the oil.

2-37E. The load on a 2-in.-diameter hydraulic cylinder increases from 10,000 lb to 15,000 lb. Due to the compressibility of the oil, the piston retracts 0.01 in. If the volume of oil under compression is 10 in³, what is the bulk modulus of the oil in units of psi?

2-38M. Convert the data of Exercise 2-37 to metric units and determine the bulk modulus of the oil in units of MPa.

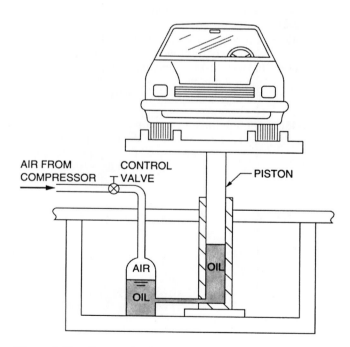

Figure 2-21. System for Exercise 2-34.

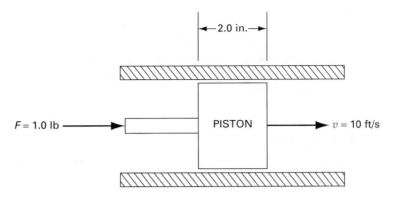

Figure 2-22.   System for Exercise 2-45.

Viscosity and Viscosity Index

2-39M. An oil (SG = 0.9) has a viscosity of 200 SUS at 120°F. Find the corresponding viscosity in units of cS and cP.

2-40E. A sample of oil with a VI of 70 is tested with a 0-VI oil and a 100-VI oil whose viscosity values at 100°F are 375 and 125 SUS, respectively. What is the viscosity of the sample oil at 100°F in units of SUS?

2-41. Derive the conversion factor between viscosity in units of lb · s/ft² and N · s/m².

2-42. Derive the conversion factor between viscosity in units of ft²/s and m²/s.

2-43E. A fluid has a viscosity of 12.0 P and a specific gravity of 0.89. Find the kinematic viscosity of this fluid in units of ft²/s.

2-44M. In Figure 2-14 the moving plate is 0.7 m on a side (in contact with the oil) and the oil film is 4 mm thick. A 6-N force is required to move the plate at a velocity of 1 m/s. If the oil has a specific gravity of 0.9, find the kinematic viscosity of the oil in units of cS.

2-45E. A 1.0-lb force moves a piston inside a cylinder at a velocity of 10 ft/s, as shown in Figure 2-22. The 4.0-in. diameter piston is centrally located in the 4.004-in inside-diameter cylinder. An oil film separates the piston from the cylinder. Find the absolute viscosity of the oil in units of lb · s/ft².

3 Energy and Power in Hydraulic Systems

Learning Objectives

Upon completing this chapter, you should be able to:

1. Differentiate between hydraulic energy and hydraulic power.
2. Define the term *efficiency*.
3. Describe the operation of an air-to-hydraulic pressure booster.
4. Explain the conservation of energy law.
5. Calculate fluid flow rates and velocities using the continuity equation.
6. Evaluate the power delivered by a hydraulic cylinder.
7. Determine the speed of a hydraulic cylinder.
8. Describe the differences among elevation energy, pressure energy, and kinetic energy.
9. Describe the operation of a hydraulic jack.
10. Apply Bernoulli's equation to determine the energy transfer within a hydraulic system.
11. Understand the meaning of the terms *elevation head, pressure head*, and *velocity head*.
12. Differentiate between the terms *pump head, motor head*, and *head loss*.

3.1 INTRODUCTION

Energy is defined as the ability to perform work, and thus the transfer of energy is a key consideration in the operation of hydraulic systems. Figure 3-1 provides a block diagram illustrating how energy is transferred throughout a hydraulic system (contained within the dashed lines). As shown, the prime mover (such as an electric

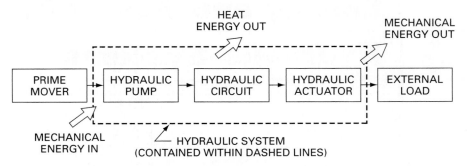

Figure 3-1. Block diagram of hydraulic system showing major components along with energy input and output terms.

motor or an internal combustion engine) delivers input energy to a pump of the hydraulic system via a rotating shaft. The pump converts this mechanical energy into hydraulic energy by increasing the fluid's pressure and velocity. The fluid flows to an actuator via a hydraulic circuit that consists of pipelines containing valves and other control components. The hydraulic circuit controls the pressures and flow rates throughout the hydraulic system. The actuator (either a hydraulic cylinder or motor) converts the hydraulic energy from the fluid into mechanical energy to drive the external load via a force or torque in an output shaft.

Some of the hydraulic energy of the fluid is lost due to friction as the fluid flows through pipes, valves, fittings, and other control components. These frictional energy losses, which are analyzed in Chapter 4, show up as heat energy which is transferred to the environment and/or to the fluid as evident by an increase in fluid temperature. Energy transformed into heat (due to friction) is lost because the heat energy is not available to perform useful work.

All these forms of energy are accounted for by the conservation of energy law, which states that energy can be neither created nor destroyed. This is expressed by the following equation for the system of Figure 3-1:

$$\text{Input } ME - \text{Lost } HE = \text{Output } ME$$

where ME = mechanical energy,
         HE = heat energy.

Power is defined as the rate of doing work or expending energy. Thus the rate at which the prime mover adds energy to the pump equals the power input to the hydraulic system. Likewise, the rate at which the actuator delivers energy to the external load equals the power output of the hydraulic system. The power output is determined by the requirements of the external load. The greater the force or torque required to move the external load and the faster this work must be done, the greater must be the power output of the hydraulic system.

Figure 3-2. Hydraulic power brush drive. *(Courtesy of Eaton Corp., Fluid Power Division, Eden Prairie, Minnesota.)*

    A hydraulic system is not a source of energy. The energy source is the prime mover, which drives the pump. Thus, in reality, a hydraulic system is merely an energy transfer system. Why not, then, eliminate hydraulics and simply couple the load directly to the prime mover? The answer is that a hydraulic system is much more versatile in its ability to transmit power. This versatility includes advantages of variable speed, reversibility, overload protection, high power-to-weight ratio, and immunity to damage under stalled conditions.

    Figure 3-2 shows a hydraulically powered brush drive that is used for cleaning roads and floors in various industrial locations. Mounted directly at the hub of the front and side sweep-scrub brushes are compact hydraulic motors that deliver power right where it's needed. The hydraulic system eliminates bulky mechanical linkages for efficient, lightweight machine operation. The result is continuous, rugged industrial cleaning action at the flip of a simple valve.

3.2 REVIEW OF MECHANICS

Introduction

Since fluid power deals with the generation of forces to accomplish useful work, it is essential that the basic laws of mechanics be clearly understood. Let's, therefore, have a brief review of mechanics as it relates to fluid power systems.

    Forces are essential to the production of work. No motion can be generated and hence no power can be transmitted without the application of some force. It was in the late seventeenth century when Sir Isaac Newton formulated the three laws of motion dealing with the effect a force has on a body:

1. A force is required to change the motion of a body.
2. If a body is acted on by a force, the body will have an acceleration proportional to the magnitude of the force and inverse to the mass of the body.
3. If one body exerts a force on a second body, the second body must exert an equal but opposite force on the first body.

The motion of a body can be either linear or angular depending on whether the body travels along a straight line or rotates about a fixed point.

Linear Motion

If a body experiences linear motion, it has a *linear velocity* (or simply *velocity*), which is defined as the distance traveled divided by the corresponding time.

$$v = \frac{s}{t} \tag{3-1}$$

where in the English system of units: s = distance (in or ft),
t = time (s or min),
v = velocity (in/s, in/min, ft/s, or ft/min).

If the body's velocity changes, the body has an acceleration, which is defined as the change in velocity divided by the corresponding change in time ($a = \Delta v/\Delta t$). In accordance with Newton's first law of motion, a force is required to produce this change in velocity. Per Newton's second law, we have

$$F = ma \tag{3-2}$$

where F = force (lb),
a = acceleration (ft/s^2),
m = mass (slugs).

This brings us to the concept of energy, which is defined as the ability to perform work. Let's assume that a force acts on a body and moves the body through a specified distance in the direction of the applied force. Then, by definition, work has been done on the body. The amount of this work equals the product of the force and distance where both the force and distance are measured in the same direction:

$$W = FS \tag{3-3}$$

where F = force (lb),
S = distance (in or ft),
W = work (in · lb or ft · lb).

This leads us to a discussion of power, which is defined as the rate of doing work or expending energy. Thus, power equals work divided by time:

$$power = \frac{FS}{t}$$

But since S/t equals v we can rewrite the power equation as follows:

$$power = Fv \tag{3-4}$$

where F = force (lb),

v = velocity (in/s, in/min, ft/s, or ft/min),

power has units of in · lb/s, in · lb/min, ft · lb/s, or ft · lb/min.

Power is a measure of how fast work is done and (in the English system of units) is usually measured in units of horsepower (hp). By definition, 1 hp equals 550 ft · lb/s or 33,000 ft · lb/min. Thus, we have

$$\text{horsepower} = \text{HP} = \frac{F(\text{lb}) \times v(\text{ft/s})}{550} = \frac{F(\text{lb}) \times v(\text{ft/min})}{33,000} \qquad \textbf{(3-5)}$$

The unit of horsepower was created by James Watt at the end of the nineteenth century, when he attempted to compare the rate of doing work by a horse in comparison with a steam engine. During a test he showed that a horse could raise a 150-lb weight (using a block-and-tackle) at an average velocity of 3.67 ft/s. This rate of doing work equals 150 lb × 3.67 ft/s or 550 ft · lb/s, which he defined as 1 horsepower.

EXAMPLE 3-1

A person exerts a 30-lb force to move a hand truck 100 ft in 60 s.

a. How much work is done?

b. What is the power delivered by the person?

Solution

a.
$$W = FS = (30 \text{ lb})(100 \text{ ft}) = 3000 \text{ ft} \cdot \text{lb}$$

b.
$$\text{power} = \frac{FS}{t} = \frac{(30 \text{ lb})(100 \text{ ft})}{60 \text{ s}} = 50 \text{ ft} \cdot \text{lb/s}$$

$$\text{HP} = \frac{50 \text{ ft} \cdot \text{lb/s}}{(550 \text{ ft} \cdot \text{lb/s})/\text{hp}} = 0.091 \text{ hp}$$

Angular Motion

Just as in the case of linear motion, angular motion is caused by the application of a force. Consider, for example, a force F applied to a wrench to tighten a bolt as shown in Figure 3-3. The force F has a moment arm R relative to the center of the bolt. Thus, the force F creates a torque T about the center of the bolt. It is the torque T that causes the wrench to rotate the bolt through a given angle until it is tightened. Note that the moment arm is measured from the center of the bolt (center of rotation)

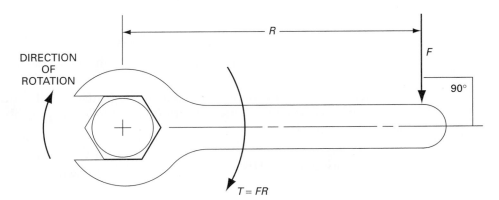

Figure 3-3. Force F applied to wrench creates torque T to tighten bolt.

perpendicularly to the line of action of the force. The resulting torque is a clockwise torque because it rotates the bolt clockwise as shown. The magnitude of the torque equals the product of the applied force F and its moment arm R.

$$T = FR \qquad (3\text{-}6)$$

where F = force (lb),
          R = moment arm (in or ft),
          T = torque (in · lb or ft · lb).

A second example of angular motion is illustrated in Figure 3-4. In this case we have a force F applied to a flat belt wrapped around the periphery of a circular disk mounted on a shaft. The shaft is mounted in bearings so that it can rotate as it drives the connecting shaft of a pump. The force F creates a torque T whose magnitude equals the product of the force F and its moment arm R that equals the radius of the disk. The resulting torque causes the disk and thus the connecting shaft to rotate at some angular speed measured in units of revolutions per minute (rpm).

The resulting torque is clockwise because it rotates the shaft clockwise as shown. As the shaft rotates and overcomes the load resistance of the pump, work is done and power is transmitted to the pump. The amount of horsepower transmitted can be found from

$$\mathrm{HP} = \frac{TN}{63{,}000} \qquad (3\text{-}7)$$

where T = torque (in · lb),
          N = rotational speed (rpm),
          HP = torque horsepower or brake horsepower.

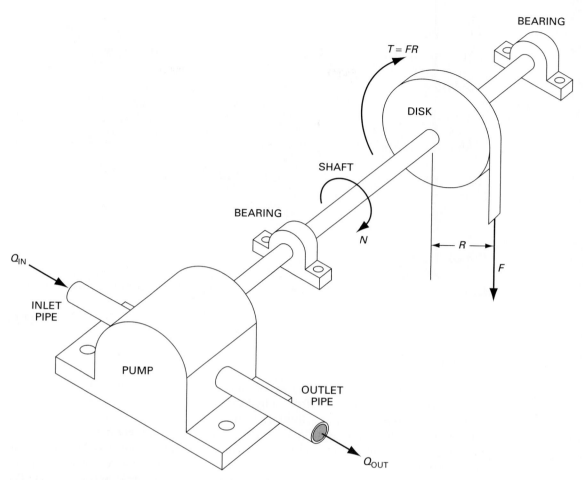

Figure 3-4. Force F applied to periphery of disk creates torque T in shaft to drive pump at rotational speed N. Q is the flow rate of oil (volume per unit time) produced by pump.

In fluid power applications, Eq. (3-7) is used to calculate the horsepower delivered by a prime mover (such as an electric motor) to drive a pump or by a hydraulic motor to drive an external load. The system in Figure 3-4 uses a disk with a flat belt wrapped around its periphery to illustrate how a force can create a torque to deliver power via a rotating shaft, per Eq. (3-7). Hence, the combination of the disk and its force-loaded belt represents the prime mover.

Since the horsepower represented by Eq. (3-7) is transmitted by torque in a rotating shaft, it is called *torque horsepower*. It is also commonly called brake horsepower (BHP) because a prony brake is a mechanical device used to measure the amount of horsepower transmitted by a torque-driven rotating shaft.

The derivation of Eq. (3-7) shows that the value of 63,000 for the constant in the denominator is valid only when the torque T has units of in · lb and the rotational speed N has units of rpm. Eq. (3-7) is derived in Appendix H.

EXAMPLE 3-2

How much torque is delivered by a 2-hp, 1800-rpm hydraulic motor?

Solution

$$HP = \frac{TN}{63,000}$$

Substituting known values, we have

$$2 = \frac{T(1800)}{63,000}$$

$$T = 70 \text{ in} \cdot \text{lb}$$

Efficiency

Efficiency is defined as output power divided by input power. Mathematically we have

$$\eta = \frac{\text{output power}}{\text{input power}} \tag{3-8}$$

where η = Greek letter eta = efficiency.

The efficiency of any system or component is always less than 100% and is calculated to determine power losses. In hydraulic systems these losses are due to fluid leakage past close-fitting parts, fluid friction due to fluid movement, and mechanical friction due to the rubbing of mating parts. Efficiency determines the amount of power that is actually delivered in comparison to the power received. The power difference (input power – output power) represents loss power since it is transformed into heat due to frictional effects and thus is not available to perform useful work. The output power is usually computed from force and linear velocity (or torque and angular velocity) associated with the load. The input power is normally computed from the same parameters associated with the prime mover.

EXAMPLE 3-3

An elevator raises a 3000-lb load through a distance of 50 ft in 10 s. If the efficiency of the entire system is 80% (0.80 in decimal fraction form for use in equations), how much input horsepower is required by the elevator hoist motor?

Solution

$$\text{output power} = \frac{FS}{t} = \frac{(3000 \text{ lb})(50 \text{ ft})}{(10 \text{ s})} = 15,000 \text{ ft} \cdot \text{lb/s}$$

$$\text{output HP} = \frac{15{,}000}{550} = 27.3 \text{ hp}$$

$$\eta = \frac{\text{output power}}{\text{input power}}$$

$$0.80 = \frac{27.3 \text{ hp}}{\text{input power}}$$

$$\text{input power} = 34.1 \text{ hp}$$

3.3 MULTIPLICATION OF FORCE (PASCAL'S LAW)

Introduction

Pascal's law reveals the basic principle of how fluid power systems perform useful work. This law can be stated as follows: *Pressure applied to a confined fluid is transmitted undiminished in all directions throughout the fluid and acts perpendicular to the surfaces in contact with the fluid.* Pascal's law explains why a glass bottle, filled with a liquid, can break if a stopper is forced into its open end. The liquid transmits the pressure, created by the force of the stopper, throughout the container, as illustrated in Figure 3-5. Let's assume that the area of the stopper is 1 in^2 and that the area of the bottom is 20 in^2. Then a 10-lb force applied to the stopper produces a pressure of 10 psi. This pressure is transmitted undiminished to the bottom of the bottle, acting on the full 20-in^2 area and producing a 200-lb force. Therefore, it is possible to break out the bottom by pushing on the stopper with a moderate force.

Pascal's Law Applied to Simple Hydraulic Jack

The glass bottle example of Figure 3-5 shows how a small force exerted on a small area can create, via Pascal's law, a proportionally larger force on a larger area. Figure 3-6(a) shows how Pascal's law can be applied to produce a useful amplified output force in a simple hydraulic jack which is a device used for lifting large weights. An input force of 10 lb is applied to a 1-in^2 area piston. This develops a 10-psi pressure throughout the oil within the housing of the jack. This 10-psi pressure acts on a 10-in^2 area piston, producing a 100-lb output force. This output force performs useful work as it lifts the 100-lb weight.

It is interesting to note that there is a similarity between the simple hydraulic jack of Figure 3-6(a) and the mechanical lever system of Figure 3-6(b). Note in Figure 3-6(b) that the 10-lb input force has a moment arm (distance from force to pivot) that is 10 times as long as the moment arm of the output force. As a result, the output force is 10 times as large as the input force and thus equals 100 lb. This force multiplication

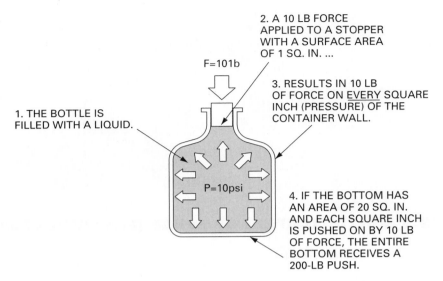

2. A 10 LB FORCE
APPLIED TO A STOPPER
WITH A SURFACE AREA
OF 1 SQ. IN. ...

F=10lb

3. RESULTS IN 10 LB
OF FORCE ON <u>EVERY</u> SQUARE
INCH (PRESSURE) OF THE
CONTAINER WALL.

1. THE BOTTLE IS
FILLED WITH A LIQUID.

P=10psi

4. IF THE BOTTOM HAS
AN AREA OF 20 SQ. IN.
AND EACH SQUARE INCH
IS PUSHED ON BY 10 LB
OF FORCE, THE ENTIRE
BOTTOM RECEIVES A
200-LB PUSH.

Figure 3-5. Demonstration of Pascal's law. *(Courtesy of Sperry Vickers, Sperry Rand Corp., Troy, Michigan.)*

in mechanical systems due to levers is similar to that produced in fluid power systems due to Pascal's law.

Analysis of Simple Hydraulic Jack

An interesting question relative to the hydraulic jack of Figure 3-6(a) is, Are we getting something for nothing? That is to say, does a hydraulic jack produce more energy than it receives? A fluid power system (like any other power system) cannot create energy. This is in accordance with the conservation of energy law presented in Section 3.6. To answer this question, let's analyze the hydraulic jack illustrated in Figure 3-7.

As shown, a downward input force F_1 is applied to the small-diameter piston 1, which has an area A_1. This produces an oil pressure p_1 at the bottom of piston 1. This pressure is transmitted through the oil to the large-diameter piston 2, which has an area A_2. The pressure p_2 at piston 2 pushes up on the piston to create an output force F_2.

By Pascal's law, $p_1 = p_2$. Since pressure equals force divided by area, we have

$$\frac{F_1}{A_1} = \frac{F_2}{A_2}$$

or

$$\frac{F_2}{F_1} = \frac{A_2}{A_1} \tag{3-9}$$

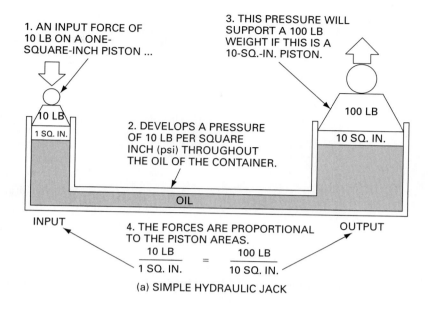

1. AN INPUT FORCE OF 10 LB ON A ONE-SQUARE-INCH PISTON ...

3. THIS PRESSURE WILL SUPPORT A 100 LB WEIGHT IF THIS IS A 10-SQ.-IN. PISTON.

10 LB

1 SQ. IN.

100 LB

10 SQ. IN.

2. DEVELOPS A PRESSURE OF 10 LB PER SQUARE INCH (psi) THROUGHOUT THE OIL OF THE CONTAINER.

OIL

INPUT

OUTPUT

4. THE FORCES ARE PROPORTIONAL TO THE PISTON AREAS.

$$\frac{10\ LB}{1\ SQ.\ IN.} = \frac{100\ LB}{10\ SQ.\ IN.}$$

(a) SIMPLE HYDRAULIC JACK

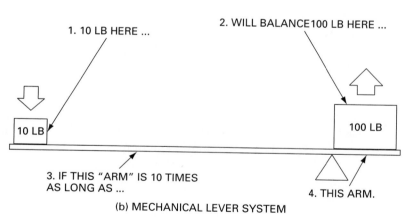

1. 10 LB HERE ...

2. WILL BALANCE 100 LB HERE ...

10 LB

100 LB

3. IF THIS "ARM" IS 10 TIMES AS LONG AS ...

4. THIS ARM.

(b) MECHANICAL LEVER SYSTEM

Figure 3-6. Force multiplication in a simple hydraulic jack and in a mechanical lever system. *(Courtesy of Sperry Vickers, Sperry Rand Corp., Troy, Michigan.)*

Thus, a force multiplication occurs from the input to the output of the jack if the output piston area is greater than the input piston area. The force multiplication ratio F_2/F_1 equals the piston area ratio A_2/A_1. However, we will next show that the output piston does not move as far as does the input piston. The ratio of the piston movements can be determined by assuming the oil to be incompressible. Thus, the

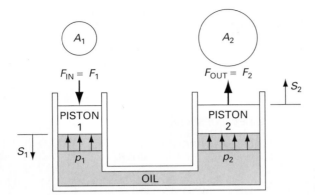

Figure 3-7. Operation of simple hydraulic jack.

cylindrical volume of oil displaced by the input piston equals the cylindrical volume displaced by the output piston:

$$V_1 = V_2$$

Since the volume of a cylinder equals the product of its cross-sectional area and its height we have

$$A_1 S_1 = A_2 S_2$$

where S_1 = the downward movement of piston 1,
S_2 = the upward movement of piston 2.

Thus,

$$\frac{S_2}{S_1} = \frac{A_1}{A_2} \qquad\qquad \textbf{(3-10)}$$

Equation 3-10 shows that the large output piston does not travel as far as the small input piston. Note that the piston stroke ratio S_2/S_1 equals the piston area ratio A_1/A_2. Thus, for a piston area ratio of 2, the output force increases by a factor of 2, but the output motion decreases by a factor of 2. Hence, in a hydraulic jack we do not get something for nothing. The output force is greater than the input force, but the output movement is less than the input movement. Combining Eqs. (3-9) and (3-10) yields the corresponding relationship

$$\frac{F_2}{F_1} = \frac{S_1}{S_2}$$

Hence,

$$F_1 S_1 = F_2 S_2 \qquad\qquad (3\text{-}11)$$

Recall that work energy equals the product of force and the distance moved by the force. Thus Eq. (3-11) states that the energy input to the hydraulic jack equals the energy output from the jack. This result occurs because the force-multiplication factor equals the motion-reduction factor. It should be noted that in a real hydraulic jack, friction between the piston and cylindrical bore surface will produce frictional energy losses. This causes the actual output energy to be less than the input energy. This loss of energy is taken into account in Chapter 4 where the energy equation is used to solve hydraulic system problems.

EXAMPLE 3-4

For the hydraulic jack of Figure 3-6, the following data are given:

$$A_1 = 2 \text{ in}^2 \quad A_2 = 20 \text{ in}^2$$

$$S_1 = 1 \text{ in}$$

$$F_1 = 100 \text{ lb}$$

Determine each of the following.

a. F_2

b. S_2

c. The energy input

d. The energy output

Solution

a. From Eq. (3-9) we have

$$F_2 = \frac{A_2}{A_1} \times F_1 = \frac{20}{2} \times 100 = 1000 \text{ lb}$$

b. From Eq. (3-10) we have

$$S_2 = \frac{A_1}{A_2} \times S_1 = \frac{2}{20} \times 1 = 0.1 \text{ in}$$

c. Energy input $= F_1 S_1 = (100 \text{ lb}) \times (1 \text{ in}) = 100 \text{ in} \cdot \text{lb}$

d. Energy output $= F_2 S_2 = (1000 \text{ lb}) \times (0.1 \text{ in}) = 100 \text{ in} \cdot \text{lb}$

3.4 APPLICATIONS OF PASCAL'S LAW

In this section we examine two applications of Pascal's law: the hand-operated hydraulic jack and the air-to-hydraulic booster.

Hand-Operated Hydraulic Jack

This system uses a piston-type hand pump to power a hydraulic load cylinder for lifting loads, as illustrated in Figure 3-8. The operation is as follows.

A hand force is applied at point A of handle ABC, which pivots about point C. The piston rod of the pump cylinder is pinned to the input handle at point B. The pump cylinder contains a small-diameter piston, which is free to move up and down. The piston and rod are rigidly connected together. When the handle is pulled up, the piston rises and creates a vacuum in the space below it. As a result, atmospheric pressure forces oil to leave the oil tank and flow through check valve 1 to fill the void created below the pump piston. This is the suction process. A check valve allows flow to pass in only one direction, as indicated by the arrow.

When the handle is then pushed down, oil is ejected from the small-diameter pump cylinder and flows through check valve 2 and enters the bottom end of the large-diameter load cylinder. The load cylinder is similar in construction to the pump cylinder and contains a piston connected to a rod. Pressure builds up below the load piston as oil ejected from the pump cylinder meets resistance in finding a place to go. From Pascal's law we know that the pressure acting on the load piston equals the pressure generated by the pump piston. The pressure generated by the pump piston equals the force applied to the pump piston rod divided by the area of the pump piston. The load that can be lifted equals the product of the pressure and the area of the load piston. Also, each time the input handle is cycled up and down, a specified volume

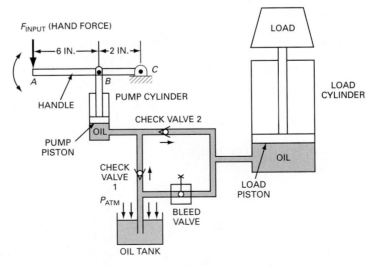

Figure 3-8. Hand-operated hydraulic jack.

of oil is ejected from the pump to raise the load cylinder a given distance. The bleed valve is a hand-operated valve, which, when opened, allows the load to be lowered by bleeding oil from the load cylinder back to the oil tank.

It should be noted that oil enters and exits from each cylinder at only one end. Such a cylinder is called *single acting* because it is hydraulically powered in only one direction. In the hydraulic jack of Figure 3-8, the retraction of the load piston is accomplished by the weight of the load pushing downward on the load piston rod when the bleed valve is opened.

Figure 3-9 shows three sizes of hand-operated hydraulic jacks having rated load capacities of 12, 20, and 30 tons respectively. The handle effort required at rated load capacity is 60, 70, and 50 lb respectively. The jack piston strokes are $3^3/_4$, $3^3/_8$, and $3^1/_8$ in and the number of pump strokes required to achieve the piston strokes are 26, 22, and 35 respectively. These jacks operate both vertically and horizontally for use in a variety of lifting, pushing, and spreading applications. Figure 3-10 shows a hand-operated hydraulic jack in which the pump and cylinder are separate units with an interconnecting hydraulic hose. These hydraulic jacks come in cylinder and pump combinations that provide strokes up to $14^1/_4$ in and rated load capacities up to 100 tons.

EXAMPLE 3-5

An operator makes one complete cycle per second interval using the hydraulic jack of Figure 3-8. Each complete cycle consists of two pump cylinder strokes (intake and power). The pump cylinder has a 1-in-diameter piston and the

Figure 3-9. Three sizes of hand-operated hydraulic jacks. *(Courtesy of SPX Fluid Power, Rockford, Illinois.)*

Figure 3-10. Hand-operated hydraulic jack with separate pump and cylinder. *(Courtesy of SPX Fluid Power, Rockford, Illinois.)*

load cylinder has a 3.25-in-diameter piston. If the average hand force is 25 lb during the power stroke,

a. How much load can be lifted?
b. How many cycles are required to lift the load 10 in assuming no oil leakage? The pump piston has a 2-in stroke.
c. What is the output HP assuming 100% efficiency?
d. What is the output HP assuming 80% efficiency?

Solution

a. First determine the force acting on the rod of the pump cylinder due to the mechanical advantage of the input handle:

$$F_{rod} = \frac{8}{2} \times F_{input} = \frac{8}{2}(25) = 100 \text{ lb}$$

Next, calculate the pump cylinder discharge pressure p:

$$p = \frac{\text{rod force}}{\text{piston area}} = \frac{F_{rod}}{A_{pump \ piston}} = \frac{100 \text{ lb}}{(\pi/4)(1)^2 in^2} = 127 \text{ psi}$$

Per Pascal's law this is also the same pressure acting on the load piston. We can now calculate the load-carrying capacity:

$$F_{load} = pA_{load \ piston} = (127) lb/in^2 \left[\frac{\pi}{4}(3.25)^2 \right] in^2 = 1055 \text{ lb}$$

b. To find the load displacement, assume the oil to be incompressible. Therefore, the total volume of oil ejected from the pump cylinder equals the volume of oil displacing the load cylinder:

$$(A \times S)_{pump \ piston} \times (\text{no. of cycles}) = (A \times S)_{load \ piston}$$

Substituting, we have

$$\frac{\pi}{4}(1)^2 in^2 \times 2 \text{ in} \times (\text{no. of cycles}) = \frac{\pi}{4}(3.25)^2 \ in^2 \times 10 \text{ in}$$

$$1.57 \ in^3 \times (\text{no. of cycles}) = 82.7 \ in^3$$

$$\text{no. of cycles} = 52.7$$

c.

$$\text{Power} = \frac{FS}{t} = \frac{(1055 \text{ lb})\left(\frac{10}{12} \text{ ft}\right)}{52.7 \text{ s}} = 16.7 \text{ ft} \cdot \text{lb/s}$$

$$HP = \frac{16.7}{550} = 0.030 \text{ hp}$$

The HP output value is small, as expected, since the power comes from a human being.

d.

$$HP = (0.80)(0.03) = 0.024 \text{ hp}$$

Air-to-Hydraulic Pressure Booster

This device is used for converting shop air into the higher hydraulic pressure needed for operating hydraulic cylinders requiring small to medium volumes of higher-pressure oil. It consists of a cylinder containing a large-diameter air piston driving a small-diameter hydraulic piston, which is actually a long rod connected to the piston (see Figure 3-11). Any shop equipped with an air line can obtain smooth, efficient hydraulic power from an air-to-hydraulic pressure booster hooked into the air line. The alternative would be a complete hydraulic system including expensive pumps and high-pressure valves. Other benefits include a space savings and a reduction in operating costs and maintenance.

Figure 3-11. Cutaway view of an air-to-hydraulic pressure booster. *(Courtesy of the S-P Manufacturing Corp., Cleveland, Ohio.)*

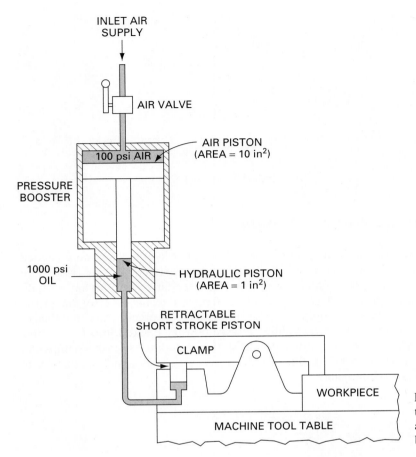

Figure 3-12. Manufacturing application of air-to-hydraulic pressure booster.

Figure 3-12 shows an application where an air-to-hydraulic pressure booster is supplying high-pressure oil to a hydraulic cylinder whose short stroke piston is used to clamp a workpiece to a machine tool table. Since shop air pressure normally operates at 100 psi, a pneumatically operated clamp would require an excessively large cylinder to rigidly hold the workpiece while it is being machined.

The air-to-hydraulic pressure booster operates as follows (see Figure 3-12): Let's assume that the air piston has a 10-in^2 area and is subjected to 100-psi air pressure. This produces a 1000-lb force on the booster's hydraulic piston. Thus, if the area of the booster's hydraulic piston is 1 in^2, the hydraulic discharge oil pressure will be 1000 psi. Per Pascal's law, this produces 1000-psi oil at the short stroke piston of the hydraulic clamping cylinder mounted on the machine tool table.

The pressure ratio of an air-to-hydraulic pressure booster can be found by using Eq. (3-12).

$$\text{pressure ratio} = \frac{\text{output oil pressure}}{\text{input air pressure}} = \frac{\text{area of air piston}}{\text{area of hydraulic piston}} \quad \textbf{(3-12)}$$

Substituting into Eq. (3-12) for the pressure booster of Figure 3-12 we have

$$\text{pressure ratio} = \frac{1000 \text{ psi}}{100 \text{ psi}} = \frac{10 \text{ in}^2}{1 \text{ in}^2} = 10$$

For a clamping cylinder piston area of 0.5 in² (diameter = 0.798 in), the clamping force equals 1000 lb/in² × 0.5 in² or 500 lb. To provide the same 500-lb clamping force without a booster would require a clamping cylinder piston area of 5 in² (diameter = 2.52 in), assuming 100-psi air pressure.

Air-to-hydraulic pressure boosters are available in a wide range of pressure ratios and can provide hydraulic pressures up to 15,000 psi using 100-psi shop air. An application circuit of a pressure booster is given in Section 11.4.

EXAMPLE 3-6

Figure 3-13 shows a pressure booster used to drive a load F via a hydraulic cylinder. The following data are given:

inlet air pressure (p_1) = 100 psi

air piston area (A_1) = 20 in²

oil piston area (A_2) = 1 in²

load piston area (A_3) = 25 in² (diameter = 5.64 in)

Find the load-carrying capacity F of the system.

Solution   First, find the booster discharge pressure p_2:

booster input force = booster output force

$$p_1 A_1 = p_2 A_2$$

$$p_2 = \frac{p_1 A_1}{A_2} = (100)\left(\frac{20}{1}\right) = 2000 \text{ psi}$$

Per Pascal's law, $p_3 = p_2 = 2000$ psi:

$$F = p_3 A_3 = (2000)(25) = 50,000 \text{ lb}$$

To produce this force without the booster would require a 500-in²-area load piston (diameter = 25.2 in), assuming 100-psi air pressure.

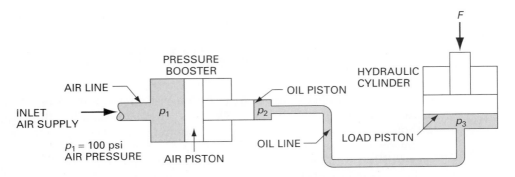

Figure 3-13. An air-to-hydraulic pressure booster system.

3.5 HYDROFORMING OF METAL COMPONENTS

Introduction

Metalworking companies frequently use a process called hydroforming to make components such as sheet metal panels or stiff and strong structural parts. Hydroforming utilizes a hydraulic liquid to form the sheet metal panel or structural part. There are two different types of hydroforming methods that are used depending on whether a sheet metal panel or structural part is to be made. The names of these two methods are given as follows:

1. Sheet hydroforming
2. Tube hydroforming

Sheet Hydroforming

Figure 3-14(a) through (f) shows the basic configuration and operation of sheet hydroforming. As shown in Figure 3-14(a), the system consists of an upper section and a lower section that are currently separated from each other. The upper section (called a die) contains an internal cavity into which a liquid can enter via a pipe at the top. The bottom of this die consists of a flexible member (rubber diaphragm). Figure 3-14(a) also shows that in the lower section, a circular, flat sheet metal blank is placed on the top of a member called a draw ring or blank holder. The draw ring has a central opening through which a punch can be pushed upward to bear on the sheet metal blank. In Figure 3-14(b) the die is lowered until its diaphragm makes contact with the blank and then the die is locked into place. Figure 3-14(c) shows liquid pressure established on the top surface of the diaphragm to a predetermined setting. In Figure 3-14(d) we see the punch being pushed upward into the sheet metal blank and diaphragm. As the punch moves upward, the liquid pressure pushes down on the diaphragm uniformly along the surface of the deforming blank. As this occurs the originally flat blank takes on the shape of the punch. In this example the punch is a hemisphere to produce the desired part. Figure 3-14(e) shows the die raised until the diaphragm is fully

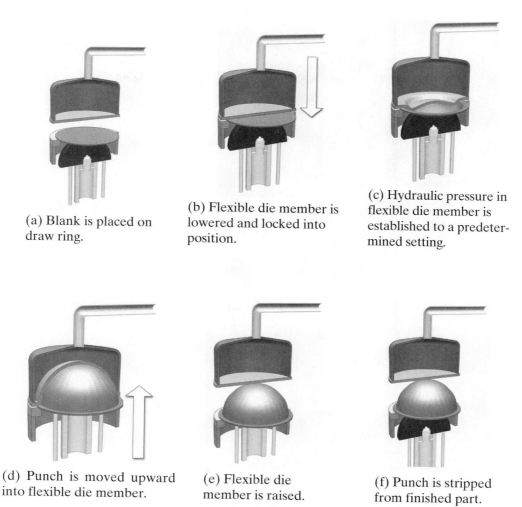

(a) Blank is placed on draw ring.

(b) Flexible die member is lowered and locked into position.

(c) Hydraulic pressure in flexible die member is established to a predetermined setting.

(d) Punch is moved upward into flexible die member.

(e) Flexible die member is raised.

(f) Punch is stripped from finished part.

Figure 3-14. Configuration and operation of sheet hydroforming. *(Courtesy of Jones Metal Products Company, West Lafayette, Ohio.)*

separated from the finished part. Finally, Figure 3-14(f) shows the punch lowered from the finished part which is a semi-spherical shell. The part is then removed from the system.

Figure 3-15 is a photograph showing a technician examining pump containment shells produced by the hydroforming machine shown in the background. This machine can operate with hydraulic pressures up to 10,000 psi. It can handle blanks with thicknesses up to 3/8 in for steel; 1 in for aluminum; and blanks made of other materials such as stainless steel, brass, copper, titanium, and nickel alloys. Production runs can accommodate blank sizes up to 32.5 in to form parts with depths to 14 in and diameters to 27 in. Figure 3-16 displays a variety of complex shaped parts, consisting of different materials, which have been produced by the hydroforming machine of Figure 3-15.

Figure 3-15. Technician examining pump containment shells produced by hydroforming machine in background. *(Courtesy of Jones Metal Products Company, West Lafayette, Ohio.)*

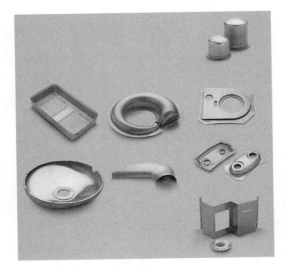

Figure 3-16. Variety of complex shaped parts consisting of different materials, produced by the hydroforming machine of Figure 3-15. *(Courtesy of Jones Metal Products Company, West Lafayette, Ohio.)*

Tube Hydroforming

Tube hydroforming is the second type of hydroforming process and it is used to shape metal tubes into lightweight, structurally stiff and strong parts. The components of a typical tube hydroforming system are shown in Figure 3-17. Figure 3-17(a) shows the original metal tube. In Figure 3-17(b) we see the deformed tube as it is located inside a die that has the shape of the desired part to be produced. As shown, hydraulic liquid enters the inside of the tube through a central opening in the top of the die. At

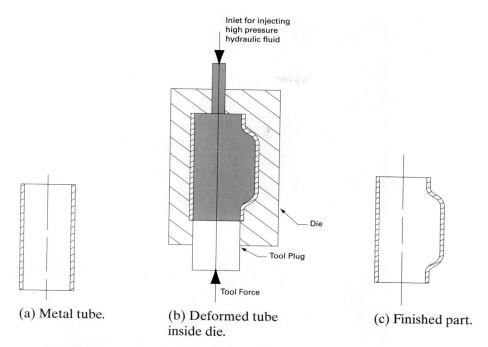

Inlet for injecting
high pressure
hydraulic fluid

Die

Tool Plug

Tool Force

(a) Metal tube.

(b) Deformed tube
inside die.

(c) Finished part.

Figure 3-17. Components of tube hydroforming system.

the bottom of the die is a central hole through which a tool plug is forced in to hold the tube in place while it is being deformed by the increasing hydraulic pressure caused by the injection of liquid into the tube by a pump. Figure 3-17(c) shows the finished part that is removed from the die. The maximum hydraulic pressure required for most tube hydroforming applications is about 50,000 psi.

One of the applications of tube hydroforming is in the automotive industry, which makes use of the complex shapes made possible by hydroforming to produce stronger, lighter, and more rigid unibody structures for vehicles. Lighter vehicles offer the benefits of material and energy savings, and so are more environmentally friendly. Tube hydroforming is particularly popular with the high-end sports car industry and is also frequently employed in the shaping of aluminum tube bicycle frames. One such automotive example is the Corvette, which is shown in Figure 3-18. This Corvette has hydroformed aluminum frame rails, which are the foundation to its agile handling because its frame rails are lighter and stiffer than the nonhydroformed type.

Figure 3-19 shows Corvette aluminum frame rails being produced in a factory using the tube hydroforming method. In Figure 3-20 we see the location of the hydroformed frame rails relative to the rear and front wheel axles of the Corvette. Figure 3-21 shows the entire Corvette chassis, which contains the hydroformed frame rails.

Figure 3-18. The Corvette convertible sports car. *(Courtesy of General Motors Corp. Used with permission, GM Media Archives, Detroit, Michigan.)*

Figure 3-19. Corvette frame rails being produced in factory using tube hydro-forming. *(Courtesy of General Motors Corp. Used with permission, GM Media Archives, Detroit, Michigan.)*

3.6 CONSERVATION OF ENERGY

The conservation of energy law states that energy can be neither created nor destroyed. This means that the total energy of a system remains constant. The total energy includes potential energy due to elevation and pressure and also kinetic energy due to velocity. Let's examine each of the three types of energy.

Figure 3-20. Location of hydroformed frame rails relative to rear and front wheel axles of Corvette. *(Courtesy of General Motors Corp. Used with permission, GM Media Archives, Detroit, Michigan.)*

Figure 3-21. Entire Corvette chassis, which contains the hydro-formed frame rails. *(Courtesy of General Motors Corp. Used with permission, GM Media Archives, Detroit, Michigan.)*

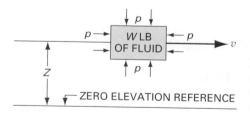

Figure 3-22. The three forms of energy as established by elevation (Z), pressure (p), and velocity (v).

    1. Potential energy due to elevation (EPE): Figure 3-22 shows a chunk of fluid of weight W lb at an elevation Z with respect to a reference plane. The weight has potential energy (EPE) relative to the reference plane because work would have to be done on the fluid to lift it through a distance Z:

$$EPE = WZ \qquad\qquad \textbf{(3-13)}$$

The units of EPE are ft · lb.

2. Potential energy due to pressure (PPE): If the W lb of fluid in Figure 3-22 possesses a pressure p, it contains pressure energy as represented by

$$\text{PPE} = W\frac{p}{\gamma} \tag{3-14}$$

where γ is the specific weight of the fluid. PPE has units of ft $\cdot$ lb.

3. Kinetic energy (KE): If the W lb of fluid in Figure 3-22 is moving with a velocity v, it contains kinetic energy, which can be found using

$$\text{KE} = \frac{1}{2}\frac{W}{g}v^2 \tag{3-15}$$

where g = acceleration of gravity.
KE has units of ft $\cdot$ lb.

Per the conservation of energy law we can make the following statement about the W-lb chunk fluid in Figure 3-22: The total energy E_T possessed by the W-lb chunk of fluid remains constant (unless energy is added to the fluid via pumps or removed from the fluid via hydraulic motors or friction) as the W-lb chunk flows through a pipeline of a hydraulic system. Mathematically we have

$$E_T = WZ + W\frac{p}{\gamma} + \frac{1}{2}\frac{W}{g}v^2 = \text{constant} \tag{3-16}$$

Of course, energy can change from one form to another. For example, the chunk of fluid may lose elevation as it flows through the hydraulic system and thus have less potential energy. This, however, would result in an equal increase in either the fluid's pressure energy or kinetic energy. The energy equation (presented in Section 3.9) takes into account the fact that energy is added to the fluid via pumps and that energy is removed from the fluid via hydraulic motors and friction as the fluid flows through actual hydraulic systems.

3.7 THE CONTINUITY EQUATION

Use of Weight Flow Rate

The continuity equation states that for steady flow in a pipeline, the weight flow rate (weight of fluid passing a given station per unit time) is the same for all locations of the pipe.

To illustrate the significance of the continuity equation, refer to Figure 3-23, which shows a pipe in which fluid is flowing at a weight flow rate w that has units of weight per unit time. The pipe has two different-sized cross-sectional areas identified by stations 1 and 2. The continuity equation states that if no fluid is added or withdrawn from the pipeline between stations 1 and 2, then the weight flow rate at stations 1 and 2 must be equal.

$$w_1 = w_2$$

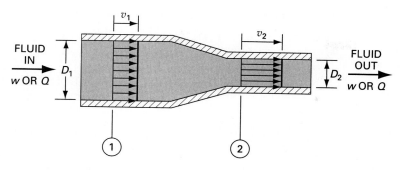

Figure 3-23. Continuity of flow.

or

$$\gamma_1 A_1 v_1 = \gamma_2 A_2 v_2 \tag{3-17}$$

where γ = specific weight of fluid (lb/ft³),

A = cross-sectional area of pipe (ft²),

v = velocity of fluid (ft/s).

Checking units, we have

$$(lb/ft^3)(ft^2)(ft/s) = (lb/ft^3)(ft^2)(ft/s)$$

or

$$lb/s = lb/s$$

A derivation of the continuity equation (Eq. 3-17) is presented in Appendix H.

Use of Volume Flow Rate

If the fluid is a liquid we can cancel out the specific weight terms from the continuity equation. This is because a liquid is essentially incompressible, and hence $\gamma_1 = \gamma_2$. The result is the continuity equation for hydraulic systems:

$$Q_1 = A_1 v_1 = A_2 v_2 = Q_2 \tag{3-18}$$

where Q is the volume flow rate (volume of fluid passing a given station per unit time). Checking units, we have

$$(ft^2)(ft/s) = (ft^2)(ft/s)$$

or

$$ft^3/s = ft^3/s$$

Hence, for hydraulic systems, the volume flow rate is also constant in a pipeline.

The continuity equation for hydraulic systems can be rewritten as follows:

$$\frac{v_1}{v_2} = \frac{A_2}{A_1} = \frac{(\pi/4)D_2^2}{(\pi/4)D_1^2}$$

where D_1 and D_2 are the pipe diameters at stations 1 and 2, respectively. The final result is

$$\frac{v_1}{v_2} = \left(\frac{D_2}{D_1}\right)^2 \tag{3-19}$$

Equation (3-19) shows that the smaller the pipe size, the greater the velocity, and vice versa. It should be noted that the pipe diameters and areas are inside values and do not include the pipe wall thickness.

EXAMPLE 3-7

For the pipe in Figure 3-23, the following data are given:

$$D_1 = 4 \text{ in} \quad D_2 = 2 \text{ in}$$

$$v_1 = 4 \text{ ft/s}$$

Find:

    a. The volume flow rate Q

    b. The fluid velocity at station 2

Solution

    a.
$$Q = Q_1 = A_1 v_1$$

$$A_1 = \frac{\pi}{4} D_1^2 = \frac{\pi}{4}\left(\frac{4}{12} \text{ ft}\right)^2 = 0.0873 \text{ ft}^2$$

$$Q = (0.0873 \text{ ft}^2)(4 \text{ ft/s}) = 0.349 \text{ ft}^3/\text{s}$$

    b. Solving Eq. (3-19) for v_2, we have

$$v_2 = v_1 \left(\frac{D_1}{D_2}\right)^2 = 4\left(\frac{4}{2}\right)^2 = 16 \text{ ft/s}$$

3.8 HYDRAULIC POWER

Hydraulic Cylinder Example

Now that we have established the concepts of flow rate and pressure, we can find the power delivered by a hydraulic fluid to a load-driving device such as a hydraulic cylinder. This power is called *hydraulic power*. Let's analyze the hydraulic cylinder

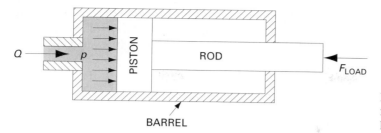

Figure 3-24. Cylinder example for determining hydraulic horsepower.

of Figure 3-24 by developing equations that will allow us to answer the following three questions:

1. How do we determine how large a piston diameter is required for the cylinder?
2. What is the pump flow rate required to drive the cylinder through its stroke in a specified time?
3. How much hydraulic horsepower does the fluid deliver to the cylinder?

Note that the horsepower delivered by the fluid to the cylinder is called hydraulic horsepower (HHP). The output horsepower delivered by the cylinder to the load equals the hydraulic horsepower minus any horsepower loss due to friction and fluid leakage between the piston and bore of the cylinder. The horsepower delivered by the cylinder to the load is called *output horsepower* (OHP). Output horsepower is always less than hydraulic horsepower due to friction and leakage losses. This is consistent with the fact that the efficiency of any component is always less than 100%.

Answer to Question 1.   A pump receives fluid on its inlet side at about atmospheric pressure (0 psig) and discharges the fluid on the outlet side at some elevated pressure p sufficiently high to overcome the load. This pressure p acts on the area of the piston A to produce the force required to overcome the load:

$$pA = F_{\text{load}}$$

Solving for the piston area A, we obtain

$$A = \frac{F_{\text{load}}}{p} \tag{3-20}$$

The load is known from the application, and the maximum allowable pressure is established based on the pump design. Thus, Eq. (3-20) allows us to calculate the required piston area if the friction between the piston and cylinder bore is negligibly small.

Answer to Question 2.   The volumetric displacement V_D of the hydraulic cylinder equals the fluid volume swept out by the piston traveling through its stroke S:

$$V_D(\text{ft}^3) = A(\text{ft}^2) \times S(\text{ft})$$

If there is negligibly small leakage between the piston and cylinder bore, the required pump volume flow rate Q equals the volumetric displacement of the cylinder divided by the time t required for the piston to travel through its stroke.

$$Q(\text{ft}^3/\text{s}) = \frac{V_D(\text{ft}^3)}{t(\text{s})}$$

but since $V_D = AS$, we have

$$Q(\text{ft}^3/\text{s}) = \frac{A(\text{ft}^2) \times S(\text{ft})}{t(\text{s})} \qquad \textbf{(3-21)}$$

Since stroke S and time t are basically known from the particular application, Eq. (3-21) permits the calculation of the required pump flow.

Recall that for a pipe we determined that $Q = Av$ where v equals the fluid velocity. Shouldn't we obtain the same equation for a hydraulic cylinder since it is essentially a pipe that contains a piston moving at velocity v? The answer is yes, as can be verified by noting that S/t can be replaced by v in Eq. (3-21) to obtain the expected result:

$$Q(\text{ft}^3/\text{s}) = A(\text{ft}^2) \times v(\text{ft}/\text{s}) \qquad \textbf{(3-22)}$$

where v = piston velocity.

Note that the larger the piston area and velocity, the greater must be the pump flow rate.

Answer to Question 3. It has been established that energy equals force times distance:

$$\text{energy} = (F)(S) = (pA)(S)$$

Since power is the rate of doing work, we have

$$\text{power} = \frac{\text{energy}}{\text{time}} = \frac{(pA)(S)}{t} = p(Av)$$

Since $Q = Av$, the final result is

$$\text{hydraulic power (ft} \cdot \text{lb/s)} = p(\text{lb/ft}^2) \times Q(\text{ft}^3/\text{s}) \qquad \textbf{(3-23)}$$

Recalling that 1 hp = 550 ft · lb/s, we obtain

$$\text{hydraulic horsepower} = \text{HHP} = p(\text{lb/ft}^2) \times Q(\text{ft}^3/\text{s}) \times \frac{1 \text{ hp}}{550 \text{ ft} \cdot \text{lb/s}}$$

As expected, all the units cancel out except the units of hp, yielding the following result:

$$\text{HHP} = \frac{p(\text{lb/ft}^2) \times Q(\text{ft}^3/\text{s})}{550} \qquad \textbf{(3-24)}$$

Hydraulic Horsepower in Terms of psi and gpm Units

Pressure in psi (lb/in²) and flow rate in gallons per minute (gpm) are the most common English units used for hydraulic systems. Thus a hydraulic horsepower equation using these most common English units is developed as follows, noting that hydraulic power equals the product of pressure and volume flow rate:

$$\text{hydraulic power} = p \times Q$$

$$= p\left(\frac{\text{lb}}{\text{in}^2}\right) \times Q\left(\frac{\text{gal}}{\text{min}}\right) \times \frac{231 \text{ in}^3}{1 \text{ gal}} \times \frac{1 \text{ min}}{60 \text{ s}} \times \frac{1 \text{ ft}}{12 \text{ in}} \times \frac{1 \text{ hp}}{550 \text{ ft} \cdot \text{lb/s}}$$

Since all the units cancel out except hp, we have the final result:

$$\text{HHP} = \frac{p(\text{psi}) \times Q(\text{gpm})}{1714} \qquad \textbf{(3-25)}$$

Note that the constants 550 and 1714 in Eqs. (3-24) and (3-25) contain the proper units to make these two equations have units of hp on the right side of the equal sign.

Observe the following power analogy among mechanical, electrical, and hydraulic systems:

$$\text{mechanical power} = \text{force} \times \text{linear velocity}$$
$$= \text{torque} \times \text{angular velocity}$$
$$\text{electrical power} = \text{voltage} \times \text{electric current}$$
$$\text{hydraulic power} = \text{pressure} \times \text{volume flow rate}$$

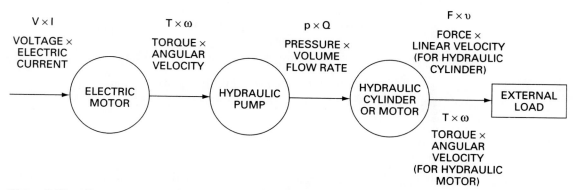

Figure 3-25. Conversion of power from input electrical to mechanical to hydraulic to output mechanical in a hydraulic system.

All three types of power (mechanical, electrical and hydraulic) are typically involved in hydraulic systems, as illustrated in Figure 3-25. In this diagram an electric motor is used as the prime mover to drive the pump. As shown, the electric motor converts electric power into mechanical power via a rotating shaft. The pump converts this mechanical power into hydraulic power. Finally, a hydraulic cylinder or hydraulic motor transforms the hydraulic power back into mechanical power to drive the load. A hydraulic cylinder delivers a force to move the load in a straight-line path with a linear velocity. Conversely a hydraulic motor delivers a torque to move the load in a rotary motion with an angular velocity.

EXAMPLE 3-8

A hydraulic cylinder is to compress a car body down to bale size in 10 s. The operation requires a 10-ft stroke and a 8000-lb force. If a 1000-psi pump has been selected, and assuming the cylinder is 100% efficient, find

  a. The required piston area
  b. The necessary pump flow rate
  c. The hydraulic horsepower delivered to the cylinder
  d. The output horsepower delivered by the cylinder to the load

Solution

a.
$$A = \frac{F_{load}}{p} = \frac{8000 \text{ lb}}{1000 \text{ lb/in}^2} = 8 \text{ in}^2$$

b.
$$Q(\text{ft}^3/\text{s}) = \frac{A(\text{ft}^2) \times S(\text{ft})}{t(\text{s})} = \frac{\left(\dfrac{8}{144}\right)(10)}{10} = 0.0556 \text{ ft}^3/\text{s}$$

Per Appendix E, 1 ft³/s = 449 gpm. Thus,

$$Q(\text{gpm}) = 449Q(\text{ft}^3/\text{s}) = (449)(0.0556) = 24.9 \text{ gpm}$$

c.
$$\text{HHP} = \frac{(1000)(24.9)}{1714} = 14.5 \text{ hp}$$

d.
$$\text{OHP} = \text{HHP} \times \eta = 14.5 \times 1.0 = 14.5 \text{ hp}$$

Thus, assuming a 100% efficient cylinder (losses are negligibly small), the hydraulic horsepower equals the output horsepower.

EXAMPLE 3-9

Solve the problem of Example 3-8 assuming a frictional force of 100 lb and a leakage of 0.2 gpm.

Solution

a.
$$A = \frac{F_{load} + F_{friction}}{p} = \frac{8000 \text{ lb} + 100 \text{ lb}}{1000 \text{ lb/in}^2} = 8.10 \text{ in}^2$$

b.
$$Q_{theoretical} = \frac{A(\text{ft}^2) \times S(\text{ft})}{t(\text{s})} = \frac{\left(\frac{8.10}{144}\right)(10)}{10} = 0.0563 \text{ ft}^3/\text{s} = 25.2 \text{ gpm}$$

$$Q_{actual} = Q_{theoretical} + Q_{leakage} = 25.2 + 0.2 = 25.4 \text{ gpm}$$

c.
$$\text{HHP} = \frac{1000 \times 25.4}{1714} = 14.8 \text{ hp}$$

d.
$$\text{OHP} = \frac{F(\text{lb}) \times v(\text{ft/s})}{550} = \frac{8000 \times 1}{550} = 14.5 \text{ hp}$$

Thus, a hydraulic horsepower of 14.8 must be delivered by the fluid to the cylinder to produce an output horsepower at 14.5 for driving the load. The efficiency of the cylinder is

$$\eta = \frac{\text{OHP}}{\text{HHP}} = \frac{14.5}{14.8} = 0.980 = 98.0\%$$

3.9 BERNOULLI'S EQUATION

Derivation of Bernoulli's Equation

Bernoulli's equation is one of the most useful relationships for performing hydraulic circuit analysis. Its application allows us to size components such as pumps, valves, and piping for proper system operation. The original Bernoulli equation can be derived by applying the conservation of energy law to a hydraulic pipeline, as shown in Figure 3-26. At station 1 we have W lb of fluid possessing an elevation Z_1, a pressure p_1, and a velocity v_1. When this W lb of fluid arrives at station 2, its elevation, pressure, and velocity have become Z_2, p_2, and v_2, respectively.

With respect to a common zero elevation reference plane, we can identify the various energy terms as follows:

Type of Energy	Station 1	Station 2
Elevation	WZ_1	WZ_2
Pressure	$W\dfrac{p_1}{\gamma}$	$W\dfrac{p_2}{\gamma}$
Kinetic	$\dfrac{Wv_1^2}{2g}$	$\dfrac{Wv_2^2}{2g}$

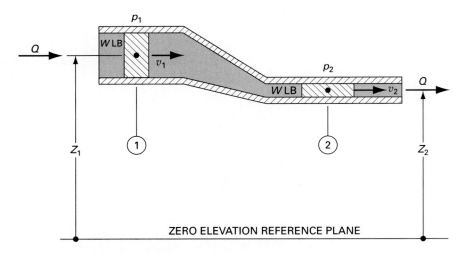

Figure 3-26. Pipeline for deriving Bernoulli's equation.

Daniel Bernoulli, the eighteenth-century Swiss scientist, formulated his equation by noting that the total energy possessed by the W lb of fluid at station 1 equals the total energy possessed by the same W lb of fluid at station 2, provided frictional losses are negligibly small:

$$WZ_1 + W\frac{p_1}{\gamma} + \frac{Wv_1^2}{2g} = WZ_2 + W\frac{p_2}{\gamma} + \frac{Wv_2^2}{2g} \qquad \text{(3-26)}$$

If we divide both sides of Eq. (3-26) by W, we are examining the energy possessed by 1 lb of fluid rather than W lb. This yields the original Bernoulli equation for an ideal frictionless system that does not contain pumps and hydraulic motors between stations 1 and 2: *The total energy per pound of fluid at station 1 equals the total energy per pound of fluid at station 2:*

$$Z_1 + \frac{p_1}{\gamma} + \frac{v_1^2}{2g} = Z_2 + \frac{p_2}{\gamma} + \frac{v_2^2}{2g} \qquad \text{(3-27)}$$

Checking units, we find that each term has units of length (ft · lb/lb = ft). This is as expected, since each term represents energy per pound of fluid:

$$Z = \text{ft}$$

$$\frac{p}{\gamma} = \frac{\text{lb/ft}^2}{\text{lb/ft}^3} = \text{ft}$$

$$\frac{v^2}{2g} = \frac{(\text{ft/s})^2}{\text{ft/s}^2} = \text{ft}$$

Since each term of Bernoulli's equation has units of length, the use of the expression *head* has picked up widespread use as follows:

Z is called *elevation head*.

p/γ is called *pressure head*.

$v^2/2g$ is called *velocity head*.

The Energy Equation

Bernoulli modified his original equation to take into account that frictional losses (H_L) take place between stations 1 and 2. H_L (called head loss) represents the energy per pound of fluid loss due to friction in going from station 1 to station 2. In addition, he took into account that a pump (which adds energy to fluid) or a hydraulic motor (which removes energy from fluid) may exist between stations 1 and 2. H_P (pump head) represents the energy per pound of fluid added by a pump, and H_m (motor head) represents the energy per pound of fluid removed by a hydraulic motor.

The modified Bernoulli equation (also called the energy equation), is stated as follows for a fluid flowing in a pipeline from station 1 to station 2:

The total energy possessed by a 1-lb chunk of fluid at station 1 plus the energy added to it by a pump minus the energy removed from it by a hydraulic motor minus the energy it loses due to friction, equals the total energy possessed by the 1-lb chunk of fluid when it arrives at station 2.

The energy equation is given as follows, where each term represents a head and thus has units of length:

$$Z_1 + \frac{p_1}{\gamma} + \frac{v_1^2}{2g} + H_p - H_m - H_L = Z_2 + \frac{p_2}{\gamma} + \frac{v_2^2}{2g} \qquad \textbf{(3-28)}$$

The method used for determining the head loss (H_L) for a given application is presented in Chapter 4. The equation for determining the pump head (H_P) can be derived by using Eq. (2-9):

$$p = \gamma H.$$

Thus we have $p = \gamma H_p$. Then, as shown in Appendix H, substituting $p = \gamma H_p$ into Eq. (3-25) yields

$$H_p(\text{ft}) = \frac{3950 \times (\text{HHP})}{Q(\text{gpm}) \times \text{SG}} \qquad \textbf{(3-29)}$$

The motor head can also be calculated using Eq. (3-29), where the H_p term is replaced by H_m. The HHP and Q terms then represent the motor hydraulic horsepower and gpm flow rate, respectively. Note that pump hydraulic horsepower is the power

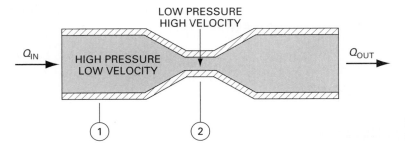

Figure 3-27. Pressure reduction in a venturi.

the pump delivers to the fluid, whereas motor hydraulic horsepower is the power the fluid delivers to the motor.

Venturi Application

The use of a venturi in an automobile engine carburetor is an application of Bernoulli's equation.

Figure 3-27 shows a venturi, which is a special pipe whose diameter is gradually reduced until a constant-diameter throat is reached. Then the pipe gradually increases in diameter until it reaches the original size. We know that inlet station 1 has a lower velocity than does the throat station 2 due to the continuity equation. Therefore, v_2 is greater than v_1.

Let's write Bernoulli's equation between stations 1 and 2 assuming ideal flow and equal elevations:

$$\frac{p_1}{\gamma} + \frac{v_1^2}{2g} = \frac{p_2}{\gamma} + \frac{v_2^2}{2g}$$

Solving for $p_1 - p_2$, we have

$$p_1 - p_2 = \frac{\gamma}{2g} (v_2^2 - v_1^2)$$

Since v_2 is greater than v_1, we know that p_1 must be greater than p_2. The reason is simple: In going from station 1 to 2 the fluid gained kinetic energy due to the continuity theorem. As a result, the fluid had to lose pressure energy in order not to create or destroy energy. This venturi effect is commonly called Bernoulli's principle.

Figure 3-28 shows how the venturi effect is used in an automobile carburetor. The volume of airflow is determined by the opening position of the butterfly valve. As the air flows through the venturi, it speeds up and loses some of its pressure. The pressure in the fuel bowl is equal to the pressure in the air horn above the venturi. This differential pressure between the fuel bowl and the venturi throat causes gasoline to flow into the airstream. The reduced pressure in the venturi helps the gasoline to vaporize.

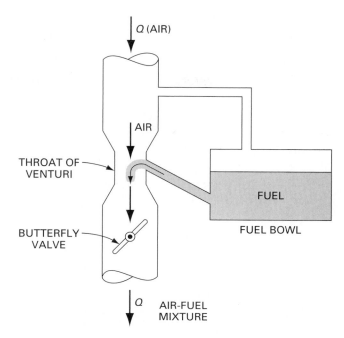

Figure 3-28. Use of venturi to produce Bernoulli effect in automobile carburetor.

EXAMPLE 3-10

For the hydraulic system of Figure 3-29, the following data are given:

a. The pump is adding 5 hp to the fluid (pump hydraulic horsepower = 5).
b. Pump flow is 30 gpm.
c. The pipe has a 1-in. inside diameter.
d. The specific gravity of the oil is 0.9.

Find the pressure available at the inlet to the hydraulic motor (station 2). The pressure at station 1 in the hydraulic tank is atmospheric (0 psig). The head loss H_L due to friction between stations 1 and 2 is 30 ft of oil.

Solution   Writing the energy equation between stations 1 and 2, we have

$$Z_1 + \frac{p_1}{\gamma} + \frac{v_1^2}{2g} + H_p - H_m - H_L = Z_2 + \frac{p_2}{\gamma} + \frac{v_2^2}{2g}$$

Since there is no hydraulic motor between stations 1 and 2, $H_m = 0$. Also, $v_1 = 0$ because the cross section of an oil tank is large. Thus, the velocity of the oil surface is negligible, and the value of v_1 approaches zero. Per Figure 3-29, $Z_2 - Z_1 = 20$ ft. Also, $H_L = 30$ ft, and $p_1 = 0$ gage pressure per the given input data.

Substituting known values, we have

$$Z_1 + 0 + 0 + H_p - 0 - 30 = Z_2 + \frac{p_2}{\gamma} + \frac{v_2^2}{2g}$$

Solving for p_2/γ, we have

$$\frac{p_2}{\gamma} = (Z_1 - Z_2) + H_p - \frac{v_2^2}{2g} - 30$$

Since $Z_2 - Z_1 = 20$ ft, we have

$$\frac{p_2}{\gamma} = H_p - \frac{v_2^2}{2g} - 50$$

Using Eq. (3-29) yields

$$H_p = \frac{(3950)(5)}{(30)(0.9)} = 732 \text{ ft}$$

Then, solve for $v_2^2/2g$ as follows

$$Q(\text{ft}^3/\text{s}) = \frac{Q(\text{gpm})}{449} = \frac{30}{449} = 0.0668$$

$$v_2(\text{ft/s}) = \frac{Q(\text{ft}^3/\text{s})}{A(\text{ft}^2)}$$

$$A(\text{ft}^2) = \frac{\pi}{4}(D \text{ ft})^2 = \frac{\pi}{4}\left(\frac{1}{12}\text{ ft}\right)^2 = 0.00546 \text{ ft}^2$$

$$v_2 = \frac{0.0668}{0.00546} \frac{\text{ft}^3/\text{s}}{\text{ft}^2} = 12.2 \text{ ft/s}$$

$$\frac{v_2^2}{2g} = \frac{(12.2 \text{ ft/s})^2}{64.4 \text{ ft/s}^2} = 2.4 \text{ ft}$$

On final substitution, we have

$$\frac{p_2}{\gamma} = 732 - 2.4 - 50 = 679.2 \text{ ft}$$

Solving for p_2 yields

$$p_2 (\text{lb/ft}^2) = (679.2 \text{ ft}) \, \gamma \, (\text{lb/ft}^3)$$

where $\gamma = (SG) \, \gamma_{\text{water}}$,
$\gamma = (0.9)(62.4) = 56.2 \text{ lb/ft}^3$,
$p_2 = (679.2)(56.2) = 38{,}200 \text{ lb/ft}^2$.

Changing to units of psi yields

$$p_2 = \frac{38{,}200}{144} = 265 \text{ psig}$$

The answer, 265 obtained for p_2 is a gage pressure since a gage pressure value of zero was used in the energy equation for p_1.

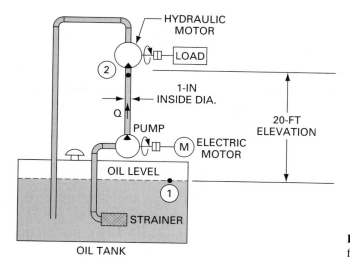

Figure 3-29. Hydraulic system for Example 3-10.

3.10 TORRICELLI'S THEOREM

Torricelli's theorem states that ideally the velocity of a free jet of fluid is equal to the square root of the product of two times the acceleration of gravity times the head producing the jet. As such, Torricelli's theorem is essentially a special case of Bernoulli's equation for the system of Figure 3-30. In this system we have a tank with an opening in its side. The tank is filled with a liquid to a height h above the centerline of the opening. As a result, a jet of fluid flows through the opening.

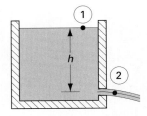

Figure 3-30. System for Torricelli's equation.

To derive Torricelli's equation we shall consider a reference point (1) at the top surface of liquid in the tank and a second reference point (2) in the jet just beyond its opening. Since the fluid jet flows into the atmosphere, it is called a "free jet."

Writing the energy equation between points 1 and 2, we have

$$Z_1 + \frac{p_1}{\gamma} + \frac{v_1^2}{2g} + H_p - H_m - H_L = Z_2 + \frac{p_2}{\gamma} + \frac{v_2^2}{2g}$$

For the system of Figure 3-30, we can make the following observations:

1. $p_1 = p_2$ = atmospheric pressure = 0 psig.
2. The area of the surface of liquid in the tank is large so that the velocity v_1 equals essentially zero.
3. There is no pump or motor ($H_p = H_m = 0$).
4. The fluid is ideal, and therefore there are no frictional losses ($H_L = 0$).
5. Z_2 can be taken as the reference zero elevation plane ($Z_2 = 0$).

Substituting known values yields

$$h + 0 + 0 + 0 - 0 - 0 = 0 + 0 + \frac{v_2^2}{2g}$$

Solving for v_2, we have

$$v_2 = \sqrt{2gh} \tag{3-30}$$

where v_2 = jet velocity (ft/s),
       g = acceleration of gravity (ft/s²),
       h = pressure head (ft).

If we do not assume an ideal fluid, frictional head losses occur, and therefore H_L does not equal zero. The solution becomes

$$v_2 = \sqrt{2g(h - H_L)} \tag{3-31}$$

This shows that the jet velocity is reduced if the fluid is not ideal because there is less net pressure head to push the fluid through the opening. Thus, the actual velocity of the jet depends on the viscosity of the fluid.

EXAMPLE 3-11

For the system of Figure 3-30, $h = 36$ ft and the diameter of the opening is 2 in. Assuming an ideal fluid, find

  a. The jet velocity in units of ft/s
  b. The flow rate in units of gpm
  c. The answers to parts a and b if $H_L = 10$ ft

Solution

  a. Substituting into Eq. (3-30) yields

$$v_2 = \sqrt{(2)(32.2 \text{ ft/s}^2) \times (36 \text{ ft})} = 48.3 \text{ ft/s}$$

  b. Solving for Q we have

$$Q(\text{ft}^3/\text{s}) = A(\text{ft}^2) \times v(\text{ft/s}) = \frac{\pi}{4}\left(\frac{2}{12}\text{ ft}\right)^2 \times 48.3 \text{ ft/s} = 1.05 \text{ ft}^3/\text{s}$$

$$Q(\text{gpm}) = 449Q(\text{ft}^3/\text{s}) = 449 \times 1.05 = 471 \text{ gpm}$$

  c. Use Eq. (3-31):

$$v_2 = \sqrt{(64.4)(36 - 10)} = 40.9 \text{ ft/s}$$

Since flow rate is proportional to velocity, we have

$$Q = \frac{40.9}{48.3} \times 471 \text{ gpm} = 399 \text{ gpm}$$

As expected, there is a greater flow rate for the ideal fluid as compared to a real fluid possessing viscosity.

3.11 THE SIPHON

The siphon is a familiar hydraulic device. It is commonly used to cause a liquid to flow from one container in an upward direction over an obstacle and then flow downward into a second lower container. As shown in Figure 3-31, a siphon consists of a U-tube with one end submerged below the surface of the liquid in the container. The center portion of the U-tube lies above the level of the liquid surface, and the free end lies below it on the outside of the container.

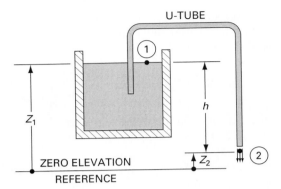

Figure 3-31. Siphon.

For the fluid to flow out of the free end, two conditions must be met:

1. The elevation of the free end must be lower than the elevation of the liquid surface inside the container.
2. The fluid must initially be forced to flow up from the container into the center portion of the U-tube. This is normally done by temporarily providing a suction pressure at the free end of the siphon. For example, when siphoning gasoline from an automobile gas tank, a person can develop this suction pressure by momentarily sucking on the free end of the hose. This allows atmospheric pressure in the tank to push the gasoline up the U-tube hose, as required. For continuous flow operation, the free end of the U-tube hose must lie below the gasoline level in the tank.

We can analyze the flow through a siphon by applying the energy equation using points 1 and 2 in Figure 3-31:

$$Z_1 + \frac{p_1}{\gamma} + \frac{v_1^2}{2g} + H_p - H_m - H_L = Z_2 + \frac{p_2}{\gamma} + \frac{v_2^2}{2g}$$

The following conditions apply for the siphon:

1. $p_1 = p_2 =$ atmospheric pressure $= 0$ psig.
2. The area of the surface of the liquid in the container is large so that the velocity v_1 equals essentially zero.
3. There is no pump or motor ($H_p = H_m = 0$).
4. $Z_1 - Z_2 = h =$ differential head between liquid level and free end of U-tube.

Substituting known values, we have

$$Z_1 + 0 + 0 + 0 - 0 - H_L = Z_2 + 0 + \frac{v_2^2}{2g}$$

Solving for v_2 yields

$$v_2 = \sqrt{2g(Z_1 - Z_2 - H_L)} = \sqrt{2g(h - H_L)} \qquad \text{(3-32)}$$

Equation (3-32) is identical to Torricelli's equation, and, as expected, the velocity inside the siphon tube is reduced as the head loss H_L due to viscosity increases. The flow rate of the siphon can be found by using the continuity equation.

EXAMPLE 3-12

For the siphon of Figure 3-31, the following information is given:

$$h = 30 \text{ ft}$$
$$H_L = 10 \text{ ft}$$
$$\text{U-tube inside diameter} = 1 \text{ in.}$$

Find the velocity and flow rate through the siphon.

Solution   Substitute known values into Eq. (3-32):

$$v_2 = \sqrt{(2)(32.2)(30 - 10)} = 35.8 \text{ ft/s}$$

Solving for the flow rate we have

$$Q(\text{ft}^3/\text{s}) = A(\text{ft}^2) \times v(\text{ft/s}) = \frac{\pi}{4}\left(\frac{1}{12}\text{ft}\right)^2 \times 35.8 \text{ ft/s} = 0.195 \text{ ft}^3/\text{s}$$

$$Q(\text{gpm}) = 449Q(\text{ft}^3/\text{s}) = 449 \times 0.195 = 87.6 \text{ gpm}$$

3.12 ENERGY, POWER, AND FLOW RATE IN THE SI METRIC SYSTEM

In this section we provide the necessary information to calculate energy, power, and flow rate in the SI metric system.

Energy

In the SI system, the joule (J) is the work done when a force of 1 N acts through a distance of 1 m. Since work equals force times distance, we have

$$1 \text{ J} = 1 \text{ N} \times 1 \text{ m} = 1 \text{ N} \cdot \text{m}$$

Thus, we have

$$\text{energy (J)} = F(\text{N}) \times S(\text{m}) \qquad \text{(3–33)}$$

Power

Power is the rate of doing work. In the SI system, 1 watt (W) of power is the rate of 1 J of work per second:

$$\text{power} = \frac{\text{work}}{\text{time}}$$

$$1 \text{ W} = \frac{1 \text{ J}}{\text{s}} = 1 \text{N} \cdot \text{m/s}$$

Thus, we have

$$\text{power (W)} = \frac{\text{work}(\text{N} \cdot \text{m})}{\text{time(s)}} \tag{3-34}$$

$$\text{hydraulic power (W)} = p(\text{N/m}^2) \times Q(\text{m}^3/\text{s}) \tag{3-35}$$

In the SI metric system all forms of power are expressed in watts. Appendix E, which provides tables of conversion factors, shows that 1 hp = 746 W = 0.746 kW.

Pump head H_p in units of meters can be related to pump power in units of watts by using Eq. (2-9): $p = \gamma H$. Thus, we have $p = \gamma H_p$. Substituting $p = \gamma H_p$ into Eq. (3-35) yields

$$H_p(\text{m}) = \frac{\text{pump hydraulic power (W)}}{\gamma(\text{N/m}^3) \times Q(\text{m}^3/\text{s})} \tag{3-36}$$

Equation (3-36) can be used to solve for the pump head for use in the energy equation. The motor head can also be calculated using Eq. (3-36), where the H_p term is replaced by H_m. The pump hydraulic power is replaced by the motor hydraulic power and Q represents the motor flow rate.

The mechanical output power (brake power or torque power) delivered by a hydraulic motor can be found from Eq. (3-37), where T is torque and ω or N is angular speed.

$$\text{power(kW)} = \frac{T(\text{N} \cdot \text{m}) \times \omega(\text{rad/s})}{1000} = \frac{T(\text{N} \cdot \text{m}) \times N(\text{rpm})}{9550} \tag{3-37}$$

Equation (3-37) is derived in Appendix H.

Flow Rate

Volume flow rate within a pipeline equals the product of the pipe cross-sectional area and the fluid velocity:

$$Q(\text{m}^3/\text{s}) = A(\text{m}^2) \times v(\text{m/s}) \tag{3-38}$$

A flow rate of 1 m³/s is an extremely large flow rate (1 m³/s = 15,800 gpm). Thus flow rates are frequently specified in units of liters per second (Lps) or liters per minute (Lpm). Since 1 liter = 1 L = 0.001 m³, we have

$$Q(\text{Lps}) = Q(\text{L/s}) = 1000 \, Q(\text{m}^3/\text{s})$$

3.13 ILLUSTRATIVE EXAMPLES USING THE SI METRIC SYSTEM

In this section we show how to solve hydraulic system problems using the SI metric system units.

EXAMPLE 3-13

For the hydraulic jack of Figure 3-7 the following data are given:

$$A_1 = 25 \text{ cm}^2 \qquad A_2 = 100 \text{ cm}^2$$
$$F_1 = 200 \text{ N}$$
$$S_1 = 5 \text{ cm}$$

Determine
  a. F_2
  b. S_2

Solution

a.
$$F_2 = \frac{A_2}{A_1} \times F_1 = \frac{100}{25} \times 200 = 800 \text{ N}$$

b.
$$S_2 = \frac{A_1}{A_2} \times S_1 = \frac{25}{100} \times 5 = 1.25 \text{ cm}$$

EXAMPLE 3-14

Oil is flowing through a 30-mm diameter pipe at 60 liters per minute (Lpm). Determine the velocity.

Solution Per Eq. (3-38) we have

$$v(\text{m/s}) = \frac{Q(\text{m}^3/\text{s})}{A(\text{m}^2)}$$

where $Q(\text{m}^3/\text{s}) = Q(\text{L/min}) \times \dfrac{0.001 \text{ m}^3}{1 \text{ L}} \times \dfrac{1 \text{ min}}{60 \text{ s}} = 0.0000167 \, Q \text{ (Lpm)}$

$$= 0.0000167 \times 60 = 0.0010 \text{ m}^3/\text{s}$$

$$A = \frac{\pi D^2}{4} = \frac{\pi}{4}(0.03 \text{ m})^2 = 0.000707 \text{ m}^2$$

Substituting values yields

$$v = \frac{0.0010}{0.000707} = 1.41 \text{ m/s}$$

EXAMPLE 3-15

A hydraulic pump delivers oil at 50 Lpm and 10,000 kPa. How much hydraulic power does the pump deliver?

Solution   Per Eq. (3-35), we have

$$\text{hydraulic power (kW)} = \frac{p(\text{Pa}) \times Q(\text{m}^3/\text{s})}{1000} = p(\text{kPa}) \times Q(\text{m}^3/\text{s})$$

where $Q(\text{m}^3/\text{s}) = Q(\text{L/min}) \times \dfrac{0.001 \text{ m}^3}{1 \text{ L}} \times \dfrac{1 \text{ min}}{60 \text{ s}} = 0.0000167Q \text{ (Lpm)}$

$$= 0.0000167 \times 50 = 0.000835 \text{ m}^3/\text{s}$$

Substituting values yields

$$\text{Hydraulic power (kW)} = 10,000 \times 0.000835 = 8.35 \text{ kW}$$

EXAMPLE 3-16

Determine the torque delivered by a hydraulic motor if the speed is 1450 rpm and the mechanical output power is 10 kW.

Solution   Solving Eq. (3-37) for torque yields

$$T(\text{N} \cdot \text{m}) = \frac{9550 \times \text{power(kW)}}{N(\text{rpm})}$$

$$= \frac{9550 \times 10}{1450} = 65.9 \text{ N} \cdot \text{m}$$

EXAMPLE 3-17

For the hydraulic system of Figure 3-29, the following SI metric data are given:

a. The pump is adding 3.73 kW (pump hydraulic power = 3.73 kW) to the fluid.

b. Pump flow is 0.001896 m³/s.

c. The pipe has a 0.0254-m inside diameter. Note that this size can also be represented in units of centimeters or millimeters as 2.54 cm or 25.4 mm, respectively.

d. The specific gravity of the oil is 0.9.

e. The elevation difference between stations 1 and 2 is 6.096 m.

Find the pressure available at the inlet to the hydraulic motor (station 2). The pressure at station 1 in the hydraulic tank is atmospheric (0 Pa or 0 N/m² gage). The head loss H_L due to friction between stations 1 and 2 is 9.144 m of oil.

Solution   Writing the energy equation between stations 1 and 2, we have

$$Z_1 + \frac{p_1}{\gamma} + \frac{v_1^2}{2g} + H_p - H_m - H_L = Z_2 + \frac{p_2}{\gamma} + \frac{v_2^2}{2g}$$

Since there is no hydraulic motor between stations 1 and 2, $H_m = 0$. Also, $v_1 = 0$ because the cross section of an oil tank is large. Also, $H_L = 9.144$ m and $p_1 = 0$ per the given input data.

Substituting known values, we have

$$Z_1 + 0 + 0 + H_p - 0 - 9.144 = Z_2 + \frac{p_2}{\gamma} + \frac{v_2^2}{2g}$$

Solving for p_2/γ, we have

$$\frac{p_2}{\gamma} = (Z_1 - Z_2) + H_p - \frac{v_2^2}{2g} - 9.144$$

Since $Z_2 - Z_1 = 6.096$ m, we have

$$\frac{p_2}{\gamma} = H_p - \frac{v_2^2}{2g} - 15.24$$

From Eq. (3-36) we solve for the pump head:

$$H_p(m) = \frac{\text{pump hydraulic power (W)}}{\gamma(N/m^3) \times Q(m^3/s)}$$

where

$$\gamma_{\text{oil}} = (\text{SG})\, \gamma_{\text{water}} = 0.9 \times 9797 \text{ N/m}^3 = 8817 \text{ N/m}^3$$

$$H_p(\text{m}) = \frac{3730}{8817 \times 0.001896} = 223.1 \text{ m.}$$

Next, we solve for v_2 and $v_2^2/2g$:

$$v_2(\text{m/s}) = \frac{Q(\text{m}^3/\text{s})}{A(\text{m}^2)} = \frac{0.001896}{(\pi/4)(0.0254 \text{ m})^2} = 3.74 \text{ m/s}$$

$$\frac{v_2^2}{2g} = \frac{(3.74 \text{ m/s})^2}{(2)(9.80 \text{ m/s}^2)} = 0.714 \text{ m}$$

On final substitution, we have

$$\frac{p_2}{\gamma} = 223.1 - 0.714 - 15.24 = 207.1 \text{ m}$$

Solving for p_2 yields

$$p_2(\text{N/m}^2) = (207.1 \text{ m})\gamma(\text{N/m}^3)$$

$$p_2 = (207.1)(8817) = 1,826,000 \text{ Pa} = 1826 \text{ kPa gage}$$

3.14 KEY EQUATIONS

Newton's law of motion:	$F = ma$	**(3-2)**
Work:	$W = FS$	**(3-3)**
Mechanical power (linear):	power $= Fv$	**(3-4)**
Mechanical horsepower (linear):	$\text{HP} = \dfrac{F(\text{lb}) \times v(\text{ft/s})}{550}$	**(3-5)**
Torque:	$T = FR$	**(3-6)**

Mechanical power
(rotary)

Brake horsepower:

$$HP = \frac{T(\text{in} \cdot \text{lb}) \times N(\text{rpm})}{63{,}000}$$ **(3-7)**

Brake power
(metric units):

$$\text{Power(kW)} = \frac{T(\text{N} \cdot \text{m}) \times \omega(\text{rad/s})}{1000}$$

$$= \frac{T(\text{N} \cdot \text{m}) \times N(\text{rpm})}{9550}$$ **(3-37)**

Efficiency:

$$\eta = \frac{\text{output power}}{\text{input power}}$$ **(3-8)**

Force multiplication
ratio in hydraulic jack:

$$\frac{F_2}{F_1} = \frac{A_2}{A_1}$$ **(3-9)**

Potential energy due
to elevation:

$$EPE = WZ$$ **(3-13)**

Potential energy
due to pressure:

$$PPE = W\frac{p}{\gamma}$$ **(3-14)**

Kinetic energy:

$$KE = \frac{1}{2}\frac{W}{g}v^2$$ **(3-15)**

Continuity equation:

$$Q_1 = A_1 v_1 = A_2 v_2 = Q_2$$ **(3-18)**

Hydraulic power

English units:

$$\text{hydraulic power (ft} \cdot \text{lb/s)} = p(\text{lb/ft}^2) \times Q(\text{ft}^3/\text{s})$$ **(3-23)**

Metric units:

$$\text{hydraulic power (W)} = p(\text{N/m}^2) \times Q(\text{m}^3/\text{s})$$ **(3-35)**

Hydraulic
horsepower:

$$HHP = \frac{p(\text{psi}) \times Q(\text{gpm})}{1714}$$ **(3-25)**

Bernoulli's equation:

$$Z_1 + \frac{p_1}{\gamma} + \frac{v_1^2}{2g} = Z_2 + \frac{p_2}{\gamma} + \frac{v_2^2}{2g}$$ **(3-27)**

Energy equation
(modified
Bernoulli's equation):

$$Z_1 + \frac{p_1}{\gamma} + \frac{v_1^2}{2g} + H_p - H_m - H_L = Z_2 + \frac{p_2}{\gamma} + \frac{v_2^2}{2g} \quad \textbf{(3-28)}$$

Pump head

Special English
units:

$$H_p(\text{ft}) = \frac{3950 \times (\text{HHP})}{Q(\text{gpm}) \times (\text{SG})} \quad \textbf{(3-29)}$$

Metric units:

$$H_p(\text{m}) = \frac{\text{pump hydraulic power(W)}}{\gamma(\text{N/m}^3) \times Q(\text{m}^3/\text{s})} \quad \textbf{(3-36)}$$

Fluid velocity:

$$v = \frac{Q}{A} \quad \textbf{(3-22 and 3-38)}$$

Energy:

$$\text{Energy} = FS \quad \textbf{(3-33)}$$

EXERCISES

Questions, Concepts, and Definitions

3-1. State Pascal's law.
3-2. Explain the meaning of Bernoulli's equation and how it affects the flow of a liquid in a hydraulic circuit.
3-3. What is the continuity equation, and what are its implications relative to fluid flow?
3-4. State Torricelli's theorem in your own words.
3-5. Explain how a siphon operates.
3-6. State the law of conservation of energy.
3-7. Explain how a venturi is used to produce the Bernoulli effect in an automobile carburetor.
3-8. What is meant by the terms *elevation head, pressure head*, and *velocity head?*
3-9. State Newton's three laws of motion.
3-10. Differentiate between energy and power.
3-11. Define the term *torque.*
3-12. What is meant by the term *efficiency?*
3-13. Relative to power, there is an analogy among mechanical, electrical, and hydraulic systems. Describe this analogy.
3-14. What is the significance of each term in the energy equation?

Problems

Note: The letter *E* following an exercise number means that English units are used. Similarly, the letter *M* indicates metric units.

Pascal's Law

3-15M. In the hydraulic jack shown in Figure 3-32, a force of 100 N is exerted on the small piston. Determine the upward force on the large piston. The area of the small piston is 50 cm², and the area of the large piston is 500 cm².

3-16M. For the system in Exercise 3-15 (shown in Figure 3-32), if the small piston moves 10 cm, how far will the large piston move? Assume the oil to be incompressible.

3-17M. Figure 3-33 shows a pneumatic/hydraulic system used to lift a load. If the inlet air pressure is 500 kPa, determine the maximum load that can be lifted.

3-18E. A pump delivers oil to a cylindrical storage tank, as shown in Figure 3-34. A faulty electric pressure switch, which controls the electric motor driving the pump, allows the pump to fill the tank completely. This causes the pressure p_1 near the base of the tank to build up to 15 psig. What force is exerted on the top of the tank? What does the pressure difference between the tank top and point 1 say about Pascal's law? What must be true about the magnitude of system pressure if the change in pressure due to elevation changes can be ignored in a fluid power system?

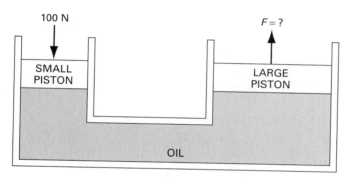

Figure 3-32. System for Exercise 3-15.

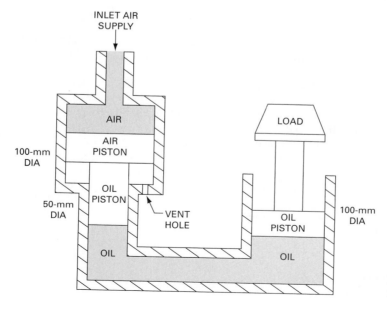

Figure 3-33. System for Exercise 3-17.

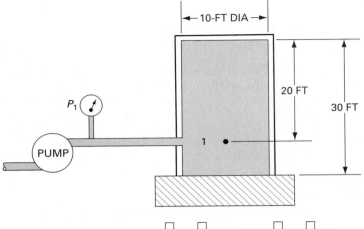

Figure 3-34. System for Exercise 3-18.

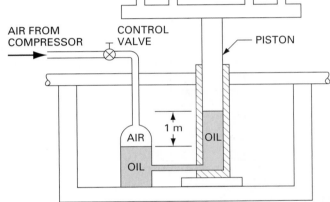

Figure 3-35. System for Exercise 3-21.

3-19E. For the hydraulic jack of Figure 3-32, a force of 100 lb is exerted on the small piston. Determine the upward force on the large piston. The diameter of the small piston is 2 in and the diameter of the large piston is 6 in.

3-20E. For the system in Exercise 3-19 (shown in Figure 3-32), if the small piston moves 1.5 in, how far will the large piston move? Assume the oil to be incompressible.

3-21M. For the fluid power automobile lift system of Figure 3-35, the air pressure equals 550 kPa gage. If the hydraulic piston has a 250-mm diameter, what is the maximum weight of an automobile that can be lifted? The specific gravity of the oil is 0.90. What percent error in the answer to this problem occurs by ignoring the 1-m head of oil between the air-oil interface surface and the bottom surface of the piston?

3-22E. Two hydraulic cylinders are connected at their piston ends (cap ends rather than rod ends) by a single pipe. Cylinder A has a 2-in diameter and cylinder B has a 4-in diameter. A 500-lb retraction force is applied to the piston rod of cylinder A. Determine the

    a. Pressure in cylinder A
    b. Pressure in cylinder B
    c. Pressure in the connecting pipe
    d. Output force of cylinder B

3-23E. An operator makes 20 complete cycles during a 15-sec interval using the hand pump of Figure 3-8. Each complete cycle consists of two pump strokes (intake and power). The pump has a 2-in-diameter piston and the load cylinder has a 4-in-diameter piston. The average hand force is 20 lb during each power stroke.

    a. How much load can be lifted?

    b. Through what distance will the load be moved during the 15-sec interval assuming no oil leakage? The pump piston has a 3-in stroke.

    c. What is the output HP assuming 90% efficiency?

3-24M. For the system in Exercise 3-23, change the data to metric units and solve parts a, b, and c.

3-25E. For the pressure booster of Figure 3-13, the following data are given:

$$\text{inlet air pressure } (p_1) = 125 \text{ psi}$$
$$\text{air piston area } (A_1) = 20 \text{ in}^2$$
$$\text{oil piston area } (A_2) = 1 \text{ in}^2$$
$$\text{load-carrying capacity } (F) = 75{,}000 \text{ lb}$$

Find the required load piston area A_3.

3-26M. For the system in Exercise 3-25, change the data to metric units and solve for the required load piston area A_3.

3-27M. For the pressure booster of Figure 3-13, the following data are given:

$$\text{inlet oil pressure } (p_1) = 1 \text{ MPa}$$
$$\text{air piston area } (A_1) = 0.02 \text{ m}^2$$
$$\text{oil piston area } (A_2) = 0.001 \text{ m}^2$$
$$\text{load-carrying capacity } (F) = 300{,}000 \text{ N}$$

Find the required load piston area A_3.

3-28E. A hydraulic system has a 100-gal reservoir mounted above the pump to produce a positive pressure (above atmospheric) at the pump inlet, as shown in Figure 3-36. The purpose of the positive pressure is to prevent the pump from cavitating when operating, especially at start-up. If the pressure at the pump inlet is to be 5 psi prior to turning the pump on and the oil has a specific gravity of 0.90, what should the oil level be above the pump inlet?

3-29E. For the hydraulic system of Figure 3-37, what would be the pressure at the pump inlet if the reservoir were located below the pump so that the oil level were 4 ft below the pump inlet? The specific gravity of the oil is 0.90. Ignore frictional losses and changes in kinetic energy. What effect would frictional losses and changes in kinetic energy have on the pressure at the pump inlet? Why? Would this increase or decrease the chances for having pump cavitation? Why?

3-30E. The hydraulic jack shown in Figure 3-38 is filled with oil. The large and small pistons have diameters of 3 in and 1 in, respectively. What force F on the handle is required to support the 2000-lb weight? If the force moves down 5 in, how far will the weight be lifted?

3-31M. For the system in Exercise 3-30, as shown in Figure 3-38, change the data to metric units. Then calculate the force F in units of N and determine how far the weight will be lifted in units of cm.

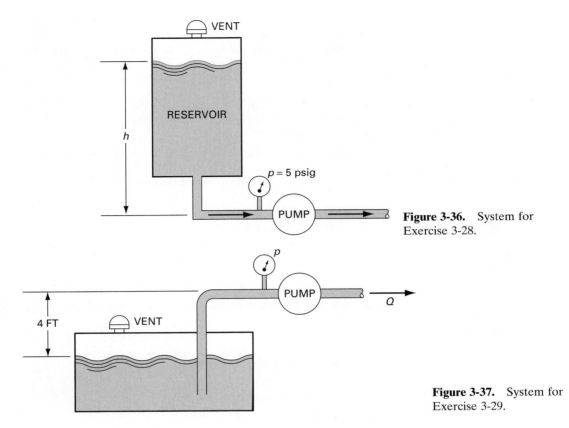

Figure 3-36. System for Exercise 3-28.

Figure 3-37. System for Exercise 3-29.

3-32M. Figure 3-39 shows a mechanical/hydraulic system used for clamping a cylindrical workpiece during a machining operation. If the machine operator applies a 100-N force to the lever as shown, what clamping force is applied to the workpiece?

Continuity Equation

3-33E. At a velocity of 10 ft/s, how many gpm of fluid will flow through a 1-in-inside-diameter pipe?

3-34E. How large an inside-diameter pipe is required to keep the velocity at 15 ft/s if the flow rate is 20 gpm?

3-35M. For the system in Exercise 3-34, change the data to metric units and solve for the size of inside-diameter pipe required.

3-36M. A hydraulic pump delivers fluid at 40 Lpm through a 25-mm-diameter pipe. Determine the fluid velocity.

3-37M. The following relationship containing metric units is analogous to Eq. (3-38):

$$v(\text{m/s}) = \frac{C Q(\text{m}^3/\text{s})}{[D(\text{m})]^2}$$

For this equation, determine the numerical value of C. Using this value of C, calculate v_2 (m/s) for the system of Example 3-17 and resolve any numerical difference.

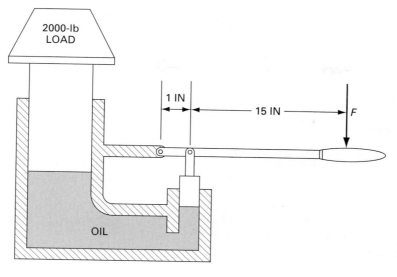

Figure 3-38. System for Exercise 3-30.

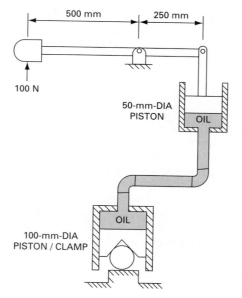

Figure 3-39. System for Exercise 3-32.

3-38M. At a velocity of 3 m/s, how many m³/s of fluid will flow through a 0.10-m-inside-diameter pipe?

3-39E. How long would it take a 20-gpm pump to extend a 4-in-diameter cylinder through a 20-in stroke?

3-40E. In Exercise 3-39, if the rod diameter is 2 in, how long will it take to retract the cylinder?

3-41E. A double-rod cylinder is one in which the rod extends out of the cylinder at both ends. Such a cylinder, with a 3-in-diameter piston and 2-in-diameter rod, cycles through a 10-in stroke at 60 cycles per minute. What gpm size pump is required?

3-42M. A cylinder with a 8-cm-diameter piston and 3-cm-diameter rod receives fluid at 30 Lpm. If the cylinder has a stroke of 35 cm, what is the maximum cycle rate that can be accomplished?

3-43M. Oil with specific gravity 0.9 enters a tee, as shown in Figure 3-40, with velocity $v_1 = 5$ m/s. The diameter at section 1 is 10 cm, the diameter at section 2 is 7 cm, and the diameter at section 3 is 6 cm. If equal flow rates are to occur at sections 2 and 3, find v_2 and v_3.

Hydraulic Power

3-44M. A hydraulic pump delivers a fluid at 50 Lpm and 10,000 kPa. How much hydraulic power does the pump produce?

3-45E. A hydraulic system is powered by a 5-hp motor and operates at 1000 psi. Assuming no losses, what is the flow rate through the system in gpm?

3-46E. A hydraulic cylinder is to compress a car body down to bale size in 8 sec. The operation requires an 8-ft stroke and a 10,000-lb force. If a 1000-psi pump has been selected, and assuming the cylinder is 100% efficient, find

   a. The required piston area
   b. The necessary pump flow rate
   c. The hydraulic horsepower delivered to the cylinder
   d. The output horsepower delivered by the cylinder to the load
   e. Solve parts a, b, c, and d assuming a 100-lb friction force and a leakage of 0.3 gpm. What is the efficiency of the cylinder with the given friction force and leakage?

3-47M. Show that Eq. (3-36) can be rewritten as follows:

$$H_p(\text{m}) = \frac{0.102 \times \text{pump hydraulic power (kW)}}{Q(\text{m}^3/\text{s}) \times (\text{SG})}$$

3-48E. The power and load-carrying capacity of a hydraulic cylinder (extension direction) are 10 hp and 5000 lb, respectively. Find the piston velocity in units of ft/s.

3-49M. The power and load-carrying capacity of a hydraulic cylinder (extension direction) are 10 kW and 20,000 N, respectively. Find the piston velocity in units of m/s.

3-50M. A hydraulic system is powered by a 5-kW motor and operates at 10 MPa. Assuming no losses, what is the flow rate through the system in units of m^3/s?

3-51M. A hydraulic cylinder is to compress a car body down to bale size in 8 s. The operation requires a 3-m stroke and a 40,000-N force. If a 10-MPa pump has been selected, and assuming the cylinder is 100% efficient, find

   a. The required piston area (m^2)
   b. The necessary pump flow rate (m^3/s)
   c. The hydraulic power (kW) delivered to the cylinder
   d. The output power (kW) delivered by the cylinder to the load
   e. Solve parts a, b, c, and d assuming a 400-N friction force and a leakage of 1.0 Lpm. What is the efficiency of the cylinder with the given friction force and leakage?

3-52E. An automotive lift raises a 3500-lb car 7 ft above floor level. If the hydraulic cylinder contains an 8-in-diameter piston and 4-in-diameter rod, determine the

   a. Work necessary to lift the car
   b. Required pressure
   c. Power if the lift raises the car in 10 s
   d. Descending speed of the lift for a 10-gpm flow rate
   e. Flow rate for the auto to descend in 10 s

3-53M. For the system in Exercise 3-52, change the data to metric units and solve parts a–e.

Bernoulli's Equation

3-54E. Fluid is flowing horizontally at 100 gpm from a 2-in-diameter pipe to a 1-in-diameter pipe, as shown in Figure 3-41. If the pressure at point 1 is 10 psi, find the pressure at point 2. The specific gravity of the fluid is 0.9.

3-55M. For the system in Exercise 3-54, change the data to metric units and solve for the pressure at point 2.

3-56E. What is the potential energy of 1000 gal of water at an elevation of 100 ft?

3-57E. What is the kinetic energy of 1 gal of water traveling at 20 ft/s?

3-58E. A siphon made of a 1-in-inside-diameter pipe is used to maintain a constant level in a 20-ft-deep tank. If the siphon discharge is 30 ft below the top of the tank, what will be the flow rate if the fluid level is 5 ft below the top of the tank?

3-59M. For the system in Exercise 3-58, change the data to metric units and solve for the flow rate.

3-60E. For the hydraulic system of Figure 3-29, the following data are given:
  1. The pump is adding 4 hp to the fluid.
  2. Pump flow is 25 gpm.
  3. The pipe has been changed to an 0.75-in-inside diameter size.
  4. The specific gravity of the oil is 0.9.
  5. The oil tank is vented to the atmosphere.
  6. The head loss H_L between stations 1 and 2 is 40 ft of oil.
  Find the pressure available at the inlet to the hydraulic motor (station 2).

3-61E. The oil tank for the hydraulic system of Figure 3-42 is air-pressurized at 10 psig. The inlet line to the pump is 10 ft below the oil level. The pump-flow rate is 30 gpm. Find the pressure at station 2 if
  a. There is no head loss between stations 1 and 2
  b. There is a 25-ft head loss between stations 1 and 2

3-62M. Solve Exercise 3-60 using metric units. The equivalent metric data are given as follows:
  1. The pump is adding 2.984 kW to the fluid.
  2. Pump flow is 0.00158 m³/s.

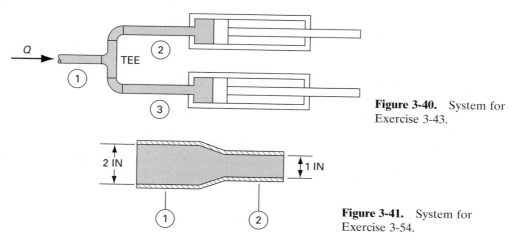

Figure 3-40. System for Exercise 3-43.

Figure 3-41. System for Exercise 3-54.

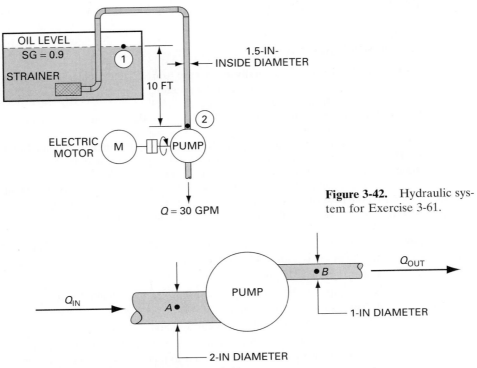

Figure 3-42. Hydraulic system for Exercise 3-61.

Figure 3-43. System for Exercise 3-64.

3. The pipe has a 0.01905-m inside diameter.
4. The specific gravity of the oil is 0.9.
5. The oil tank is vented to the atmosphere.
6. The head loss H_L between stations 1 and 2 is 12.19 m of oil.
7. The elevation difference between stations 1 and 2 is 6.096 m.
Find the pressure available at the inlet to the hydraulic motor (station 2).

3-63M. Solve Exercise 3-61 using metric units. The equivalent metric data are given as follows:
   1. Pump flow is 0.001896 m³/s.
   2. The air pressure at station 1 in the hydraulic tank is 68.97-kPa gage pressure.
   3. The inlet line to the pump is 3.048 m below the oil level.
   4. The pipe has a 0.0381-m inside diameter.
   Find the pressure at station 2 if
   a. There is no head loss between stations 1 and 2
   b. There is a 7.622-m head loss between stations 1 and 2

3-64E. For the pump in Figure 3-43, Q_{out} = 30 gpm of oil having a specific gravity of 0.9. Points A and B are at the same elevation. What is Q_{in}? Assuming the pump efficiency is 100%, find the pressure difference between points A and B if
   a. The pump is turned off
   b. The input power to the pump is 2 hp

3-65M. For the system in Exercise 3-64, change the data to metric units and solve parts a and b.

4 Frictional Losses in Hydraulic Pipelines

Learning Objectives

Upon completing this chapter, you should be able to:

1. Differentiate between laminar and turbulent flow.
2. Understand the significance of the Reynolds number.
3. Determine the Reynolds number at any location in a pipeline.
4. Explain the meaning of the term *friction factor*.
5. Determine friction factors for laminar and turbulent flow.
6. Evaluate the head loss in a pipeline undergoing laminar or turbulent flow.
7. Calculate frictional losses in valves and fittings.
8. Explain the meaning of the term *K factor*.
9. Discuss the significance of the term *equivalent length*.
10. Perform an energy analysis of a complete hydraulic circuit.

4.1 INTRODUCTION

Up to now we have not investigated the mechanism of energy losses due to friction associated with the flow of a fluid inside a pipeline. It is intuitive that liquids, such as water or gasoline, flow much more readily than do other liquids, such as oil. The resistance to flow is essentially a measure of the viscosity of the fluid. The greater the viscosity of a fluid, the less readily it flows and the more energy is required to move it. This energy is lost because it is dissipated into heat and thus represents wasted energy.

Energy losses also occur in valves and fittings. A fitting is a component (other than a straight pipe) that is used to carry the fluid. Examples are bends, couplings, tees, elbows, filters, and strainers. The nature of the flow path through valves and fittings determines the amount of energy losses. Generally speaking, the more tortuous the path, the greater the losses. In many fluid power applications, energy losses due to flow in valves and fittings exceed those due to flow in pipes.

It is very important to keep all energy losses in a fluid power system to a minimum acceptable level. This requires the proper selection of the sizes of the pipes, valves, and fittings that make up the system. In general, the smaller the pipe diameter as well as valve and fitting size, the greater the losses. However, using large-diameter pipes, valves, and fittings results in greater cost and poor space utilization. Thus, the selection of component sizes represents a compromise between energy losses and component cost and space requirements.

The resistance to flow of pipes, valves, and fittings can be determined using empirical formulas that have been developed by experimentation. This permits the calculation of energy losses for any system component. The energy equation and the continuity equation can then be used to perform a complete analysis of a fluid power system. This includes calculating the pressure drops, flow rates, and power losses for all components of the fluid power system.

Figure 4-1 depicts a hydraulic system (shown inside the dashed lines) whose pump receives mechanical energy from an electric motor. The hydraulic system delivers mechanical energy to an external load via a hydraulic cylinder while it also

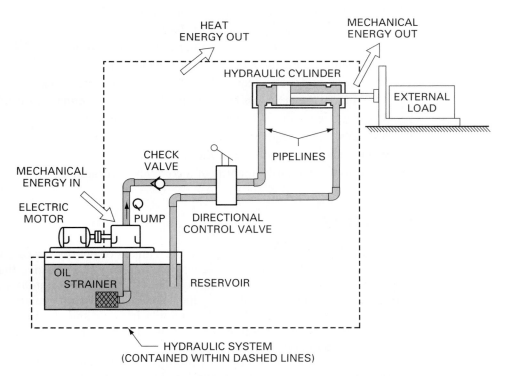

Figure 4-1. Energy transfer into and out of a hydraulic system includes energy loss (in the form of heat) due to frictional fluid flow through pipes, valves, and fittings.

Figure 4-2. Hydraulic-powered backhoe loader. *(Courtesy of Caterpillar Inc., Peoria, Illinois.)*

transfers heat energy (such as that created by frictional fluid flow through pipes, valves, and fittings) to the surroundings. Within the hydraulic system, the pump receives low-pressure oil from the reservoir and delivers it at high pressure to the hydraulic cylinder. The hydraulic cylinder uses this flow of high-pressure oil to drive the external load as low-pressure oil returns to the reservoir.

Figure 4-2 shows a backhoe loader that uses a variable displacement, pressure-compensated, axial-piston pump to provide optimum performance in both backhoe and loader operations. The backhoe portion of the machine performs operations such as digging a trench. The front loader portion can then be used to load a dump truck with the earth removed from the trench dug by the backhoe. The pump delivers the desired flow to the hydraulic cylinders at the required pressure to fulfill implement demands. At an operating speed of 2200 rpm, the pump produces a maximum flow of 43 gpm (163 Lpm) at a system pressure of 3300 psi (22,700 kPa).

4.2 LAMINAR AND TURBULENT FLOW

In our discussions of fluid flow in pipes in Chapter 3, we assumed a constant velocity over the cross section of a pipe at any one station (see Figure 3-23). However, when a fluid flows through a pipe, the layer of fluid at the wall has zero velocity. This is due to viscosity, which causes fluid particles to cling to the wall. Layers of fluid at the progressively greater distances from the pipe surface have higher velocities, with the maximum velocity occurring at the pipe centerline, as illustrated in Figure 4-3.

There are two types of flow in pipes. The first type is *laminar flow*, which is characterized by the fluid flowing in smooth layers or laminae. In this type of flow, a particle of fluid in a given layer stays in that layer, as shown in Figure 4-4. This type

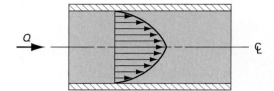

Figure 4-3. Velocity profile in pipe.

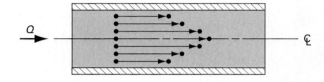

Figure 4-4. Straight-line path of fluid particles in laminar flow.

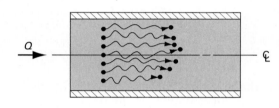

Figure 4-5. Random fluctuation of fluid particles in turbulent flow.

of fluid motion is called *streamline flow* because all the particles of fluid are moving in parallel paths. Therefore, laminar flow is smooth with essentially no collision of particles. For laminar flow, the friction is caused by the sliding of one layer or particle of fluid over another in a smooth continuous fashion.

If the velocity of flow reaches a high enough value, the flow ceases to be laminar and becomes turbulent. As shown in Figure 4-5, in turbulent flow the movement of a particle becomes random and fluctuates up and down in a direction perpendicular as well as parallel to the mean flow direction. This mixing action generates turbulence due to the colliding fluid particles. This causes considerably more resistance to flow and thus greater energy losses than that produced by laminar flow.

The difference between laminar and turbulent flow can be seen when using a water faucet. When the faucet is turned only partially open, with just a small amount of flow, the flow pattern observed is a smooth laminar one, as depicted in Figure 4-6(a). However, when the faucet is opened wide, the flow mixes and becomes turbulent, as illustrated in Figure 4-6(b).

4.3 REYNOLDS NUMBER

It is important to know whether the flow pattern inside a pipe is laminar or turbulent. This brings us to the experiments performed by Osborn Reynolds in 1833 to determine the conditions governing the transition from laminar to turbulent flow.

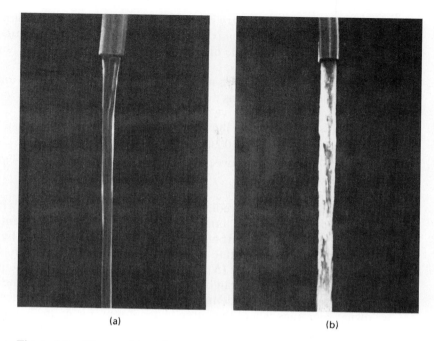

Figure 4-6. Flow patterns from water faucet. (a) Laminar flow,
(b) turbulent flow. *(Reprinted from Introduction to Fluid Mechanics by J. E.
John and W. L. Haberman, Prentice Hall, Englewood Cliffs, NJ, 1988.)*

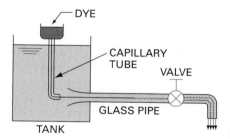

Figure 4-7. Reynolds' experiment.

Using the test setup shown in Figure 4-7, Reynolds allowed the fluid in the large
tank to flow through a bell-mouthed entrance and along a smooth glass pipe. He controlled the flow rate by means of a valve at the end of the pipe. A capillary tube,
connected to a reservoir of dye, allowed the flow of a fine jet of dye into the main
flow stream.

If the flow in the pipe was laminar, the dye jet flowed smoothly. However, when
turbulent flow occurred in the pipe, the dye jet would mix with the main fluid.

Reynolds came to a very significant conclusion as a result of his experiments: *The nature of the flow depends on the dimensionless parameter* $\upsilon D\rho/\mu$, where υ = fluid velocity, D = pipe inside diameter, ρ = fluid density, and μ = absolute viscosity of the fluid.

This parameter has been named the *Reynolds number* (N_R) and (as Reynolds discovered from his tests) has the following significance:

1. If N_R is less than 2000, the flow is laminar.
2. If N_R is greater than 4000, the flow is turbulent.
3. Reynolds numbers between 2000 and 4000 cover a critical zone between laminar and turbulent flow.

It is not possible to predict the type of flow that will exist within the critical zone. Thus if the Reynolds number lies in the critical zone, turbulent flow should be assumed. However, since turbulent flow results in greater losses, hydraulic systems should normally be designed to operate in the laminar flow region.

The Reynolds number ($N_R = \upsilon D\rho/\mu = \upsilon D/\nu$), which is a dimensionless parameter (has no overall units), can be calculated in several ways depending on the units used for each of its dependent variables (υ, D, ρ, μ, and ν). This is shown in the following equations, where the letter M denotes metric units.

$$N_R = \frac{\upsilon(\text{ft/s}) \times D(\text{ft}) \times \rho(\text{slugs/ft}^3)}{\mu(\text{lb} \cdot \text{s/ft}^2)} = \frac{\upsilon(\text{ft/s}) \times D(\text{ft})}{\nu(\text{ft}^2/\text{s})} \qquad \textbf{(4-1)}$$

$$N_R = \frac{\upsilon(\text{m/s}) \times D(\text{m}) \times \rho(\text{kg/m}^3)}{\mu(\text{N} \cdot \text{s/m}^2)} = \frac{\upsilon(\text{m/s}) \times D(\text{m})}{\nu(\text{m}^2/\text{s})} \qquad \textbf{(4-1M)}$$

Using absolute viscosity in units of cP gives

$$N_R = \frac{7740\upsilon(\text{ft/s}) \times D(\text{in}) \times \text{SG}}{\mu(\text{cP})} \qquad \textbf{(4-2)}$$

$$N_R = \frac{1000\upsilon(\text{m/s}) \times D(\text{mm}) \times \text{SG}}{\mu(\text{cP})} \qquad \textbf{(4-2M)}$$

A final relationship using kinematic viscosity in units of cS is:

$$N_R = \frac{7740\upsilon(\text{ft/s}) \times D(\text{in})}{\nu(\text{cS})} \qquad \textbf{(4-3)}$$

$$N_R = \frac{1000\upsilon(\text{m/s}) \times D(\text{mm})}{\nu(\text{cS})} \qquad \textbf{(4-3M)}$$

Note that in Eqs. (4-2), (4-2M), (4-3), and (4-3M) the constant contains the proper units to cause the Reynolds number to be dimensionless.

If turbulent flow is allowed to exist, higher fluid temperatures will occur due to greater frictional energy losses. Therefore, turbulent flow systems suffering from excessive fluid temperatures can be helped by increasing the pipe diameter to establish laminar flow.

EXAMPLE 4-1

The kinematic viscosity of a hydraulic oil is 100 cS. If the fluid is flowing in a 1-in-diameter pipe at a velocity of 10 ft/s, what is the Reynolds number?

Solution   Substitute directly into Eq. (4-3):

$$N_R = \frac{(7740)(10)(1)}{100} = 774$$

EXAMPLE 4-2

Oil ($v = 0.001$ m²/s) is flowing in a 50-mm-diameter pipe at a velocity of 5 m/s. What is the Reynolds number?

Solution   Per Eq. (4-1M), we have

$$N_R = \frac{5 \times 0.050}{0.001} = 250$$

4.4 DARCY'S EQUATION

Friction is the main cause of energy losses in fluid power systems. The energy loss due to friction is transferred into heat, which is given off to the surrounding air. The result is a loss of potential energy in the system, and this shows up as a loss in pressure or head. In Chapter 3 we included this head loss in the energy equation as an H_L term. However, we did not discuss how the magnitude of this head loss term could be evaluated. The head loss (H_L) in a system actually consists of two components:

1. Losses in pipes
2. Losses in valves and fittings

Head losses in pipes can be found by using Darcy's equation:

$$H_L = f\left(\frac{L}{D}\right)\left(\frac{v^2}{2g}\right) \tag{4-4}$$

where f = friction factor (dimensionless),
         L = length of pipe (ft, m),
         D = pipe inside diameter (ft, m),
         υ = average fluid velocity (ft/s, m/s),
         g = acceleration of gravity (ft/s², m/s²).

Darcy's equation can be used to calculate the head loss due to friction in pipes for both laminar and turbulent flow. The difference between the two lies in the evaluation of the friction factor f. The technique is discussed for laminar and turbulent flow in Sections 4.5 and 4.6, respectively.

4.5 FRICTIONAL LOSSES IN LAMINAR FLOW

Darcy's equation can be used to find head losses in pipes experiencing laminar flow by noting that for laminar flow the friction factor equals the constant 64 divided by the Reynolds number:

$$f = \frac{64}{N_R} \tag{4-5}$$

Substituting Eq. (4-5) into Eq. (4-4) yields the Hagen-Poiseuille equation, which is valid for laminar flow only:

$$H_L = \frac{64}{N_R}\left(\frac{L}{D}\right)\left(\frac{\upsilon^2}{2g}\right) \tag{4-6}$$

Examples 4-3 and 4-4 illustrate the use of the Hagen-Poiseuille equation.

EXAMPLE 4-3

For the system of Example 4-1, find the head loss due to friction in units of psi for a 100-ft length of pipe. The oil has a specific gravity of 0.90.

Solution   From Eq. (4-6) we solve for the head loss in units of feet of oil:

$$H_L = \frac{64}{774}\left(\frac{100}{\frac{1}{12}}\right)\left(\frac{10^2}{64.4}\right) = 154 \text{ ft}$$

Note that the units for H_L are really ft · lb/lb. Thus, we can conclude that 154 ft · lb of energy is lost by each pound of oil as it flows through the 100-ft length of pipe.

Using Eq. (2-9), we convert head loss in units of feet of oil to pressure loss in units of psi:

$$p_L = \gamma H_L = (SG \times \gamma_{H_2O}) \times H_L = (0.9 \times 0.0361) \text{lb/in}^3$$

$$\times (12 \times 154) \text{in} = 60 \text{ psi}$$

Thus, there is a 60-psi pressure loss as the oil flows through the 100-ft length of pipe. This pressure loss is due to friction.

EXAMPLE 4-4

For the system of Example 4-2, find the head loss due to friction for a 50-m length of pipe. $\gamma_{oil} = 8800$ N/m³.

Solution   From Eq. (4-6), we have

$$H_L = \frac{64}{250}\left(\frac{50}{0.050}\right)\left(\frac{5^2}{2 \times 9.80}\right) = 326 \text{ m of oil}$$

The pressure drop is

$$\Delta p = \gamma(\text{N/m}^3) \times H_L(\text{m}) = 8800 \times 326 = 28,700 \text{ Pa} = 28.7 \text{ kPa}$$

4.6 FRICTIONAL LOSSES IN TURBULENT FLOW

Effect of Pipe Roughness

Darcy's equation will be used for calculating energy losses in turbulent fluid flow. However, the friction factor cannot be represented by a simple formula as was the case for laminar flow. This is due to the random and fluctuating movement of the fluid particles.

For turbulent flow, experiments have shown that the friction factor is a function of not only the Reynolds number but also the relative roughness of the pipe. The relative roughness is defined as the pipe inside surface roughness, ε (Greek letter epsilon), divided by the pipe inside diameter D:

$$\text{relative roughness} = \frac{\varepsilon}{D} \tag{4-7}$$

Figure 4-8 illustrates the physical meaning of the pipe inside surface roughness ε, which is called the absolute roughness.

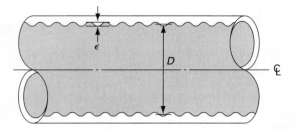

Figure 4-8. Pipe absolute roughness (ε).

TYPE OF PIPE	ABSOLUTE ROUGHNESS	
	ε (in)	ε (mm)
GLASS OR PLASTIC	SMOOTH	SMOOTH
DRAWN TUBING	0.00006	0.0015
COMMERCIAL STEEL OR WROUGHT IRON	0.0018	0.046
ASPHALTED CAST IRON	0.0048	0.12
GALVANIZED IRON	0.006	0.15
CAST IRON	0.0102	0.26
RIVETED STEEL	0.072	1.8

Figure 4-9. Typical values of absolute roughness.

Pipe roughness values depend on the pipe material as well as the method of manufacture. Figure 4-9 gives typical values of absolute roughness for various types of pipes.

It should be noted that the values given in Figure 4-9 are average values for new clean pipe. After the pipes have been in service for a time, the roughness values may change significantly due to the buildup of deposits on the pipe walls.

The Moody Diagram

To determine the value of the friction factor for use in Darcy's equation, we use the Moody diagram shown in Figure 4-10. This diagram contains curves that were determined by data taken by L. F. Moody. Each curve represents values of friction factor as a function of the Reynolds number for a given value of relative roughness. Thus, if we know the Reynolds number and relative roughness, we can quickly determine the friction factor.

The following important characteristics should be noted about the Moody diagram:

1. It is plotted on logarithmic paper because of the large range of values encountered for f and N_R.
2. At the left end of the chart (Reynolds numbers less than 2000) the straight-line curve gives the relationship for laminar flow: $f = 64/N_R$.

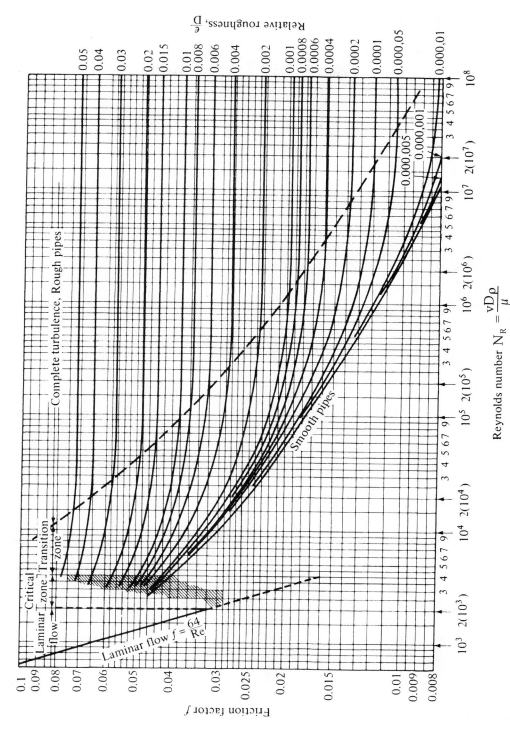

Figure 4-10. The Moody diagram. *(Reprinted from Introduction to Fluid Mechanics by J. E. John and W. L. Haberman, Prentice Hall, Englewood Cliffs, NJ, 1988.)*

3. No curves are drawn in the critical zone ($2000 < N_R < 4000$), because it is not possible to predict whether the flow is laminar or turbulent in this region.

4. For Reynolds numbers greater than 4000, each curve plotted represents a particular value of ε/D. For intermediate values of ε/D, interpolation is required.

5. Once complete turbulence is reached (region to the right of the dashed line), increasing values of N_R have no effect on the value of f.

Example 4-5 illustrates the use of the Moody diagram for finding values of friction factor f for laminar and turbulent flow.

EXAMPLE 4-5

The kinematic viscosity of a hydraulic oil is 50 cS. If the oil flows in a 1-in-diameter commercial steel pipe, find the friction factor if

a. The velocity is 10 ft/s

b. The velocity is 40 ft/s

Solution

a. Find N_R from Eq. (4-3):

$$N_R = \frac{(7740)(10)(1)}{50} = 1548 = 1.548 \times 10^3$$

Since the flow is laminar, we do not need to know the relative roughness value. To find f, locate 1.548×10^3 on the N_R axis of the Moody diagram (approximate value $= 1.5 \times 10^3$). Then project vertically up until the straight-line curve ($f = 64/N_R$) is reached. Then project horizontally to the f axis to obtain a value of 0.042.

b.
$$N_R = \frac{(7740)(40)(1)}{50} = 6192 = 6.192 \times 10^3$$

Since the flow is turbulent, we need the value of ε/D. First, the relative roughness (a dimensionless parameter) is found using Figure 4-9 to get the value of ε:

$$\frac{\varepsilon}{D} = \frac{0.0018 \text{ in}}{1 \text{ in}} = 0.0018$$

Now locate the value 6.192×10^3 on the N_R axis of the Moody diagram (approximate value = 6.2×10^3). Then project vertically up until you are between the ε/D curves of 0.0010 and 0.0020 (where the 0.0018 curve would approximately exist if it were drawn). Then project horizontally to the f axis to obtain a value of 0.036.

Since approximate values of N_R are used and interpolation of ε/D values is required, variations in the determined value of f are expected when using the Moody diagram. However, this normally produces variations of less than ±0.001, which is generally acceptable for calculating frictional losses in piping systems. This type of analysis is done for a complete system in Section 4.9.

4.7 LOSSES IN VALVES AND FITTINGS

The K Factor

In addition to losses due to friction in pipes, there also are energy losses in valves and fittings such as tees, elbows, and bends. For many fluid power applications, the majority of the energy losses occur in these valves and fittings in which there is a change in the cross section of the flow path and a change in the direction of flow. Thus, the nature of the flow through valves and fittings is very complex.

As a result, experimental techniques are used to determine losses. Tests have shown that head losses in valves and fittings are proportional to the square of the velocity of the fluid:

$$H_L = K\left(\frac{v^2}{2g}\right) \tag{4-8}$$

The constant of proportionality (K) is called the K factor (also called loss coefficient) of the valve or fitting. Figure 4-11 gives typical K-factor values for several common types of valves and fittings.

Common Valves and Fittings

Illustrations of several common valves and fittings are given as follows:

1. **Globe valve.** See Figure 4-12. In this design, the fluid changes direction when flow occurs between the globe and seat. This construction increases resistance to fluid flow but also permits close regulation. Figure 4-12 shows the globe valve in its fully closed position. A threaded stem (attached to the globe by a swivel

VALVE OR FITTING		K FACTOR
GLOBE VALVE:	WIDE OPEN	10.0
	1/2 OPEN	12.5
GATE VALVE:	WIDE OPEN	0.19
	3/4 OPEN	0.90
	1/2 OPEN	4.5
	1/4 OPEN	24.0
RETURN BEND		2.2
STANDARD TEE		1.8
STANDARD ELBOW?		0.9
45° ELBOW		0.42
90° ELBOW		0.75
BALL CHECK VALVE		4.0

Figure 4-11. *K* factors of common valves and fittings.

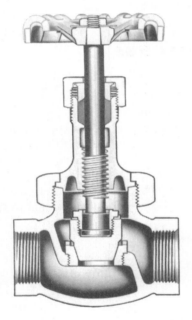

Figure 4-12. Globe valve. *(Courtesy of Crane Co., New York, New York.)*

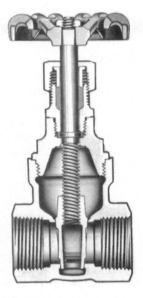

Figure 4-13. Gate valve. *(Courtesy of Crane Co., New York, New York.)*

connection and rotated by the handwheel) not only raises and lowers the globe, but also helps guide it squarely to its seat.

2. **Gate valve.** See Figure 4-13. Fluids flow through gate valves in straight-line paths; thus, there is little resistance to flow and the resulting pressure drops are small. A gatelike disk (actuated by a stem screw and handwheel) moves up and down at right angles to the path of flow and seats against two seat

Figure 4-14. 45° elbow. *(Courtesy of Crane Co., New York, New York.)*

Figure 4-15. 90° elbow. *(Courtesy of Crane Co., New York, New York.)*

Figure 4-16. Tee. *(Courtesy of Crane Co., New York, New York.)*

Figure 4-17. Return bend. *(Courtesy of Crane Co., New York, New York.)*

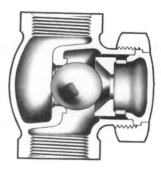

Figure 4-18. Ball check valve. *(Courtesy of Crane Co., New York, New York.)*

faces to shut off flow. Gate valves are best for services that require infrequent valve operation and where the disk is kept either fully opened or closed. They are not practical for throttling. With the usual type of gate valve, close regulation is impossible. Velocity of flow against a partly opened disk may cause vibration and chattering and result in damage to the seating surfaces.

3. **45° elbow.** See Figure 4-14.
4. **90° elbow.** See Figure 4-15.
5. **Tee.** See Figure 4-16.
6. **Return bend.** See Figure 4-17.
7. **Ball check valve.** See Figure 4-18. The function of a check valve is to allow flow to pass through in only one direction. Thus, check valves are used to prevent backflow in hydraulic lines.

Pressure Drop versus Flow Rate Curves

For some fluid power valves, in addition to specifying K factors, empirical curves of pressure drop versus flow rate are given by the valve manufacturer. Thus, if the flow rate through the valve is known, the pressure drop can be determined by referring to the curve. This is normally done for directional control valves and also for flow control valves for various opening positions.

Figure 4-19 shows a cutaway of a directional control valve whose pressure drop versus flow rate characteristics are provided by curves (see Figure 4-20) as well as by K-factor values. As expected from Eq. (4-8), the curves show that the pressure drop increases approximately as the square of the flow rate.

Most fluid power valves (including the valve of Figure 4-19) are of the spool design. A spool is a circular shaft containing lands that are larger-diameter sections machined to slide in a very close fitting bore of the valve body. The grooves between the lands provide the desired flow paths through the valve for unique positions of the sliding spool. The design and operation of hydraulic valves are discussed in detail in Chapter 8.

4.8 EQUIVALENT-LENGTH TECHNIQUE

Darcy's equation shows that the head loss in a pipe, due to fluid friction, is proportional not only to the square of the fluid velocity but also to the length of the pipe. There is a similarity between Darcy's equation and Eq. (4-8), which states that the head loss in a valve or fitting is proportional to the square of the fluid velocity.

This suggests we can find a length of pipe that for the same flow rate would produce the same head loss as a valve or fitting. This length of pipe, which is called *the equivalent length* of a valve or fitting, can be found by equating the head losses across the valve or fitting and the pipe:

$$H_{L(\text{valve or fitting})} = H_{L(\text{pipe})}$$

Figure 4-19.   Cutaway of a directional control valve. *(Courtesy of Continental Hydraulics, Division of Continental Machines, Inc., Savage, Minnesota.)*

EXAMPLE 4-6

What is the head loss across a 1-in wide-open globe valve when oil (SG = 0.9) flows through it at a rate of 30 gpm?

Solution   Find the fluid velocity using Eq. (3-22):

$$v(\text{ft/s}) = \frac{Q(\text{ft}^3/s)}{A(\text{ft}^2)} = \frac{30/449}{\frac{\pi}{4}(1/12)^2} = 12.2 \text{ ft/s}$$

From Figure 4-11, K for a wide-open globe valve equals 10. Thus, per Eq. (4-8), we have

$$H_L = \frac{(10)(12.2)^2}{64.4} = 23.1 \text{ ft of oil}$$

The pressure drop (Δp) across the valve can now be found:

$$\Delta p = \gamma H_L = (SG \times \gamma_{H_2O}) \times H_L = (0.9 \times 0.0361) \text{ lb/in}^3$$

$$\times (12 \times 23.1) \text{ in} = 9.01 \text{ psi}$$

EXAMPLE 4-7

What is the head loss across a 50-mm wide-open gate valve when oil ($v = 0.001$ m²/s, $\gamma = 8800$ N/m³) flows through it at a rate of 0.02 m³/s?

Solution   The velocity is found first.

$$v(\text{m/s}) = \frac{Q(\text{m}^3/s)}{A(\text{m}^2)} = \frac{0.02}{\frac{\pi}{4}(0.050)^2} = 10.2 \text{ m/s}$$

From Figure 4-11, $K = 0.19$. Thus, per Eq. (4-8), we have

$$H_L = \frac{0.19(10.2)^2}{2 \times 9.80} = 1.01 \text{ m of oil}$$

The pressure drop across the valve can now be found:

$$\Delta p(\text{N/m}^2) = \gamma(\text{N/m}^3) \times H_L(\text{m}) = 8800(1.01) = 8890 \text{ Pa} = 8.89 \text{ kPa}$$

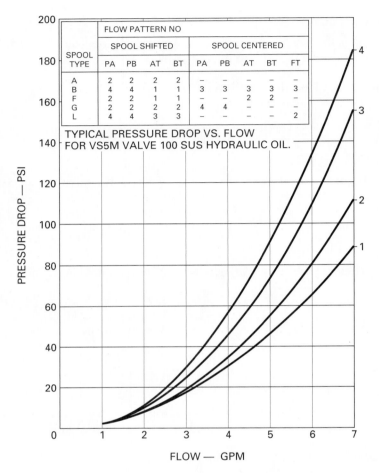

Figure 4-20. Pressure drop (vs.) flow curves for a directional control valve. *(Courtesy of Continental Hydraulics, Division of Continental Machines, Inc., Savage, Minnesota.)*

Substituting the corresponding expressions, we have

$$K\left(\frac{v^2}{2g}\right) = f\left(\frac{L_e}{D}\right)\left(\frac{v^2}{2g}\right)$$

Since the velocities are equal, we can cancel the $v^2/2g$ terms from both sides of the equation. The result is

$$L_e = \frac{KD}{f} \tag{4-9}$$

where L_e is the equivalent length of a valve or fitting whose K factor is K. Note that parameters K and f are both dimensionless. Therefore, L_e and D will have the same dimensions. For example, if D is measured in units of feet, then L_e will be calculated in units of feet.

Equation (4-9) permits the convenience of examining each valve or fitting of a fluid power system as though it were a pipe of length L_e. This provides a convenient method of analyzing hydraulic circuits where frictional energy losses are to be taken into account. Section 4.9 deals with this type of problem. Example 4-8 shows how to find the equivalent length of a hydraulic component.

EXAMPLE 4-8

Hydraulic oil ($\nu = 100$ cS) flows through a 1-in-diameter commercial steel pipe at a rate of 30 gpm. What is the equivalent length of a 1-in wide-open globe valve placed in the line?

Solution We need to find the friction factor f, so let's first find the velocity υ from Eq. (3-22):

$$\upsilon(\text{ft/s}) = \frac{Q(\text{ft}^3/\text{s})}{A(\text{ft}^2)} = \frac{30/449}{\frac{\pi}{4}(1/12)^2} = 12.2 \text{ ft/s}$$

Using Eq. (4-3), we find the Reynolds number:

$$N_R = \frac{7740(12.2)(1)}{100} = 944$$

Since the flow is laminar, we do not need to know the relative roughness to find the friction factor.

$$f = \frac{64}{N_R} = \frac{64}{944} = 0.0678$$

Finally, we determine the equivalent length using Eq. (4-9):

$$L_e = \frac{KD}{f} = \frac{(10)(\frac{1}{12} \text{ ft})}{0.0678} = 12.3 \text{ ft}$$

Thus, a 1-in-diameter pipe of length 12.3 ft would produce the same frictional energy loss as a 1-in wide-open globe valve for a flow rate of 30 gpm.

4.9 HYDRAULIC CIRCUIT ANALYSIS

We are now ready to perform a complete analysis of a hydraulic circuit, taking into account energy losses due to friction. Let's analyze the hydraulic system of Figure 4-21 by doing an example problem.

EXAMPLE 4-9

For the hydraulic system of Figure 4-21, the following data are given:

a. The pump is adding 5 hp to the fluid (pump hydraulic horsepower = 5).
b. Pump flow is 30 gpm.
c. The pipe has a 1-in inside diameter.
d. The specific gravity of oil is 0.9.
e. The kinematic viscosity of oil is 100 cS.

Find the pressure available at the inlet to the hydraulic motor (station 2). The pressure at the oil top surface level in the hydraulic tank is atmospheric (0 psig). The head loss H_L due to friction between stations 1 and 2 is not given.

Solution   Writing the energy equation between stations 1 and 2, we have

$$Z_1 + \frac{p_1}{\gamma} + \frac{v_1^2}{2g} + H_p - H_m - H_L = Z_2 + \frac{p_2}{\gamma} + \frac{v_2^2}{2g}$$

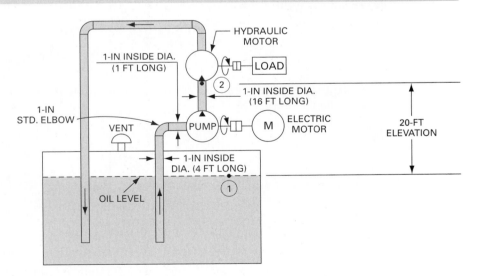

Figure 4-21.   Hydraulic system for Example 4-9.

Since there is no hydraulic motor between stations 1 and 2, $H_m = 0$. Also, $v_1 = 0$ and $p_1/\gamma = 0$ (the oil tank is vented to the atmosphere). Also, $Z_2 - Z_1 = 20$ ft, per Figure 4-21.

To make use of the energy equation, let's first solve for v_2 using Eq. (3-22):

$$v_2(\text{ft/s}) = \frac{Q(\text{ft}^3/\text{s})}{A(\text{ft}^2)} = \frac{30/449}{\frac{\pi}{4}(1/12)^2} = 12.2 \text{ ft/s}$$

Next, let's evaluate the velocity head at station 2:

$$\frac{v_2^2}{2g} = \frac{(12.2)^2}{64.4} = 2.4 \text{ ft}$$

The Reynolds number can now be found:

$$N_R = \frac{7740v(\text{ft/s}) \times D(\text{in})}{\nu(\text{cS})} = \frac{7740(12.2)(1)}{100} = 944$$

Since the flow is laminar, the friction factor can be found directly from the Reynolds number:

$$f = \frac{64}{N_R} = \frac{64}{944} = 0.0678$$

We can now determine the head loss due to friction between stations 1 and 2:

$$H_L = f\left(\frac{L_{eTOT}}{D}\right)\frac{v^2}{2g}$$

where

$$L_{eTOT} = L_{TOT} + L_e(\text{std elbow})$$

$$L_{eTOT} = 16 + 1 + 4 + \left(\frac{KD}{f}\right)_{\text{std elbow}}$$

$$L_{eTOT} = 21 + \frac{(0.9)(\frac{1}{12}\text{ ft})}{0.0678} = 21 + 1.1 + 22.1 \text{ ft}$$

Substituting values gives the head loss.

$$H_L = (0.0678)\frac{(22.1)}{\frac{1}{12}}(2.4) = 43 \text{ ft}$$

We can now substitute into the energy equation to solve for p_2/γ:

$$\frac{p_2}{\gamma} = (Z_1 - Z_2) + H_p + \frac{p_1}{\gamma} - H_L - \frac{v_2^2}{2g}$$

$$\frac{p_2}{\gamma} = -20 + H_p + 0 - 43 - 2.4$$

$$\frac{p_2}{\gamma} = H_p - 65.4$$

Using Eq. (3-29) allows us to solve for the pump head:

$$H_p(\text{ft}) = \frac{3950(\text{HHP})}{Q(\text{gpm}) \times (\text{SG})} = \frac{(3950)(5)}{(30)(0.9)} = 732 \text{ ft}$$

Thus, we can solve for the pressure head at station 2:

$$\frac{p_2}{\gamma} = 732 - 65.4 = 667 \text{ ft}$$

Finally, we solve for the pressure at station 2:

$$p_2(\text{lb/ft}^2) = (667 \text{ ft}) \times \gamma(\text{lb/ft}^3)$$

where

$$\gamma = (\text{SG})\gamma_{\text{water}} = (0.9)(62.4) = 56.2 \text{ lb/ft}^3$$

Thus,

$$p_2 = (667)(56.2) = 37,500 \text{ lb/ft}^2 = 260 \text{ psi}$$

Note that the pressure increase across the pump (from inlet to discharge) can be found as follows:

$$\Delta p_{\text{pump}} = \gamma H_p = 56.2 \text{ lb/ft}^3 \times 732 \text{ ft} = 41,100 \text{ lb/ft}^2 = 286 \text{ psi}$$

Thus, the pump has to produce a pressure increase of 286 psi (from inlet to outlet) to provide a pressure of 260 psi at the inlet to the hydraulic motor.

4.10 CIRCUIT ANALYSIS USING THE SI METRIC SYSTEM

In this section we perform a complete analysis of a hydraulic circuit using the SI metric system of units and taking into account energy losses due to friction.

EXAMPLE 4-10

For the hydraulic system of Figure 4-21, the following metric data are given:

  a. The pump is adding 3.73 kW to the fluid.
  b. Pump flow is 0.00190 m³/s.
  c. The pipe has a 0.0254-m inside diameter.
  d. The specific gravity of oil is 0.9.
  e. The kinematic viscosity of oil is 100 cS.
  f. The elevation difference between stations 1 and 2 is 6.10 m.
  g. Pipe lengths are as follows: pump inlet pipe length = 1.53 m and pump outlet pipe length up to hydraulic motor = 4.88 m.

Find the pressure available at the inlet to the hydraulic motor (station 2). The pressure at the oil top surface level in the hydraulic tank is atmospheric (0 Pa gage). The head loss, H_L, due to friction between stations 1 and 2 is not given.

Solution   In the SI metric system, absolute viscosity is given in units of newton-seconds per meter squared. Thus, we have $\mu = N \cdot s/m^2 = Pa \cdot s$. Our problem is to convert viscosity in cS to the appropriate units in the SI metric system. This is accomplished as follows: The conversion between $dyn \cdot s/cm^2$ or poise and $N \cdot s/m^2$ is found first:

$$\mu(N \cdot s/m^2) = \mu(dyn \cdot s/cm^2) \times \frac{1N}{10^5\, dyn} \times \left(\frac{100\ cm}{1\ m}\right)^2$$

This yields a useful conversion equation dealing with absolute viscosities:

$$\mu(N \cdot s/m^2) = \frac{\mu(dyn \cdot s/cm^2)}{10} = \frac{\mu(poise)}{10} \qquad \textbf{(4-10)}$$

Since a centipoise is one-hundredth of a poise, we can develop a conversion equation between $N \cdot s/m^2$ and centipoise:

$$\mu(P) = \frac{\mu(cP)}{100} \qquad \textbf{(4-11)}$$

Substituting Eq. (4-11) into Eq. (4-10) yields the desired result:

$$\mu(N \cdot s/m^2) = \frac{\mu(cP)}{100(10)} = \frac{\mu(cP)}{1000} \qquad \textbf{(4-12)}$$

However, in this problem, we were given a kinematic viscosity value rather than an absolute viscosity value. Therefore, we need to convert viscosity in cS to the appropriate units in the SI metric system. Kinematic viscosity in the SI metric system is given in units of meters squared per second. Thus, we have $\nu = m^2/s$. The conversion between cm^2/s or stokes and m^2/s is found as follows:

$$\nu(m^2/s) = \nu(cm^2/s) \times \left(\frac{1 \text{ m}}{100 \text{ cm}}\right)^2 = \frac{\nu(cm^2/s)}{10,000} = \frac{\nu(\text{stokes})}{10,000} \quad \textbf{(4-13)}$$

Since a centistoke is one-hundredth of a stoke, we can now develop the desired conversion between m^2/s and centistokes:

$$\nu(\text{stokes}) = \frac{\nu(cS)}{100} \quad \textbf{(4-14)}$$

Substituting Eq. (4-14) into Eq. (4-13) yields

$$\nu(m^2/s) = \frac{\nu(\text{stokes})}{10,000} = \frac{\nu(cS)}{10,000 \times 100} = \frac{\nu(cS)}{1,000,000} \quad \textbf{(4-15)}$$

Now back to the problem at hand. We write the energy equation between stations 1 and 2 as:

$$Z_1 + \frac{p_1}{\gamma} + \frac{v_1^2}{2g} + H_p - H_m - H_L = Z_2 + \frac{p_2}{\gamma} + \frac{v_2^2}{2g}$$

Since there is no hydraulic motor between stations 1 and 2, $H_m = 0$. Also $v_1 = 0$ and $p_1/\gamma = 0$ (the oil tank is vented to the atmosphere). Also, $Z_2 - Z_1 = 6.10$ m per given input data. To make use of the energy equation, let's first solve for v_2:

$$v_2(m/s) = \frac{Q(m^3/s)}{A(m^2)} = \frac{0.00190}{(\pi/4)(0.0254)^2} = 3.74 \text{ m/s}$$

Next, let's evaluate the velocity head at station 2:

$$\frac{v_2^2}{2g} = \frac{(3.74 \text{ m/s})^2}{2(9.80 \text{ m/s}^2)} = 0.174 \text{ m}$$

The Reynolds number can now be found:

$$N_R = \frac{v(m/s) \times D(m)}{\nu(m^2/s)} = \frac{(3.74)(0.0254)}{100/1,000,000} = 944$$

Since the flow is laminar, the friction factor can be found directly from the Reynolds number.

$$f = \frac{64}{N_R} = \frac{64}{944} = 0.0678$$

We can now determine the head loss due to friction between stations 1 and 2:

$$H_L = f\left(\frac{L_{eTOT}}{D}\right)\frac{v^2}{2g}$$

where

$$L_{eTOT} = 4.88 + 1.53 + \left(\frac{KD}{f}\right)_{std\ elbow}$$

$$L_{eTOT} = 6.41 + \frac{(0.9)(0.0254)}{0.0678} = 6.41 + 0.34 = 6.75\ m$$

$$H_L = (0.0678)\frac{(6.75)}{0.0254}(0.714) = 12.9\ m$$

We can now substitute into the energy equation to solve for p_2/γ:

$$\frac{p_2}{\gamma} = (Z_1 - Z_2) + H_p + \frac{p_1}{\gamma} - H_L - \frac{v_2^2}{2g}$$

$$\frac{p_2}{\gamma} = -6.10 + H_p + 0 - 12.9 - 0.714 = H_p - 19.7$$

Using Eq. (3-36) allows us to solve for the pump head:

$$H_p(m) = \frac{pump\ hydraulic\ power\ (W)}{\gamma(N/m^3) \times Q(m^3/s)}$$

where $\gamma = (SG)\ \gamma_{water} = 0.9 \times 9800\ N/m^3 = 8820\ N/m^3$

$$H_p(m) = \frac{3730}{8820 \times 0.00190} = 223.1\ m$$

Thus, we can now solve for the pressure head at station 2:

$$\frac{p_2}{\gamma} = 223.1 - 19.7 = 203.4\ m$$

Solving for p_2 yields

$$p_2(N/m^2) = (203.4\ m) \times \gamma(N/m^3)$$

$$p_2(N/m^2) = 203.4 \times 8820 = 1,790,000\ Pa = 1790\ kPa$$

4.11 KEY EQUATIONS

Reynolds number

English or metric units:
$$N_R = \frac{vDp}{\mu} = \frac{vD}{v} \tag{4-1}$$

Special English units using cP:
$$N_R = \frac{7740v\,(\text{ft>s}) \times D(\text{in}) \times \text{SG}}{\mu\,(\text{cP})} \tag{4-2}$$

Special metric units using cP:
$$N_R = \frac{1000v\,(\text{m/s}) \times D(\text{mm}) \times \text{SG}}{\mu\,(\text{cP})} \tag{4-2M}$$

Special English units using cS:
$$N_R = \frac{7740v\,(\text{ft>s}) \times D(\text{in})}{v\,(\text{cS})} \tag{4-3}$$

Special metric units using cS:
$$N_R = \frac{1000v\,(\text{m/s}) \times D(\text{mm})}{v\,(\text{cS})} \tag{4-3M}$$

Darcy's equation:
$$H_L = f\left(\frac{L}{D}\right)\left(\frac{v^2}{2g}\right) \tag{4-4}$$

Laminar friction factor:
$$f = \frac{64}{N_R} \tag{4-5}$$

Hagen-Poiseuille equation:
$$H_L = \frac{64}{N_R}\left(\frac{L}{D}\right)\left(\frac{v^2}{2g}\right) \tag{4-6}$$

Relative roughness:
$$\text{relative roughness} = \frac{\varepsilon}{D} \tag{4-7}$$

Head loss in valves and fittings:
$$H_L = K\left(\frac{v^2}{2g}\right) \tag{4-8}$$

Equivalent length of valves and fittings:
$$L_e = \frac{KD}{f} \tag{4-9}$$

EXERCISES

Questions, Concepts, and Definitions

4-1. Why is it important to select properly the size of pipes, valves, and fittings in hydraulic systems?

4-2. What is the physical difference between laminar and turbulent flow?

4-3. What are the important conclusions resulting from Reynolds' experiment?

4-4. Define the term *relative roughness*.

4-5. What is meant by the K factor of a valve or fitting?

4-6. What is meant by the equivalent length of a valve or fitting?

4-7. To minimize pressure losses, the K factor of a valve should be made as small as possible. True or false?

4-8. Name two causes of turbulence in fluid flow.

Problems

Note: The letter E following an exercise number means that English units are used. Similarly, the letter M indicates metric units.

Reynolds Number

4-9E. The kinematic viscosity of a hydraulic oil is 75 cS. If it is flowing in a $1\frac{1}{2}$-in-diameter pipe at a velocity of 20 ft/s, what is the Reynolds number? Is the flow laminar or turbulent?

4-10M. The kinematic viscosity of a hydraulic oil is 0.0001 m^2/s. If it is flowing in a 30-mm-diameter pipe at a velocity of 6 m/s, what is the Reynolds number? Is the flow laminar or turbulent?

4-11. A hydraulic system is operating at a Reynolds number of 1000. If the temperature increases so that the oil viscosity decreases, would the Reynolds number increase, decrease, or remain the same?

Frictional Losses in Pipelines

4-12E. For the system in Exercise 4-9, find the head loss due to friction in units of psi for a 100-ft length of smooth pipe. The oil has a specific gravity of 0.90.

4-13E. The kinematic viscosity of a hydraulic oil is 100 cS. If it is flowing in a 3/4-in-diameter commercial steel pipe, find the friction factor in each case.
   a. The velocity is 15 ft/s.
   b. The velocity is 45 ft/s.

4-14M. For the system in Exercise 4-10, find the head loss due to friction in units of bars for a 100-m length of smooth pipe. The oil has a specific gravity of 0.90.

4-15M. The kinematic viscosity of a hydraulic oil is 0.0001 m^2/s. If it is flowing in a 20-mm-diameter commercial steel pipe, find the friction factor in each case.
   a. The velocity is 2 m/s.
   b. The velocity is 10 m/s.

4-16. For laminar flow of a liquid in a pipe, frictional pressure losses are ____ to the liquid velocity.

4-17. For fully turbulent flow of a liquid in a pipe, frictional pressure losses vary as the ____ of the velocity.

Losses in Valves and Fittings

4-18E. What is the head loss across a 1½-in-wide open gate valve when oil (SG = 0.90) flows through it at a rate of 100 gpm?

4-19M. What is the head loss (in units of bars) across a 30-mm-wide open gate valve when oil (SG = 0.90) flows through at a rate of 0.004 m³/s?

4-20. If the volume flow rate through a valve is doubled, by what factor does the pressure drop increase?

4-21E. A directional control valve with an effective area of 0.5 in² provides a pressure drop of 40 psi at 60 gpm. If the fluid has a specific gravity of 0.90, what are the flow coefficient and K factor for the valve?

4-22M. For the directional control valve in Exercise 4-21, if the data were converted to metric units, how would the values of the flow coefficient and K factor calculated compare to those determined using English units? Explain your answer.

Equivalent Length Method

4-23E. Oil (SG = 0.90, ν = 75 cS) flows at a rate of 30 gpm through a ¾-in-diameter commercial steel pipe. What is the equivalent length of a ¾-in-wide open gate valve placed in the line?

4-24M. Oil (SG = 0.90, ν = 0.0001 m²/s) flows at a rate of 0.002 m³/s through a 20-mm-diameter commercial steel pipe. What is the equivalent length of a 20-mm-wide open gate valve placed in the line?

Hydraulic Circuit Analysis

4-25E. For the hydraulic system of Figure 4-21, the following data are given:
  1. The pump is adding 4 hp to the fluid (pump hydraulic horsepower = 4).
  2. Pump flow is 25 gpm.
  3. The pipe has a 0.75-in inside diameter.
  4. The specific gravity of oil is 0.90.
  5. The kinematic viscosity of oil is 75 cS.

Find the pressure available at the inlet to the hydraulic motor (station 2).

4-26E. The oil tank for the hydraulic system of Figure 4-22 is air-pressurized at 10 psig. The inlet line to the pump is 10 ft below the oil level. The pump flow rate is 30 gpm. Find the pressure at station 2. The specific gravity of the oil is 0.90, and the kinematic viscosity of the oil is 100 cS. Assume that the pressure drop across the strainer is 1 psi.

4-27M. Solve Exercise 4-25 using metric units. The equivalent metric data are given as follows:
  1. The pump is adding 2.984 kW to the fluid (pump hydraulic power = 2.984 kW).
  2. Pump flow is 0.00158 m³/s.
  3. The pipe has a 19.05-mm inside diameter.
  4. The specific gravity of oil is 0.90.
  5. The kinematic viscosity of oil is 75 cS.
  6. The elevation difference between stations 1 and 2 is 6.096 m.
  7. Pipe lengths are as follows: 1-ft length = 0.305 m, 4-ft length = 1.22 m, and 16-ft length = 4.88 m.

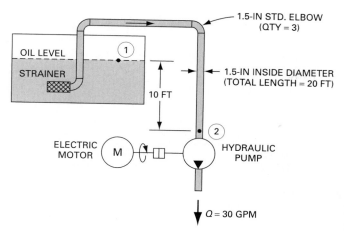

Figure 4-22.   Hydraulic system for Exercise 4-26.

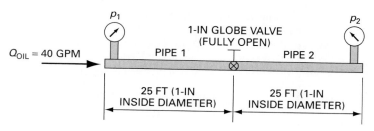

Figure 4-23.   Hydraulic system for Exercise 4-29.

Find the pressure available at the inlet to the hydraulic motor (station 2).

4-28M. Solve Exercise 4-26 using metric units. The equivalent metric data are given as follows:
  1. The oil tank is air-pressurized at 68.97 kPa gage pressure.
  2. The inlet line to the pump is 3.048 m below the oil level.
  3. The pump flow rate is 0.001896 m³/s.
  4. The specific gravity of the oil is 0.90.
  5. The kinematic viscosity of the oil is 100 cS.
  6. Assume that the pressure drop across the strainer is 6.897 kPa.
  7. The pipe has a 38.1-mm inside diameter.
  8. The total length of pipe is 6.097 m.

Find the pressure at station 2.

4-29E. For the system of Figure 4-23, solve for $p_2 - p_1$ in units of psi. The kinematic viscosity (v) of the oil is 100 cS and the specific gravity (SG) is 0.90.

4-30M. For the system in Exercise 4-29, the following new data are applicable:
                 pipe 1: length = 8 m, ID = 25 mm
                 pipe 2: length = 8 m, ID = 25 mm
  The globe valve is 25 mm in size and is wide open.

   SG = 0.90, v = 0.0001 m²/s,   Q = 0.0025 m³/s

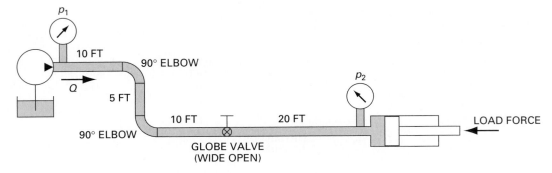

Figure 4-24.   Hydraulic system for Exercise 4-31.

Find $p_2 - p_1$ in units of bars.

4-31E. For the system of Figure 4-24, if $p_1 = 100$ psi, solve for p_2 in units of psi. The pipe is 45 ft long, has a 1.5-in ID throughout, and lies in a horizontal plane. $Q = 30$ gpm of oil (SG = 0.90 and $v = 100$ cS).

4-32M. For the system in Exercise 4-31, the following new data are applicable:

$$p_1 = 7 \text{ bars}, \quad Q = 0.002 \text{ m}^3/\text{s}$$

$$\text{pipe: } L \text{ (total)} = 15 \text{ m and ID} = 38 \text{ mm}$$

$$\text{oil: SG} = 0.90 \text{ and } v = 0.0001 \text{ m}^2/\text{s}$$

Solve for p_2 in units of bars.

4-33E. For the fluid power system shown in Figure 4-25, determine the external load F that the hydraulic cylinder can sustain while moving in the extending direction. Take frictional pressure losses into account. The pump produces a pressure increase of 1000 psi from the inlet port to the discharge port and a flow rate of 40 gpm. The following data are applicable:

$$\text{kinematic viscosity of oil} = 0.001 \text{ ft}^2/\text{s}$$

$$\text{weight density of oil} = 50 \text{ lb/ft}^3$$

$$\text{cylinder piston diameter} = 8 \text{ in}$$

$$\text{cylinder rod diameter} = 4 \text{ in}$$

All elbows are 90° with K factor = 0.75. Pipe lengths and inside diameters are given in Figure 4-25.

4-34E. For the system in Exercise 4-33, as shown in Figure 4-25, determine the heat-generation rate due to frictional pressure losses.

4-35E. For the system in Exercise 4-33, as shown in Figure 4-25, determine the extending and retracting speeds of the cylinder.

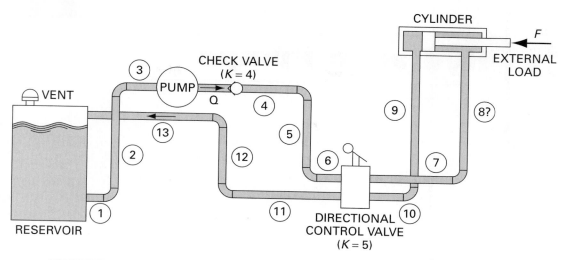

Figure 4-25 System for Exercise 4-33.

Pipe No.	Length (ft)	Dia (in)	Pipe No.	Length (ft)	Dia (in)
1	2	1.5	8	5	1.0
2	6	1.5	9	5	0.75
3	2	1.5	10	5	0.75
4	50	1.0	11	60	0.75
5	10	1.0	12	10	0.75
6	5	1.0	13	20	0.75
7	5	1.0			

4-36M. For the system in Exercise 4-33, as shown in Figure 4-25, change the data to metric units and solve for the external load F that the cylinder can sustain while moving in the extending direction.

4-37M. For the system in Exercise 4-36, determine the heat-generation rate.

4-38M. For the system in Exercise 4-36, determine the extending and retracting speeds of the cylinder.

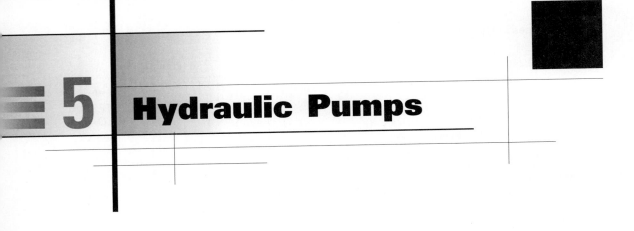

5 Hydraulic Pumps

Learning Objectives

Upon completing this chapter, you should be able to:

1. Distinguish between positive displacement and dynamic (nonpositive displacement) pumps.
2. Describe the pumping action of pumps.
3. Explain the operation of gear, vane, and piston pumps.
4. Determine the flow rate delivered by positive displacement pumps.
5. Differentiate between axial and radial piston pumps.
6. Understand the difference between fixed displacement and variable displacement pumps.
7. Explain the operation of pressure-compensated pumps.
8. Distinguish between bent-axis-type piston pumps and the swash plate design.
9. Differentiate between internal and external gear pumps.
10. Evaluate the performance of pumps by determining the volumetric, mechanical, and overall efficiencies.
11. Explain the phenomenon of pump cavitation and identify ways to eliminate its occurrence.
12. Compare the various performance factors of gear, vane, and piston pumps.
13. Understand the significance of sound intensity levels in decibels.
14. Explain the causes of pump noise and identify ways to reduce noise levels.
15. Describe the sequence of operations used to select a pump for a given application.

5.1 INTRODUCTION

A pump, which is the heart of a hydraulic system, converts mechanical energy into hydraulic energy. The mechanical energy is delivered to the pump via a prime mover such as an electric motor. Due to mechanical action, the pump creates a partial vacuum at its inlet. This permits atmospheric pressure to force the fluid through the inlet line and into the pump. The pump then pushes the fluid into the hydraulic system.

There are two broad classifications of pumps as identified by the fluid power industry.

1. Dynamic (nonpositive displacement) pumps. This type is generally used for low-pressure, high-volume flow applications. Because they are not capable of withstanding high pressures, they are of little use in the fluid power field. Normally their maximum pressure capacity is limited to 250–300 psi. This type of pump is primarily used for transporting fluids from one location to another. The two most common types of dynamic pumps are the centrifugal and the axial flow propeller pumps.

2. Positive displacement pumps. This type is universally used for fluid power systems. As the name implies, a positive displacement pump ejects a fixed amount of fluid into the hydraulic system per revolution of pump shaft rotation. Such a pump is capable of overcoming the pressure resulting from the mechanical loads on the system as well as the resistance to flow due to friction. These are two features that are desired of fluid power pumps. These pumps have the following advantages over nonpositive displacement pumps:

   a. High-pressure capability (up to 12,000 psi)
   b. Small, compact size
   c. High volumetric efficiency
   d. Small changes in efficiency throughout the design pressure range
   e. Great flexibility of performance (can operate over a wide range of pressure requirements and speed ranges)

There are three main types of positive displacement pumps: gear, vane, and piston. Many variations exist in the design of each of these main types of pumps. For example, vane and piston pumps can be of either fixed or variable displacement. A fixed displacement pump is one in which the amount of fluid ejected per revolution (displacement) cannot be varied. In a variable displacement pump, the displacement can be varied by changing the physical relationships of various pump elements. This change in pump displacement produces a change in pump flow output even though pump speed remains constant.

It should be understood that pumps do not pump pressure. Instead they produce fluid flow. The resistance to this flow, produced by the hydraulic system, is what determines the pressure. For example, if a positive displacement pump has its discharge line open to the atmosphere, there will be flow, but there will be no discharge pressure above atmospheric because there is essentially no resistance to flow. However, if the discharge line is blocked, then we have theoretically infinite resistance to flow. Hence, there is no place for the fluid to go. The pressure will therefore rise until some component breaks unless pressure relief is provided. This is the reason a pressure relief valve is needed when a positive displacement pump is used. When the pressure reaches a set value, the relief valve will open to allow flow back to the oil tank. Thus, a pressure relief valve determines the maximum pressure level that the system will experience regardless of the magnitude of the load resistance. A discussion of pressure relief valves is provided in Chapter 8.

Some pumps are made with variable displacement, pressure compensation capability. Such pumps are designed so that as system pressure builds up, they produce less flow. Finally, at some predetermined maximum pressure level, the flow output goes to zero due to zero displacement. This prevents any additional pressure buildup. Pressure relief valves are not needed when pressure-compensated pumps are used.

The hydraulic power developed by pumps is converted back into mechanical power by hydraulic cylinders and motors, which produce the useful work output. Hydraulic cylinders and motors are presented in Chapters 6 and 7, respectively.

5.2 WATERJET CUTTING

In this section we examine a manufacturing application of hydraulics involving the use of pumps, before delving into a discussion of the design and operation of pumps that follows in Sections 5.3 through 5.12. Figure 5-1 illustrates this hydraulics application, which is called waterjet cutting. In simple terms, a waterjet cutting system pressurizes water up to about 60,000 psi and then allows the high-pressure water to flow through a diamond orifice and nozzle. The orifice and nozzle allow for a

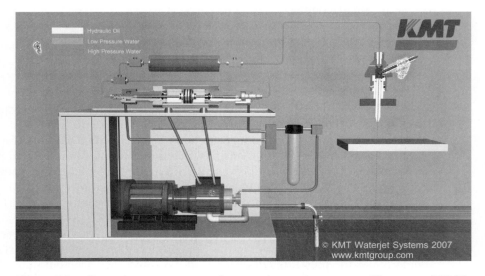

Figure 5-1. Operating components of a waterjet cutting system. *(Courtesy of KMT Waterjet Systems Inc., Baxter Springs, Kansas.)*

precise jet of water (with a diameter between 0.005 and 0.010 in) to exit from the system at the nozzle tip at speeds of up to three times the speed of sound. The high speed and precise location of the waterjet allows it to readily and accurately cut through raw components to produce finished parts. Any material with a consistent density can be cut, including steel, aluminum, rubber, glass, plastics, composites, and ceramics. Cutting very hard materials such as titanium requires adding a fine mesh abrasive to the cutting stream just upstream of the nozzle.

The hydraulics portion of the system includes a main pump that delivers oil at pressures in the range of 2000 to 3000 psi, to a pressure intensifier. The pressure intensifier contains a reciprocating large-diameter piston that has two small-diameter rod ends. Reciprocation occurs because the oil alternately enters and leaves at either side of the piston. The oil pressure bears on the piston over an area equal to the piston area minus the rod area. Ultra-high-pressure water is discharged from the intensifier at the rod ends as a result of the reciprocating action of the rods. This is done when the area of either rod end pushes on low-pressure water that has entered the cavity near either rod end of the intensifier. The low-pressure water is delivered to the intensifier via a small booster pump and filter. The large vessel shown above the pressure intensifier in Figure 5-1 is a shock attenuator that dampens the water pressure fluctuations to ensure the water exiting the cutting head is steady and consistent.

The ratio of the ultra-high water pressure at the outlet of the intensifier to the moderate oil pressure at the inlet of the intensifier equals the ratio of the piston area minus the rod area to the rod area. Thus for an area ratio of 20 to 1, in order to produce a water pressure of 60,000 psi, the pump discharge oil pressure would have to be 3000 psi. Conversely for this same area ratio, the oil flow rate would equal 20 times the water flow rate. Thus to produce a waterjet flow rate of 1 gpm, the pump would have to deliver oil at a flow rate of 20 gpm.

Figure 5-2 shows a waterjet cutting a gear from a metal plate. Notice that the cutting head needs to be very close to the material being cut. Figure 5-3 is a photograph showing an actual waterjet system that includes a cutting table on which the raw component to be cut is mounted. The cutting head moves according to outputs given by the computer shown to the right of the table. The head cutting speeds and direction of motion are programmed to optimize the production of parts of different shapes and materials.

5.3 PUMPING THEORY

Pumps operate on the principle whereby a partial vacuum is created at the pump inlet due to the internal operation of the pump. This allows atmospheric pressure to push the fluid out of the oil tank (reservoir) and into the pump intake. The pump then mechanically pushes the fluid out the discharge line.

This type of operation can be visualized by referring to the simple piston pump of Figure 5-4. Note that this pump contains two ball check valves, which are described as follows:

- Check valve 1 is connected to the pump inlet line and allows fluid to enter the pump only at this location.

Figure 5-2. Waterjet cutting a gear from a metal plate. *(Courtesy of KMT Waterjet Systems Inc., Baxter Springs, Kansas.)*

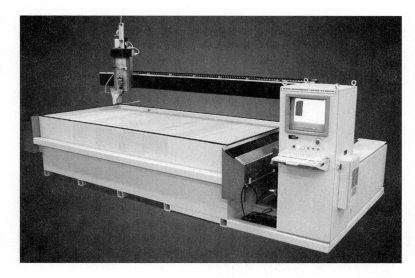

Figure 5-3. Waterjet system with cutting table and computer. *(Courtesy of KMT Waterjet Systems Inc., Baxter Springs, Kansas.)*

- Check valve 2 is connected to the pump discharge line and allows fluid to leave the pump only at this location.

As the piston is pulled to the left, a partial vacuum is generated in pump cavity 3, because the close tolerance between the piston and cylinder (or the use of piston ring seals) prevents air inside cavity 4 from traveling into cavity 3. This flow of air, if allowed to occur, would destroy the vacuum. This vacuum holds the ball of check valve 2 against its seat (lower position) and allows atmospheric pressure to push fluid from the reservoir

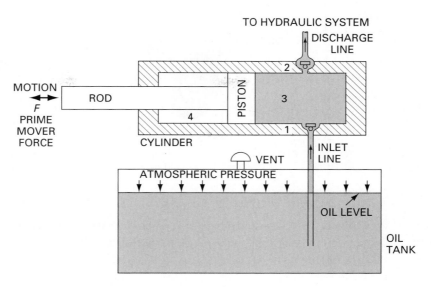

Figure 5-4.   Pumping action of a simple piston pump.

into the pump via check valve 1. This inlet flow occurs because the force of the fluid pushes the ball of check valve 1 off its seat.

When the piston is pushed to the right, the fluid movement closes inlet valve 1 and opens outlet valve 2. The quantity of fluid, displaced by the piston, is forcibly ejected out the discharge line leading to the hydraulic system. The volume of oil displaced by the piston during the discharge stroke is called the *displacement volume* of the pump.

5.4 PUMP CLASSIFICATION

Dynamic Pumps

The two most common types of dynamic pumps are the centrifugal (impeller) and axial (propeller) pumps shown in Figure 5-5. Although these pumps provide smooth continuous flow, their flow output is reduced as circuit resistance is increased and thus are rarely used in fluid power systems. In dynamic pumps there is a great deal of clearance between the rotating impeller or propeller and the stationary housing. Thus as the resistance of the external system starts to increase, some of the fluid slips back into the clearance spaces, causing a reduction in the discharge flow rate. This slippage is due to the fact that the fluid follows the path of least resistance. When the resistance of the external system becomes infinitely large (for example, a valve is closed in the outlet line), the pump will produce no flow.

These pumps are typically used for low-pressure, high-volume flow applications. Also, since there is a great deal of clearance between the rotating and stationary elements, dynamic pumps are not self-priming unlike positive displacement pumps.

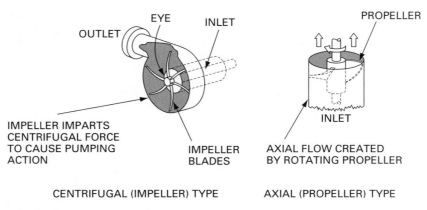

CENTRIFUGAL (IMPELLER) TYPE   AXIAL (PROPELLER) TYPE

Figure 5-5. Dynamic (nonpositive displacement) pumps. *(Courtesy of Sperry Vickers, Sperry Rand Corp., Troy, Michigan.)*

This is because the large clearance space does not permit a suction pressure to occur at the inlet port when the pump is first turned on. Thus if the fluid is being pumped from a reservoir located below the pump, priming is required. Priming is the prefilling of the pump housing and inlet pipe with fluid so that the pump can initially draw in the fluid and pump it efficiently.

Figure 5-6(a) shows the construction features of a centrifugal pump, the most commonly used type of dynamic pump. The operation of a centrifugal pump is as follows. The fluid enters at the center of the impeller and is picked up by the rotating impeller. As the fluid rotates with the impeller, the centrifugal force causes the fluid to move radially outward. This causes the fluid to flow through the outlet discharge port of the housing. One of the interesting characteristics of a centrifugal pump is its behavior when there is no demand for fluid. In such a case, there is no need for a pressure relief valve to prevent pump damage. The tips of the impeller blades merely slosh through the fluid, and the rotational speed maintains a fluid pressure corresponding to the centrifugal force established. The fact that there is no positive internal seal against leakage is the reason that the centrifugal pump is not forced to produce flow against no demand. When demand for the fluid occurs (for example, the opening of a valve), the pressure delivers the fluid to the source of the demand. This is why centrifugal pumps are so desirable for pumping stations used for delivering water to homes and factories. The demand for water may go to near zero during the evening and reach a peak sometimes during the daytime. The centrifugal pump can readily handle these large changes in fluid demand.

Although dynamic pumps provide smooth continuous flow (when a demand exists), their output flow rate is reduced as resistance to flow is increased. This is shown for centrifugal pumps in Figure 5-6(b), where pump pressure is plotted versus pump flow. The maximum pressure is called the *shutoff head* because an external circuit valve is closed, which shuts off the flow. As the external resistance decreases due to the valve being opened, the flow increases at the expense of reduced pressure.

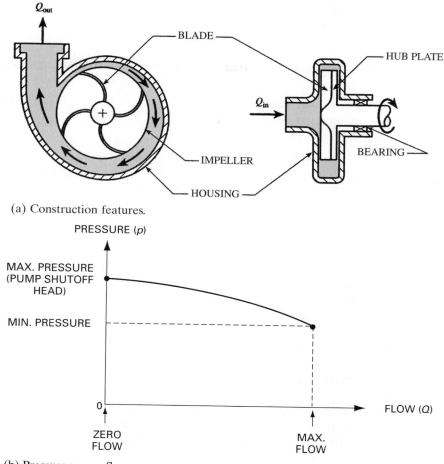

(a) Construction features.

(b) Pressure versus flow curve.

Figure 5-6. Centrifugal pump. (a) Construction features. (b) Pressure versus flow curve.

Positive Displacement Pumps

This type of pump ejects a fixed quantity of fluid per revolution of the pump shaft. As a result, pump output flow, neglecting changes in the small internal leakage, is constant and not dependent on system pressure. This makes them particularly well suited for fluid power systems. However, positive displacement pumps must be protected against overpressure if the resistance to flow becomes very large. This can happen if a valve is completely closed and there is no physical place for the fluid to go. The reason for this is that a positive displacement pump continues to eject fluid (even though it has no place to go), causing an extremely rapid buildup in pressure as the fluid is compressed. A pressure relief valve is used to protect the pump against overpressure by diverting pump flow back to the hydraulic tank, where the fluid is stored for system use.

Positive displacement pumps can be classified by the type of motion of internal elements. The motion may be either rotary or reciprocating. Although these pumps come in a wide variety of different designs, there are essentially three basic types:

1. Gear pumps (fixed displacement only by geometrical necessity)

   a. External gear pumps
   b. Internal gear pumps
   c. Lobe pumps
   d. Screw pumps

2. Vane pumps

   a. Unbalanced vane pumps (fixed or variable displacement)
   b. Balanced vane pumps (fixed displacement only)

3. Piston pumps (fixed or variable displacement)

   a. Axial design
   b. Radial design

In addition, vane pumps can be of the balanced or unbalanced design. The unbalanced design can have pressure compensation capability, which automatically protects the pump against overpressure. In Sections 5.5, 5.6, and 5.7, we discuss the details of the construction and operation of gear, vane, and piston pumps, respectively.

5.5 GEAR PUMPS

External Gear Pump

Figure 5-7 illustrates the operation of an external gear pump, which develops flow by carrying fluid between the teeth of two meshing gears. One of the gears is connected to a drive shaft connected to the prime mover. The second gear is driven as it meshes with the driver gear. Oil chambers are formed between the gear teeth, the pump housing, and the side wear plates. The suction side is where teeth come out of mesh, and it is here that the volume expands, bringing about a reduction in pressure to below atmospheric pressure. Fluid is pushed into this void by atmospheric pressure because the oil supply tank is vented to the atmosphere. The discharge side is where teeth go into mesh, and it is here that the volume decreases between mating teeth. Since the pump has a positive internal seal against leakage, the oil is positively ejected into the outlet port.

Volumetric Displacement and Theoretical Flow Rate

The following analysis permits us to evaluate the theoretical flow rate of a gear pump using specified nomenclature:

D_o = outside diameter of gear teeth (in, m)
D_i = inside diameter of gear teeth (in, m)

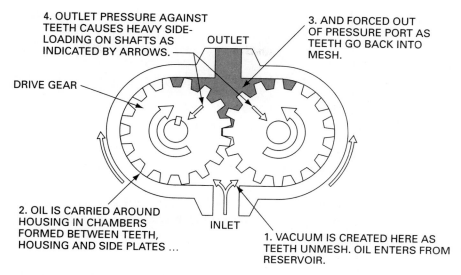

4. OUTLET PRESSURE AGAINST TEETH CAUSES HEAVY SIDE-LOADING ON SHAFTS AS INDICATED BY ARROWS.

3. AND FORCED OUT OF PRESSURE PORT AS TEETH GO BACK INTO MESH.

OUTLET

DRIVE GEAR

2. OIL IS CARRIED AROUND HOUSING IN CHAMBERS FORMED BETWEEN TEETH, HOUSING AND SIDE PLATES ...

INLET

1. VACUUM IS CREATED HERE AS TEETH UNMESH. OIL ENTERS FROM RESERVOIR.

Figure 5-7. External gear pump operation. *(Courtesy of Sperry Vickers, Sperry Rand Corp., Troy, Michigan.)*

L = width of gear teeth (in, m)
V_D = displacement volume of pump (in³/rev, m³/rev)
N = rpm of pump
Q_T = theoretical pump flow rate

The volumetric displacement of a gear pump can be found by calculating the volume of a hollow cylinder of outside diameter D_o and inside diameter D_i, where the length of the cylinder is L. There are actually two such cylinder volumes (because there are two gears) where oil could fill the inside of the pump if there were no gear teeth. However one half of these two volumes is taken up by the gear teeth of both gears. Thus, the volumetric displacement can be represented by Eq. (5-1).

$$V_D = \frac{\pi}{4}(D_o^2 - D_i^2)L \qquad \textbf{(5-1)}$$

The theoretical flow rate (in English units) is determined next:

$$Q_T(\text{in}^3/\text{min}) = V_D(\text{in}^3/\text{rev}) \times N \text{ (rev/min)}$$

Since 1 gal = 231 in³, we have

$$Q_T(\text{gpm}) = \frac{V_D(\text{in}^3/\text{rev}) \times N(\text{rev/min})}{231} \qquad \textbf{(5-2)}$$

Using metric units, we have

$$Q_T(\text{m}^3/\text{min}) = V_D(\text{m}^3/\text{rev}) \times N(\text{rev/min}) \qquad \textbf{(5-2M)}$$

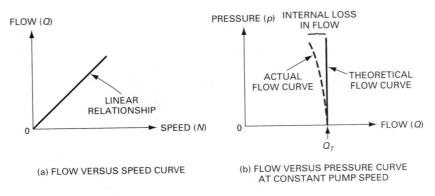

Figure 5-8. Positive displacement pump Q versus N and p versus Q curves. (a) Flow versus speed curve. (b) Flow versus pressure curve at constant pump speed.

Equations (5-2) and (5-2M) show that the pump flow varies directly with speed as graphically illustrated in Figure 5-8(a). Hence, the theoretical flow is constant at a given speed, as shown by the solid line in Figure 5-8(b).

Volumetric Efficiency

There must be a small clearance (about 0.001 in) between the teeth tip and pump housing. As a result, some of the oil at the discharge port can leak directly back toward the suction port. This means that the actual flow rate Q_A is less than the theoretical flow rate Q_T, which is based on volumetric displacement and pump speed. This internal leakage, called *pump slippage*, is identified by the term *volumetric efficiency*, η_v, which equals about 90% for positive displacement pumps, operating at design pressure:

$$\eta_v = \frac{Q_A}{Q_T} \tag{5-3}$$

The higher the discharge pressure, the lower the volumetric efficiency because internal leakage increases with pressure. This is shown by the dashed line in Figure 5-8(b). Pump manufacturers usually specify volumetric efficiency at the pump rated pressure. The rated pressure of a positive displacement pump is that pressure below which no mechanical damage due to overpressure will occur to the pump and the result will be a long, reliable service life. Too high a pressure not only produces excessive leakage but also can damage a pump by distorting the casing and overloading the shaft bearings. This brings to mind once again the need for overpressure protection. High pressures occur when a high resistance to flow is encountered, such as a large actuator load or a closed valve in the pump outlet line.

EXAMPLE 5-1

A gear pump has a 3-in outside diameter, a 2-in inside diameter, and a 1-in width. If the actual pump flow at 1800 rpm and rated pressure is 28 gpm, what is the volumetric efficiency?

Solution   Find the displacement volume using Eq. (5-1):

$$V_D = \frac{\pi}{4}[(3)^2 - (2)^2](1) = 3.93 \text{ in}^3$$

Next, use Eq. (5-2) to find the theoretical flow rate:

$$Q_T = \frac{V_D N}{231} = \frac{(3.93)(1800)}{231} = 30.6 \text{ gpm}$$

The volumetric efficiency is then found using Eq. (5-3):

$$\eta_v = \frac{28}{30.6} = 0.913 = 91.3\%$$

EXAMPLE 5-2

A gear pump has a 75-mm outside diameter, a 50-mm inside diameter, and a 25-mm width. If the volumetric efficiency is 90% at rated pressure, what is the corresponding actual flow rate? The pump speed is 1000 rpm.

Solution   The volume displacement is

$$V_D = \frac{\pi}{4}[(0.075)^2 - (0.050)^2](0.025) = 0.0000614 \text{ m}^3/\text{rev}$$

Since 1 L = 0.001 m³, $V_D = 0.0614$ L.
   Next, combine Eqs. (5-2M) and (5-3) to find the actual flow rate:

$$Q_A = \eta_v Q_T = \eta_v V_D(\text{m}^3/\text{rev}) \times N(\text{rev/min})$$

$$= 0.90 \times 0.0000614 \times 1000 = 0.0553 \text{ m}^3/\text{min}$$

Since 1 L = 0.001 m³, we have

$$Q_A = 55.3 \text{ Lpm}$$

Figure 5-9 shows detailed features of an external gear pump. Also shown is the hydraulic symbol used to represent fixed displacement pumps in hydraulic circuits. This external gear pump uses spur gears (teeth are parallel to the axis of the gear), which are noisy at relatively high speeds. To reduce noise and provide smoother

Figure 5-9. Detailed features of an external gear pump. *(Courtesy of Webster Electric Company, Inc., subsidiary of STA-RITE Industries, Inc., Racine, Wisconsin.)*

operation, helical gears (teeth inclined at a small angle to the axis of the gear) are sometimes used. However, these helical gear pumps are limited to low-pressure applications (below 200 psi) because they develop excessive end thrust due to the action of the helical gears. Herringbone gear pumps eliminate this thrust action and thus can be used to develop much higher pressures (up to 3000 psi). Herringbone gears consist basically of two rows of helical teeth cut into one gear. One of the rows of each gear is right-handed and the other is left-handed to cancel out the axial thrust force. Herringbone gear pumps operate as smoothly as helical gear pumps and provide greater flow rates with much less pulsating action.

Internal Gear Pump

Figure 5-10 illustrates the configuration and operation of the internal gear pump. This design consists of an internal gear, a regular spur gear, a crescent-shaped seal, and an external housing. As power is applied to either gear, the motion of the gears draws fluid from the reservoir and forces it around both sides of the crescent seal, which acts as a seal between the suction and discharge ports. When the teeth mesh on the side opposite to the crescent seal, the fluid is forced to enter the discharge port of the pump.

Figure 5-11 provides a cutaway view of an internal gear pump that contains its own built-in safety relief valve.

Lobe Pump

Also in the general family of gear pumps is the lobe pump, which is illustrated in Figure 5-12. This pump operates in a fashion similar to the external gear pump. But

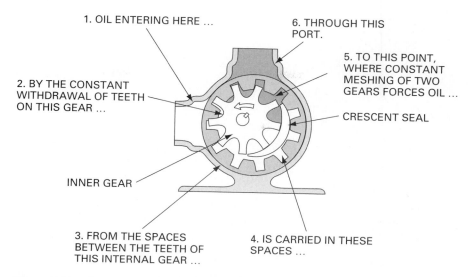

1. OIL ENTERING HERE ...

6. THROUGH THIS PORT.

5. TO THIS POINT, WHERE CONSTANT MESHING OF TWO GEARS FORCES OIL ...

2. BY THE CONSTANT WITHDRAWAL OF TEETH ON THIS GEAR ...

CRESCENT SEAL

INNER GEAR

3. FROM THE SPACES BETWEEN THE TEETH OF THIS INTERNAL GEAR ...

4. IS CARRIED IN THESE SPACES ...

Figure 5-10. Operation of an internal gear pump. *(Courtesy of Sperry Vickers, Sperry Rand Corp., Troy, Michigan.)*

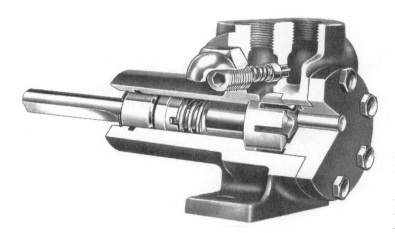

Figure 5-11. Cutaway view of an internal gear pump with built-in safety relief valve. *(Courtesy of Viking Pump Division of Houdaille Industries Inc., Cedar Falls, Iowa.)*

unlike the external gear pump, both lobes are driven externally so that they do not actually contact each other. Thus, they are quieter than other types of gear pumps. Due to the smaller number of mating elements, the lobe pump output will have a somewhat greater amount of pulsation, although its volumetric displacement is generally greater than that for other types of gear pumps.

Gerotor Pump

The Gerotor pump, shown in Figure 5-13, operates very much like the internal gear pump. The inner gear rotor (Gerotor element) is power-driven and draws the outer

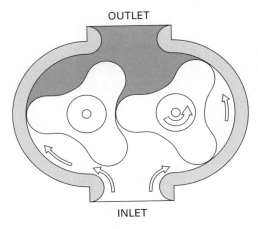

OUTLET

INLET

Figure 5-12. Operation of the lobe pump. *(Courtesy of Sperry Vickers, Sperry Rand Corp., Troy, Michigan.)*

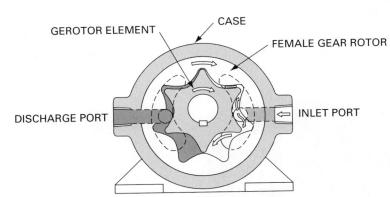

GEROTOR ELEMENT   CASE   FEMALE GEAR ROTOR

DISCHARGE PORT   INLET PORT

Figure 5-13. Operation of the Gerotor pump. *(Courtesy of Sperry Vickers, Sperry Rand Corp., Troy, Michigan.)*

gear rotor around as they mesh together. This forms inlet and discharge pumping chambers between the rotor lobes. The tips of the inner and outer rotors make contact to seal the pumping chambers from each other. The inner gear has one tooth less than the outer gear, and the volumetric displacement is determined by the space formed by the extra tooth in the outer rotor.

Figure 5-14 shows an actual Gerotor pump. As can be seen, this is a simple type of pump since there are only two moving parts.

Screw Pump

The screw pump (see Figure 5-15 for construction and nomenclature) is an axial flow positive displacement unit. Three precision ground screws, meshing within a close-fitting housing, deliver nonpulsating flow quietly and efficiently. The two symmetrically opposed idler rotors act as rotating seals, confining the fluid in a succession of closures or stages. The idler rotors are in rolling contact with the central power rotor and are free to float in their respective housing bores on a hydrodynamic oil

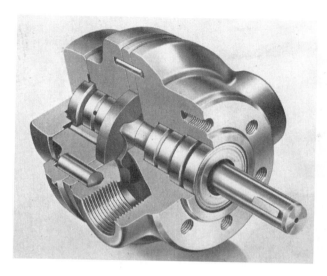

Figure 5-14. Gerotor pump. *(Courtesy of Brown & Sharpe Mfg. Co., Manchester, Michigan.)*

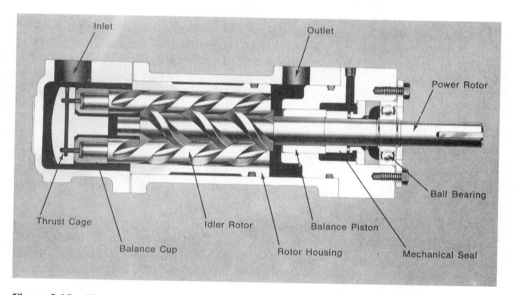

Figure 5-15. Nomenclature of a screw pump. *(Courtesy of DeLaval, IMO Pump Division, Trenton, New Jersey.)*

film. There are no radial bending loads. Axial hydraulic forces on the rotor set are balanced, eliminating any need for thrust bearings.

In Figure 5-16, we see a cutaway view of an actual screw pump. It is rated at 500 psi and can deliver up to 123 gpm. High-pressure designs are available for 3500-psi operation with output flow rates up to 88 gpm.

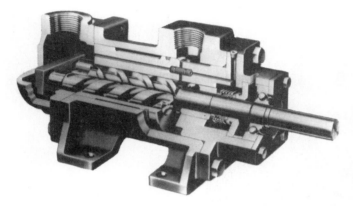

Figure 5-16. Screw pump. *(Courtesy of DeLaval, IMO Pump Division, Trenton, New Jersey.)*

5.6 VANE PUMPS

Basic Design and Operation

Figure 5-17 illustrates the operation of a vane pump. The rotor, which contains radial slots, is splined to the drive shaft and rotates inside a cam ring. Each slot contains a vane designed to mate with the surface of the cam ring as the rotor turns. Centrifugal force keeps the vanes out against the surface of the cam ring. During one-half revolution of rotor rotation, the volume increases between the rotor and cam ring. The resulting volume expansion causes a reduction of pressure. This is the suction process, which causes fluid to flow through the inlet port and fill the void. As the rotor rotates through the second half revolution, the surface of the cam ring pushes the vanes back into their slots, and the trapped volume is reduced. This positively ejects the trapped fluid through the discharge port.

Analysis of Volumetric Displacement

Careful observation of Figure 5-17 will reveal that there is an eccentricity between the centerline of the rotor and the centerline of the cam ring. If the eccentricity is zero, there will be no flow. The following analysis and nomenclature is applicable to the vane pump:

$$D_C = \text{diameter of cam ring (in, m)}$$
$$D_R = \text{diameter of rotor (in, m)}$$
$$L = \text{width of rotor (in, m)}$$
$$V_D = \text{pump volumetric displacement (in}^3, \text{m}^3)$$
$$e = \text{eccentricity (in, m)}$$
$$e_{\max} = \text{maximum possible eccentricity (in, m)}$$
$$V_{D\max} = \text{maximum possible volumetric displacement (in}^3, \text{m}^3)$$

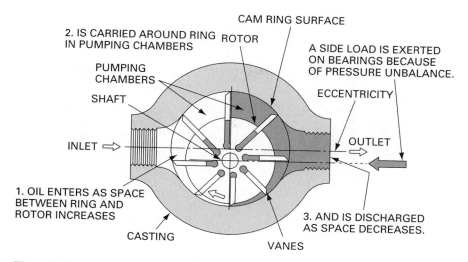

CAM RING SURFACE

2. IS CARRIED AROUND RING ROTOR
IN PUMPING CHAMBERS

A SIDE LOAD IS EXERTED
ON BEARINGS BECAUSE
OF PRESSURE UNBALANCE.

PUMPING
CHAMBERS

ECCENTRICITY

SHAFT

INLET ⇒

OUTLET
⇒

1. OIL ENTERS AS SPACE
BETWEEN RING AND
ROTOR INCREASES

3. AND IS DISCHARGED
AS SPACE DECREASES.

CASTING

VANES

Figure 5-17. Vane pump operation. *(Courtesy of Sperry Vickers, Sperry Rand Corp., Troy, Michigan.)*

From geometry, we can find the maximum possible eccentricity:

$$e_{\max} = \frac{D_C - D_R}{2}$$

This maximum value of eccentricity produces a maximum volumetric displacement:

$$V_{D_{\max}} = \frac{\pi}{4}(D_C^2 - D_R^2)L$$

Noting that we have the difference between two squared terms yields

$$V_{D_{\max}} = \frac{\pi}{4}(D_C + D_R)(D_C - D_R)L$$

Substituting the expression for $e_{\max}$ yields

$$V_{D_{\max}} = \frac{\pi}{4}(D_C + D_R)(2e_{\max})L$$

The actual volumetric displacement occurs when $e_{\max} = e$:

$$V_D = \frac{\pi}{2}(D_C + D_R)eL \tag{5-4}$$

Some vane pumps have provisions for mechanically varying the eccentricity. Such a design is called a *variable displacement pump* and is illustrated in Figure 5-18. A handwheel or a pressure compensator can be used to move the cam ring to change the eccentricity. The direction of flow through the pump can be reversed by movement of the cam ring on either side of center.

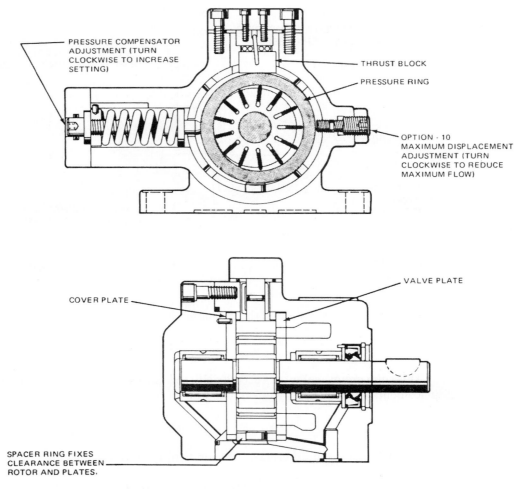

Figure 5-18. Variable displacement, pressure-compensated vane pump. *(Courtesy of Brown & Sharpe Mfg. Co., Manchester, Michigan.)*

Pressure-Compensated Vane Pump

The design we see in Figure 5-18 is a pressure-compensated one in which system pressure acts directly on the cam ring via a hydraulic piston on the right side (not shown). This forces the cam ring against the compensator spring-loaded piston on the left side of the cam ring. If the discharge pressure is large enough, it overcomes the compensator spring force and shifts the cam ring to the left. This reduces the eccentricity, which is maximum when discharge pressure is zero. As the discharge pressure continues to increase, zero eccentricity is finally achieved, and the pump flow becomes zero. Such a pump basically has its own protection against excessive pressure buildup, as shown in Figure 5-19. When the pressure reaches a value called

p_{cutoff}, the compensator spring force equals the hydraulic piston force. As the pressure continues to increase, the compensator spring is compressed until zero eccentricity is achieved. The maximum pressure achieved is called p_{deadhead}, at which point the pump is protected because it produces no more flow. As a result, there is no power wasted and fluid heating is reduced.

Figure 5-20 shows the internal configuration of an actual pressure-compensated vane pump. This design contains a cam ring that rotates slightly during use, thereby distributing wear over the entire inner circumference of the ring.

Note in Figures 5-17 and 5-18 that a side load is exerted on the bearings of the vane pump because of pressure unbalance. This same undesirable side load exists for the gear pump of Figure 5-7. Such pumps are hydraulically unbalanced.

Balanced Vane Pump

A balanced vane pump is one that has two intake and two outlet ports diametrically opposite each other. Thus, pressure ports are opposite each other, and a

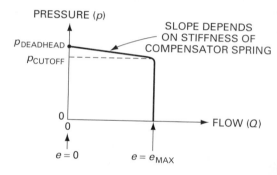

Figure 5-19. Pressure versus flow for pressure-compensated vane pump.

Figure 5-20. Cutaway photograph of pressure-compensated vane pump. *(Courtesy of Continental Hydraulics, Division of Continental Machines, Inc., Savage, Minnesota.)*

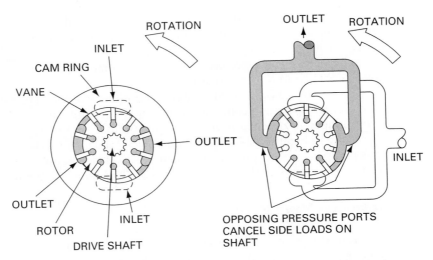

Figure 5-21. Balanced vane pump principles. *(Courtesy of Sperry Vickers, Sperry Rand Corp., Troy, Michigan.)*

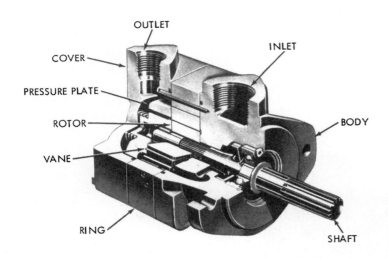

Figure 5-22. Cutaway view of balanced vane pump. *(Courtesy of Sperry Vickers, Sperry Rand Corp., Troy, Michigan.)*

complete hydraulic balance is achieved. One disadvantage of a balanced vane pump is that it cannot be designed as a variable displacement unit. Instead of having a circular cam ring, a balanced design vane pump has an elliptical housing, which forms two separate pumping chambers on opposite sides of the rotor. This eliminates the bearing side loads and thus permits higher operating pressures. Figure 5-21 shows the balanced vane pump principle of operation.

Figure 5-22 is a cutaway view of a balanced vane pump containing 12 vanes and a spring-loaded end plate. The inlet port is in the body, and the outlet port is in the cover, which may be assembled in any of four positions for convenience in piping.

EXAMPLE 5-3

A vane pump is to have a volumetric displacement of 5 in³. It has a rotor diameter of 2 in, a cam ring diameter of 3 in, and a vane width of 2 in. What must be the eccentricity?

Solution   Use Eq. (5-4):

$$e = \frac{2V_D}{\pi(D_C + D_R)L} = \frac{(2)(5)}{\pi(2 + 3)(2)} = 0.318 \text{ in}$$

EXAMPLE 5-4

A vane pump has a rotor diameter of 50 mm, a cam ring diameter of 75 mm, and a vane width of 50 mm. If the eccentricity is 8 mm, determine the volumetric displacement.

Solution   Substituting values into Eq. (5-4) yields

$$V_D = \frac{\pi}{2}(0.050 + 0.075)(0.008)(0.050) = 0.0000785 \text{ m}^3$$

Since 1 L = 0.001 m³, V_D = 0.0785 L.

EXAMPLE 5-5

A fixed displacement vane pump delivers 1000 psi oil to an extending hydraulic cylinder at 20 gpm. When the cylinder is fully extended, oil leaks past its piston at a rate of 0.7 gpm. The pressure relief valve setting is 1200 psi. If a pressure-compensated vane pump were used it would reduce pump flow from 20 gpm to 0.7 gpm when the cylinder is fully extended to provide the leakage flow at the pressure relief valve setting of 1200 psi. How much hydraulic horsepower would be saved by using the pressure-compensated pump?

Solution   The fixed displacement pump produces 20 gpm at 1200 psi when the cylinder is fully extended (0.7 gpm leakage flow through the cylinder and 19.3 gpm through the relief valve). Thus, we have

$$\text{hydraulic HP lost} = \frac{pQ}{1714} = \frac{1200 \times 20}{1714} = 14.0 \text{ hp}$$

A pressure-compensated pump would produce only 0.7 gpm at 1200 psi when the cylinder is fully extended. For this case we have

$$\text{hydraulic HP lost} = \frac{pQ}{1714} = \frac{1200 \times 0.7}{1714} = 0.49 \text{ hp}$$

Hence, the hydraulic horsepower saved = 14.0 − 0.49 = 13.51 hp. This horsepower savings occurs only while the cylinder is fully extended because either pump would deliver 1000 psi oil at 20 gpm while the cylinder is extending.

5.7 PISTON PUMPS

Introduction

A piston pump works on the principle that a reciprocating piston can draw in fluid when it retracts in a cylinder bore and discharge it when it extends. The basic question is how to mechanize a series of reciprocating pistons. There are two basic types of piston pumps. One is the axial design, having pistons that are parallel to the axis of the cylinder block. Axial piston pumps can be either of the bent axis configuration or of the swash plate design. The second type of piston pump is the radial design, which has pistons arranged radially in a cylinder block.

Axial Piston Pump (Bent-Axis Design)

Figure 5-23 shows an axial piston pump (bent-axis type) that contains a cylinder block rotating with the drive shaft. However, the centerline of the cylinder block is set at an offset angle relative to the centerline of the drive shaft. The cylinder block contains a number of pistons arranged along a circle. The piston rods are connected to the drive shaft flange by ball-and-socket joints. The pistons are forced in and out of their bores as the distance between the drive shaft flange and cylinder block changes. A universal link connects the block to the drive shaft to provide alignment and positive drive.

The volumetric displacement of the pump varies with the offset angle θ, as shown in Figure 5-24. No flow is produced when the cylinder block centerline is parallel to the drive shaft centerline. θ can vary from 0° to a maximum of about 30°. Fixed displacement units are usually provided with 23° or 30° offset angles.

Variable displacement units are available with a yoke and some external control to change the offset angle such as a stroking cylinder. Some designs have controls that move the yoke over the center position to reverse the direction of flow through the pump. Figure 5-25 is a cutaway of a variable displacement piston pump in which an external handwheel can be turned to establish the desired offset angle. Also shown is the hydraulic symbol used to represent variable displacement pumps in hydraulic circuits.

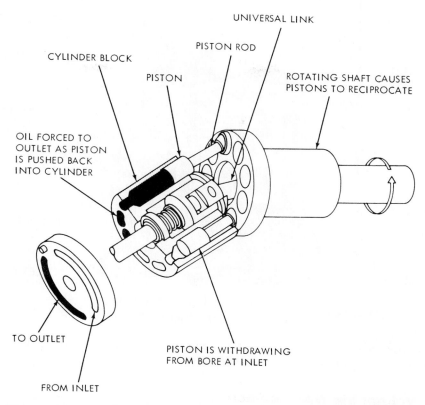

UNIVERSAL LINK

PISTON ROD

CYLINDER BLOCK

PISTON

ROTATING SHAFT CAUSES
PISTONS TO RECIPROCATE

OIL FORCED TO
OUTLET AS PISTON
IS PUSHED BACK
INTO CYLINDER

TO OUTLET

PISTON IS WITHDRAWING
FROM BORE AT INLET

FROM INLET

Figure 5-23. Axial piston pump (bent-axis type). *(Courtesy of Sperry Vickers, Sperry Rand Corp., Troy, Michigan.)*

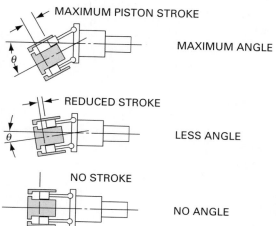

MAXIMUM PISTON STROKE

MAXIMUM ANGLE

θ

REDUCED STROKE

LESS ANGLE

θ

NO STROKE

NO ANGLE

Figure 5-24. Volumetric displacement changes with offset angle. *(Courtesy of Sperry Vickers, Sperry Rand Corp., Troy, Michigan.)*

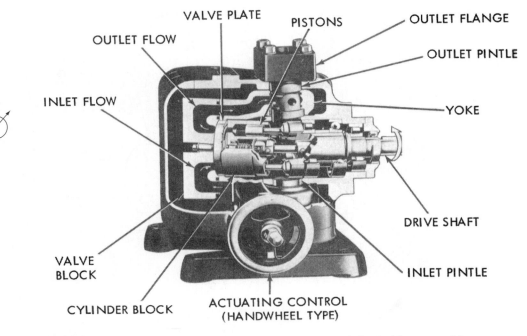

Figure 5-25. Variable displacement piston pump with handwheel. *(Courtesy of Sperry Vickers, Sperry Rand Corp., Troy, Michigan.)*

Volumetric Displacement and Theoretical Flow Rate

The following nomenclature and analysis are applicable to an axial piston pump:

θ = offset angle (°)
S = piston stroke (in, m)
D = piston circle diameter (in, m)
Y = number of pistons
A = piston area (in², m²)
N = pump speed (rpm)
Q_T = theoretical flow rate (gpm, m³/min)

From trigonometry we have

$$\tan(\theta) = \frac{S}{D}$$

or

$$S = D \tan(\theta)$$

The total displacement volume equals the number of pistons multiplied by the displacement volume per piston:

$$V_D = YAS$$

Substituting, we have

$$V_D = YAD \tan (\theta) \tag{5-5}$$

From Eqs. (5-2) and (5-5) we obtain a relationship for the theoretical flow rate using English units.

$$Q_T(\text{gpm}) = \frac{DANY \tan (\theta)}{231} \tag{5-6}$$

Similarly, using Eqs. (5-2M) and (5-5), we obtain a relationship for the theoretical flow rate in metric units.

$$Q_T(\text{m}^3/\text{min}) = DANY \tan (\theta) \tag{5-6M}$$

EXAMPLE 5-6

Find the offset angle for an axial piston pump that delivers 16 gpm at 3000 rpm. The pump has nine $\frac{1}{2}$-in-diameter pistons arranged on a 5-in-diameter piston circle. The volumetric efficiency is 95%.

Solution From Eq. (5-3) we calculate the theoretical flow rate:

$$Q_T = \frac{Q_A}{\eta_v} = \frac{16 \text{ gpm}}{0.95} = 16.8 \text{ gpm}$$

Using Eq. (5-6) yields

$$\tan (\theta) = \frac{231 Q_T}{DANY} = \frac{231 \times 16.8}{5[\pi/4(1/2)^2] \times 3000 \times 9} = 0.146$$

$$\theta = 8.3°$$

EXAMPLE 5-7

Find the flow rate in units of L/s that an axial piston pump delivers at 1000 rpm. The pump has nine 15-mm-diameter pistons arranged on a 125-mm-diameter piston circle. The offset angle is set at 10° and the volumetric efficiency is 94%.

Solution   Substituting directly into Eq. (5-6M) yields

$$Q_T(\text{m}^3/\text{min}) = D(\text{m}) \times A(\text{m}^2) \times N(\text{rev}/\text{min}) \times Y \tan(\theta)$$

$$= 0.125 \left(\frac{\pi}{4} \times 0.015^2\right) \times 1000 \times 9 \times \tan 10° = 0.0351 \text{ m}^3/\text{min}$$

From Eq. (5-3) we calculate the actual flow rate:

$$Q_A = Q_T \eta_v = 0.0351 \text{ m}^3/\text{min} \times 0.94 = 0.0330 \text{ m}^3/\text{min}$$

To convert to flow rate in units of L/s, we perform the following manipulation of units:

$$Q_A(\text{L/s}) = Q_A(\text{m}^3/\text{min}) \times \frac{1 \text{ min}}{60 \text{ s}} \times \frac{1 \text{ L}}{0.001 \text{ m}^3}$$

$$= 0.0330 \times \frac{1}{60} \times \frac{1}{0.001} = 0.550 \text{ L/s}$$

In-Line Piston Pump (Swash Plate Design)

Figure 5-26 provides a photograph and sketch illustrating the swash plate design in-line piston pump. In this type, the cylinder block and drive shaft are located on the same centerline. The pistons are connected to a shoe plate, which bears against an angled swash plate. As the cylinder rotates (see Figure 5-27), the pistons reciprocate because the piston shoes follow the angled surface of the swash plate. The outlet and inlet ports are located in the valve plate so that the pistons pass the inlet as they are being pulled out and pass the outlet as they are being forced back in.

This type of pump can also be designed to have variable displacement capability. In such a design, the swash plate is mounted in a movable yoke, as depicted in Figure 5-28. The swash plate angle can be changed by pivoting the yoke on pintles (see Figure 5-29 for the effect of swash plate angle on piston stroke). Positioning of the yoke can be accomplished by manual operation, servo control, or a compensator control, as shown in Figure 5-28. The maximum swash plate angle is limited to $17\frac{1}{2}°$ by construction.

Radial Piston Pump

The operation and construction of a radial piston pump is illustrated in Figure 5-30. This design consists of a pintle to direct fluid in and out of the cylinders, a cylinder barrel with pistons, and a rotor containing a reaction ring. The pistons remain in constant contact with the reaction ring due to centrifugal force and back pressure

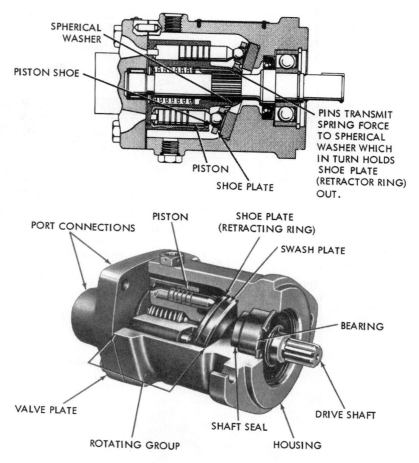

SPHERICAL WASHER

PISTON SHOE

PINS TRANSMIT SPRING FORCE TO SPHERICAL WASHER WHICH IN TURN HOLDS SHOE PLATE (RETRACTOR RING) OUT.

PISTON

SHOE PLATE

PORT CONNECTIONS

PISTON

SHOE PLATE (RETRACTING RING)

SWASH PLATE

BEARING

VALVE PLATE

DRIVE SHAFT

SHAFT SEAL

ROTATING GROUP

HOUSING

Figure 5-26. In-line design piston pump. *(Courtesy of Sperry Vickers, Sperry Rand Corp., Troy, Michigan.)*

on the pistons. For pumping action, the reaction ring is moved eccentrically with respect to the pintle or shaft axis. As the cylinder barrel rotates, the pistons on one side travel outward. This draws in fluid as each cylinder passes the suction ports of the pintle. When a piston passes the point of maximum eccentricity, it is forced inward by the reaction ring. This forces the fluid to enter the discharge port of the pintle. In some models, the displacement can be varied by moving the reaction ring to change the piston stroke.

Figure 5-31 provides a cutaway view of an actual radial piston pump that has variable displacement, pressure-compensated discharge. This pump is available in three sizes (2.40-, 3.00-, and 4.00-in^3 volumetric displacements) and weighs approximately 60 lb. Variable displacement is accomplished by hydraulic rather than mechanical means and is responsive to discharge line pressure.

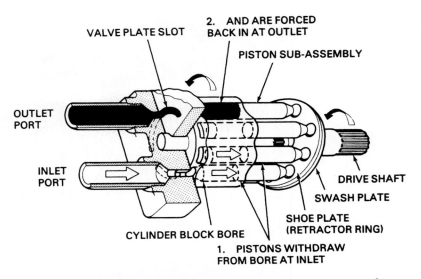

VALVE PLATE SLOT

2. AND ARE FORCED
BACK IN AT OUTLET

PISTON SUB-ASSEMBLY

OUTLET
PORT

INLET
PORT

DRIVE SHAFT

SWASH PLATE

SHOE PLATE
(RETRACTOR RING)

CYLINDER BLOCK BORE

1. PISTONS WITHDRAW
FROM BORE AT INLET

Figure 5-27. Swash plate causes pistons to reciprocate. *(Courtesy of Sperry Vickers, Sperry Rand Corp., Troy, Michigan.)*

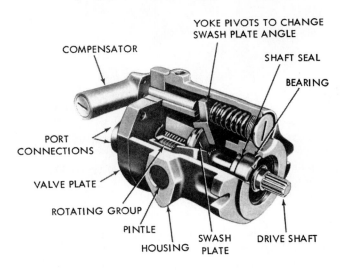

YOKE PIVOTS TO CHANGE
SWASH PLATE ANGLE

COMPENSATOR

SHAFT SEAL

BEARING

PORT
CONNECTIONS

VALVE PLATE

ROTATING GROUP

PINTLE

SWASH
PLATE

DRIVE SHAFT

HOUSING

Figure 5-28. Variable displacement version of in-line piston pump. *(Courtesy of Sperry Vickers, Sperry Rand Corp., Troy, Michigan.)*

5.8 PUMP PERFORMANCE

Introduction

The performance of a pump is primarily a function of the precision of its manufacture. Components must be made to close tolerances, which must be maintained while the pump is operating under design conditions. The maintenance of close tolerances is accomplished by designs that have mechanical integrity and balanced pressures.

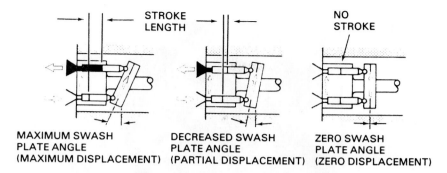

Figure 5-29. Variation in pump displacement. *(Courtesy of Sperry Vickers, Sperry Rand Corp., Troy, Michigan.)*

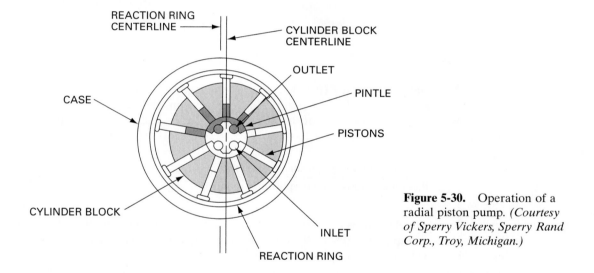

Figure 5-30. Operation of a radial piston pump. *(Courtesy of Sperry Vickers, Sperry Rand Corp., Troy, Michigan.)*

Theoretically the ideal pump would be one having zero clearance between all mating parts. Although this is not feasible, working clearances should be as small as possible while maintaining proper oil films for lubrication between rubbing parts.

Pump Efficiencies

Pump manufacturers run tests to determine performance data for their various types of pumps. The overall efficiency of a pump can be computed by comparing the hydraulic power output of the pump to the mechanical input power supplied by the prime mover. Overall efficiency can be broken into two distinct components called *volumetric efficiency* and *mechanical efficiency*. These three efficiencies are discussed on the following pages.

Figure 5-31. Cutaway view of
a radial piston pump. *(Courtesy
of Deere & Co., Moline,
Illinois.)*

1. Volumetric efficiency (η_v). *Volumetric efficiency* indicates the amount of
leakage that takes place within the pump. This involves considerations such as manufacturing tolerances and flexing of the pump casing under design pressure operating conditions:

$$\eta_v = \frac{\text{actual flow-rate produced by pump}}{\text{theoretical flow-rate pump should produce}} = \frac{Q_A}{Q_T} \tag{5-7}$$

Volumetric efficiencies typically run from 80% to 90% for gear pumps, 82% to
92% for vane pumps, and 90% to 98% for piston pumps. Note that when substituting efficiency values into equations, decimal fraction values should be used instead
of percentage values. For example, an efficiency value of 90% would be represented
by a value of 0.90.

2. Mechanical efficiency (ι_m). *Mechanical efficiency* indicates the amount of
energy losses that occur for reasons other than leakage. This includes friction in
bearings and between other mating parts. It also includes energy losses due to fluid
turbulence. Mechanical efficiencies typically run from 90% to 95%.

$$\eta_m = \frac{\text{pump output power assuming no leakage}}{\text{actual power delivered to pump}}$$

Using English units and horsepower for power yields

$$\eta_m = \frac{pQ_T/1714}{T_AN/63,000} \tag{5-8}$$

In metric units, using watts for power,

$$\eta_m = \frac{pQ_T}{T_A N} \tag{5-8M}$$

The parameters of Eqs. (5-8) and (5-8M) are defined as follows in conjunction with Figure 5-32.

p = pump discharge pressure (psi, Pa)
Q_T = pump theoretical flow rate (gpm, m³/s)
T_A = actual torque delivered to pump (in · lb, N · m)
N = pump speed (rpm, rad/s)

Mechanical efficiency can also be computed in terms of torques:

$$\eta_m = \frac{\text{theoretical torque required to operate pump}}{\text{actual torque delivered to pump}} = \frac{T_T}{T_A} \tag{5-9}$$

Note that the theoretical torque required to operate a pump (T_T) is the torque that would be required if there were no leakage.

Equations for evaluating the theoretical torque and the actual torque are as follows:

Theoretical Torque

$$T_T(\text{in} \cdot \text{lb}) = \frac{V_D(\text{in}^3) \times p(\text{psi})}{2\pi} \tag{5-10}$$

or

$$T_T(\text{N} \cdot \text{m}) = \frac{V_D(\text{m}^3) \times p(\text{Pa})}{2\pi} \tag{5-10M}$$

Actual Torque

$$T_A = \frac{\text{actual horsepower delivered to pump} \times 63,000}{N(\text{rpm})} \tag{5-11}$$

or

$$T_A = \frac{\text{actual power delivered to pump (W)}}{N(\text{rad/s})} \tag{5-11M}$$

where

$$N(\text{rad/s}) = \frac{2\pi}{60} N(\text{rpm})$$

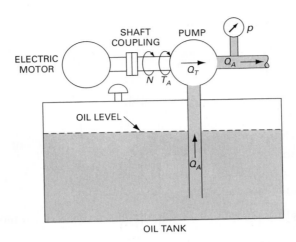

Figure 5-32. Terms involving pump mechanical efficiency.

3. Overall efficiency (η_o). The overall efficiency considers all energy losses and hence is defined as follows:

$$\text{overall efficiency} = \frac{\text{actual power delivered by pump}}{\text{actual power delivered to pump}} \tag{5-12}$$

The overall efficiency can also be represented mathematically as follows:

$$\eta_o = \eta_v \times \eta_m \tag{5-13}$$

Substituting Eq. (5-7) and (5-8) into Eq. (5-13), we have (for English units):

$$\eta_o = \eta_v \times \eta_m = \frac{Q_A}{Q_T} \times \frac{pQ_T/1714}{T_A N/63,000}$$

Canceling like terms yields the desired result showing the equivalency of Eq. (5-12) and Eq. (5-13).

$$\eta_o = \frac{pQ_A/1714}{T_A N/63,000} = \frac{\text{actual horsepower delivered by pump}}{\text{actual horsepower delivered to pump}} \tag{5-14}$$

Repeating this substitution for metric units using Eqs. (5-7), (5-8M), and (5-13) yields:

$$\eta_o = \frac{pQ_A}{T_A N} = \frac{\text{actual power delivered by pump}}{\text{actual power delivered to pump}} \tag{5-14M}$$

Note that the actual power delivered to a pump from a prime mover via a rotating shaft is called *brake power* and the actual power delivered by a pump to the fluid is called *hydraulic power*.

EXAMPLE 5-8

A pump has a displacement volume of 5 in³. It delivers 20 gpm at 1000 rpm and 1000 psi. If the prime mover input torque is 900 in · lb,

a. What is the overall efficiency of the pump?

b. What is the theoretical torque required to operate the pump?

Solution

a. Use Eq. (5-2) to find the theoretical flow rate:

$$Q_T = \frac{V_D N}{231} = \frac{(5)(1000)}{231} = 21.6 \text{ gpm}$$

Next, solve for the volumetric efficiency:

$$\eta_v = \frac{Q_A}{Q_T} = \frac{20}{21.6} = 0.926 = 92.6\%$$

Then solve for the mechanical efficiency:

$$\eta_m = \frac{p Q_T / 1714}{T_A N / 63,000} = \frac{[(1000)(21.6)]/1714}{[(900)(1000)]/63,000} = 0.881 = 88.1\%$$

Finally, we solve for the overall efficiency:

$$\eta_o = \eta_v \eta_m = 0.926 \times 0.881 = 0.816 = 81.6\%$$

b. Using Eq. (5-9) to solve for the theoretical torque we have

$$T_T = T_A \eta_m = 900 \times 0.881 = 793 \text{ in} \cdot \text{lb}$$

Thus, due to mechanical losses within the pump, 900 in · lb of torque are required to drive the pump instead of 793 in · lb.

Pump Performance Curves

Pump manufacturers specify pump performance characteristics in the form of graphs. Test data are obtained initially in tabular form and then put in graphical form for better visual interpretation. Figure 5-33 represents typical performance curves obtained for a 6-in³ variable displacement pump operating at full displacement. The upper graph gives curves of overall and volumetric efficiencies as a function of pump speed (rpm) for pressure levels of 3000 and 5000 psi. The lower graph gives curves of pump input horsepower (hp) and pump output flow (gpm) as a function of pump speed for the same two pressure levels.

Performance curves for the radial piston pump of Figure 5-31 are presented in Figure 5-34. Recall that this pump comes in three different sizes:

PR24: 2.40-in³ displacement

PR30: 3.00-in³ displacement

PR40: 4.00-in³ displacement

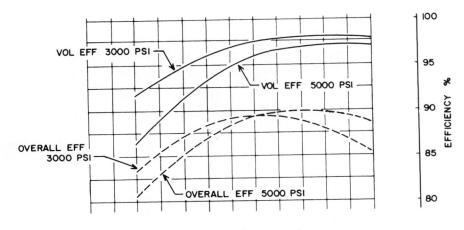

THESE CURVES INCLUDES LOSSES FROM INTEGRAL
SERVO/CHARGE PUMP 8 TRANSMISSION VALVE PACKAGE

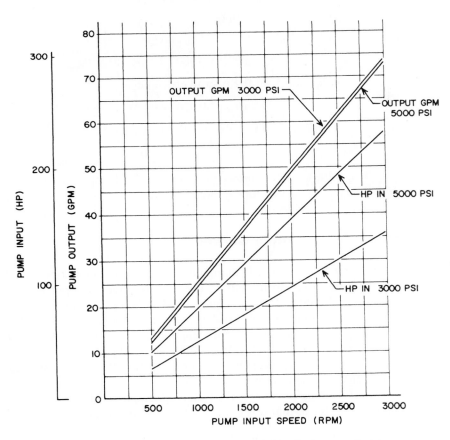

Figure 5-33. Performance curves for 6-in³ variable displacement piston pump.
(Courtesy of Abex Corp., Dennison Division, Columbus, Ohio.)

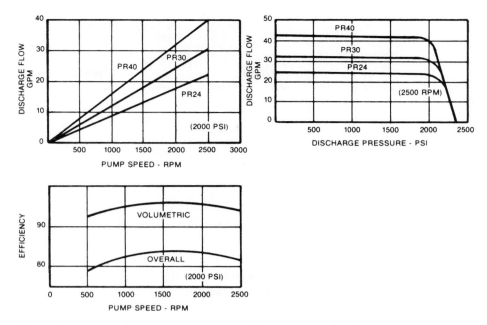

Figure 5-34. Performance curves of radial piston pumps. *(Courtesy of Deere & Co., Moline, Illinois.)*

Thus, there are three curves on two of the graphs. Observe the linear relationship between discharge flow (gpm) and pump speed (rpm). Also note that the discharge flow of these pumps is nearly constant over a broad pressure range. Discharge flow is infinitely variable between the point of inflection on the constant-discharge portion of the curve and zero flow. The volumetric and overall efficiency curves are based on a 2000-psi pump pressure.

Pump Performance Comparison Factors

Figure 5-35 compares various performance factors for hydraulic pumps. In general, gear pumps are the least expensive but also provide the lowest level of performance. In addition, gear pump efficiency is rapidly reduced by wear, which contributes to high maintenance costs. The volumetric efficiency of gear pumps is greatly affected by the following leakage losses, which can rapidly accelerate due to wear:

1. Leakage around the outer periphery of the gears
2. Leakage across the faces of the gears
3. Leakage at the points where the gear teeth make contact

Gear pumps are simple in design and compact in size. Therefore, they are the most common type of pump used in fluid power systems. The greatest number of applications of gear pumps are in the mobile equipment and machine tool fields.

PUMP TYPE	PRESSURE RATING (PSI)	SPEED RATING (RPM)	OVERALL EFFICIENCY (PERCENT)	HP PER LB RATIO	FLOW CAPACITY (GPM)	COST (DOLLARS PER HP)
EXTERNAL GEAR	2000–3000	1200–2500	80–90	2	1–150	4–8
INTERNAL GEAR	500–2000	1200–2500	70–85	2	1–200	4–8
VANE	1000–2000	1200–1800	80–95	2	1–80	6–30
AXIAL PISTON	2000–12,000	1200–3000	90–98	4	1–200	6–50
RADIAL PISTON	3000–12,000	1200–1800	85–95	3	1–200	5–35

Figure 5-35. Comparison of various performance factors for pumps.

Vane pump efficiencies and costs fall between those of gear and piston pumps. Vane pumps have good efficiencies and last for a reasonably long time. However, continued satisfactory performance necessitates clean oil with good lubricity. Excessive shaft speeds can cause operating problems. Leakage losses in vane pumps occur across the faces of the rotor and between the bronze wear plates and the pressure ring.

Piston pumps are the most expensive and provide the highest level of overall performance. They can be driven at high speeds (up to 5000 rpm) to provide a high horsepower-to-weight ratio. They produce essentially a nonpulsating flow and can operate at the highest pressure levels. Due to very close-fitting pistons, they have the highest efficiencies. Since no side loads occur to the pistons, the pump life expectancy is at least several years. However, because of their complex design, piston pumps cannot normally be repaired in the field.

5.9 PUMP NOISE

Introduction

Noise is sound that people find undesirable. For example, prolonged exposure to loud noise can result in loss of hearing. In addition, noise can mask sounds that people want to hear, such as voice communication between people and warning signals emanating from safety equipment.

The sounds that people hear come as pressure waves through the surrounding air medium. The pressure waves, which possess an amplitude and frequency, are generated by a vibrating object such as a pump, hydraulic motor, or pipeline. The human ear receives the sound waves and converts them into electrical signals that are transmitted to the brain. The brain translates these electrical signals into the sensation of sound.

	140	JET TAKEOFF AT CLOSE RANGE
	130	HYDRAULIC PRESS
THRESHOLD OF PAIN	120	NEARBY RIVETER
DEAFENING	110	AMPLIFIED ROCK BAND
	100	NOISY CITY TRAFFIC
VERY LOUD	90	NOISY FACTORY, GEAR PUMP
	80	VACUUM CLEANER, VANE PUMP
LOUD	70	NOISY OFFICE, PISTON PUMP
	60	AVERAGE FACTORY, SCREW PUMP
MODERATE	50	AVERAGE OFFICE
	40	PRIVATE OFFICE
FAINT	30	QUIET CONVERSATION
	20	RUSTLE OF LEAVES
VERY FAINT	10	WHISPER
	0	THRESHOLD OF HEARING

Figure 5-36. Common sound levels (dB).

Sound Intensity Levels (dB)

The strength of a sound wave, which depends on the pressure amplitude, is described by its intensity. *Intensity* is defined as the rate at which sound energy is transmitted through a unit area. Note that this is the definition of power per unit area. As such, intensity is typically represented in units of W/m^2. However, it is general practice to express this energy-transfer rate in units of decibels (dB). Decibels give the relative magnitudes of two intensities by comparing the one under consideration to the intensity of a sound at the threshold of hearing (the weakest intensity that the human ear can hear). One bel (1 B = 10 dB) represents a very large change in sound intensity. Thus, it has become standard practice to express sound intensity in units of decibels.

Note that intensity and loudness are not the same, because loudness depends on each person's sense of hearing. The loudness of a sound may differ for two people sitting next to each other and listening to the same sound. However, the intensity of a sound, which represents the amount of energy possessed by the sound, can be measured and thus does not depend on the person hearing it.

One decibel equals approximately the smallest change in intensity that can be detected by most people. The weakest sound intensity that the human ear can hear is designated as 0 dB. In contrast, sound intensities of 120 dB or greater produce pain and may cause permanent loss of hearing.

Figure 5-36 gives examples of some common sounds and the corresponding intensity levels in decibels. As shown in Figure 5-36, a sound level of 90 dB is considered very loud and is represented as the sound level in a noisy factory. The Occupational Safety and Health Administration (OSHA) stipulates that 90 dB(A) is the maximum sound level that a person may be exposed to during an 8-hr period in the workplace. The letter A following the symbol dB signifies that the sound level–measuring equipment uses a filtering system that more closely simulates the sensitivity of the human ear.

Calculating Sound Intensity Levels

The sound level in dB is obtained by taking the logarithm to the base 10 of the ratio of the intensity under consideration to the threshold of hearing intensity. The logarithm is used because the intensity of even a moderate sound (50 dB, per Figure 5-36) is actually 100,000 (which equals 10^5) times the smallest intensity that can be detected by the human ear (0 dB). Using a logarithmic scale reduces this huge factor to a more comprehensible one, as shown in the following equation:

$$I(\text{B}) = \log \frac{I}{I(\text{hear. thrsh.})} \tag{5-15}$$

where
$$I = \text{the intensity of a sound under consideration in units of W/m}^2,$$
$$I(\text{hear. thrsh.}) = \text{the intensity of a sound at the threshold of hearing in units of W/m}^2,$$
$$I(\text{B}) = \text{the intensity of a sound under consideration in units of bels.}$$

Thus, for a moderate sound the intensity in bels is

$$I(\text{moderate sound}) = \log 10^5 = 5 \text{ B}$$

This means that if bels are used, the intensity varies from 0 to only 12 for the entire sound range from threshold of hearing to threshold of pain. This is a fairly restricted range. To increase this range by a factor of 10, the decibel is used instead of the bel per the following equation:

$$I(\text{dB}) = 10 \log \frac{I}{I(\text{hear. thrsh.})} \tag{5-16}$$

Thus for a moderate sound, the decibel level is

$$I(\text{moderate sound}) = 10 \log 10^5 = 50 \text{ dB}$$

Likewise, for the threshold of hearing, the sound level in dB is

$$I(\text{hear. thrsh.}) = 10 \log 1 = 0 \text{ db}$$

These values check with those provided in Figure 5-36.

Equation (5-16) can be rewritten to determine the amount that the intensity of sound increases in units of dB if its intensity in W/m² increases by a given factor. The applicable equation is

$$dB \text{ increase} = 10 \log \frac{I(\text{final})}{I(\text{initial})} \qquad \textbf{(5-17)}$$

Thus, for example, if the intensity (in units of W/m²) of a sound doubles, the dB increase becomes

$$dB \text{ increase} = 10 \log 2 = 3.01 \text{ dB}$$

This result means that if the intensity of a sound increases by only 3.01 dB, the intensity doubles in power units of W/m². Hence the change in the intensity of a sound of only a few dB is significant.

Control of Noise Levels

Controlling noise levels is critically important in terms of preventing human accidents due to noise masking as well as protecting against permanent loss of hearing. Masking describes the ability of one sound to make the human ear incapable of hearing a second one, such as a safety warning signal. In general, noise reduction can be accomplished as follows:

1. Make changes to the source of the noise, such as a noisy pump. Problems here include misaligned pump/motor couplings, improperly installed pump/motor mounting plates, pump cavitation, and excess pump speed or pressure.
2. Modify components connected to the primary source of the noise. An example is the clamping of hydraulic piping at specifically located supports.
3. Use sound-absorption materials in nearby screens or partitions. This practice will reduce the reflection of sound waves to other areas of the building where noise can be a problem.

Pump Noise as a Performance Parameter

Noise is a significant parameter used to determine the performance of a pump. Any increase in the noise level normally indicates increased wear and imminent pump failure. Pumps are good generators but poor radiators of noise. As such, pumps are one of the main contributors to noise in a fluid power system. However, the noise we hear is not just the sound coming directly from the pump. It includes the vibration and fluid pulsations produced by the pump as well. Pumps are compact, and because of their relatively small size, they are poor radiators of noise, especially at lower frequencies. Reservoirs, electric motors, and piping, being large, are better radiators. Therefore, pump-induced vibrations or pulsations can cause them to radiate audible noise greater than that coming from the pump. In general, fixed displacement

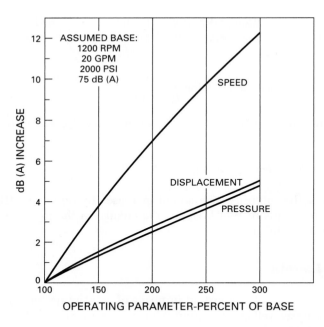

Figure 5-37. Data showing effect of changing size, pressure, and speed on noise. *(Courtesy of Sperry Vickers, Sperry Rand Corp., Troy, Michigan.)*

pumps are less noisy than variable displacement units because they have a more rigid construction.

As illustrated in Figure 5-37, pump speed has a strong effect on noise, whereas pressure and pump size have about equal but smaller effects. Since these three factors determine horsepower, they provide a trade-off for noise. To achieve the lowest noise levels, use the lowest practical speed (1000 or 2000 rpm where electric motors are used, a reducer gear for engine prime movers) and select the most advantageous combination of size and pressure to provide the needed horsepower.

Pump Cavitation

Still another noise problem, called *pump cavitation*, can occur due to entrained air bubbles in the hydraulic fluid or vaporization of the hydraulic fluid. This occurs when pump suction lift is excessive and the pump inlet pressure falls below the vapor pressure of the fluid (usually about 5-psi suction). As a result, air or vapor bubbles, which form in the low-pressure inlet region of the pump, are collapsed when they reach the high-pressure discharge region. This produces high fluid velocity and impact forces, which can erode the metallic components and shorten pump life.

The following rules will control or eliminate cavitation of a pump by keeping the suction pressure above the saturation pressure of the fluid:

1. Keep suction line velocities below 4 ft/s (1.2 m/s).
2. Keep pump inlet lines as short as possible.
3. Minimize the number of fittings in the inlet line.

PUMP DESIGN	NOISE LEVEL (dB-A)
GEAR	80–100
VANE	65–85
PISTON	60–80
SCREW	50–70

Figure 5-38. Noise levels for various pump designs.

4. Mount the pump as close as possible to the reservoir.

5. Use low-pressure drop inlet filters or strainers. Use indicating-type filters and strainers so that they can be replaced at proper intervals as they become dirty.

6. Use the proper oil as recommended by the pump manufacturer. The importance of temperature control lies in the fact that increased temperatures tend to accelerate the liberation of air or vapor bubbles. Therefore, operating oil temperatures should be kept in the range of 120°F to 150°F (50°C to 65°C) to provide an optimum viscosity range and maximum resistance to liberation of air or vapor bubbles to reduce the possibility of cavitation.

Pump noise is created as the internal rotating components abruptly increase the fluid pressure from inlet to outlet. The abruptness of the pressure increases plays a big role in the intensity of the pump noise. Thus, the noise level at which a pump operates depends greatly on the design of the pump. Gear and vane pumps generate a much higher noise level than do screw pumps. Figure 5-38 provides the approximate noise levels associated with various pump designs.

5.10 PUMP SELECTION

Pumps are selected by taking into account a number of considerations for a complete hydraulic system involving a particular application. Among these considerations are flow-rate requirements (gpm), operating speed (rpm), pressure rating (psi), performance, reliability, maintenance, cost, and noise. The selection of a pump typically entails the following sequence of operations:

1. Select the actuator (hydraulic cylinder or motor) that is appropriate based on the loads encountered.

2. Determine the flow-rate requirements. This involves the calculation of the flow rate necessary to drive the actuator to move the load through a specified distance within a given time limit.

3. Select the system pressure. This ties in with the actuator size and the magnitude of the resistive force produced by the external load on the system. Also involved here is the total amount of power to be delivered by the pump.

4. Determine the pump speed and select the prime mover. This, together with the flow-rate calculation, determines the pump size (volumetric displacement).

5. Select the pump type based on the application (gear, vane, or piston pump and fixed or variable displacement).

6. Select the reservoir and associated plumbing, including piping, valving, filters and strainers, and other miscellaneous components such as accumulators.

7. Consider factors such as noise levels, horsepower loss, need for a heat exchanger due to generated heat, pump wear, and scheduled maintenance service to provide a desired life of the total system.

8. Calculate the overall cost of the system.

Normally the sequence of operation is repeated several times with different sizes and types of components. After the procedure is repeated for several alternative systems, the best overall system is selected for the given application. This process is called *optimization*. It means determining the ultimate selection of a combination of system components to produce the most efficient overall system at minimum cost commensurate with the requirements of a particular application.

5.11 PUMP PERFORMANCE RATINGS IN METRIC UNITS

Performance data for hydraulic pumps are measured and specified in metric units as well as English units. Figure 5-39 shows actual performance data curves for a variable displacement, pressure-compensated vane pump operating at 1200 rpm. The curves give values of flow rate (gpm), efficiency, and power (hp and kW) versus output pressure (psi and bars). This particular pump (see Figure 5-39) can operate at speeds between 1000 and 1800 rpm, is rated at 2540 psi (175 bars), and has a nominal displacement volume of 1.22 in³ (20 cm³ or 0.02 L). Although the curves give flow rates in gpm, metric flow rates of liters per minute (Lpm) are frequently specified.

EXAMPLE 5-9

A pump has a displacement volume of 100 cm³. It delivers 0.0015 m³/s at 1000 rpm and 70 bars. If the prime mover input torque is 120 N · m,

a. What is the overall efficiency of the pump?

b. What is the theoretical torque required to operate the pump?

Solution

a. Using Eq. (5-2M), where the volumetric displacement is

$$V_D = 100 \text{ cm}^3/\text{rev} \times \left(\frac{1 \text{ m}}{100 \text{ cm}}\right)^3 = 0.000100 \text{ m}^3/\text{rev}$$

we have

$$Q_T = V_D N = (0.000100 \text{ m}^3/\text{rev})\left(\frac{1000}{60} \text{ rev/s}\right) = 0.00167 \text{ m}^3/\text{s}$$

Next, solve for the volumetric efficiency:

$$\eta_v = \frac{Q_A}{Q_T} = \frac{0.0015}{0.00167} = 0.898 = 89.8\%$$

Model VVB020

1200 RPM

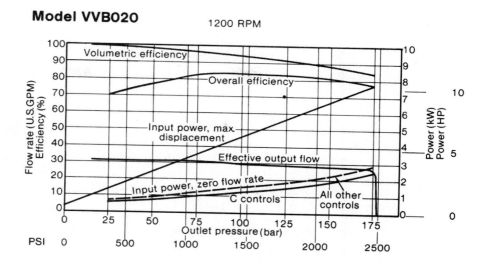

Figure 5-39. English/metric performance curves for variable displacement pressure-compensated vane pump at 1200 rpm. *(Courtesy of Vickers, Inc., Troy, Michigan.)*

Then solve for the mechanical efficiency:

$$\eta_m = \frac{pQ_T}{T_A N} = \frac{(70 \times 10^5 \text{ N/m}^2)(0.00167 \text{ m}^3/\text{s})}{(120 \text{ N} \cdot \text{m})\left(1000 \times \frac{2\pi}{60} \text{ rad/s}\right)}$$

$$\eta_m = \frac{11{,}690 \text{ N} \cdot \text{m/s}}{12{,}570 \text{ N} \cdot \text{m/s}} = 0.930 = 93.0\%$$

Note that the product $T_A N$ gives power in units of N · m/s (W) where torque (T_A) has units of N · m and shaft speed has units of rad/s. Finally, we solve for the overall efficiency:

$$\eta_o = \eta_v \eta_m = 0.898 \times 0.930 = 0.835 = 83.5\%$$

b. $\qquad\qquad T_T = T_A \eta_m = (120)(0.93) = 112\,\text{N} \cdot \text{m}$

Thus, due to mechanical losses within the pump, 120 N · m of torque are required to drive the pump instead of 112 N · m.

EXAMPLE 5-10

The pump in Example 5-9 is driven by an electric motor having an overall efficiency of 85%. The hydraulic system operates 12 hours per day for 250 days per year. The cost of electricity is \$0.11 per kilowatt hour. Determine

    a. The yearly cost of electricity to operate the hydraulic system
    b. The amount of the yearly cost of electricity that is due to the inefficiencies of the electric motor and pump

Solution

    a. First, we calculate the mechanical input power the electric motor delivers to the pump. Per Eq. (3-37), we have

$$\text{Pump input power (kW)} = \frac{T_A(\text{N} \cdot \text{m}) \times N(\text{rpm})}{9550} = \frac{120 \times 1000}{9550} = 12.6\,\text{kW}$$

Next, we calculate the electrical input power the electric motor receive:

$$\text{Electric motor input power} = \frac{\text{Electric motor output power}}{\text{Electric motor overall efficiency}}$$

Since the electric motor output power equals the pump input power we have

$$\text{Electric motor input power} = \frac{12.6\,\text{kW}}{0.85} = 14.8\,\text{kW}$$

Finally, we determine the yearly cost of electricity:

$$\text{yearly cost} = \text{power rate} \times \text{time per year} \times \text{unit cost of electricity}$$

$$= 14.8\,\text{kW} \times 12\,\frac{\text{hr}}{\text{day}} \times 250\,\frac{\text{days}}{\text{year}} \times \frac{\$0.11}{\text{kw hr}} = \$4884/\text{yr}$$

    b. The total kW loss equals the kW loss due to the electric motor plus the kW loss due to the pump. Thus, we have

$$\text{Total kW loss} = (1 - 0.85) \times 14.8 + (1 - 0.835) \times 12.6$$
$$= 2.22 + 2.08 = 4.30\,\text{kW}$$

$$\text{Yearly cost due to inefficiencies} = \frac{4.3}{14.8} \times \$4884/\text{yr} = \$1419/\text{yr}$$

Since $4.3/14.8 = 0.29$, we conclude that 29% of the total cost of electricity is due to the inefficiencies of the electric motor and pump. This also means that only 71% of the electrical power entering the electric motor is transferred into hydraulic power at the pump outlet port.

5.12 KEY EQUATIONS

Gear pump volumetric displacement:

$$V_D = \frac{\pi}{4}(D_o^2 - D_i^2)L \tag{5-1}$$

Theoretical flow rate of a pump
(flow rate a no-leak pump would deliver):

English or metric units:

$$Q_T = V_D N \tag{5-2}$$

English units:

$$Q_T\,(\text{ft}^3/\text{min}) = V_D(\text{ft}^3/\text{rev}) \times N(\text{rev}/\text{min}) \tag{5-2}$$

Special English units:

$$Q_T(\text{gpm}) = \frac{V_D(\text{in}^3/\text{rev}) \times N(\text{rpm})}{231} \tag{5-2}$$

Metric units:

$$Q_T\,(\text{m}^3/\text{min}) = V_D(\text{m}^3/\text{rev}) \times N(\text{rev}/\text{min}) \tag{5-2M}$$

Pump volumetric efficiency:

$$\eta_v = \frac{Q_A}{Q_T} \tag{5-3}$$

Vane pump volumetric displacement:

$$V_D = \frac{\pi}{2}(D_C + D_R)eL \tag{5-4}$$

Piston pump volumetric displacement:

$$V_D = YAD \tan(\theta) \tag{5-5}$$

Piston pump theoretical flow rate

Special
English
units:

$$Q_T(\text{gpm}) = \frac{D(\text{in}) \times A(\text{in}^2) \times N(\text{rpm}) \times Y\tan(\theta)}{231}$$

(5-6)

Metric units:

$$Q_T\left(\frac{\text{m}^3}{\text{min}}\right) = D(\text{m}) \times A(\text{m}^2) \times N\left(\frac{\text{rev}}{\text{min}}\right) \times Y\tan(\theta)$$

(5-6M)

Pump mechanical efficiency

Special English
units:

$$\eta_m = \frac{p(\text{psi}) \times Q_T(\text{gpm})/1714}{T_A(\text{in}\cdot\text{lb}) \times N(\text{rpm})/63,000}$$

(5-8)

Metric units:

$$\eta_m = \frac{p(\text{Pa}) \times Q_T(\text{m}^3/\text{s})}{T_A(\text{N}\cdot\text{m}) \times N(\text{rad/s})}$$

(5-8M)

English or
metric units:

$$\eta_m = \frac{T_T}{T_A}$$

(5-9)

Theoretical torque of a pump
(input torque a frictionless
pump would require
from the prime mover)

English units:

$$T_T(\text{in}\cdot\text{lb}) = \frac{V_D(\text{in}^3) \times p(\text{psi})}{2\pi}$$

(5-10)

Metric units:

$$T_T(\text{N}\cdot\text{m}) = \frac{V_D(\text{m}^3) \times p(\text{Pa})}{2\pi}$$

(5-10M)

Actual torque a pump
receives from the prime mover

Special
English units:

$$T_A(\text{in}\cdot\text{lb}) = \frac{\text{actual horsepower delivered to pump} \times 63,000}{N(\text{rpm})}$$

(5-11)

Metric units:

$$T_A(\text{N}\cdot\text{m}) = \frac{\text{actual power delivered to pump}(\text{W})}{N(\text{rad/s})}$$

(5-11M)

Pump overall efficiency

Definition:

$$\eta_o = \frac{\text{actual power delivered by pump}}{\text{actual power delivered to pump}}$$

(5-14)

Special English units:

$$\eta_o = \frac{p(\text{psi}) \times Q_A(\text{gpm})/1714}{T_A(\text{in} \cdot \text{lb}) \times N(\text{rpm})/63,000}$$

(5-14)

Metric units:

$$\eta_o = \frac{p(\text{Pa}) \times Q_A(\text{m}^3/\text{s})}{T_A(\text{N} \cdot \text{m}) \times N(\text{rad}/\text{s})}$$

(5-14M)

No units involved:

$$\eta_o = \eta_v \times \eta_m$$

(5-13)

Sound intensity:

$$I(\text{dB}) = 10 \log \frac{I(\text{W/m}^2)}{I(\text{hear. thrsh.})(\text{W/m}^2)}$$

(5-16)

Increase in sound intensity:

$$\text{dB increase} = 10 \log \frac{I \, \text{final} \, (\text{W/m}^2)}{I \, \text{initial} (\text{W/m}^2)}$$

(5-17)

EXERCISES

Questions, Concepts, and Definitions

5-1. Name the three popular construction types of positive displacement pumps.

5-2. What is a positive displacement pump? In what ways does it differ from a centrifugal pump?

5-3. How is the pumping action in positive displacement pumps accomplished?

5-4. How is the volumetric efficiency of a positive displacement pump determined?

5-5. How is the mechanical efficiency of a positive displacement pump determined?

5-6. How is the overall efficiency of a positive displacement pump determined?

5-7. Explain how atmospheric pressure pushes hydraulic oil up into the inlet port of a pump.

5-8. What is the difference between a fixed displacement pump and a variable displacement pump?

5-9. Name three designs of external gear pumps.

5-10. Name two designs of internal gear pumps.

5-11. Why is the operation of a screw pump quiet?

5-12. Name the important considerations when selecting a pump for a particular application.

5-13. What is a pressure-compensated vane pump, and how does it work?

5-14. What is pump cavitation, and what is its cause?

5-15. How is pressure developed in a hydraulic system?

5-16. Why should the suction head of a pump not exceed 5 psi?

5-17. Why must positive displacement pumps be protected by relief valves?

5-18. Why are centrifugal pumps so little used in fluid power systems?

5-19. What is meant by a balanced-design hydraulic pump?

5-20. Name the two basic types of piston pumps.

5-21. What parameters affect the noise level of a positive displacement pump?

5-22. What is meant by the pressure rating of a positive displacement pump?

5-23. Name four rules that will control or eliminate cavitation of a pump.

5-24. Comment on the relative comparison in performance among gear, vane, and piston pumps.

5-25. What are two ways of expressing pump size?

5-26. What types of pumps are available in variable displacement designs?

5-27. Explain the principle of a balanced vane design pump.

5-28. How can displacement be varied in an axial piston pump?

5-29. Explain how the size of the pumping chamber of a variable displacement vane pump is changed.

5-30. How is the capability of a variable displacement pump affected by the addition of pressure compensation?

5-31. What distinction is made between the terms *sound* and *noise?*

5-32. What is the difference between the terms *sound intensity* and *sound loudness?*

5-33. What is a decibel? Why are decibels used instead of bels to measure sound intensity?

5-34. What two human-safety concerns exist due to excessive noise levels?

5-35. Name the three principal ways in which noise reduction can be accomplished.

Problems

Note: The letter *E* following an exercise number means that English units are used. Similarly, the letter *M* indicates metric units.

Pump Flow Rates

5-36E. What is the theoretical flow rate from a fixed displacement axial piston pump with a nine-bore cylinder operating at 2000 rpm? Each bore has a 0.5-in diameter and the stroke is 0.75 in.

5-37E. A vane pump is to have a volumetric displacement of 7 in^3. It has a rotor diameter of $2\frac{1}{2}$ in, a cam ring diameter of $3\frac{1}{2}$ in, and a vane width of 2 in. What must be the eccentricity?

5-38E. Find the offset angle for an axial piston pump that delivers 30 gpm at 3000 rpm. The volumetric efficiency is 96%. The pump has nine $\frac{5}{8}$-in-diameter pistons arranged on a 5-in piston circle diameter.

5-39M. What is the theoretical flow rate from a fixed displacement axial piston pump with a nine-bore cylinder operating at 2000 rpm? Each bore has a 15-mm diameter and the stroke is 20 mm.

5-40M. A vane pump is to have a volumetric displacement of 115 cm^3. It has a rotor diameter of 63.5 mm, a cam ring diameter of 88.9 mm, and a vane width of 50.8 mm. What must be the eccentricity?

5-41M. Find the offset angle for an axial piston pump that delivers 0.0019 m^3/s at 3000 rpm. The pump has nine 15.9-mm-diameter pistons arranged on a 127-mm piston circle diameter. The volumetric efficiency is 95%.

5-42M. A pump having a 96% volumetric efficiency delivers 29 Lpm of oil at 1000 rpm. What is the volumetric displacement of the pump?

5-43. What is the metric equivalent of the following equation?

$$Q_T = \frac{V_D N}{231}$$

Pump Efficiencies

5-44. A positive displacement pump has an overall efficiency of 88% and a volumetric efficiency of 92% What is the mechanical efficiency?

5-45E. A gear pump has a $3\frac{1}{4}$-in outside diameter, a $2\frac{1}{4}$-in inside diameter, and a 1-in width. If the actual pump flow rate at 1800 rpm and rated pressure is 29 gpm, what is the volumetric efficiency?

5-46M. A gear pump has a 82.6-mm outside diameter, a 57.2-mm inside diameter, and a 25.4-mm width. If the actual pump flow rate at 1800 rpm and rated pressure is 0.00183 m^3/s, what is the volumetric efficiency?

5-47E. A pump has an overall efficiency of 88% and a volumetric efficiency of 92% while consuming 8 hp. Determine the mechanical efficiency and the horsepower loss due to friction (not usable to pump fluid).

5-48M. Determine the overall efficiency of a pump driven by a 10-hp prime mover if the pump delivers fluid at 40 Lpm at a pressure of 10 MPa.

5-49. The intensity (in units of W/m^2) of the noise of a pump increases by a factor of 10 due to cavitation. What is the corresponding increase in noise level in decibels?

Pump Power

5-50E. How much hydraulic horsepower would a pump produce when operating at 2000 psi and delivering 10 gpm? What HP electric motor would be selected to drive this pump if its overall efficiency is 85%?

5-51E. A pump has a displacement volume of 6 in^3. It delivers 24 gpm at 1000 rpm and 1000 psi. If the prime mover input torque is 1100 in · lb,

    a. What is the overall efficiency of the pump?

    b. What is the theoretical torque required to operate the pump?

5-52M. How much hydraulic power would a pump produce when operating at 140 bars and delivering 0.001 m^3/s of oil? What power-rated electric motor would be selected to drive this pump if its overall efficiency is 85%?

5-53M. A pump has a displacement volume of 98.4 cm^3. It delivers 0.0152 m^3/s of oil at 1000 rpm and 70 bars. If the prime mover input torque is 124.3 N · m,

    a. What is the overall efficiency of the pump?

    b. What is the theoretical torque required to operate the pump?

5-54E. A pump operates at 3000 psi and delivers 5 gpm. It requires 10 hp to drive the pump. Determine the overall efficiency of the pump. If the pump is driven at 1000 rpm, what is the input torque to the pump?

5-55E. A pump has a displacement volume of 6 in^3 and delivers 29 gpm at 1200 rpm and 500 psi. If the overall efficiency of the pump is 88%, find the actual torque required to operate the pump.

5-56E. A 1.5-in^3/rev displacement pump delivers 10 gpm at a pressure of 2000 psi. At 100% overall efficiency, what prime mover hp and speed are required?

System Problems

5-57E. For the fluid power system of Figure 5-40, the following data are given:

    cylinder piston diameter   = 8 in
    cylinder rod diameter   = 4 in
    extending speed of cylinder = 3 in/s

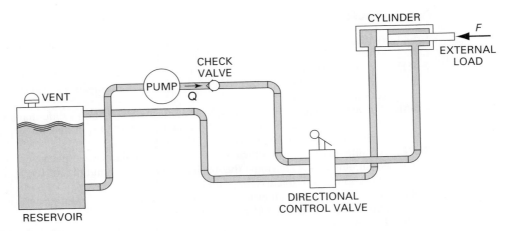

Figure 5-40. System for Exercise 5-57.

external load on cylinder	= 40,000 lb
pump volumetric efficiency	= 92%
pump mechanical efficiency	= 90%
pump speed	= 1800 rpm
pump inlet pressure	= −4.0 psi

The total pressure drop in the line from the pump discharge port to the blank end of the cylinder is 75 psi. The total pressure drop in the return line from the rod end of the cylinder is 50 psi. Determine the

a. Volumetric displacement of the pump
b. Input HP required to drive the pump
c. Input torque required to drive the pump
d. Percentage of pump input power delivered to the load

5-58M. For the system of Exercise 5-57, change the data to metric units and solve parts a, b, c, and d.

5-59E. The system of Exercise 5-57 contains a fixed displacement pump with a pressure relief valve set at 1000 psi. The system operates 20 hours per day for 250 days per year. The cylinder is stalled in its fully extended position 70% of the time. When the cylinder is fully extended, 1.0 gpm leaks past its piston.

a. If the electric motor driving the pump has an efficiency of 85% and the cost of electricity is $0.10 per kilowatt hour, find the annual cost of electricity for powering the system.

b. It is being considered to replace the fixed displacement pump with a pressure-compensated pump (compensator set at 1000 psi) that costs $2,500 more. How long will it take for the pressure-compensated pump to pay for itself if its overall efficiency is the same as the fixed displacement pump?

5-60M. For the system of Exercise 5-59, change the data to metric units and solve parts a and b.

6 Hydraulic Cylinders and Cushioning Devices

Learning Objectives

Upon completing this chapter, you should be able to:

1. Describe the construction and design features of hydraulic cylinders.
2. Identify the various types of hydraulic cylinder mountings and mechanical linkages for transmitting power.
3. Calculate the load-carrying capacity, speed, and power of hydraulic cylinders during the extending and retracting strokes.
4. Determine the maximum pressure developed by cushions at the ends of a hydraulic cylinder.
5. Explain the operation and features of double-rod cylinders and telescoping cylinders.
6. Describe the purpose, construction, and operation of hydraulic shock absorbers.
7. Analyze hydraulic cylinder piston rod forces resulting from driving external loads via various mechanical linkages.

6.1 INTRODUCTION

Pumps perform the function of adding energy to the fluid of a hydraulic system for transmission to some output location. Hydraulic cylinders (covered in this chapter) and hydraulic motors (covered in Chapter 7) do just the opposite. They extract energy from the fluid and convert it to mechanical energy to perform useful work. *Hydraulic cylinders* (also called *linear actuators*) extend and retract a piston rod to provide a push or pull force to drive the external load along a straight-line path. On the other hand, *hydraulic motors* (also called *rotary actuators*) rotate a shaft to provide a torque to drive the load along a rotary path.

Figure 6-1 shows a materials handling application in which hydraulic cylinders are used to drive the boom of a forklift truck operating on a rough terrain. On the rough ground around building construction, operating a forklift truck with a high reach requires a high degree of skill and care. One of the most difficult operations in this setting involves inserting loads horizontally onto scaffolds. In this system, automatic leveling and insertion of the fork is accomplished with computer-controlled electrohydraulics. Pure horizontal motion of the fork is achieved by hydraulic cylinders continuously changing the boom angle and length. This is accomplished by position sensors located on the boom. These sensors transmit signals to the on-board computer that respond by sending commands to the servo valves for controlling the oil flow rate to the hydraulic cylinders.

A hydraulic shock absorber is a device that brings a moving load to a gentle rest through the use of metered hydraulic fluid. Typically hydraulic shock absorbers use orifices to meter internal oil flow to accomplish a uniform gentle deceleration of the moving load. Two common applications of hydraulic shock absorbers are in moving cranes and automobile suspension systems.

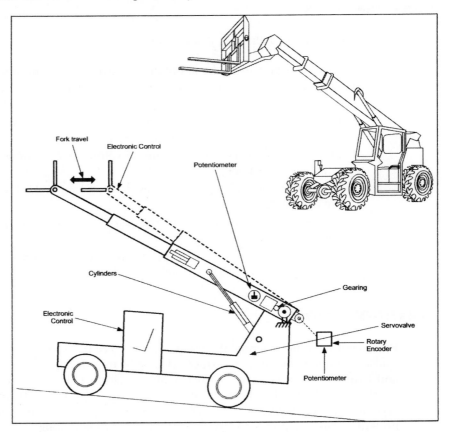

Figure 6-1.   Rough terrain forklift driven by hydraulic cylinders. *(Courtesy of National Fluid Power Association, Milwaukee, Wisconsin.)*

6.2 HYDRAULIC CYLINDER OPERATING FEATURES

The simplest type of hydraulic cylinder is the single-acting design, which is shown schematically in Figure 6-2(a). It consists of a piston inside a cylindrical housing called a *barrel*. Attached to one end of the piston is a rod, which extends outside one end of the cylinder (rod end). At the other end (blank end) is a port for the entrance and exit of oil. A single-acting cylinder can exert a force in only the extending direction as fluid from the pump enters the blank end of the cylinder. Single-acting cylinders do not retract hydraulically. Retraction is accomplished by using gravity or by the inclusion of a compression spring in the rod end. Figure 6-2(b) shows the graphic symbol of a single-acting cylinder.

A graphic symbol implies how a component operates without showing any of its construction details. In drawing hydraulic circuits (as done in Chapter 9) graphic symbols of all components are used. This facilitates circuit analysis and troubleshooting. The symbols, which are merely combinations of simple geometric figures such as circles, rectangles, and lines, make no attempt to show the internal configuration of a component. However, symbols must clearly show the function of each component.

Figure 6-3 shows the design of a double-acting hydraulic cylinder. Such a cylinder can be extended and retracted hydraulically. Thus, an output force can be applied in two directions (extension and retraction). This particular cylinder has a working pressure rating of 2000 psi for its smallest bore size of $1\frac{1}{8}$ in and 800 psi for its largest bore size of 8 in.

The nomenclature of a double-acting cylinder is provided in Figure 6-3. In this design, the barrel is made of seamless steel tubing, honed to a fine finish on the inside. The piston, which is made of ductile iron, contains U-cup packings to seal against leakage between the piston and barrel. The ports are located in the end caps, which are secured to the barrel by tie rods. The tapered cushion plungers provide smooth deceleration at both ends of the stroke. Therefore, the piston does not bang into the

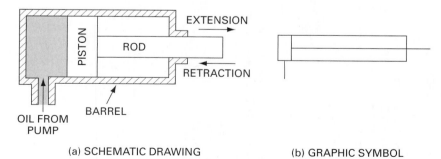

(a) SCHEMATIC DRAWING   (b) GRAPHIC SYMBOL

Figure 6-2.   Single-acting hydraulic cylinder.

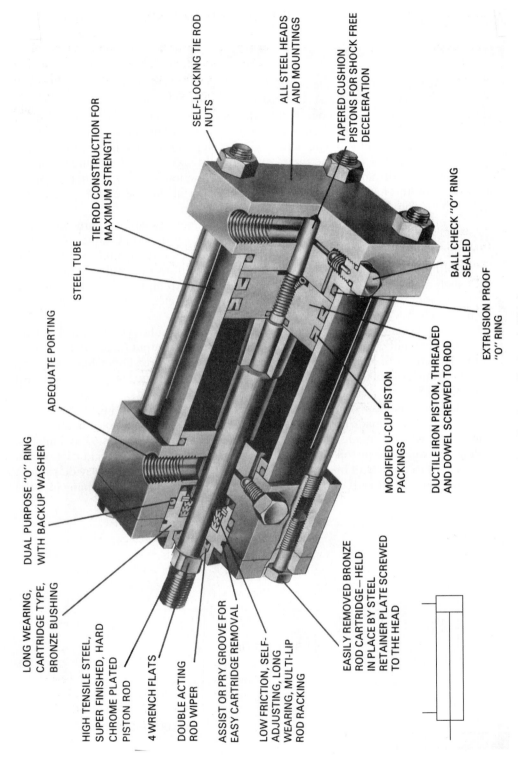

SELF-LOCKING TIE ROD NUTS

ALL STEEL HEADS AND MOUNTINGS

TAPERED CUSHION PISTONS FOR SHOCK FREE DECELERATION

TIE ROD CONSTRUCTION FOR MAXIMUM STRENGTH

STEEL TUBE

BALL CHECK "O" RING SEALED

EXTRUSION PROOF "O" RING

ADEQUATE PORTING

MODIFIED U-CUP PISTON PACKINGS

DUCTILE IRON PISTON, THREADED AND DOWEL SCREWED TO ROD

LONG WEARING, CARTRIDGE TYPE, BRONZE BUSHING

DUAL PURPOSE "O" RING WITH BACKUP WASHER

HIGH TENSILE STEEL, SUPER FINISHED, HARD CHROME PLATED PISTON ROD

4 WRENCH FLATS

DOUBLE ACTING ROD WIPER

ASSIST OR PRY GROOVE FOR EASY CARTRIDGE REMOVAL

LOW FRICTION, SELF-ADJUSTING, LONG WEARING, MULTI-LIP ROD RACKING

EASILY REMOVED BRONZE ROD CARTRIDGE—HELD IN PLACE BY STEEL RETAINER PLATE SCREWED TO THE HEAD

Figure 6-3. Double-acting cylinder design. (*Courtesy of Sheffer Corp., Cincinnati, Ohio.*)

end caps with excessive impact, which could damage the hydraulic cylinder after a given number of cycles. The graphic symbol for a double-acting cylinder is also shown in Figure 6-3.

6.3 CYLINDER MOUNTINGS AND MECHANICAL LINKAGES

Various types of cylinder mountings are in existence, as illustrated in Figure 6-4. This permits versatility in the anchoring of cylinders. The rod ends are usually threaded so that they can be attached directly to the load, a clevis, a yoke, or some other mating device.

Through the use of various mechanical linkages, the applications of hydraulic cylinders are limited only by the ingenuity of the fluid power designer. As illustrated in Figure 6-5, these linkages can transform a linear motion into either an oscillating or rotary motion. In addition, linkages can also be employed to increase or decrease the effective leverage and stroke of a cylinder.

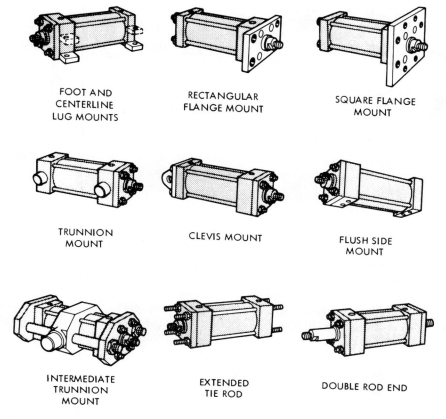

Figure 6-4. Various cylinder mountings. (*Courtesy of Sperry Vickers, Sperry Rand Corp., Troy, Michigan.*)

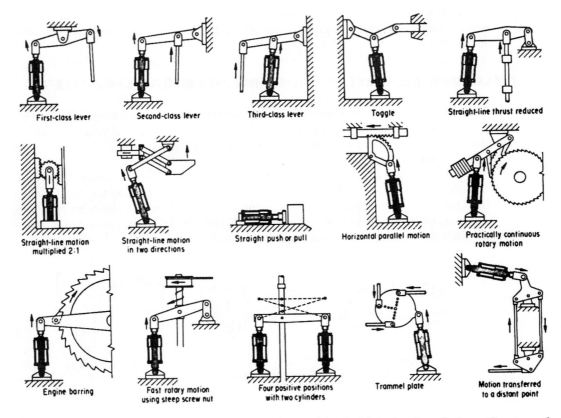

Figure 6-5. Typical mechanical linkages that can be combined with hydraulic cylinders. *(Courtesy of Rexnord Industries, Hydraulic Components Division, Racine, Wisconsin.)*

Much effort has been made by manufacturers of hydraulic cylinders to reduce or eliminate the side loading of cylinders created as a result of misalignment. It is almost impossible to achieve perfect alignment even though the alignment of a hydraulic cylinder has a direct bearing on its life.

A universal alignment mounting accessory designed to reduce misalignment problems is illustrated in Figure 6-6. By using one of these accessory components and a mating clevis at each end of the cylinder (see Figure 6-6), the following benefits are obtained:

1. Freer range of mounting positions
2. Reduced cylinder binding and side loading
3. Allowance for universal swivel
4. Reduced bearing and tube wear
5. Elimination of piston blow-by caused by misalignment

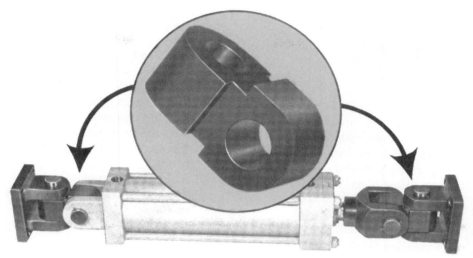

Figure 6-6. Universal alignment mounting accessory for fluid cylinders. *(Courtesy of Sheffer Corp., Cincinnati, Ohio.)*

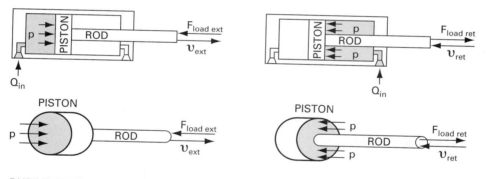

DURING EXTENSION, THE ENTIRE PISTON AREA (A_p) WHICH IS SHOWN SHADED IN THE LOWER VIEW, IS EXPOSED TO FLUID PRESSURE.

(a) EXTENSION STROKE

DURING RETRACTION, ONLY THE ANNULAR AREA AROUND THE ROD (A_p-A_r) WHICH IS SHOWN SHADED IN THE LOWER VIEW, IS EXPOSED TO FLUID PRESSURE.

(b) RETRACTION STROKE

Figure 6-7. The effective area of double-acting cylinders is greater for the extension stroke than it is for the retraction stroke.

6.4 CYLINDER FORCE, VELOCITY, AND POWER

The output force (F) and piston velocity (v) of double-acting cylinders are not the same for extension and retraction strokes. This is explained as follows (in conjunction with Figure 6-7): During the extension stroke, fluid enters the blank end of the cylinder through the entire circular area of the piston (A_p). However, during the retraction stroke, fluid enters the rod end through the smaller annular area between the rod and cylinder bore $(A_p - A_r)$, where A_p equals the piston area and A_r equals the rod area. This difference in flow-path cross-sectional area accounts for the difference in piston velocities. Since A_p is greater than $(A_p - A_r)$, the retraction velocity is greater than the extension velocity for the same input flow rate.

Similarly, during the extension stroke, fluid pressure bears on the entire circular area of the piston. However, during the retraction stroke, fluid pressure bears only on the smaller annular area between the piston rod and cylinder bore. This difference in area accounts for the difference in output forces. Since A_p is greater than $(A_p - A_r)$, the extension force is greater than the retraction force for the same operating pressure.

Equations (6-1) through (6-4) allow for the calculation of the output force and velocity for the extension and retraction strokes of 100% efficient double-acting cylinders.

Extension Stroke

$$F_{ext}(\text{lb}) = p(\text{psi}) \times A_p(\text{in}^2) \qquad \text{(6-1)}$$

$$F_{ext}(\text{N}) = p(\text{Pa}) \times A_p(\text{m}^2) \qquad \text{(6-1M)}$$

$$v_{ext}(\text{ft/s}) = \frac{Q_{in}(\text{ft}^3/\text{s})}{A_p(\text{ft}^2)} \qquad \text{(6-2)}$$

$$v_{ext}(\text{m/s}) = \frac{Q_{in}(\text{m}^3/\text{s})}{A_p(\text{m}^2)} \qquad \text{(6-2M)}$$

Retraction Stroke

$$F_{ret}(\text{lb}) = p(\text{psi}) \times (A_p - A_r)\text{in}^2 \qquad \text{(6-3)}$$

$$F_{ret}(\text{N}) = p(\text{Pa}) \times (A_p - A_r)\text{m}^2 \qquad \text{(6-3M)}$$

$$v_{ret}(\text{ft/s}) = \frac{Q_{in}(\text{ft}^3/\text{s})}{(A_p - A_r)\text{ft}^2} \qquad \text{(6-4)}$$

$$v_{ret}(\text{m/s}) = \frac{Q_{in}(\text{m}^3/\text{s})}{(A_p - A_r)\text{m}^2} \qquad \text{(6-4M)}$$

The power developed by a hydraulic cylinder equals the product of its force and velocity during a given stroke. Using this relationship and Eqs. (6-1) and (6-2) for the extending stroke and Eqs. (6-3) and (6-4) for the retraction stroke, we arrive at the same result: Power = $p \times Q_{in}$. Thus, we conclude that the power developed equals the product of pressure and cylinder input volume flow rate for both the extension and retraction strokes.

The horsepower developed by a hydraulic cylinder for either the extension or retraction stroke can be found using Eq. (6-5).

$$\text{Power (HP)} = \frac{v_p(\text{ft/s}) \times F(\text{lb})}{550} = \frac{Q_{in}(\text{gpm}) \times p(\text{psi})}{1714} \qquad \textbf{(6-5)}$$

Using metric units, the kW power developed for either the extension or retraction stroke can be found using Eq. (6-5M).

$$\text{Power (kW)} = v_p(\text{m/s}) \times F(\text{kN}) = Q_{in}(\text{m}^3/\text{s}) \times p(\text{kPa}) \qquad \textbf{(6-5M)}$$

Note that the equating of the input hydraulic power and the output mechanical power in Eqs. (6-5) and (6-5M) assumes a 100% efficient hydraulic cylinder.

EXAMPLE 6-1

A pump supplies oil at 20 gpm to a 2-in-diameter double-acting hydraulic cylinder. If the load is 1000 lb (extending and retracting) and the rod diameter is 1 in, find

a. The hydraulic pressure during the extending stroke
b. The piston velocity during the extending stroke
c. The cylinder horsepower during the extending stroke
d. The hydraulic pressure during the retraction stroke
e. The piston velocity during the retraction stroke
f. The cylinder horsepower during the retraction stroke

Solution

a.
$$p_{ext} = \frac{F_{ext}(\text{lb})}{A_p(\text{in}^2)} = \frac{1000}{(\pi/4)(2)^2} = \frac{1000}{3.14} = 318 \text{ psi}$$

b.
$$v_{ext} = \frac{Q_{in}(\text{ft}^3/\text{s})}{A_p(\text{ft}^2)} = \frac{20/449}{3.14/144} = \frac{0.0446}{0.0218} = 2.05 \text{ ft/s}$$

c.
$$\text{HP}_{ext} = \frac{v_{ext}(\text{ft/s}) \times F_{ext}(\text{lb})}{550} = \frac{2.05 \times 1000}{550} = 3.72 \text{ hp}$$

or
$$\text{HP}_{ext} = \frac{Q_{in}(\text{gpm}) \times p_{ext}(\text{psi})}{1714} = \frac{20 \times 318}{1714} = 3.72 \text{ hp}$$

d.
$$p_{ret} = \frac{F_{ret}(\text{lb})}{(A_p - A_r)\text{in}^2} = \frac{1000}{3.14 - (\pi/4)(1)^2} = \frac{1000}{2.355} = 425 \text{ psi}$$

Therefore, as expected, more pressure is required to retract than to extend the same load due to the effect of the rod.

e.
$$v_{ret} = \frac{Q_{in}(\text{ft}^3/\text{s})}{(A_p - A_r)\text{ft}^2} = \frac{0.0446}{2.355/144} = 2.73 \text{ ft/s}$$

Hence, as expected (for the same pump flow), the piston retraction velocity is greater than that for extension due to the effect of the rod.

f.
$$HP_{ret} = \frac{v_{ret}(\text{ft/s}) \times F_{ret}(\text{lb})}{550} = \frac{2.73 \times 1000}{550} = 4.96 \text{ hp}$$

or
$$HP_{ret} = \frac{Q_{in}(\text{gpm}) \times p_{ret}(\text{psi})}{1714} = \frac{20 \times 425}{1714} = 4.96 \text{ hp}$$

Thus, more horsepower is supplied by the cylinder during the retraction stroke because the piston velocity is greater during retraction and the load force remained the same during both strokes. This, of course, was accomplished by the greater pressure level during the retraction stroke. Recall that the pump output flow rate is constant, with a value of 20 gpm.

6.5 CYLINDER LOADS DUE TO MOVING OF WEIGHTS

The force a cylinder must produce equals the load the cylinder is required to overcome. In many cases the load is due to the weight of an object the cylinder is attempting to move. In the case of a vertical cylinder, the load simply equals the weight of the object because gravity acts in a downward, vertical direction.

Sometimes a cylinder is used to slide an object along a horizontal surface. In this case, the cylinder load is theoretically zero. This is because there is no component of the object's weight acting along the axis of the cylinder (a horizontal direction). However, as the object slides across the horizontal surface, the cylinder must overcome the frictional force created between the object and the horizontal surface. This frictional force, which equals the load acting on the cylinder, opposes the direction of motion of the moving object.

If the cylinder is mounted in neither a vertical nor horizontal direction, the cylinder load equals the component of the object's weight acting along the axis of the cylinder, plus a frictional force if the object is sliding along an inclined surface. Thus for an inclined cylinder, the load the cylinder must overcome is less than the weight of the object to be moved if the object is not sliding on an inclined surface.

The cylinder loads described up to now are based on moving an object at a constant velocity. However, an object to be moved at a given velocity is initially at rest. Thus the object has to be accelerated from zero velocity up to a steady state (constant) velocity as determined by the pump flow rate entering the cylinder. This acceleration represents an additional force (called an *inertial force*) that must be added to the weight component and any frictional force involved.

Examples 6-2, 6-3, and 6-4 illustrate how to determine loads acting on cylinders that are moving objects in any given direction, taking into account sliding friction and acceleration.

EXAMPLE 6-2

Find the cylinder force F required to move a 6000-lb weight W along a horizontal surface at a constant velocity. The coefficient of friction (CF) between the weight and horizontal support surface equals 0.14.

Solution

The frictional force f between the weight and its horizontal supporting surface equals CF times W. Thus, we have

$$F = f = (CF) \times W = 0.14 \times 6000 \text{ lb} = 840 \text{ lb}$$

EXAMPLE 6-3

Find the cylinder force F required to lift the 6000-lb weight W of Example 6-2 along a direction which is 30° from the horizontal, as shown in Figure 6-8(a). The weight is moved at constant velocity.

Solution

As shown in Figure 6-8(b), the cylinder force F must equal the component of the weight acting along the centerline of the cylinder. Thus, we have a right triangle with the hypotenuse W and the side F forming a 60° angle. From trigonometry we have sin 30° = F/W. Solving for F yields

$$F = W \sin 30° = 6000 \text{ lb} \times \sin 30° = 3000 \text{ lb}$$

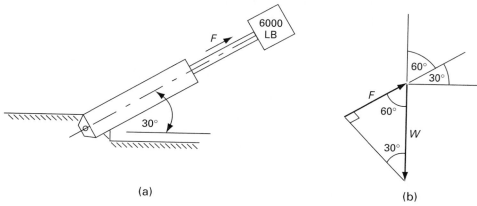

Figure 6-8. Inclined cylinder lifting a load for Example 6-3.

EXAMPLE 6-4

The 6000-lb weight of Example 6-2 is to be lifted upward in a vertical direction. Find the cylinder force required to

   a. Move the weight at a constant velocity of 8 ft/s
   b. Accelerate the weight from zero velocity to a velocity of 8 ft/s in 0.50 s

Solution

   a. For a constant velocity the cylinder force simply equals the weight of 6000 lb.
   b. From Newton's law of motion, the force required to accelerate a mass m equals the product of the mass m and its acceleration a. Noting that mass equals weight divided by the acceleration of gravity g, we have

$$a = \frac{8 \text{ ft/s} - 0 \text{ ft/s}}{0.50 \text{ s}} = 16 \text{ ft/s}^2$$

Thus, the force required to accelerate the weight is

$$F_{accel} = \frac{6000 \text{ lb}}{32.2 \text{ ft/s}^2} \times 16 \text{ ft/s}^2 = 2980 \text{ lb}$$

The cylinder force F_{cyl} required equals the sum of the weight and the acceleration force.

$$F_{cyl} = 6000 \text{ lb} + 2980 \text{ lb} = 8980 \text{ lb}$$

6.6 SPECIAL CYLINDER DESIGNS

Figure 6-9 illustrates a double-rod cylinder in which the rod extends out of the cylinder at both ends. For such a cylinder, the words *extend* and *retract* have no meaning. Since the force and speed are the same for either end, this type of cylinder is

Figure 6-9. Double-rod cylinder. *(Courtesy of Allenair Corp., Mineola, New York.)*

Figure 6-10. Telescopic cylinder. *(Courtesy of Commercial Shearing, Inc., Youngstown, Ohio.)*

typically used when the same task is to be performed at either end. Since each end contains the same size rod, the velocity of the piston is the same for both strokes.

Figure 6-10 illustrates the internal design features of a telescopic cylinder. This type actually contains multiple cylinders that slide inside each other. They are used where long work strokes are required but the full retraction length must be minimized. One application for a telescopic cylinder is the high-lift fork truck, illustrated in Figure 6-11. As shown, this lift truck is in the process of accessing materials located inside a second-story storage area of a warehouse, from the outside.

6.7 CYLINDER LOADINGS THROUGH MECHANICAL LINKAGES

In many applications, the load force that a hydraulic cylinder must overcome does not act along the axis of the hydraulic cylinder. Thus, the load force and the hydraulic cylinder force are in general not equal. The following is an analysis on how to determine the hydraulic cylinder force required to drive nonaxial loads using the first-class, second-class, and third-class lever systems of Figure 6-5. Note from Figure 6-5 that in first-class, second-class, and third-class lever systems, the cylinder rod and load rod are pin connected by a lever that can rotate about a fixed hinge pin. A similar analysis can be made of any of the other types of linkage arrangements shown in Figure 6-5.

Figure 6-11. High-lift truck. *(Courtesy of Mitsubishi Caterpillar Forklift America, Houston, Texas.)*

First-Class Lever System

Figure 6-12 shows a first-class lever system, which is characterized by the lever fixed-hinge pin being located between the cylinder and load rod pins. Note that the length of the lever portion from the cylinder rod pin to the fixed hinge is L_1, whereas the length of the lever portion from the load rod pin to the fixed hinge is L_2.

To determine the cylinder force F_{cyl} required to drive a load force F_{load}, we equate moments about the fixed hinge, which is the pivot point of the lever. The cylinder force attempts to rotate the lever counterclockwise about the pivot, and this creates a counterclockwise moment. Similarly, the load force creates a clockwise moment about the pivot. At equilibrium, these two moments are equal in magnitude:

$$\text{Counterclockwise moment} = \text{clockwise moment}$$

$$F_{cyl}(L_1 \cos\theta) = F_{load}(L_2 \cos\theta)$$

or

$$F_{cyl} = \frac{L_2}{L_1} F_{load} \qquad\qquad \textbf{(6-6)}$$

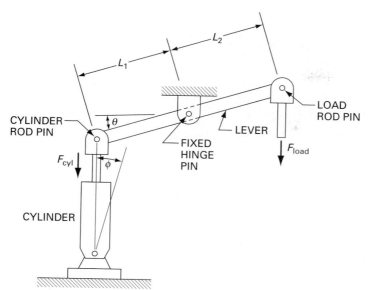

Figure 6-12. Use of a first-class lever to drive a load.

It should be noted that the cylinder is clevis-mounted (see Figure 6-4 for the clevis-mount design) to allow the rod-pinned end to travel along the circular path of the lever as it rotates about its fixed-hinge pin. If the centerline of the hydraulic cylinder becomes offset by an angle ϕ from the vertical, as shown in Figure 6-12, the relationship becomes

$$F_{cyl}(L_1 \cos\theta \times \cos\phi) = F_{load}(L_2 \cos\theta)$$

or

$$F_{cyl} = \frac{L_2}{L_1 \cos\phi} F_{load} \qquad \textbf{(6-7)}$$

Examination of Eq. (6-6) shows that when L_1 (distance from cylinder rod to hinge pin) is greater than L_2, the cylinder force is less than the load force. Of course, this results in a load stroke that is less than the cylinder stroke, as required by the conservation of energy law. When ϕ is 10° or less, the value of cos ϕ is very nearly unity (cos 0° = 1 and cos 10° = 0.985) and thus Eq. (6-6) can be used instead of Eq. (6-7).

The length of the moment arm for either the cylinder force or the load force is the perpendicular distance from the hinge pin to the line of action of the force. Thus, for the development of Eq. (6-6), the moment arms are $L_1 \cos\theta$ and $L_2 \cos\theta$ rather than simply L_1 and L_2. Similarly, for the development of Eq. (6-7), the moment arm for the cylinder force is $L_1 \cos\theta \times \cos\phi$, rather than simply $L_1 \cos\theta$, which is based on the assumption that $\phi = 0°$.

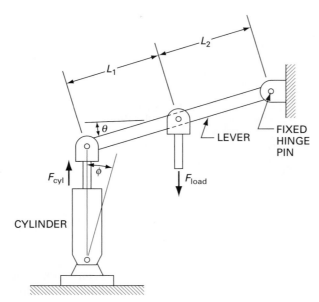

Figure 6-13. Use of a second-class lever system to drive a load.

Second-Class Lever System

Figure 6-13 shows a second-class lever system, which is characterized by the load rod pin being located between the fixed-hinge pin and cylinder rod pin of the lever.

   The analysis is accomplished by equating moments about the fixed-hinge pin, as follows:

$$F_{cyl} \cos \phi (L_1 + L_2) \cos \theta = F_{load}(L_2 \cos \theta)$$

or

$$F_{cyl} = \frac{L_2}{(L_1 + L_2) \cos \phi} F_{load} \tag{6-8}$$

   Comparing Eq. (6-7) to Eq. (6-8) shows that a smaller cylinder force is required to drive a given load force for a given lever length if a second-class lever is used instead of a first-class lever. Thus, using a second-class lever rather than a first-class lever reduces the required cylinder piston area for a given application. Of course, using a second-class lever also results in a smaller load stroke for a given cylinder stroke.

Third-Class Lever System

As shown in Figure 6-14, for a third-class lever system the cylinder rod pin lies between the load rod pin and fixed-hinge pin of the lever.

   Equating moments about the fixed-hinge pin yields

$$F_{cyl} \cos \phi (L_2 \cos \theta) = F_{load}(L_1 + L_2) \cos \theta$$

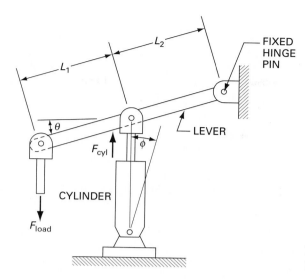

Figure 6-14. Use of a third-class lever system to drive a load.

or

$$F_{cyl} = \frac{L_1 + L_2}{L_2 \cos\phi} F_{load} \qquad \textbf{(6-9)}$$

Examination of Eq. (6-9) reveals that for a third-class lever, the cylinder force is greater than the load force. The reason for using a third-class lever system would be to produce a load stroke that is greater than the cylinder stroke, at the expense of requiring a larger cylinder diameter.

EXAMPLE 6-5

For the first-, second-, and third-class lever systems of Figures 6-12, 6-13, and 6-14 the following data are given:

$$L_1 = L_2 = 10 \text{ in}$$

$$\phi = 0°$$

$$F_{load} = 1000 \text{ lb}$$

Find the cylinder force required to overcome the load force for the

a. First-class lever
b. Second-class lever
c. Third-class lever

Solution

a. Per Eq. (6-7), we have

$$F_{cyl} = \frac{L_2}{L_1 \cos\phi} F_{load} = \frac{10}{10 \times 1}(1000) = 1000 \text{ lb}$$

b. Using Eq. (6-8) yields

$$F_{cyl} = \frac{L_2}{(L_1 + L_2)\cos\phi} F_{load} = \frac{10}{(10 + 10) \times 1}(1000) = 500 \text{ lb}$$

c. Substituting into Eq. (6-9), we have

$$F_{cyl} = \frac{L_1 + L_2}{L_2\cos\phi} F_{load} = \frac{(10 + 10)}{10 \times 1}(1000) = 2000 \text{ lb}$$

Thus, as expected, the second-class lever requires the smallest cylinder force, whereas the third-class lever requires the largest cylinder force.

Figure 6-15 shows an excavator lifting a huge concrete pipe at a construction site, via a chain connecting the pipe to the pinned end of the hydraulically actuated bucket. An excavator is a good example of an industrial machine in which the hydraulic cylinder loadings occur through mechanical linkages. Observe that there are a total of four hydraulic cylinders used to drive the three pin-connected members called the boom, stick, and bucket. The boom is the member that is pinned at one end to the cab frame. The stick is the central member that is pin connected at one end to the boom and pin connected at the other end to the bucket. Two of the cylinders connect the

Figure 6-15. Excavator contains hydraulic cylinders whose loadings occur through mechanical linkages. *(Courtesy of Caterpillar, Inc., Peoria, Illinois.)*

cab frame to the boom. A third cylinder connects the boom to the stick, and the fourth cylinder connects the stick to the bucket.

The problem is to determine the load on each cylinder for given positions of the boom, stick, and bucket. In order to do this, it is necessary to make a force analysis using the resulting mechanical linkage configuration and the given external load applied to the bucket. To determine the load on all four cylinders also requires knowing the weights and center of gravity locations of the boom, stick, and bucket. For the excavator of Figure 6-15, the maximum size bucket is 1.1 cubic yards and the maximum lifting capacity at ground level is 16,000 lb.

6.8 HYDRAULIC CYLINDER CUSHIONS

Double-acting cylinders sometimes contain cylinder cushions at the ends of the cylinder to slow the piston down near the ends of the stroke. This prevents excessive impact when the piston is stopped by the end caps, as illustrated in Figure 6-16. As shown, deceleration starts when the tapered plunger enters the opening in the cap. This restricts the exhaust flow from the barrel to the port. During the last small portion of the stroke, the oil must exhaust through an adjustable opening. The cushion design also incorporates a check valve to allow free flow to the piston during direction reversal.

The maximum pressure developed by cushions at the ends of a cylinder must be considered since excessive pressure buildup would rupture the cylinder. Example 6-6 illustrates how to calculate this pressure, which decelerates the piston at the ends of its extension and retraction strokes.

EXAMPLE 6-6

A pump delivers oil at a rate of 18.2 gpm into the blank end of the 3-in-diameter hydraulic cylinder shown in Figure 6-17. The piston contains a 1-in-diameter cushion plunger that is 0.75 in long, and therefore the piston decelerates over a distance of 0.75 in at the end of its extension stroke. The cylinder drives a 1500-lb weight, which slides on a flat horizontal surface having a coefficient of friction (CF) equal to 0.12. The pressure relief valve setting equals 750 psi. Therefore, the maximum pressure (p_1) at the blank end of the cylinder equals 750 psi while the cushion is decelerating the piston. Find the maximum pressure (p_2) developed by the cushion.

Solution
Step 1: Calculate the steady-state piston velocity v prior to deceleration:

$$v = \frac{Q_{\text{pump}}}{A_{\text{piston}}} = \frac{(18.2/449)\text{ft}^3/\text{s}}{[(\pi/4)(3)^2/144]\text{ft}^2} = \frac{0.0406}{0.049} = 0.83 \text{ ft/s}$$

Step 2: Calculate the deceleration a of the piston during the 0.75-in displacement S using the constant acceleration (or deceleration) equation:

$$v^2 = 2aS$$

Solving for deceleration, we have

$$a = \frac{v^2}{2S}$$

(6-10)

Substituting known values, we obtain the value of deceleration:

$$a = \frac{(0.83 \text{ ft/s})^2}{2(0.75/12 \text{ ft})} = 5.51 \text{ ft/s}^2$$

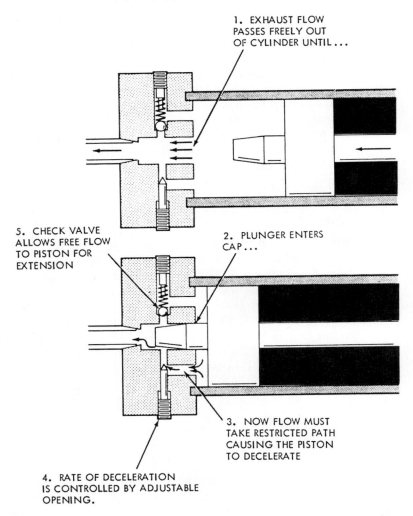

1. EXHAUST FLOW PASSES FREELY OUT OF CYLINDER UNTIL...

5. CHECK VALVE ALLOWS FREE FLOW TO PISTON FOR EXTENSION

2. PLUNGER ENTERS CAP...

3. NOW FLOW MUST TAKE RESTRICTED PATH CAUSING THE PISTON TO DECELERATE

4. RATE OF DECELERATION IS CONTROLLED BY ADJUSTABLE OPENING.

Figure 6-16. Operation of cylinder cushions. *(Courtesy of Sperry Vickers, Sperry Rand Corp., Troy, Michigan.)*

Step 3: Use Newton's law of motion: The sum of all external forces ΣF acting on a mass m equals the product of the mass m and its acceleration or deceleration a:

$$\Sigma F = ma$$

When substituting into Newton's equation, we shall consider forces that tend to slow down the piston as being positive forces. Also, the mass m equals the mass of all the moving members (piston, rod, and load). Since the weight of the piston and rod is small compared to the weight of the load, the weight of the piston and rod will be ignored. Also note that mass m equals weight W divided by the acceleration of gravity g. The frictional retarding force f between the weight W and its horizontal support surface equals CF times W. This frictional force is the external load force acting on the cylinder while it is moving the weight.

Substituting into Newton's equation yields

$$p_2(A_{\text{piston}} - A_{\text{cushion plunger}}) + (\text{CF})W - p_1(A_{\text{piston}}) = \frac{W}{g}a$$

Solving for p_2 yields a usable equation:

$$p_2 = \frac{(W/g)a + p_1(A_{\text{piston}}) - (\text{CF})W}{A_{\text{piston}} - A_{\text{cushion plunger}}} \tag{6-11}$$

Substituting known values produces the desired result:

$$p_2 = \frac{[(1500)(5.51)/32.2] + 750(\pi/4)(3)^2 - (0.12)(1500)}{(\pi/4)(3)^2 - (\pi/4)(1)^2}$$

$$p_2 = \frac{257 + 5303 - 180}{7.07 - 0.785} = \frac{5380}{6.285} = 856 \text{ psi}$$

Thus, the hydraulic cylinder must be designed to withstand an operating pressure of 856 psi rather than the pressure relief valve setting of 750 psi.

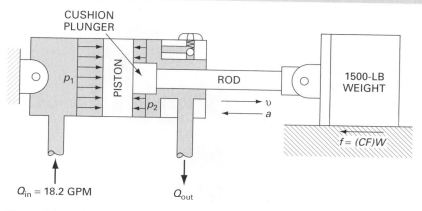

Figure 6-17. Cylinder cushion problem for Example 6-6.

6.9 HYDRAULIC SHOCK ABSORBERS

Introduction

A hydraulic shock absorber is a device that brings a moving load to a gentle rest through the use of metered hydraulic fluid. Figure 6-18 shows a hydraulic shock absorber that can provide a uniform gentle deceleration of any moving load from 25 to 25,000 lb or where the velocity and weight combination equals 3300 in · lb. Heavy-duty units are available with load capacities of over 11 million in · lb and strokes up to 20 in.

Construction and Operation Features

The construction and operation of the shock absorber of Figure 6-18 is as follows:

These shock absorbers are filled completely with oil. Therefore, they may be mounted in any position or at any angle. The spring-return units are entirely self-contained, extremely compact types that require no external hoses, valves, or fittings. In this spring-returned type a built-in cellular accumulator accommodates oil displaced by the piston rod as the rod moves inward. See Figure 6-19 for a cutaway view. Since the shock absorber is always filled with oil, there are no air pockets to cause spongy or erratic action.

These shock absorbers are multiple-orifice hydraulic devices. The orifices are simply holes through which a fluid can flow. When a moving load strikes the bumper of the shock absorber, it sets the rod and piston in motion. The moving piston pushes oil through a series of holes from an inner high-pressure chamber to an outer low-pressure chamber.

The resistance to the oil flow caused by the holes (restrictions) creates a pressure that acts against the piston to oppose the moving load. Holes are spaced geometrically according to a proven formula that produces constant pressure on the side of the piston opposite the load (constant resisting force) from the beginning to nearly

Figure 6-18. Hydraulic shock absorber. *(Courtesy of EGD Inc., Glenview, Illinois.)*

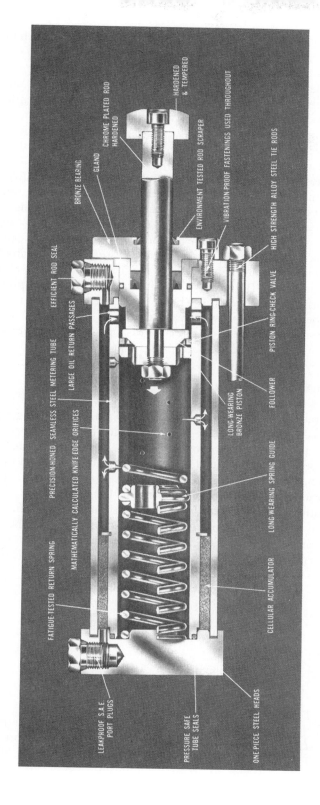

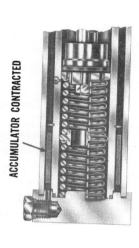

ACCUMULATOR CONTRACTED

Labels (left figure):
HARDENED & TEMPERED
CHROME PLATED ROD
HARDENED
GLAND
BRONZE BEARING
ENVIRONMENT TESTED ROD SCRAPER
VIBRATION-PROOF FASTENINGS USED THROUGHOUT
HIGH STRENGTH ALLOY STEEL TIE RODS
PISTON RING-CHECK VALVE
FOLLOWER
LONG-WEARING BRONZE PISTON
LARGE OIL RETURN PASSAGES
EFFICIENT ROD SEAL
PRECISION-HONED SEAMLESS STEEL METERING TUBE
MATHEMATICALLY CALCULATED KNIFE EDGE ORIFICES
FATIGUE-TESTED RETURN SPRING
LONG WEARING SPRING GUIDE
CELLULAR ACCUMULATOR
LEAKPROOF S.A.E. PORT PLUGS
PRESSURE SAFE TUBE SEALS
ONE-PIECE STEEL HEADS

Figure 6-19. Cutaway view of hydraulic shock absorber. (*Courtesy of E.G.D. Inc., Glenview, Illinois.*)

the end of the stroke. The piston progressively shuts off these orifices as the piston and rod move inward. Therefore, the total orifice area continually decreases and the load decelerates uniformly. At the end of the stroke, the load comes to a gentle rest and pressure drops to zero gage pressure. This results in uniform deceleration and gentle stopping, with no bounce back. In bringing a moving load to a stop, the shock absorber converts work and kinetic energy into heat, which is dissipated to the surroundings or through a heat exchanger.

Figure 6-20 illustrates various methods of decelerating the same weight from the same velocity over the same distance. The area under each curve represents the energy absorbed. The snubber or dashpot produces a high peak force at the beginning of the stroke; then the resistance is sharply reduced during the remainder of the stopping distance. The snubber, being a single-orifice device, provides a nonuniform deceleration, and the initial peak force can produce damaging stresses on the moving load and structural frame. Compression springs have a low initial stopping force and build to a peak at the end of the stroke. The springs store the energy during compression only to return it later, causing bounce-back. The rising force deflection curve requires a longer stroke to stay below a given maximum deceleration force. Liquid springs rely on the slight compressibility of the hydraulic fluid to stop a load. The reaction of a liquid spring is similar to that of a mechanical spring.

Crane and Automotive Applications

One application for hydraulic shock absorbers is energy dissipation of moving cranes, as illustrated in Figure 6-21. Shock absorbers prevent bounce-back of the bridge or trolley and thus provide protection for the operator, crane, and building structure.

Perhaps the most common application of hydraulic shock absorbers is for the suspension systems of automobiles. Figure 6-22(a) and (b) provide external and cutaway views respectively of an actual design. When an automobile tire hits a road irregularity such as a bump or pothole, the piston rod, which extends out the top of the shock absorber, is driven downward into its cylindrical housing. This is called the bump or compression stroke. During piston rod movement in the opposite direction (called the rebound stroke), energy stored in the automobile suspension springs is dissipated

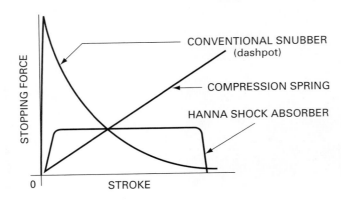

Figure 6-20. Hydraulic shock absorbers create a uniform stopping force. *(Courtesy of EGD Inc., Glenview, Illinois.)*

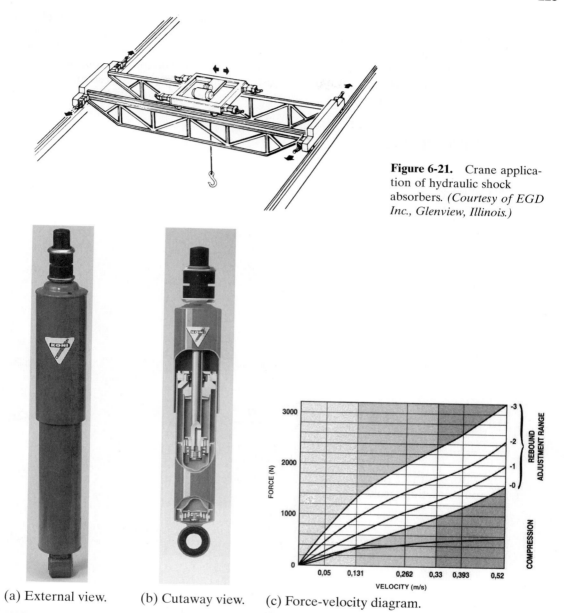

Figure 6-21. Crane application of hydraulic shock absorbers. *(Courtesy of EGD Inc., Glenview, Illinois.)*

(a) External view.   (b) Cutaway view.   (c) Force-velocity diagram.

Figure 6-22. Automotive hydraulic shock absorber. *(Courtesy of KONI North America, Hebron, Kentucky.)*

by the shock absorber into heat. Figure 6-22(c) provides a force-velocity diagram of this shock absorber. The curves in this diagram give the relationships between the damping force and the speed of the piston rod during operation. In this diagram the units of force (vertical axis) and velocity (horizontal axis) are N and m/s, respectively.

In essence the curves indicate how stiff or soft the automobile ride will feel to a passenger. The bottom curve represents the damping characteristics for the bump stroke. The upper four curves (labeled 0, −1, −2, and −3) are for the rebound stroke, showing four different damping characteristics depending on the ride comfort desired. This is accomplished by turning an adjustment knob through a desired angle to provide the rebound damping characteristics within the range indicated on the diagram.

6.10 KEY EQUATIONS

Cylinder extension force:

$$F_{ext} = pA_p$$

(6-1)

Cylinder extension velocity:

$$y_{ext} = \frac{Q_{in}}{A_p}$$

(6-2)

Cylinder retraction force:

$$F_{ret} = p\,(A_p - A_r)$$

(6-3)

Cylinder retraction velocity:

$$y_{ret} = \frac{Q_{in}}{A_p - A_r}$$

(6-4)

Cylinder power for extension and retraction strokes

English units:

$$HP = \frac{v\,(\text{ft/s}) \times F(\text{lb})}{550} = \frac{Q_{in}(\text{gpm}) \times p(\text{psi})}{1714}$$

(6-5)

Metric units:

$$kW\ power = v\,(\text{m/s}) \times F(\text{kN}) = Q_{in}(\text{m}^3/\text{s}) \times p(\text{kPa})$$

(6-5M)

First-class lever:

$$F_{cyl} = \frac{L_2}{L_1 \cos \phi} F_{load}$$

(6-7)

Second-class
lever:

$$F_{cyl} = \frac{L_2}{(L_1 + L_2)\cos\phi} F_{load}$$ **(6-8)**

Third-class
lever:

$$F_{cyl} = \frac{L_1 + L_2}{L_2\cos\phi} F_{load}$$ **(6-9)**

Cylinder piston
deceleration due
to cushion:

$$a = \frac{v^2}{2S}$$ **(6-10)**

Maximum
cylinder
pressure
developed
by cushion:

$$p_2 = \frac{(W/g)a + p_1(A_{piston}) - (CF)W}{A_{piston} - A_{cushion\ plunger}}$$

(6-11)

EXERCISES

Questions, Concepts, and Definitions

6-1. What is the difference between a single-acting and a double-acting hydraulic cylinder?

6-2. Name four different types of hydraulic cylinder mountings.

6-3. What is a cylinder cushion? What is its purpose?

6-4. What is a double-rod cylinder? When would it normally be used?

6-5. What is a telescoping-rod cylinder? When would it normally be used?

6-6. Why does the rod of a double-acting cylinder retract at a greater velocity than it extends for the same input flow rate?

6-7. How are single-acting cylinders retracted?

6-8. Differentiate between first-, second-, and third-class lever systems used with hydraulic cylinders to drive loads.

6-9. Relative to mechanical lever systems, define the term *moment*.

6-10. Relative to mechanical lever systems, define the term *moment arm*.

6-11. When using lever systems with hydraulic cylinders, why must the cylinder be clevis-mounted?

6-12. Using the mechanics of cylinder loadings with lever systems as an example, explain the difference between a torque and a moment.

6-13. What is the purpose of a hydraulic shock absorber? Name two applications.

Problems

Note: The letter *E* following an exercise number means that English units are used. Similarly, the letter *M* indicates metric units.

Hydraulic Cylinder Force, Velocity, and Power

6-14. An electric motor drives a pump at constant speed and delivers power to the pump at a constant rate. The pump delivers oil to a hydraulic cylinder. By what factor would the cylinder force and time to travel through full stroke change during extension if

a. The cylinder stroke is doubled, and the piston and rod diameters remain the same

b. The piston and rod diameters are both doubled and the stroke remains the same

c. The stroke, piston, and rod diameters are all doubled

6-15. Repeat exercise 6-14 for the retraction stroke.

6-16M. An 8-cm-diameter hydraulic cylinder has a 4-cm-diameter rod. If the cylinder receives flow at 100 Lpm and 12 MPa, find the

a. Extension and retraction speeds

b. Extension and retraction load-carrying capacities

6-17E. A 3-in-diameter hydraulic cylinder has a 1.5-in-diameter rod. Find the flow rate leaving the cylinder when it is extending by an entering flow rate of 8 gpm.

6-18E. A pump supplies oil at 25 gpm to a $1\frac{1}{2}$-in-diameter double-acting hydraulic cylinder. If the load is 1200 lb (extending and retracting) and the rod diameter is $\frac{3}{4}$ in, find the

a. Hydraulic pressure during the extending stroke

b. Piston velocity during the extending stroke

c. Cylinder horsepower during the extending stroke

d. Hydraulic pressure during the retracting stroke

e. Piston velocity during the retracting stroke

f. Cylinder horsepower during the retracting stroke

6-19M. A pump supplies oil at 0.0016 m³/s to a 40-mm-diameter double-acting hydraulic cylinder. If the load is 5000 N (extending and retracting) and the rod diameter is 20 mm, find the

a. Hydraulic pressure during the extending stroke

b. Piston velocity during the extending stroke

c. Cylinder kW power during the extending stroke

d. Hydraulic pressure during the retracting stroke

e. Piston velocity during the retracting stroke

f. Cylinder kW power during the retracting stroke

6-20. Determine the value of constants C_1 and C_2 in the following equations for determining the speed of a hydraulic cylinder.

$$v(\text{in/min}) = \frac{C_1 Q(\text{gpm})}{A(\text{in}^2)} \qquad v(\text{m/s}) = \frac{C_2 Q(\text{m}^3/\text{s})}{A(\text{m}^2)}$$

6-21. A hydraulic cylinder has a rod diameter equal to one-half the piston diameter. Determine the difference in load-carrying capacity between extension and retraction if the pressure is constant.

6-22. For the cylinder in Exercise 6-21, what would happen if the pressure were applied to both sides of the cylinder at the same time?

Cylinder Loads Due to Moving of Weights

6-23E. The inclined cylinder of Figure 6-23 has a 3-in-diameter piston. Determine the pressure required to extend the 5000-lb weight.

6-24M. A 10,000 N weight is to be lowered by a vertical cylinder, as shown in Figure 6-24. The cylinder has a 75-mm-diameter piston and a 50-mm-diameter rod. The weight is to decelerate from 100 m/min to a stop in 0.5 s. Determine the required pressure in the rod end of the cylinder during the deceleration motion.

6-25E. A 6000-lb weight is being pushed up an inclined surface at constant speed by a cylinder, as shown in Figure 6-25. The coefficient of friction between the weight and inclined surface equals 0.15. For a pressure of 1000 psi, determine the required cylinder piston diameter.

6-26E. Solve the problem of Exercise 6-25 if the weight is to accelerate from a zero velocity to a velocity of 5 ft/s in 0.5 s.

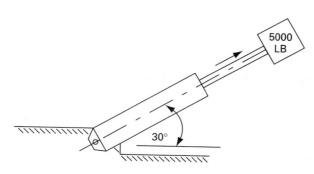

Figure 6-23. System for Exercise 6-23.

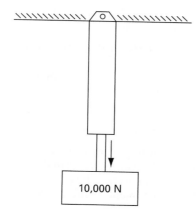

Figure 6-24. System for Exercise 6-24.

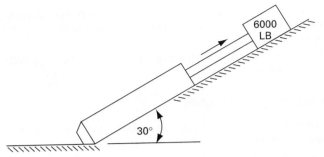

Figure 6-25. System for Exercises 6-25 and 6-26.

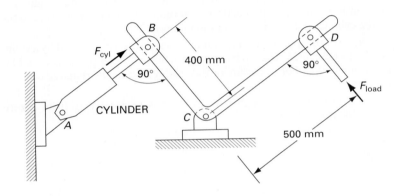

Figure 6-26. System for Exercise 6-28.

Cylinder Loadings Through Mechanical Linkages

6-27M. Change the data of Example 6-5 to metric units.

    a. Solve for the cylinder force required to overcome the load force for the first-, second-, and third-class lever systems.

    b. Repeat part a with $\theta = 10°$.

    c. Repeat part a with $\phi = 5°$ and $20°$.

6-28M. For the system of Figure 6-26, determine the hydraulic cylinder force required to drive a 1000-N load.

6-29E. For the crane system of Figure 6-27, determine the hydraulic cylinder force required to lift the 2000-lb load.

6-30E. For the toggle mechanism of Figure 6-28, determine the output load force for a hydraulic cylinder force of 1000 lb.

Hydraulic Cylinder Cushions

6-31E. A pump delivers oil at a rate of 20 gpm to the blank end of a 2-in-diameter hydraulic cylinder as shown in Figure 6-17. The piston contains a $\frac{3}{4}$-in-diameter cushion plunger that is 1 in long. The cylinder drives a 1000-lb weight, which slides on a flat, horizontal surface having a coefficient of friction equal to 0.15. The

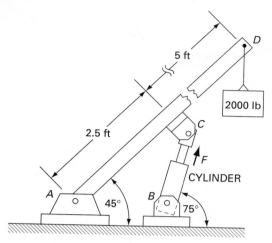

Figure 6-27. System for Exercise 6-29.

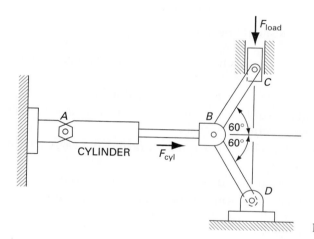

Figure 6-28. System for Exercise 6-30.

pressure relief valve setting equals 500 psi. Find the maximum pressure developed by the cushion.

6-32M. Change the data of Exercise 6-31 to metric units and solve for the maximum pressure developed by the cushion.

7 Hydraulic Motors

Learning Objectives

Upon completing this chapter, you should be able to:

1. Evaluate torque-pressure-displacement relationships for limited rotation hydraulic motors.
2. Explain the operation of gear, vane, and piston hydraulic motors.
3. Evaluate the performance of hydraulic motors by determining the volumetric, mechanical, and overall efficiencies.
4. Make a comparison of the various performance factors of gear, vane, and piston hydraulic motors.
5. Determine the torque and power delivered by hydraulic motors.
6. Analyze the operation and performance of hydrostatic transmissions.

7.1 INTRODUCTION

As in the case of hydraulic cylinders, hydraulic motors extract energy from a fluid and convert it to mechanical energy to perform useful work. Hydraulic motors can be of the limited rotation or the continuous rotation type. A limited rotation motor, which is also called a *rotary actuator* or an *oscillating motor*, can rotate clockwise and counterclockwise but through less than one complete revolution. A continuous rotation hydraulic motor, which is simply called a *hydraulic motor*, can rotate continuously at an rpm that is determined by the motor's input flow rate. In reality, hydraulic motors are pumps that have been redesigned to withstand the different forces that are involved in motor applications. As a result, hydraulic motors are typically of the gear, vane, or piston configuration.

Figure 7-1 shows a hydraulic impact wrench, which is ideal for heavy-duty drilling, loosening, and tightening operations. It delivers an infinitely adjustable range

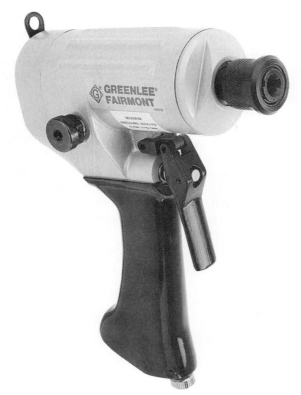

Figure 7-1. Hydraulic impact wrench. *(Courtesy of Greenlee Textron, Inc., Rockford, Illinois.)*

of speed and torque up to 5400 rpm and 400 ft · lb (540 N · m), respectively. This impact wrench weighs 7.1 lb (3.2 kg) and uses a gear motor. It operates with a flow-rate range of 4 to 12 gpm (15 to 45 Lpm) and a pressure range of 1000 to 2000 psi (70 to 140 bars).

Hydrostatic transmissions are hydraulic systems specifically designed to have a pump drive a hydraulic motor. Thus, a hydrostatic transmission simply transforms mechanical power into fluid power and then reconverts the fluid power back into shaft power. The advantages of hydrostatic transmissions include power transmission to remote areas, infinitely variable speed control, self-overload protection, reverse rotation capability, dynamic braking, and a high power-to-weight ratio. Applications include materials handling equipment, farm tractors, railway locomotives, buses, lawn mowers, and machine tools.

7.2 LIMITED ROTATION HYDRAULIC MOTORS

Introduction

A limited rotation hydraulic motor (also called *oscillation motor* or *rotary actuator*) provides rotary output motion over a finite angle. This device produces high instantaneous torque in either direction and requires only a small space and simple mountings. Rotary actuators consist of a chamber or chambers containing the working fluid and a movable surface against which the fluid acts. The movable surface is connected to an output shaft to produce the output motion. A direct-acting vane-type actuator is shown schematically along with its graphic symbol in Figure 7-2. Fluid under pressure is directed to one side of the moving vane, causing it to rotate. This type provides about 280° of rotation. Vane unit capacity ranges from 3 to 1 million in · lb.

Rotary actuators are available with working pressures up to 5000 psi. They are typically mounted by foot, flange, and end mounts. Cushioning devices are available in most designs. Figure 7-3 shows an actual rotary actuator similar to the design depicted schematically in Figure 7-2. Since it contains two vanes, the maximum angle of rotation is reduced to about 100°. However, the torque-carrying capacity is twice that obtained by a single-vane design. This particular unit can operate with either air or oil at pressures up to 1000 psi.

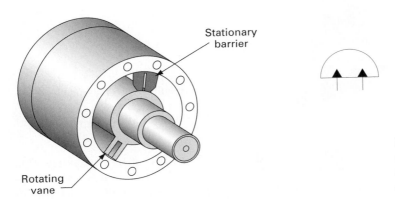

Figure 7-2. Rotary actuator. *(Courtesy of Rexnord Inc., Hydraulic Components Division, Racine, Wisconsin.)*

Figure 7-3. Rotary actuator. *(Courtesy of Ex-Cell-O Corp., Troy, Michigan.)*

Analysis of Torque Capacity

The following nomenclature and analysis are applicable to a limited rotation hydraulic motor containing a single rotating vane:

R_R = outer radius of rotor (in, m)

R_V = outer radius of vane (in, m)

L = width of vane (in, m)

p = hydraulic pressure (psi, Pa)

F = hydraulic force acting on vane (lb, N)

A = surface area of vane in contact with oil (in^2, m^2)

T = torque capacity (in $\cdot$ lb, N $\cdot$ m)

The force on the vane equals the pressure times the vane surface area:

$$F = pA = p(R_V - R_R)L$$

The torque equals the vane force times the mean radius of the vane:

$$T = p(R_V - R_R)L \frac{(R_V + R_R)}{2}$$

On rearranging, we have

$$T = \frac{pL}{2}(R_V^2 - R_R^2) \tag{7-1}$$

A second equation for torque can be developed by noting the following relationship for volumetric displacement V_D:

$$V_D = \pi (R_V^2 - R_R^2)L \tag{7-2}$$

Combining Eqs. (7-1) and (7-2) yields

$$T = \frac{pV_D}{2\pi} \tag{7-3}$$

Observe from Eq. (7-3) that torque capacity can be increased by increasing the pressure or volumetric displacement or both.

EXAMPLE 7-1

A single-vane rotary actuator has the following physical data:

outer radius of rotor = 0.5 in

outer radius of vane = 1.5 in

width of vane = 1 in

If the torque load is 1000 in · lb, what pressure must be developed to overcome the load?

Solution   Use Eq. (7-2) to solve for the volumetric displacement:

$$V_D = \pi(1.5^2 - 0.5^2)(1) = 6.28 \text{ in}^3$$

Then use Eq. (7-3) to solve for the pressure:

$$p = \frac{2\pi T}{V_D} = \frac{2\pi(1000)}{6.28} = 1000 \text{ psi}$$

As shown in Figure 7-4, applications for rotary actuators include conveyor sorting; valve turning; air bending operations; flipover between work stations; positioning for welding; lifting; rotating; and dumping. The symbol for a rotary actuator is shown in the lower right-hand side of Figure 7-4.

7.3 GEAR MOTORS

Hydraulic motors can rotate continuously and as such have the same basic configuration as pumps. However, instead of pushing on the fluid as pumps do, motors are pushed on by the fluid. In this way, hydraulic motors develop torque and produce continuous rotary motion. Since the casing of a hydraulic motor is pressurized from an outside source, most hydraulic motors have casing drains to protect shaft seals. There are three basic types of hydraulic motors: gear, vane, and piston. Let's first examine the operation and configuration of the gear motor.

A gear motor develops torque due to hydraulic pressure acting on the surfaces of the gear teeth, as illustrated in Figure 7-5. The direction of rotation of the motor can be reversed by reversing the direction of flow. As is the case for gear pumps, the volumetric displacement of a gear motor is fixed. The gear motor shown in Figure 7-5 is not balanced with respect to pressure loads. The high pressure at the inlet, coupled with the low pressure at the outlet, produces a large side load on the shaft and bearings. Gear motors are normally limited to 2000-psi operating pressures and 2400-rpm operating speeds. They are available with a maximum flow capacity of 150 gpm.

The main advantages of a gear motor are its simple design and subsequent low cost. Figure 7-6 shows a cutaway view of an actual gear motor. Also shown is the hydraulic symbol used in hydraulic circuits for representing fixed displacement motors.

Hydraulic motors can also be of the internal gear design. This type can operate at higher pressures and speeds and also has greater displacements than the external gear motor.

As in the case of pumps, screw-type hydraulic motors exist using three meshing screws (a power rotor and two idler rotors). Such a motor is illustrated in Figure 7-7. The rolling screw set results in extremely quiet operation. Torque is developed by

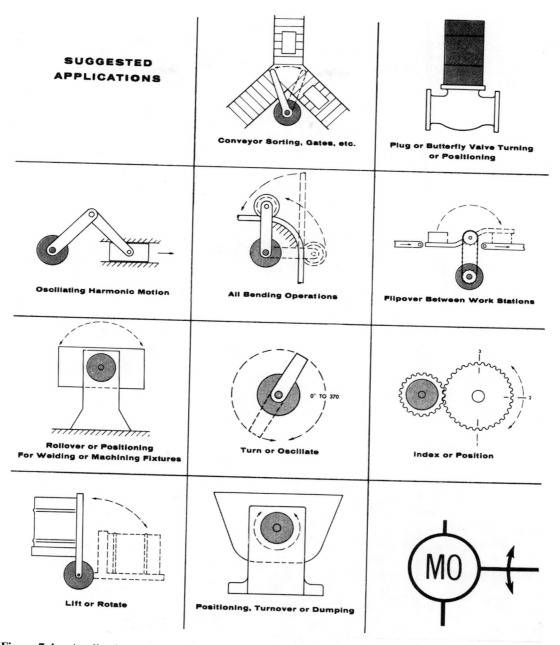

Figure 7-4. Applications of rotary actuators. *(Courtesy of Carter Controls, Inc., Lansing, Illinois.)*

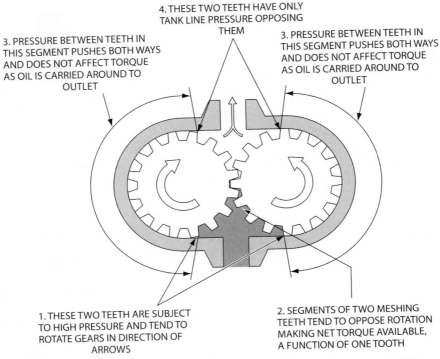

4. THESE TWO TEETH HAVE ONLY TANK LINE PRESSURE OPPOSING THEM

3. PRESSURE BETWEEN TEETH IN THIS SEGMENT PUSHES BOTH WAYS AND DOES NOT AFFECT TORQUE AS OIL IS CARRIED AROUND TO OUTLET

3. PRESSURE BETWEEN TEETH IN THIS SEGMENT PUSHES BOTH WAYS AND DOES NOT AFFECT TORQUE AS OIL IS CARRIED AROUND TO OUTLET

1. THESE TWO TEETH ARE SUBJECT TO HIGH PRESSURE AND TEND TO ROTATE GEARS IN DIRECTION OF ARROWS

2. SEGMENTS OF TWO MESHING TEETH TEND TO OPPOSE ROTATION MAKING NET TORQUE AVAILABLE, A FUNCTION OF ONE TOOTH

Figure 7-5. Torque development by a gear motor. *(Courtesy of Sperry Vickers, Sperry Rand Corp., Troy, Michigan.)*

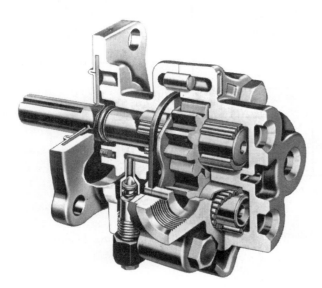

Figure 7-6. External gear motor. *(Courtesy of Webster Electric Company, Inc., subsidiary of STA-RITE Industries, Inc., Racine, Wisconsin.)*

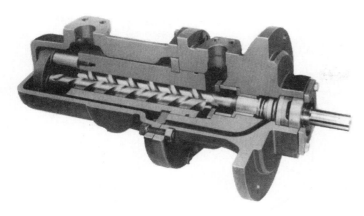

Figure 7-7. Screw motor. *(Courtesy of DeLaval, IMO Pump Division, Trenton, New Jersey.)*

differential pressure acting on the thread area of the screw set. Motor torque is proportional to differential pressure across the screw set. This particular motor can operate at pressures up to 3000 psi and can possess volumetric displacements up to 13.9 in³.

7.4 VANE MOTORS

Vane motors develop torque by the hydraulic pressure acting on the exposed surfaces of the vanes, which slide in and out of the rotor connected to the drive shaft (see Figure 7-8, view A). As the rotor revolves, the vanes follow the surface of the cam ring because springs (not shown in Figure 7-8) are used to force the vanes radially outward. No centrifugal force exists until the rotor starts to revolve. Therefore, the vanes must have some means other than centrifugal force to hold them against the cam ring. Some designs use springs, whereas other types use pressure-loaded vanes. The sliding action of the vanes forms sealed chambers, which carry the fluid from the inlet to the outlet.

Vane motors are universally of the balanced design illustrated in view B of Figure 7-8. In this design, pressure buildup at either port is directed to two interconnected cavities located 180° apart. The side loads that are created are therefore canceled out. Since vane motors are hydraulically balanced, they are fixed displacement units.

Figure 7-9 shows a design where pivoted rocker arms are attached to the rotor and serve as springs to force the vanes outward against the elliptical cam ring. This type of motor is available to operate at pressures up to 2500 psi and at speeds up to 4000 rpm. The maximum flow delivery is 250 gpm.

Four hydraulic vane motors are used to drive the paddle wheel on the river boat shown in Figure 7-10(a). Figure 7-10(b) is a photograph showing a cutaway view of one of these motors that is rated at 5000 psi and has a 250 cubic inch volumetric displacement.

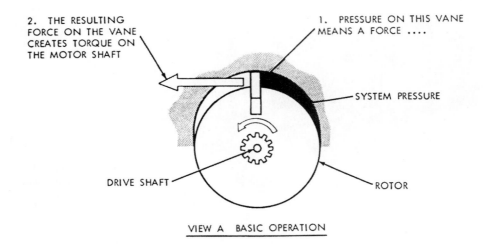

2. THE RESULTING
FORCE ON THE VANE
CREATES TORQUE ON
THE MOTOR SHAFT

1. PRESSURE ON THIS VANE
MEANS A FORCE ....

SYSTEM PRESSURE

DRIVE SHAFT

ROTOR

VIEW A   BASIC OPERATION

1. THIS VANE IS
SUBJECT TO HIGH
PRESSURE AT THE
INLET SIDE AND LOW
PRESSURE OPPOSITE

OUTLET

ROTATION

INLET

3. THE INLET
CONNECTS TO TWO
OPPOSING PRESSURE
PASSAGES TO BALANCE
SIDE LOADS ON THE
ROTOR.

2. THE RESULTING FORCE
ON THE VANE CREATES
TORQUE ON THE ROTOR
SHAFT

VIEW B   BALANCED DESIGN

Figure 7-8. Operation of a vane motor. *(Courtesy of Sperry Vickers, Sperry Rand Corp., Troy, Michigan.)*

7.5 PISTON MOTORS

In-Line Piston Motor (Swash Plate Design)

Piston motors can be either fixed or variable displacement units. They generate torque by pressure acting on the ends of pistons reciprocating inside a cylinder block. Figure 7-11 illustrates the in-line design in which the motor driveshaft and

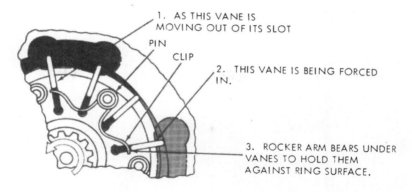

1. AS THIS VANE IS MOVING OUT OF ITS SLOT

PIN

CLIP

2. THIS VANE IS BEING FORCED IN.

3. ROCKER ARM BEARS UNDER VANES TO HOLD THEM AGAINST RING SURFACE.

NOTE: ONLY ONE ROCKER SPRING IS SHOWN FOR SIMPLICITY.

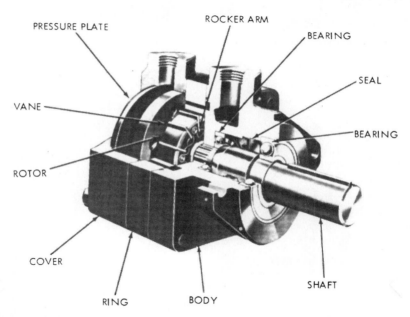

PRESSURE PLATE

ROCKER ARM

BEARING

SEAL

VANE

BEARING

ROTOR

COVER

SHAFT

RING

BODY

Figure 7-9. Vane motors with spring-loaded vanes. *(Courtesy of Sperry Vickers, Sperry Rand Corp, Troy, Michigan.)*

cylinder block are centered on the same axis. Pressure acting on the ends of the pistons generates a force against an angled swash plate. This causes the cylinder block to rotate with a torque that is proportional to the area of the pistons. The torque is also a function of the swash plate angle. The in-line piston motor is designed either as a fixed or variable displacement unit (see Figure 7-12). As illustrated in Figure 7-13, the swash plate angle determines the volumetric displacement.

In variable displacement units, the swash plate is mounted in a swinging yoke. The angle of the swash plate can be altered by various means, such as a lever,

(a) Paddle wheel–driven riverboat.

(b) Hydraulic vane motor.

Figure 7-10. Riverboat with hydraulic motor–driven paddle wheel. *(Courtesy of Rineer, Hydraulics, Inc., San Antonio, Texas.)*

handwheel, or servo control. If the swash plate angle is increased, the torque capacity is increased, but the driveshaft speed is decreased. Mechanical stops are usually incorporated so that the torque and speed capacities stay within prescribed limits.

Axial Piston Motor (Bent-Axis Design)

A bent-axis piston motor is illustrated in Figure 7-14. This type of motor also develops torque due to pressure acting on reciprocating pistons. This design, however,

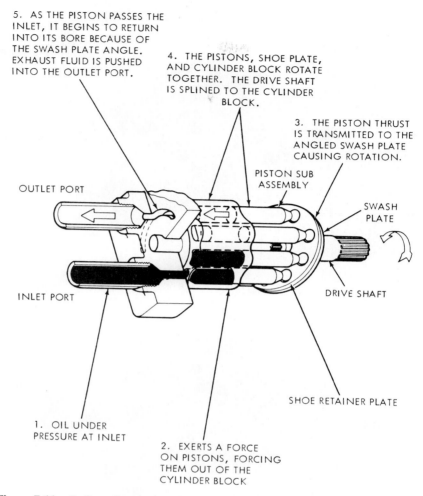

5. AS THE PISTON PASSES THE INLET, IT BEGINS TO RETURN INTO ITS BORE BECAUSE OF THE SWASH PLATE ANGLE. EXHAUST FLUID IS PUSHED INTO THE OUTLET PORT.

4. THE PISTONS, SHOE PLATE, AND CYLINDER BLOCK ROTATE TOGETHER. THE DRIVE SHAFT IS SPLINED TO THE CYLINDER BLOCK.

3. THE PISTON THRUST IS TRANSMITTED TO THE ANGLED SWASH PLATE CAUSING ROTATION.

PISTON SUB ASSEMBLY

OUTLET PORT

SWASH PLATE

DRIVE SHAFT

INLET PORT

SHOE RETAINER PLATE

1. OIL UNDER PRESSURE AT INLET

2. EXERTS A FORCE ON PISTONS, FORCING THEM OUT OF THE CYLINDER BLOCK

Figure 7-11. In-line piston motor operation. *(Courtesy of Sperry Vickers, Sperry Rand Corp., Troy, Michigan.)*

has the cylinder block and driveshaft mounted at an angle to each other so that the force is exerted on the driveshaft flange.

Speed and torque depend on the angle between the cylinder block and driveshaft. The larger the angle, the greater the displacement and torque but the smaller the speed. This angle varies from a minimum of $7\frac{1}{2}°$ to a maximum of $30°$. Figure 7-15 shows a fixed displacement bent-axis motor, whereas Figure 7-16 illustrates the variable displacement design in which the displacement is varied by a handwheel.

Piston motors are the most efficient of the three basic types and are capable of operating at the highest speeds and pressures. Operating speeds of 12,000 rpm and

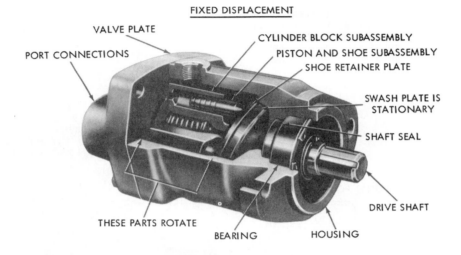

FIXED DISPLACEMENT

VALVE PLATE

PORT CONNECTIONS

CYLINDER BLOCK SUBASSEMBLY

PISTON AND SHOE SUBASSEMBLY

SHOE RETAINER PLATE

SWASH PLATE IS STATIONARY

SHAFT SEAL

DRIVE SHAFT

THESE PARTS ROTATE

BEARING

HOUSING

VARIABLE DISPLACEMENT

COMPENSATOR

CONTROL PISTON

SWASH PLATE IS MOUNTED IN SWINGING YOKE

SPRING

SHAFT SEAL

BEARING

PORT CONNECTIONS

VALVE PLATE

DRIVE SHAFT

ROTATING GROUP

PINTLE

HOUSING

YOKE

Figure 7-12. Two configurations of in-line piston motors. *(Courtesy of Sperry Vickers, Sperry Rand Corp., Troy, Michigan.)*

pressures of 5000 psi can be obtained with piston motors. Large piston motors are capable of delivering flows up to 450 gpm.

7.6 HYDRAULIC MOTOR THEORETICAL TORQUE, POWER, AND FLOW RATE

Due to frictional losses, a hydraulic motor delivers less torque than it should theoretically. The theoretical torque (which is the torque that a frictionless hydraulic

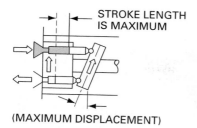

(MAXIMUM DISPLACEMENT)

MAXIMUM SWASH
PLATE ANGLE
AND
MAXIMUM TORQUE
CAPABILITY

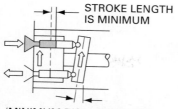

(MINIMUM DISPLACEMENT)

MINIMUM SWASH
PLATE ANGLE
AND
MINIMUM TORQUE
CAPABILITY

Figure 7-13. Motor displacement varies with swash plate angle. *(Courtesy of Sperry Vickers, Sperry Rand Corp., Troy, Michigan.)*

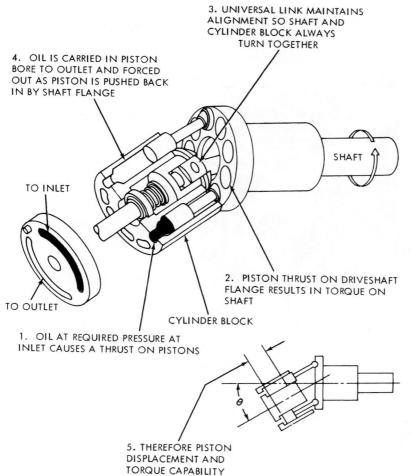

3. UNIVERSAL LINK MAINTAINS
ALIGNMENT SO SHAFT AND
CYLINDER BLOCK ALWAYS
TURN TOGETHER

4. OIL IS CARRIED IN PISTON
BORE TO OUTLET AND FORCED
OUT AS PISTON IS PUSHED BACK
IN BY SHAFT FLANGE

SHAFT

TO INLET

TO OUTLET

2. PISTON THRUST ON DRIVESHAFT
FLANGE RESULTS IN TORQUE ON
SHAFT

CYLINDER BLOCK

1. OIL AT REQUIRED PRESSURE AT
INLET CAUSES A THRUST ON PISTONS

θ

5. THEREFORE PISTON
DISPLACEMENT AND
TORQUE CAPABILITY
DEPEND ON ANGLE

Figure 7-14. Bent-axis piston motor. *(Courtesy of Sperry Vickers, Sperry Rand Corp., Troy, Michigan.)*

Figure 7-15. Fixed displacement bent-axis piston motor. *(Courtesy of Sperry Vickers, Sperry Rand Corp., Troy, Michigan.)*

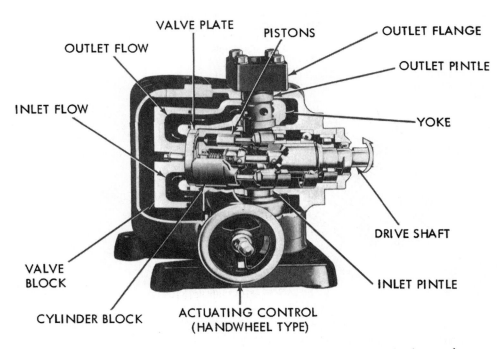

Figure 7-16. Variable displacement bent-axis piston motor with handwheel control. *(Courtesy of Sperry Vickers, Sperry Rand Corp., Troy, Michigan.)*

motor would deliver) can be determined by the following equation developed for limited rotation hydraulic actuators in Section 7.2:

$$T_T(\text{in} \cdot \text{lb}) = \frac{V_D(\text{in}^3/\text{rev}) \times p(\text{psi})}{2\pi} \tag{7-4}$$

Using metric units, we have

$$T_T(\text{N} \cdot \text{m}) = \frac{V_D(\text{m}^3/\text{rev}) \times p(\text{Pa})}{2\pi} \qquad \textbf{(7-4M)}$$

Thus, the theoretical torque is proportional not only to the pressure but also to the volumetric displacement.

The theoretical horsepower (which is the horsepower a frictionless hydraulic motor would develop) can also be mathematically expressed:

$$HP_T = \frac{T_T(\text{in} \cdot \text{lb}) \times N(\text{rpm})}{63{,}000}$$

$$= \frac{V_D(\text{in}^3/\text{rev}) \times p(\text{psi}) \times N(\text{rpm})}{395{,}000} \qquad \textbf{(7-5)}$$

In metric units,

$$\text{theoretical power (W)} = T_T(\text{N} \cdot \text{m}) \times N(\text{rad/s})$$

$$= \frac{V_D(\text{m}^3/\text{rev}) \times p(\text{Pa}) \times N(\text{rad/s})}{2\pi} \qquad \textbf{(7-5M)}$$

Also, due to leakage, a hydraulic motor consumes more flow rate than it should theoretically. The theoretical flow rate is the flow rate a hydraulic motor would consume if there were no leakage. As is the case for pumps, the following equation gives the relationship among speed, volumetric displacement, and theoretical flow rate.

$$Q_T(\text{gpm}) = \frac{V_D(\text{in}^3/\text{rev}) \times N(\text{rpm})}{231} \qquad \textbf{(7-6)}$$

or

$$Q_T(\text{m}^3/\text{s}) = V_D(\text{m}^3/\text{rev}) \times N(\text{rev/s}) \qquad \textbf{(7-6M)}$$

The use of Eqs. (7-4), (7-5), and (7-6) can be illustrated by an example.

EXAMPLE 7-2

A hydraulic motor has a 5-in³ volumetric displacement. If it has a pressure rating of 1000 psi and it receives oil from a 10-gpm theoretical flow-rate pump, find the motor

- **a.** Speed
- **b.** Theoretical torque
- **c.** Theoretical horsepower

Solution

a. From Eq. (7-6) we solve for motor speed:

$$N = \frac{231Q_T}{V_D} = \frac{(231)(10)}{5} = 462 \text{ rpm}$$

b. Theoretical torque is found using Eq. (7-4):

$$T_T = \frac{V_D p}{2\pi} = \frac{(5)(1000)}{2\pi} = 795 \text{ in} \cdot \text{lb}$$

c. Theoretical horsepower is obtained from Eq. (7-5):

$$\text{HP}_T = \frac{T_T N}{63,000} = \frac{(795)(462)}{63,000} = 5.83 \text{ HP}$$

7.7 HYDRAULIC MOTOR PERFORMANCE

Introduction

The performance of any hydraulic motor depends on the precision of its manufacture as well as the maintenance of close tolerances under design operating conditions. As in the case for pumps, internal leakage (slippage) between the inlet and outlet reduces the volumetric efficiency of a hydraulic motor. Similarly, friction between mating parts and due to fluid turbulence reduces the mechanical efficiency of a hydraulic motor.

Gear motors typically have an overall efficiency of 70 to 75% as compared to 75 to 85% for vane motors and 85 to 95% for piston motors.

Some systems require that a hydraulic motor start under load. Such systems should include a stall-torque factor when making design calculations. For example, only about 80% of the maximum torque can be expected if the motor is required to start either under load or operate at speeds below 500 rpm.

Motor Efficiencies

Hydraulic motor performance is evaluated on the same three efficiencies (volumetric, mechanical, and overall) used for hydraulic pumps. They are defined for motors as follows:

1. **Volumetric efficiency (η_v).** The volumetric efficiency of a hydraulic motor is the inverse of that for a pump. This is because a pump does not produce as much flow as it should theoretically, whereas a motor uses more flow than it should theoretically due to leakage. Thus, we have

$$\eta_v = \frac{\text{theoretical flow-rate motor should consume}}{\text{actual flow-rate consumed by motor}} = \frac{Q_T}{Q_A} \qquad \textbf{(7-7)}$$

Determination of volumetric efficiency requires the calculation of the theoretical flow rate, which is defined for a motor in Section 7.6 by Eqs. (7-6) and (7-6M). Substituting values of the calculated theoretical flow rate and the actual flow rate (which is measured) into Eq. (7-7) yields the volumetric efficiency for a given motor.

2. **Mechanical efficiency (η_m).** The mechanical efficiency of a hydraulic motor is the inverse of that for a pump. This is because due to friction, a pump requires a greater torque than it should theoretically whereas a motor produces less torque than it should theoretically. Thus, we have

$$\eta_m = \frac{\text{actual torque delivered by motor}}{\text{torque motor should theoretically deliver}} = \frac{T_A}{T_T} \qquad \textbf{(7-8)}$$

Equations (7-9) and (7-10) allow for the calculation of T_T and T_A, respectively:

$$T_T(\text{in} \cdot \text{lb}) = \frac{V_D(\text{in}^3) \times p(\text{psi})}{2\pi} \qquad \textbf{(7-9)}$$

or

$$T_T(\text{N} \cdot \text{m}) = \frac{V_D(\text{m}^3/\text{rev}) \times p(\text{Pa})}{2\pi} \qquad \textbf{(7-9M)}$$

$$T_A(\text{in} \cdot \text{lb}) = \frac{\text{actual HP delivered by motor} \times 63{,}000}{N(\text{rpm})} \qquad \textbf{(7-10)}$$

or

$$T_A(\text{N} \cdot \text{m}) = \frac{\text{actual wattage delivered by motor}}{N(\text{rad/s})} \qquad \textbf{(7-10M)}$$

3. **Overall efficiency (η_o).** As in the case for pumps, the overall efficiency of a hydraulic motor equals the product of the volumetric and mechanical efficiencies.

$$\eta_o = \eta_v \eta_m \qquad \textbf{(7-11)}$$

$$= \frac{\text{actual power delivered by motor}}{\text{actual power delivered to motor}}$$

In English units, we have

$$\eta_o = \frac{\dfrac{T_A(\text{in} \cdot \text{lb}) \times N(\text{rpm})}{63{,}000}}{\dfrac{p(\text{psi}) \times Q_A(\text{gpm})}{1714}} \qquad \textbf{(7-12)}$$

In metric units, we have

$$\eta_o = \frac{T_A(\text{N} \cdot \text{m}) \times N(\text{rad/s})}{p(\text{Pa}) \times Q_A(\text{m}^3/\text{s})}$$ (7-12M)

Note that the actual power delivered to a motor by the fluid is called *hydraulic power* and the actual power delivered to a load by a motor via a rotating shaft is called *brake power*.

Figure 7-17 represents typical performance curves obtained for a 6-in^3 variable displacement motor operating at full displacement. The upper graph gives curves of overall and volumetric efficiencies as a function of motor speed (rpm) for pressure levels of 3000 and 5000 psi. The lower graph gives curves of motor input flow (gpm) and motor output torque (in · lb) as a function of motor speed (rpm) for the same two pressure levels.

EXAMPLE 7-3

A hydraulic motor has a displacement of 10 in^3 and operates with a pressure of 1000 psi and a speed of 2000 rpm. If the actual flow rate consumed by the motor is 95 gpm and the actual torque delivered by the motor is 1500 in · lb, find

a. η_v

b. η_m

c. η_o

d. The actual horsepower delivered by the motor

Solution

a. To find η_v, we first calculate the theoretical flow rate:

$$Q_T = \frac{V_D N}{231} = \frac{(10)(2000)}{231} = 86.6 \text{ gpm}$$

$$\eta_v = \frac{Q_T}{Q_A} = \frac{86.6}{95} = 0.911 = 91.1\%$$

b. To find η_m, we need to calculate the theoretical torque:

$$T_T = \frac{V_D p}{2\pi} = \frac{(10)(1000)}{2\pi} = 1592 \text{ in} \cdot \text{lb}$$

$$\eta_m = \frac{T_A}{T_T} = \frac{1500}{1592} = 0.942 = 94.2\%$$

c.

$$\eta_o = \eta_v \eta_m = 0.911 \times 0.942 = 0.858 = 85.8\%$$

d.

$$HP_A = \frac{T_A N}{63,000} = \frac{(1500)(2000)}{63,000} = 47.6 \text{ hp}$$

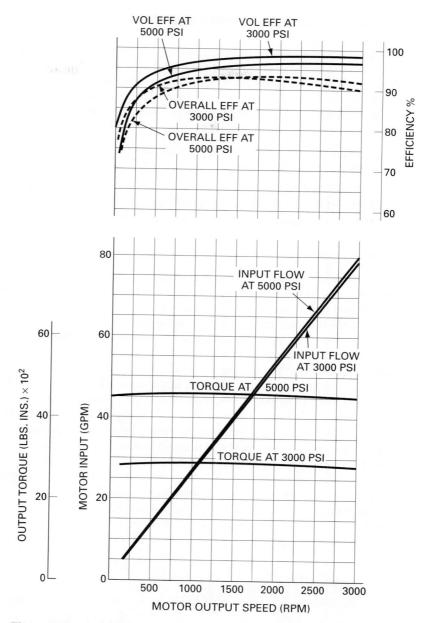

Figure 7-17. Performance curves for 6-in³ variable displacement motor. *(Courtesy of Abex Corp., Denison Division, Columbus, Ohio.)*

7.8 HYDROSTATIC TRANSMISSIONS

A system consisting of a hydraulic pump, a hydraulic motor, and appropriate valves and pipes can be used to provide adjustable-speed drives for many practical applications. Such a system is called a *hydrostatic transmission*. There must, of course, be a prime mover such as an electric motor or gasoline engine. Applications in existence include tractors, rollers, front-end loaders, hoes, and lift trucks. Some of the advantages of hydrostatic transmissions are the following:

1. Infinitely variable speed and torque in either direction and over the full speed and torque ranges
2. Extremely high power-to-weight ratio
3. Ability to be stalled without damage
4. Low inertia of rotating members, permitting fast starting and stopping with smoothness and precision
5. Flexibility and simplicity of design

Figure 7-18 shows a heavy-duty hydrostatic transmission system which uses a variable displacement piston pump and a fixed displacement piston motor. Both pump and motor are of the swash plate in-line piston design. This type of hydrostatic transmission is expressly designed for application in the agricultural, construction, materials handling, garden tractor, recreational vehicle, and industrial markets.

For the transmission of Figure 7-18, the operator has complete control of the system, with one lever for starting, stopping, forward motion, or reverse motion. Control of the variable displacement pump is the key to controlling the vehicle. Prime mover power is transmitted to the pump. When the operator moves the control lever, the swash plate in the pump is tilted from neutral. When the pump swash plate is tilted, a positive stroke of the pistons occurs. This, in turn, at any given input speed, produces a certain flow from the pump. This flow is transferred through high-pressure lines to the motor. The ratio of the volume of flow from the pump to the displacement of the motor determines the speed at which the motor will run. Moving the control lever to the opposite side of neutral causes the flow through the pump to reverse its direction. This reverses the direction of rotation of the motor. Speed of the output shaft is controlled by adjusting the displacement (flow) of the pump. Load (working pressure) is determined by the external conditions (grade, ground conditions, etc.), and this establishes the demand on the system.

EXAMPLE 7-4

A hydrostatic transmission, operating at 1000-psi pressure, has the following characteristics:

Pump	Motor
$V_D = 5 \text{ in}^3$	$V_D = ?$
$\eta_v = 82\%$	$\eta_v = 92\%$
$\eta_m = 88\%$	$\eta_m = 90\%$
$N = 500 \text{ rpm}$	$N = 400 \text{ rpm}$

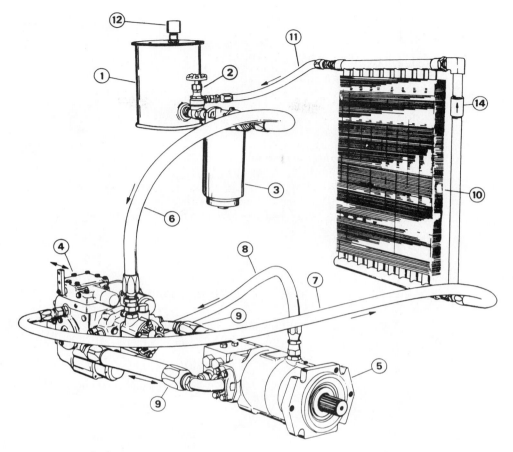

1. Reservoir
2. Shut-off Valve
3. Filter
4. Variable Displacement
   Pump
5. Fixed Displacement Motor
6. Inlet Line
7. Pump Case Drain Line
8. Motor Case Drain Line
9. High Pressure Lines
10. Heat Exchanger
11. Reservoir Return Line
12. Reservoir Fill Cap or
    Breather
14. Heat Exchanger
    By-pass Valve

Figure 7-18. Hydrostatic transmission system. *(Courtesy of Sundstrand Hydro-Transmission Division, Sundstrand Corp., Ames, Iowa.)*

Find the

    a. Displacement of the motor

    b. Motor output torque

Solution

a. $\text{pump theoretical flow rate} = \dfrac{\text{displacement of pump} \times \text{pump speed}}{231}$

$$= \frac{(5)(500)}{231} = 10.8 \text{ gpm}$$

$$\text{pump actual flow rate} = \text{pump theoretical flow rate}$$
$$\times \text{ pump volumetric efficiency}$$
$$= (10.8)(0.82) = 8.86 \text{ gpm}$$

Since the motor actual flow rate equals the pump actual flow rate we have

$$\text{motor theoretical flow rate} = \text{pump actual flow rate}$$
$$\times \text{ motor volumetric efficiency}$$
$$= (8.86)(0.92) = 8.15 \text{ gpm}$$

$$\text{motor displacement} = \frac{\text{motor theoretical flow rate} \times 231}{\text{motor speed}}$$

$$= \frac{(8.15)(231)}{400} = 4.71 \text{ in}^3$$

b. hydraulic HP delivered to motor =

$$\frac{\text{system pressure} \times \text{actual flow rate to motor}}{1714}$$

$$= \frac{(1000)(8.86)}{1714} = 5.17 \text{ hp}$$

$$\text{brake HP delivered by motor} = 5.17 \times 0.92 \times 0.90 = 4.28$$

$$\text{torque delivered by motor} = \frac{\text{HP delivered by motor} \times 63,000}{\text{motor speed}}$$

$$= \frac{4.28 \times 63,000}{400} = 674 \text{ in} \cdot \text{lb}$$

7.9 HYDRAULIC MOTOR PERFORMANCE IN METRIC UNITS

As in the case for pumps, performance data for hydraulic motors are measured and specified in metric units as well as English units. Examples 7-5, 7-6, and 7-7 show how to analyze the performance of hydraulic motors and hydrostatic transmissions using metric units.

EXAMPLE 7-5

A hydraulic motor has a 82-cm³ (0.082-L) volumetric displacement. If it has a pressure rating of 70 bars and it receives oil from a 0.0006-m³/s (0.60-Lps or 36.0-Lpm) theoretical flow-rate pump, find the motor

 a. Speed
 b. Theoretical torque
 c. Theoretical power

Solution

 a. From Eq. (7-6M) we solve for the motor speed:

$$N = \frac{Q_T}{V_D} = \frac{0.0006 \text{ m}^3/\text{s}}{0.000082 \text{ m}^3/\text{rev}} = 7.32 \text{ rev/s} = 439 \text{ rpm}$$

 b. Theoretical torque is found using Eq. (7-4M):

$$T_T = \frac{V_D p}{2\pi} = \frac{(0.000082 \text{ m}^3)(70 \times 10^5 \text{ N/m}^2)}{2\pi} = 91.4 \text{ N} \cdot \text{m}$$

 c. Theoretical power is obtained as follows:

$$\text{theoretical power} = T_T N = (91.4 \text{ N} \cdot \text{m})(7.32 \times 2\pi \text{ rad/s})$$
$$= 4200 \text{ W} = 4.20 \text{ kW}$$

EXAMPLE 7-6

A hydraulic motor has a displacement of 164 cm³ and operates with a pressure of 70 bars and a speed of 2000 rpm. If the actual flow rate consumed by the motor is 0.006 m³/s and the actual torque delivered by the motor is 170 N · m, find

 a. η_v
 b. η_m
 c. η_o
 d. The actual kW delivered by the motor

Solution

a. To find the volumetric efficiency, we first calculate the theoretical flow rate:

$$Q_T = V_D N = (0.000164 \text{ m}^3/\text{rev})\left(\frac{2000}{60} \text{ rev/s}\right) = 0.00547 \text{ m}^3/\text{s}$$

$$\eta_v = \frac{Q_T}{Q_A} = \frac{0.00547}{0.006} = 0.912 = 91.2\%$$

b. To find η_m, we need to calculate the theoretical torque:

$$T_T = \frac{V_D p}{2\pi} = \frac{(0.000164)(70 \times 10^5)}{2\pi} = 182.8 \text{ N} \cdot \text{m}$$

$$\eta_m = \frac{T_A}{T_T} = \frac{170}{182.8} = 0.930 = 93.0\%$$

c. $\eta_o = \eta_v \eta_m = 0.912 \times 0.930 = 0.848 = 84.8\%$

d. $\text{actual power} = T_A N = (170)\left(2000 \times \frac{2\pi}{60}\right) = 35{,}600 \text{ W} = 35.6 \text{ kW}$

EXAMPLE 7-7

A hydrostatic transmission, operating at 70 bars pressure, has the following characteristics:

Pump	*Motor*
$V_D = 82 \text{ cm}^3$	$V_D = ?$
$\eta_v = 82\%$	$\eta_v = 92\%$
$\eta_m = 88\%$	$\eta_m = 90\%$
$N = 500 \text{ rpm}$	$N = 400 \text{ rpm}$

Find the

a. Displacement of the motor
b. Motor output torque

Solution

a. pump theoretical flow rate = displacement of pump × pump speed

$$= (0.000082)\left(\frac{500}{60}\right) = 0.000683 \text{ m}^3/\text{s}$$

pump actual flow rate = pump theoretical flow rate × pump volumetric efficiency

$$= (0.000683)(0.82) = 0.000560 \text{ m}^3/\text{s}$$

Since the motor actual flow rate equals the pump actual flow rate, we have

motor theoretical flow rate = pump actual flow rate × motor volumetric efficiency

$$= (0.000560)(0.92) = 0.000515 \text{ m}^3/\text{s}$$

$$\text{Motor displacement} = \frac{\text{motor theoretical flow rate}}{\text{motor speed}}$$

$$= \frac{0.000515}{400/60} = 0.0000773 \text{ m}^3 = 77.3 \text{ cm}^3$$

b. hydraulic power delivered to motor = system pressure
$$\times \text{ actual flow rate to motor}$$

$$= (70 \times 10^5)(0.000560) = 3920 \text{ W}$$

brake power delivered by motor $= (3920)(0.92)(0.90) = 3246 \text{ W}$

$$\text{torque delivered by motor} = \frac{\text{power delivered by motor}}{\text{motor speed}}$$

$$= \frac{3246}{400 \times 2\pi/60} = 77.5 \text{ N} \cdot \text{m}$$

7.10 KEY EQUATIONS

Theoretical torque of a hydraulic motor or rotary actuator (torque that a frictionless hydraulic motor or rotary actuator would deliver):

English units:
$$T_T(\text{in} \cdot \text{lb}) = \frac{V_D(\text{in}^3/\text{rev}) \times p(\text{psi})}{2\pi} \qquad \textbf{(7-4 and 7-9)}$$

Metric units:
$$T_T(\text{N} \cdot \text{m}) = \frac{V_D(\text{m}^3/\text{rev}) \times p(\text{Pa})}{2\pi} \qquad \textbf{(7-4M and 7-9M)}$$

Theoretical power of a hydraulic motor (power that a frictionless hydraulic motor would deliver):

English units:
$$HP_T = \frac{T_T(\text{in} \cdot \text{lb}) \times N(\text{rpm})}{63,000} \qquad \textbf{(7-5)}$$

$$= \frac{V_D(\text{in}^3/\text{rev}) \times p(\text{psi}) \times N(\text{rpm})}{395,000}$$

Metric units:

$$\text{Theoretical power (W)} = T_T(\text{N} \cdot \text{m}) \times N(\text{rad/s}) \qquad \textbf{(7-5M)}$$

$$= \frac{V_D(\text{m}^3/\text{rev}) \times p(\text{Pa}) \times N(\text{rad/s})}{2\pi}$$

Theoretical flow rate of a hydraulic motor (flow rate that a no-leakage hydraulic motor would consume)

Special
English units:

$$Q_T \,(\text{gpm}) = \frac{V_D(\text{in}^3/\text{rev}) \times N(\text{rpm})}{231} \qquad \textbf{(7-6)}$$

Metric units:

$$Q_T \,(\text{m}^3/\text{s}) = V_D(\text{m}^3/\text{rev}) \times N(\text{rev/s}) \qquad \textbf{(7-6M)}$$

Volumetric
efficiency of a
hydraulic motor:

$$\eta_v = \frac{\text{theoretical flow rate motor should consume}}{\text{actual flow rate consumed by motor}} = \frac{Q_T}{Q_A} \qquad \textbf{(7-7)}$$

Mechanical
efficiency of a
hydraulic
motor:

$$\eta_m = \frac{\text{actual torque delivered by motor}}{\text{torque motor should theoretically deliver}} = \frac{T_A}{T_T} \qquad \textbf{(7-8)}$$

Actual torque delivered by a hydraulic motor

English units:

$$T_A(\text{in} \cdot \text{lb}) = \frac{\text{actual HP delivered by motor} \times 63{,}000}{N(\text{rpm})} \qquad \textbf{(7-10)}$$

Metric units:

$$T_A(\text{N} \cdot \text{m}) = \frac{\text{actual wattage delivered by motor}}{N(\text{rad/s})} \qquad \textbf{(7-10M)}$$

Overall efficiency of a hydraulic motor

No units
involved:

$$\eta_o = \eta_v \eta_m \qquad \textbf{(7-11)}$$

Definition:

$$\eta_o = \frac{\text{actual power delivered by motor}}{\text{actual power delivered to motor}} \qquad \textbf{(7-12)}$$

Special
English units:
$$\eta_o = \frac{T_A(\text{in} \cdot \text{lb}) \times N(\text{rpm})/63,000}{p \ (\text{psi}) \times Q_A(\text{gpm})/1714}$$

(7-12)

Metric units:
$$\eta_o = \frac{T_A(\text{N} \cdot \text{m}) \times N(\text{rad/s})}{p(\text{Pa}) \times Q_A(\text{m}^3/\text{s})}$$

(7-12M)

EXERCISES

Questions, Concepts, and Definitions

7-1. What is a limited rotation hydraulic motor? How does it differ from a hydraulic motor?

7-2. What are the main advantages of gear motors?

7-3. Why are vane motors fixed displacement units?

7-4. Name one way in which vane motors differ from vane pumps.

7-5. Can a piston pump be used as a piston motor?

7-6. For a hydraulic motor, define *volumetric, mechanical*, and *overall efficiency*.

7-7. Why does a hydraulic motor use more flow than it should theoretically?

7-8. What is a hydrostatic transmission? Name four advantages it typically possesses.

7-9. Why does a hydraulic motor deliver less torque than it should theoretically?

7-10. Explain why, theoretically, the torque output from a fixed displacement hydraulic motor operating at constant pressure is the same regardless of changes in speed.

7-11. The torque output from a fixed displacement hydraulic motor operating at constant pressure is the same regardless of changes in speed. True or false? Explain your answer.

7-12. What determines the speed of a hydraulic motor?

7-13. Define the displacement and torque ratings of a hydraulic motor.

7-14. Explain how the vanes are held in contact with the cam ring in high-performance vane motors.

7-15. How is torque developed in an in-line-type piston motor?

7-16. If a hydraulic motor is pressure compensated, what is the effect of an increase in the working load?

7-17. What type of hydraulic motor is generally most efficient?

7-18. Knowing the displacement and speed of a hydraulic motor, how do you calculate the gpm flowing through it?

Problems

Note: The letter *E* following an exercise number means that English units are used. Similarly, the letter *M* indicates metric units.

Limited Rotation Hydraulic Motors

7-19E. A rotary actuator has the following physical data:

outer radius of rotor = 0.4 in
outer radius of vane = 1.25 in
width of vane = 0.75 in

If the torque load is 750 in · lb, what pressure must be developed to overcome the load?

7-20M. A rotary actuator has the following physical data:

$$\text{outer radius of rotor} = 10 \text{ mm}$$
$$\text{outer radius of vane} = 32 \text{ mm}$$
$$\text{width of vane} = 20 \text{ mm}$$

If the torque load is 85 N · m, what pressure must be developed to overcome the load?

Hydraulic Motor Torque, Power, and Flow Rate

7-21E. A hydraulic motor receives a flow rate of 20 gpm at a pressure of 1800 psi. If the motor speed is 800 rpm, determine the actual torque delivered by the motor assuming it is 100% efficient.

7-22M. A hydraulic motor receives a flow rate of 72 Lpm at a pressure of 12,000 kPa. If the motor speed is 800 rpm, determine the actual torque delivered by the motor assuming it is 100% efficient.

7-23E. A hydraulic motor has a 6-in³ volumetric displacement. If it has a pressure rating of 2000 psi and receives oil from a 15-gpm theoretical flow-rate pump, find the motor

   a. Speed

   b. Theoretical torque

   c. Theoretical horsepower

7-24M. A hydraulic motor has a 100-cm³ volumetric displacement. If it has a pressure rating of 140 bars and receives oil from a 0.001-m³/s theoretical flow-rate pump, find the motor

   a. Speed

   b. Theoretical torque

   c. Theoretical kW power

7-25E. The pressure rating of the components in a hydraulic system is 1000 psi. The system contains a hydraulic motor to turn a 10-in-radius drum at 30 rpm to lift a 1000-lb weight W, as shown in Figure 7-19. Determine the flow rate in units of gpm and the output horsepower of the 100% efficient motor.

7-26M. The system in Exercise 7-25, as shown in Figure 7-19, has the following data using metric units:

$$\text{pressure} = 1 \times 10^5 \text{ kPa}$$
$$\text{drum radius} = 0.3 \text{ m}$$
$$\text{motor speed} = 30 \text{ rpm}$$
$$\text{weight of load} = 4000 \text{ N}$$

Determine the flow rate in units of m³/s and the output power of the 100% efficient motor in kW.

7-27E. A hydraulic system contains a pump that discharges oil at 2000 psi and 100 gpm to a hydraulic motor, as shown in Figure 7-20. The pressure at the motor inlet is 1800 psi due to a pressure drop in the line. If oil leaves the motor at 200 psi, determine the output power delivered by the 100% efficient motor.

7-28M. Change the data in Exercise 7-27 to metric units and solve for the output power delivered by the 100% efficient motor.

7-29E. If the pipeline between the pump and motor in Exercise 7-27 is horizontal and of constant diameter, what is the cause of the 200-psi pressure drop?

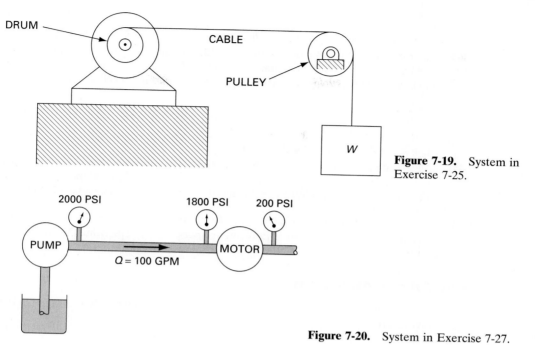

Figure 7-19. System in Exercise 7-25.

Figure 7-20. System in Exercise 7-27.

7-30M. Change the data in Exercise 7-29 to metric units and determine the cause of the pressure drop in the pipeline from the pump outlet to the motor inlet.

7-31E. What torque would a hydraulic motor deliver at a speed of 1750 rpm if it produces 4 BHP (brake horsepower)?

7-32E. In Exercise 7-31, the pressure remains constant at 2000 psi.

   a. What effect would doubling the speed have on the torque?

   b. What effect would halving the speed have on the torque?

Hydraulic Motor Efficiencies

7-33E. A hydraulic motor receives a flow rate of 20 gpm at a pressure of 1800 psi. The motor speed is 800 rpm. If the motor has a power loss of 4 HP, find the motor

   a. Actual output torque

   b. Overall efficiency

7-34M. A hydraulic motor receives a flow rate of 72 Lpm at a pressure of 12,000 kPa. The motor speed is 800 rpm. If the motor has a power loss of 3 kW, find the motor

   a. Actual output torque

   b. Overall efficiency

7-35E. A hydraulic motor has a displacement of 8 in³ and operates with a pressure of 1500 psi and a speed of 2000 rpm. If the actual flow rate consumed by the motor is 75 gpm and the actual torque delivered by the motor is 1800 in · lb, find

   a. η_v

   b. η_m

      c. η_o

      d. The horsepower delivered by the motor

7-36M. A hydraulic motor has a displacement of 130 cm^3 and operates with a pressure of 105 bars and a speed of 2000 rpm. If the actual flow rate consumed by the motor is 0.005 m^3/s and the actual torque delivered by the motor is 200 N · m, find

      a. η_v

      b. η_m

      c. η_o

      d. kW power delivered by the motor

7-37E. A hydraulic motor has volumetric efficiency of 90% and operates at a speed of 1750 rpm and a pressure of 1000 psi. If the actual flow rate consumed by the motor is 75 gpm and the actual torque delivered by the motor is 1300 in · lb, find the overall efficiency of the motor.

7-38E. A gear motor has an overall efficiency of 84% at a pressure drop of 3000 psi across its ports and when the ratio of flow rate to speed is 0.075 gpm/rpm. Determine the torque and displacement of the motor.

Hydrostatic Transmissions

7-39E. A hydrostatic transmission operating at 1500-psi pressure has the following characteristics:

Pump	*Motor*
$V_D = 6$ in^3	$V_D = ?$
$\eta_v = 85\%$	$\eta_v = 94\%$
$\eta_m = 90\%$	$\eta_m = 92\%$
$N = 1000$ rpm	$N = 600$ rpm

      Find the

      a. Displacement of the motor

      b. Motor output torque

7-40M. A hydrostatic transmission operating at 105-bars pressure has the following characteristics:

Pump	*Motor*
$V_D = 100$ in^3	$V_D = ?$
$\eta_v = 85\%$	$\eta_v = 94\%$
$\eta_m = 90\%$	$\eta_m = 92\%$
$N = 1000$ rpm	$N = 600$ rpm

      Find the

      a. Displacement of the motor

      b. Motor output torque

8 Hydraulic Valves

Learning Objectives

Upon completing this chapter, you should be able to:

1. Describe the purpose, construction, and operation of various directional control valves.
2. Differentiate among two-way, three-way, and four-way directional control valves.
3. Identify the graphic symbols used for directional, pressure, and flow control valves.
4. Explain how valves are actuated using manual, mechanical, fluid pilot, and electric solenoid methods.
5. Describe the purpose, construction, and operation of various pressure control valves.
6. Differentiate between a pressure relief valve, a pressure-reducing valve, a sequence valve, and an unloading valve.
7. Determine the cracking pressure and full pump flow pressure of pressure relief valves.
8. Calculate the power loss in pressure relief and unloading valves.
9. Describe the purpose, construction, and operation of various flow control valves.
10. Analyze how flow control valves can control the speed of hydraulic cylinders.
11. Differentiate between a noncompensated and a compensated flow control valve.
12. Describe the purpose, construction, and operation of mechanical-hydraulic and electrohydraulic servo valves.
13. Discuss the purpose, construction, and operation of cartridge valves.

8.1 INTRODUCTION

One of the most important considerations in any fluid power system is control. If control components are not properly selected, the entire system will not function as required. Fluid power is controlled primarily through the use of control devices called *valves*. The selection of these valves involves not only the type but also the size, actuating technique, and remote-control capability. There are three basic types of valves: (1) directional control valves, (2) pressure control valves, and (3) flow control valves.

Directional control valves determine the path through which a fluid traverses a given circuit. For example, they establish the direction of motion of a hydraulic cylinder or motor. This control of the fluid path is accomplished primarily by check valves; shuttle valves; and two-way, three-way, and four-way directional control valves.

Pressure control valves protect the system against overpressure, which may occur due to excessive actuator loads or due to the closing of a valve. In general pressure control is accomplished by pressure relief, pressure reducing, sequence, unloading, and counterbalance valves.

In addition, fluid flow rate must be controlled in various lines of a hydraulic circuit. For example, the control of actuator speeds depends on flow rates. This type of control is accomplished through the use of flow control valves. Noncompensated flow control valves are used where precise speed control is not required since flow rate varies with pressure drop across a flow control valve. Pressure-compensated flow control valves automatically adjust to changes in pressure drop to produce a constant flow rate. Circuit applications of the valves discussed in this chapter are presented in Chapter 9.

In Figures 8-1 and 8-2 we see a welding machine application in which a directional control valve, a check valve, and a sequence valve are used as components of a hydraulic circuit for positioning and holding parts during a welding operation. This particular application requires a sequencing system for fast and positive holding of these parts. This is accomplished by placing a sequence valve in the line leading to the second of the two hydraulic cylinders, as illustrated in Figure 8-2. When the four-way directional control valve is actuated, the first cylinder extends to the end of its stroke to complete the "positioning" cycle. Oil pressure then builds up, overcoming the sequence valve setting. This opens the sequence valve to allow oil to flow to the second cylinder so that it can extend to complete the "hold" cycle. The check valve allows the second cylinder to retract, along with the first cylinder, when the four-way valve is shifted to allow oil to flow to the rod end of both cylinders.

8.2 DIRECTIONAL CONTROL VALVES

Introduction

As the name implies, directional control valves are used to control the direction of flow in a hydraulic circuit. Any valve (regardless of its design) contains ports that are external openings through which fluid can enter and leave via connecting

Figure 8-1. Welding machine application using hydraulic control valves. *(Courtesy of Owatonna Tool Co., Owatonna, Minnesota.)*

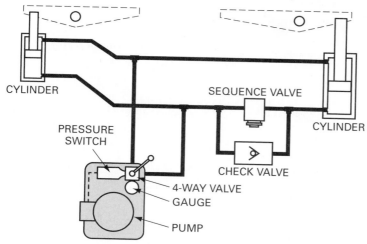

Figure 8-2. Hydraulic circuit showing control valves used for welding application. *(Courtesy of Owatonna Tool Co., Owatonna, Minnesota.)*

pipelines. The number of ports on a directional control valve (DCV) is identified using the term *way*. Thus, for example, a valve with four ports is a four-way valve.

Check Valve

The simplest type of direction control valve is a check valve (see Figure 8-3), which is a two-way valve because it contains two ports. The purpose of a check valve is to permit free flow in one direction and prevent any flow in the opposite direction.

Figure 8-4 provides two schematic drawings (one for the no-flow condition and one for the free-flow condition) showing the internal operation of a poppet check valve. A poppet is a specially shaped plug element held onto a seat (a surface surrounding the flow path opening inside the valve body) by a spring. Fluid flows through the valve in the space between the seat and poppet. As shown, a light spring holds the poppet in the closed position. In the free-flow direction, the fluid pressure overcomes the spring force at about 5 psi. If flow is attempted in the opposite direction, the fluid pressure pushes the poppet (along with the spring force) in the closed position. Therefore, no flow is permitted. The higher the pressure, the greater will be the force pushing the poppet against its seat. Thus, increased pressure will not result in any tendency to allow flow in the no-flow direction.

Figure 8-3.   Check valve. *(Courtesy of Sperry Vickers, Sperry Rand Corp., Troy, Michigan.)*

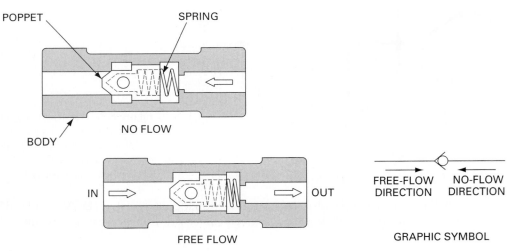

Figure 8-4.   Operation of check valve. *(Courtesy of Sperry Vickers, Sperry Rand Corp., Troy, Michigan.)*

Figure 8-4 also shows the graphic symbol of a check valve along with its no-flow and free-flow directions. Graphic symbols, which clearly show the function of hydraulic components (but without the details provided in schematic drawings), are used when drawing hydraulic circuit diagrams. Note that a check valve is analogous to a diode in electric circuits.

Pilot-Operated Check Valve

A second type of check valve is the pilot-operated check valve, shown in Figure 8-5 along with its graphic symbol. This type of check valve always permits free flow in one direction but permits flow in the normally blocked opposite direction only if pilot pressure is applied at the pilot pressure port of the valve. In the design of Figure 8-5, the check valve poppet has the pilot piston attached to the threaded poppet stem by a nut. The light spring holds the poppet seated in a no-flow condition by pushing against the pilot piston. The purpose of the separate drain port is to prevent oil from creating a pressure buildup on the bottom of the piston. The dashed line (which is part of the graphic symbol shown in Figure 8-5) represents the pilot pressure line connected to the pilot pressure port of the valve. Pilot check valves are frequently used for locking hydraulic cylinders in position.

Three-Way Valves

Three-way directional control valves, which contain three ports, are typically of the spool design rather than poppet design. A spool is a circular shaft containing lands that are large diameter sections machined to slide in a very close fitting bore of the valve body. The radial clearance between the land and bore is usually less than

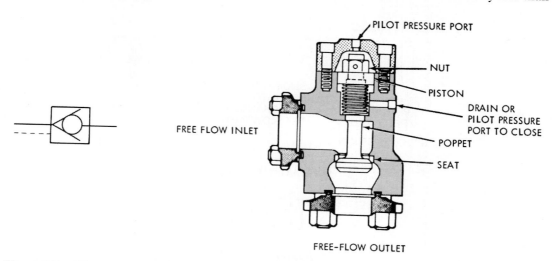

Figure 8-5. Pilot-operated check valve. *(Courtesy of Sperry Vickers, Sperry Rand Corp., Troy, Michigan.)*

0.001 in. The grooves between the lands provide the flow paths between ports. These valves are designed to operate with two or three unique positions of the spool. The spool can be positioned manually, mechanically, by using pilot pressure, or by using electrical solenoids.

Figure 8-6 shows the flow paths through a three-way valve that uses two positions of the spool. Such a valve is called a three-way, two-position directional control valve. The flow paths are shown by two schematic drawings (one for each spool position) as well as by a graphic symbol (containing two side-by-side rectangles). In discussing the operation of these valves, the rectangles are commonly called "envelopes."

The following is a description of the flow paths through the three-way valve of Figure 8-6:

Spool Position 1: Flow can go from pump port P (the port connected to the pump discharge pipe) to outlet port A as shown by the straight line and arrow in the left envelope. In this spool position, tank port T (the port connected to the pipe leading to the oil tank) is blocked.

Spool Position 2: Flow can go from port A to port T. Port P is blocked by the spool. Note that the three ports are labeled for only one of the two envelopes

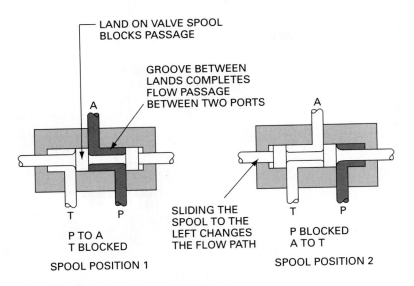

Figure 8-6. Two spool positions inside a three-way valve.

of the graphic symbol. Thus the reader must mentally identify the ports on the second envelope.

Three-way valves are typically used to control the flow directions to and from single-acting cylinders, as illustrated in Figure 8-7. As shown, the cylinder extends under hydraulic pressure (left envelope) and retracts under spring force as oil flows to the oil tank (right envelope). Observe that fluid entering the pump port of a three-way valve can be directed to only a single outlet port (in this case port *A*).

Four-Way Valves

Figure 8-8 shows the flow paths through a four-way, two-position directional control valve. Observe that fluid entering the valve at the pump port can be directed to either outlet port *A* or *B*.

The following is a description of the flow paths through this four-way valve:

Spool Position 1: Flow can go from *P* to *A* and *B* to *T*.
Spool Position 2: Flow can go from *P* to *B* and *A* to *T*.

Observe that the graphic symbol shows only one tank port *T* (for a total of four ports) even though the actual valve may have two, as shown in the schematic drawings. However, each tank port provides the same function, and thus there are only four different ports from a functional standpoint. The two internal flow-to-tank passageways can be combined inside the actual valve to provide a single tank port. Recall that the graphic symbol is concerned with only the function of a component and not its internal design.

Four-way valves are typically used to control the flow directions to and from double-acting cylinders, as shown in Figure 8-9. As shown, a four-way valve permits the cylinder to both extend (left envelope) and retract (right envelope) under hydraulic pressure.

Manually Actuated Valves

Figure 8-10 shows a cutaway of a four-way valve. Note that it is manually actuated (see hand lever). Since the spool is spring-loaded at both ends, it is a spring-centered,

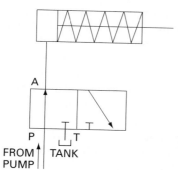

Figure 8-7. Three-way DCV controlling flow directions to and from a single-acting cylinder.

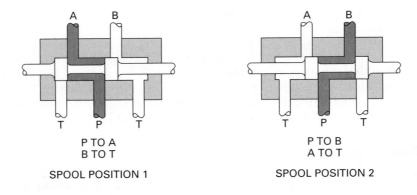

P TO A
B TO T

SPOOL POSITION 1

P TO B
A TO T

SPOOL POSITION 2

SCHEMATIC DRAWINGS

GRAPHIC SYMBOL

Figure 8-8. Two spool positions inside a four-way valve.

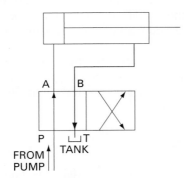

FROM PUMP

TANK

Figure 8-9. Four-way DCV controlling flow directions to and from a double-acting cylinder.

three-position directional control valve. Thus, when the valve is unactuated (no hand force on lever), the valve will assume its center position due to the balancing opposing spring forces. Figure 8-10 also provides the graphic symbol of this four-way valve.

Note in the graphic symbol that the ports are labeled on the center envelope, which represents the flow path configuration in the spring-centered position of the spool. Also observe the spring and lever actuation symbols used at the ends of the right and left envelopes. These imply a spring-centered, manually actuated valve. It should be noted that a three-position valve is used when it is necessary to stop or hold a hydraulic actuator at some intermediate position within its entire stroke range.

In Figure 8-11 we see a manually actuated, two-position, four-way valve that is spring offset. In this case the lever shifts the spool, and the spring returns the spool

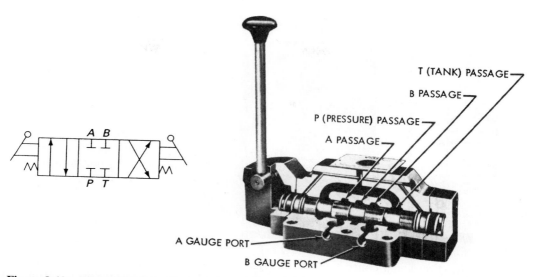

Figure 8-10. Manually actuated, spring-centered, three-position, four-way valve. *(Courtesy of Sperry Vickers, Sperry Rand Corp., Troy, Michigan.)*

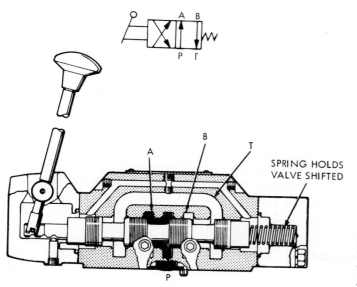

Figure 8-11. Manually actuated, two-position, spring-offset, four-way valve. *(Courtesy of Sperry Vickers, Sperry Rand Corp., Troy, Michigan.)*

to its original position when the lever is released. There are only two unique operating positions, as indicated by the graphic symbol. Note that the ports are labeled at the envelope representing the neutral (spring offset or return) or unactuated position of the spool.

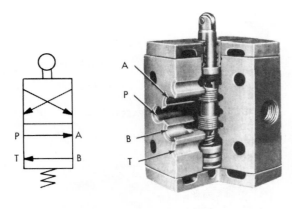

Figure 8-12. Mechanically actuated, spring-offset, two-position, four-way valve. *(Courtesy of Sperry Vickers, Sperry Rand Corp., Troy, Michigan.)*

Mechanically Actuated Valves

The directional control valves of Figures 8-10 and 8-11 are manually actuated by the use of a lever. Figure 8-12 shows a two-position, four-way, spring-offset valve that is mechanically rather than manually actuated. This is depicted in the cutaway view, with the spool end containing a roller that is typically actuated by a cam-type mechanism. Note that the graphic symbol is the same except that actuation is depicted as being mechanical (the circle represents the cam-driven roller) rather than manual.

Pilot-Actuated Valves

Directional control valves can also be shifted by applying air pressure against a piston at either end of the valve spool. Such a design is illustrated by the cutaway view of Figure 8-13. As shown, springs (located at both ends of the spool) push against centering washers to center the spool when no air is applied. When air is introduced through the left end passage, its pressure pushes against the piston to shift the spool to the right. Removal of this left end air supply and introduction of air through the right end passage causes the spool to shift to the left. Therefore, this is a four-way, three-position, spring-centered, air pilot–actuated directional control valve. In the graphic symbol in Figure 8-13, the dashed lines represent pilot pressure lines.

Solenoid-Actuated Valves

A very common way to actuate a spool valve is by using a solenoid, as illustrated in Figure 8-14. As shown, when the electric coil (solenoid) is energized, it creates a magnetic force that pulls the armature into the coil. This causes the armature to push on the push pin to move the spool of the valve.

Solenoids are actuators that are bolted to the valve housing, as shown in Figure 8-15, which gives a cutaway view of an actual solenoid-actuated directional control valve. Like mechanical or pilot actuators, solenoids work against a push pin,

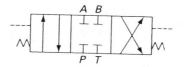

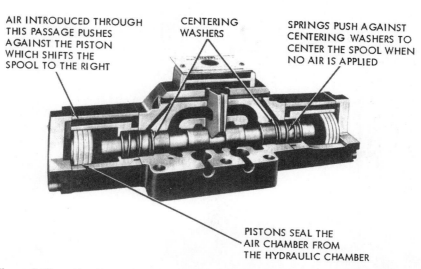

AIR INTRODUCED THROUGH
THIS PASSAGE PUSHES
AGAINST THE PISTON
WHICH SHIFTS THE
SPOOL TO THE RIGHT

CENTERING
WASHERS

SPRINGS PUSH AGAINST
CENTERING WASHERS TO
CENTER THE SPOOL WHEN
NO AIR IS APPLIED

PISTONS SEAL THE
AIR CHAMBER FROM
THE HYDRAULIC CHAMBER

Figure 8-13. Air pilot–actuated, three-position, spring-centered, four-way valve. *(Courtesy of Sperry Vickers, Sperry Rand Corp., Troy, Michigan.)*

which is sealed to prevent external leakage of oil. There are two types of solenoid designs used to dissipate the heat created by the electric current flowing in the wire of the coil. The first type simply dissipates the heat to the surrounding air and is referred to as an *air gap solenoid*. In the second type, a *wet pin solenoid*, the push pin contains an internal passageway that allows tank port oil to communicate between the housing of the valve and the housing of the solenoid. Wet pin solenoids do a better job in dissipating heat because the cool oil represents a good heat sink to absorb the heat from the solenoid. As the oil circulates, the heat is carried into the hydraulic system where it can be easily dealt with.

The solenoid valve of Figure 8-15 has a flow capacity of 12 gpm and a maximum operating pressure of 3500 psi. It has a wet pin solenoid whose armature moves in a tube that is open to the tank cavity of the valve. The fluid around the armature serves to cool it and cushion its stroke without appreciably affecting response time. There are no seals around this armature to wear or restrict its movement. This allows all the power developed by the solenoid to be transmitted to the valve spool without having to overcome seal friction. Impact loads, which frequently cause premature solenoid failure, are eliminated with this construction. This valve has a solenoid at each end of the spool. Specifically, it is a solenoid-actuated, four-way, three-position, spring-centered directional control valve. Note in the graphic symbol how the solenoid is represented at both ends of the spool.

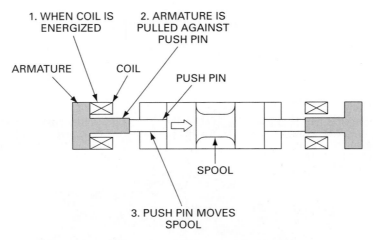

Figure 8-14. Operation of solenoid to shift spool of valve.
(Courtesy of Sperry Vickers, Sperry Rand Corp., Troy, Michigan.)

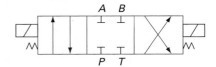

Figure 8-15. Solenoid-actuated, three-position, spring-centered, four-way directional control valve.
(Courtesy of Continental Hydraulics, Division of Continental Machines Inc., Savage, Minnesota.)

Figure 8-16 shows a single solenoid-actuated, four-way, two-position, spring-offset directional control valve. Its graphic symbol is also given in Figure 8-16.

In Figure 8-17 we see a solenoid-controlled, pilot-operated directional control valve. Note that the pilot valve is actually mounted on top of the main valve body. The upper pilot stage spool (which is solenoid actuated) controls the pilot pressure, which can be directed to either end of the main stage spool. This 35-gpm, 3000-psi valve is of the four-way, three-position, spring-centered configuration and has a manual override to shift the pilot stage mechanically when troubleshooting.

Center Flow Path Configurations for Three-Position, Four-Way Valves

Most three-position valves have a variety of possible flow path configurations. Each four-way valve has an identical flow path configuration in the actuated position but a different spring-centered flow path. This is illustrated in Figure 8-18.

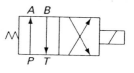

Figure 8-16. Single solenoid-actuated, four-way, two-position, spring-offset directional control valve. *(Courtesy of Continental Hydraulics, Division of Continental Machines Inc., Savage, Minnesota.)*

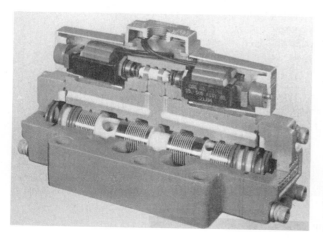

Figure 8-17. Solenoid-controlled, pilot-operated directional control valve. *(Courtesy of Continental Hydraulics, Division of Continental Machines Inc., Savage, Minnesota.)*

Note that the open-center-type connects all ports together. In this design the pump flow can return directly back to the tank at essentially atmospheric pressure. At the same time, the actuator (cylinder or motor) can be moved freely by applying an external force.

The closed-center design has all ports blocked, as is the case for the valve of Figures 8-10 and 8-13. In this way the pump flow can be used for other parts of the circuit. At the same time, the actuator connected to ports *A* and *B* is hydraulically locked. This means it cannot be moved by the application of an external force.

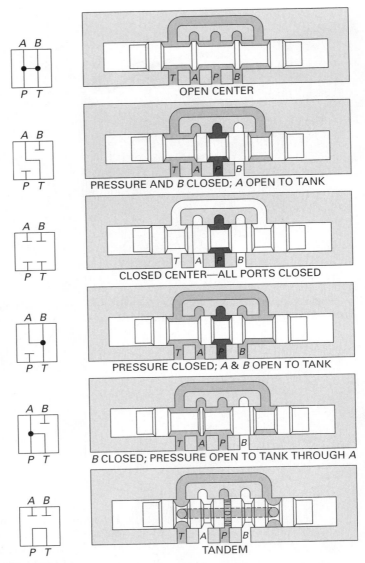

Figure 8-18. Various center flow paths for three-position, four-way valves. *(Courtesy of Sperry Vickers, Sperry Rand Corp., Troy, Michigan.)*

The tandem design also results in a locked actuator. However, it also unloads the pump at essentially atmospheric pressure. For example, the closed-center design forces the pump to produce flow at the high-pressure setting of the pressure relief valve. This not only wastes pump power but promotes wear and shortens pump life, especially if operation in the center position occurs for long periods. Another factor

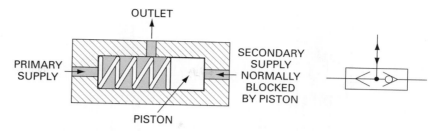

Figure 8-19. Shuttle valve (schematic and graphic symbol).

is that the wasted power shows up as heat, which raises the temperature of the oil. This promotes oil oxidation, which increases the acidity of the oil. Such an oil tends to corrode the critical metallic parts not only of the pump but also of the actuators and valves. Also affected is the viscosity of the oil. Higher temperature lowers the viscosity, which in turn increases leakage and reduces the lubricity of the oil. To keep the temperature at a safe level, an expensive oil cooler may be required.

Shuttle Valves

A shuttle valve is another type of directional control valve. It permits a system to operate from either of two fluid power sources. One application is for safety in the event that the main pump can no longer provide hydraulic power to operate emergency devices. The shuttle valve will shift to allow fluid to flow from a secondary backup pump. As shown in Figure 8-19, a shuttle valve consists of a floating piston that can be shuttled to one side or the other of the valve depending on which side of the piston has the greater pressure. Shuttle valves may be spring-loaded (biased as shown in Figure 8-19) in one direction to favor one of the supply sources or unbiased so that the direction of flow through the valve is determined by circuit conditions. A shuttle valve is essentially a direct-acting double-check valve with a cross-bleed, as shown by the graphic symbol of Figure 8-19. As shown by the double arrows on the graphic symbol, reverse flow is permitted.

8.3 PRESSURE CONTROL VALVES

Simple Pressure Relief Valves

The most widely used type of pressure control valve is the pressure relief valve, since it is found in practically every hydraulic system. It is normally a closed valve whose function is to limit the pressure to a specified maximum value by diverting pump flow back to the tank. Figure 8-20 illustrates the operation of a simple relief valve. A poppet is held seated inside the valve by the force of a stiff compression spring. When the system pressure reaches a high enough value, the resulting hydraulic force (acting on the piston-shaped poppet) exceeds the spring force and the poppet is forced off its seat. This permits flow through the outlet to the tank as

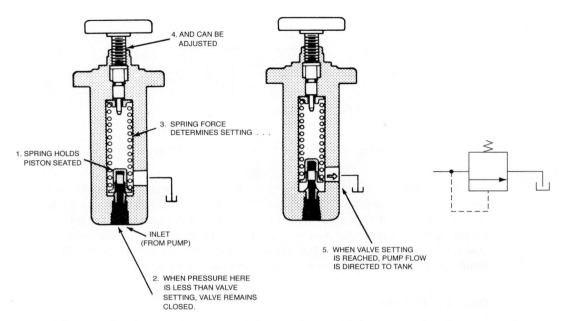

Figure 8-20. Simple pressure relief valve. *(Courtesy of Sperry Vickers, Sperry Rand Corp., Troy, Michigan.)*

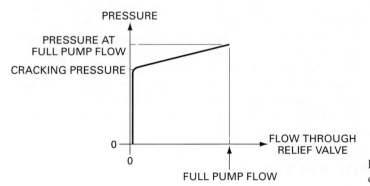

Figure 8-21. Pressure versus flow curve for simple relief valve.

long as this high pressure level is maintained. Note the external adjusting screw, which varies the spring force and, thus, the pressure at which the valve begins to open (cracking pressure). Figure 8-20 also provides the graphic symbol of a simple pressure relief valve.

It should be noted that the poppet must open sufficiently to allow full pump flow. The pressure that exists at full pump flow can be substantially greater than the cracking pressure. This is shown in Figure 8-21, where system pressure is plotted versus flow through the relief valve. The stiffness of the spring (force required to compress the spring 1 in or 1 cm) and the amount the poppet must open to permit full pump

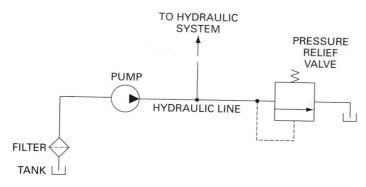

Figure 8-22. Symbolic representation of partial hydraulic circuit.

flow to determine the difference between the full pump flow pressure and the cracking pressure. The stiffness of a spring is called the spring constant and has units of lb/in or N/cm. The pressure at full pump flow is the pressure level that is specified when referring to the pressure setting of the relief valve. It is the maximum pressure level permitted by the relief valve.

Figure 8-22 shows a partial hydraulic circuit containing a pump and pressure relief valve, which are drawn symbolically. If the hydraulic system (not shown) does not accept any flow, then all the pump flow must return to the tank via the relief valve. The pressure relief valve provides protection against any overloads experienced by the actuators in the hydraulic system. Obviously one important function of a pressure relief valve is to limit the force or torque produced by hydraulic cylinders and motors.

EXAMPLE 8-1

A pressure relief valve contains a poppet with a 0.75 in² area on which system pressure acts. During assembly a spring with a spring constant of 2500 lb/in is installed to hold the poppet against its seat. The adjustment mechanism is then set so that the spring is initially compressed 0.20 in from its free-length condition. In order to pass full pump flow through the valve at the PRV pressure setting, the poppet must move 0.10 in from its fully closed position. Determine the

a. Cracking pressure
b. Full pump flow pressure (PRV pressure setting)

Solution

a. The force (F) a spring exerts equals the product of the spring constant (k) and the spring deflection (S) from its free-length condition. Thus, the spring force exerted on the poppet when it is fully closed is

$$F = kS = 2500 \text{ lb/in} \times 0.20 \text{ in} = 500 \text{ lb}$$

In order to put the poppet on the verge of opening (cracking), the hydraulic force must equal the 500-lb spring force.

$$\text{hydraulic force} = \text{spring force}$$
$$p_{\text{cracking}}A = 500 \text{ lb}$$

$$p_{\text{cracking}}(0.75 \text{ in}^2) = 500 \text{ lb} \quad \text{or} \quad p_{\text{cracking}} = 667 \text{ psi}$$

Thus, when the system pressure becomes slightly greater than 667 psi, the poppet lifts off its seat a small amount to allow fluid to begin flowing through the valve.

b. When the poppet moves 0.10 in from its fully closed position, the spring has compressed a total of 0.30 in from its free-length condition. Thus, the spring force exerted on the poppet when it is opened 0.10 in to allow full pump flow is

$$F = kS = 2500 \text{ lb/in} \times 0.30 \text{ in} = 750 \text{ lb}$$

In order to move the poppet 0.10 in from its fully closed position, the hydraulic force must equal the 750-lb spring force.

$$\text{hydraulic force} = \text{spring force}$$
$$p_{\text{full pump flow}}A = 750 \text{ lb}$$

$$p_{\text{full pump flow}}(0.75 \text{ in}^2) = 750 \text{ lb} \quad \text{or} \quad p_{\text{full pump flow}} = 1000 \text{ psi}$$

Thus, when system pressure reaches a value of 1000 psi, the poppet is lifted 0.10 in off its seat and the flow rate through the valve equals the pump flow rate. This means that the PRV pressure setting is 1000 psi and is 333 psi higher (or 50% higher) than the cracking pressure.

Compound Pressure Relief Valves

A compound pressure relief valve (see Figure 8-23 for external and cutaway views of an actual design) is one that operates in two stages. As shown in Figure 8-23, the pilot stage is located in the upper valve body and contains a pressure-limiting poppet that is held against a seat by an adjustable spring. The lower body contains the port connections. Diversion of the full pump flow is accomplished by the balanced piston in the lower body.

The operation is as follows (refer to Figure 8-24): In normal operation, the balanced piston is in hydraulic balance. Pressure at the inlet port acts under the piston and also on its top because an orifice is drilled through the large land. For pressures less than the valve setting, the piston is held on its seat by a light spring. As soon as pressure reaches the setting of the adjustable spring, the poppet is forced off its seat. This limits the pressure in the upper chamber. The restricted flow through the orifice

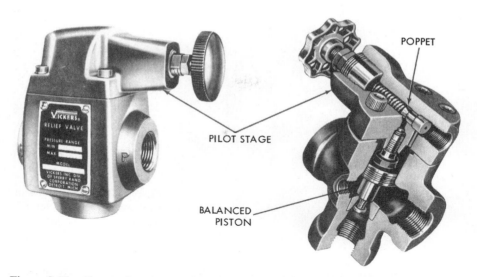

POPPET

PILOT STAGE

BALANCED
PISTON

Figure 8-23. External and cutaway views of an actual compound relief valve. *(Courtesy of Sperry Vickers, Sperry Rand Corp., Troy, Michigan.)*

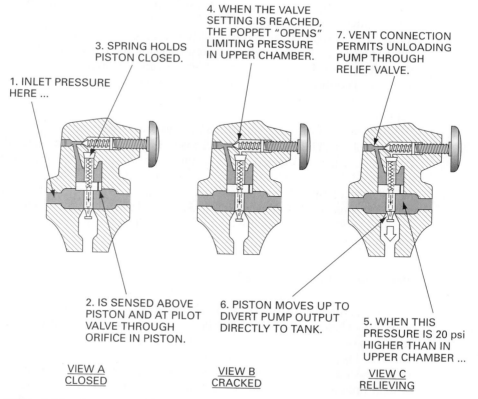

3. SPRING HOLDS
PISTON CLOSED.

4. WHEN THE VALVE
SETTING IS REACHED,
THE POPPET "OPENS"
LIMITING PRESSURE
IN UPPER CHAMBER.

7. VENT CONNECTION
PERMITS UNLOADING
PUMP THROUGH
RELIEF VALVE.

1. INLET PRESSURE
HERE ...

2. IS SENSED ABOVE
PISTON AND AT PILOT
VALVE THROUGH
ORIFICE IN PISTON.

6. PISTON MOVES UP TO
DIVERT PUMP OUTPUT
DIRECTLY TO TANK.

5. WHEN THIS
PRESSURE IS 20 psi
HIGHER THAN IN
UPPER CHAMBER ...

VIEW A
CLOSED

VIEW B
CRACKED

VIEW C
RELIEVING

Figure 8-24. Operation of compound pressure relief valve. *(Courtesy of Sperry Vickers, Sperry Rand Corp., Troy, Michigan.)*

Figure 8-25. Compound pressure relief valve with integral solenoid-actuated, two-way vent valve. *(Courtesy of Abex Corp., Denison Division, Columbus, Ohio.)*

and into the upper chamber results in an increase in pressure in the lower chamber. This causes an unbalance in hydraulic forces, which tends to raise the piston off its seat. When the pressure difference between the upper and lower chambers reaches approximately 20 psi, the large piston lifts off its seat to permit flow directly to the tank. If the flow increases through the valve, the piston lifts farther off its seat. However, this compresses only the light spring, and hence very little override occurs. Compound relief valves may be remotely operated by using the outlet port from the chamber above the piston. For example, this chamber can be vented to the tank via a solenoid directional control valve. When this valve vents the pressure relief valve to the tank, the 20-psi pressure in the bottom chamber overcomes the light spring and unloads the pump to the tank.

Figure 8-25 shows a compound pressure relief valve that has this remote operation capability. This particular model has its own built-in solenoid-actuated, two-way vent valve, which is located between the cap and body of the main valve. Manual override of the solenoid return spring is a standard feature. The pressure relief valve is vented when the solenoid is de-energized and devented when energized. This relief valve has a maximum flow capacity of 53 gpm and can be adjusted to limit system pressures up to 5000 psi. Clockwise tightening of the hex locknut prevents accidental setting changes by use of the knurled knob.

Pressure-Reducing Valves

A second type of pressure control valve is the pressure-reducing valve. This type of valve (which is normally open) is used to maintain reduced pressures in specified locations of hydraulic systems. It is actuated by downstream pressure and tends to close as this pressure reaches the valve setting. Figure 8-26 illustrates the operation of a pressure-reducing valve that uses a spring-loaded spool to control the downstream pressure. If downstream pressure is below the valve setting, fluid will flow freely from the inlet to the outlet. Note that there is an internal passageway from

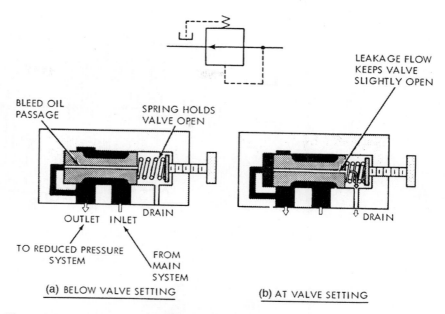

Figure 8-26.   Operation of a pressure-reducing valve. *(Courtesy of Sperry Vickers, Sperry Rand Corp., Troy, Michigan.)*

the outlet, which transmits outlet pressure to the spool end opposite the spring. When the outlet (downstream) pressure increases to the valve setting, the spool moves to the right to partially block the outlet port, as shown in Figure 8-26(b). Just enough flow is passed to the outlet to maintain its preset pressure level. If the valve closes completely, leakage past the spool could cause downstream pressure to build up above the valve setting. This is prevented from occurring because a continuous bleed to the tank is permitted via a separate drain line to the tank. Figure 8-26 also provides the graphic symbol for a pressure-reducing valve. Observe that the symbol shows that the spring cavity has a drain to the tank.

Unloading Valves

An additional pressure control device is the unloading valve. This valve is used to permit a pump to build pressure to an adjustable pressure setting and then allow it to discharge oil to the tank at essentially zero pressure as long as pilot pressure is maintained on the valve from a remote source. Hence, the pump has essentially no load and is therefore developing a minimum amount of power. This is the case in spite of the fact that the pump is delivering a full pump flow because the pressure is practically zero. This is not the same with a pressure relief valve because the pump is delivering full pump flow at the pressure relief valve setting and thus is operating at maximum power conditions.

Figure 8-27 shows a schematic of an unloading valve used to unload the pump connected to port A when the pressure at port X is maintained at the value that

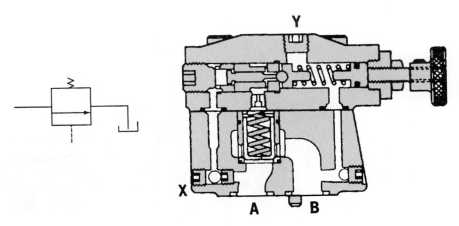

Figure 8-27. Schematic of unloading valve. *(Courtesy of Abex Corp., Denison Division, Columbus, Ohio.)*

Figure 8-28. Unloading valve. *(Courtesy of Abex Corp., Denison Division, Columbus, Ohio.)*

satisfies the valve setting. The high-flow poppet is controlled by the spring-loaded ball and the pressure applied to port X. Flow entering at port A is blocked by the poppet at low pressures. The pressure signal from A passes through the orifice in the main poppet to the topside area and on to the ball. There is no flow through these sections of the valve until the pressure rises to the maximum permitted by the adjustably set spring-loaded ball. When that occurs, the poppet lifts and flow goes from port A to port B, which is typically connected to the tank. The pressure signal to port X (sustained by another part of the system) acts against the solid control piston and forces the ball farther off the seat. This causes the topside pressure on the main poppet to go to a very low value and allows flow from A to B with a very low pressure drop as long as signal pressure at X is maintained. The ball reseats, and the main poppet closes with a snap action when the pressure at X falls to approximately 90% of the maximum pressure setting of the spring-loaded ball. Also included in Figure 8-27 is the graphic symbol of an unloading valve. Figure 8-28 shows the actual unloading valve.

EXAMPLE 8-2

A pressure relief valve has a pressure setting of 1000 psi. Compute the horse-power loss across this valve if it returns all the flow back to the tank from a 20-gpm pump.

Solution

$$HP = \frac{pQ}{1714} = \frac{(1000)(20)}{1714} = 11.7 \text{ hp}$$

EXAMPLE 8-3

An unloading valve is used to unload the pump of Example 8-2. If the pump discharge pressure (during unloading) equals 25 psi, how much hydraulic horsepower is being wasted?

Solution

$$HP = \frac{pQ}{1714} = \frac{(25)(20)}{1714} = 0.29 \text{ hp}$$

Sequence Valves

Still another pressure control device is the sequence valve, which is designed to cause a hydraulic system to operate in a pressure sequence. After the components connected to port A (see Figure 8-29) have reached the adjusted pressure of the sequence valve, the valve passes fluid through port B to do additional work in a different portion of the system. The high-flow poppet of the sequence valve is controlled by the spring-loaded cone. Flow entering at port A is blocked by the poppet at low pressures. The pressure signal at A passes through orifices to the topside of the poppet and to the cone. There is no flow through these sections until the pressure rises at A to the maximum permitted by the adjustably set spring-loaded cone. When the pressure at A reaches that value, the main poppet lifts, passing flow to port B. It maintains the adjusted pressure at port A until the pressure at B rises to the same value. A small pilot flow (about 1/4 gpm) goes through the control piston and past the pilot cone to the external drain at this time. When the pressure at B rises to the pressure at A, the control piston seats and prevents further pilot flow loss. The main poppet opens fully and allows the pressure at A and B to rise to higher values together. Flow may go either way at this time. The spring cavity of the control cone drains externally from port Y, generally to the tank. This sequence valve may be remotely controlled from vent port X. Figure 8-29 also includes the

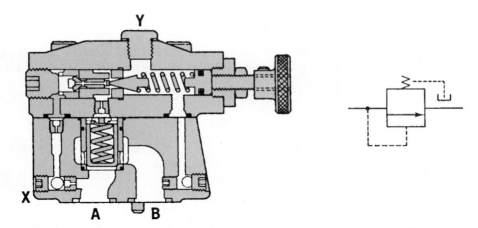

Figure 8-29. Schematic of sequence valve. *(Courtesy of Abex Corp., Denison Division, Columbus, Ohio.)*

graphic symbol for a sequence valve. The pilot line can come from anywhere in the circuit and not just from directly upstream, as shown.

Counterbalance Valves

A final pressure control valve to be presented here is the counterbalance valve (CBV). The purpose of a counterbalance valve is to maintain control of a vertical hydraulic cylinder to prevent it from descending due to the weight of its external load. As shown in Figure 8-30, the primary port of this valve is connected to the bottom of the cylinder, and the secondary port is connected to a directional control valve (DCV). The pressure setting of the counterbalance valve is somewhat higher than is necessary to prevent the cylinder load from falling due to its weight. As shown in Figure 8-30(a), when pump flow is directed (via the DCV) to the top of the cylinder, the cylinder piston is pushed downward. This causes pressure at the primary port to increase to a value above the pressure setting of the counterbalance valve and thus raise the spool of the CBV. This then opens a flow path through the counterbalance valve for discharge through the secondary port to the DCV and back to the tank.

When raising the cylinder [see Figure 8-30(b)], an integral check valve opens to allow free flow for retracting the cylinder. Figure 8-30(c) gives the graphic symbol for a counterbalance valve. Section 9.7 provides a complete circuit diagram for a counterbalance valve application.

8.4 FLOW CONTROL VALVES

Orifice as a Flow Meter or Flow Control Device

Figure 8-31 shows an orifice (a disk with a hole through which fluid flows) installed in a pipe. Such a device can be used as a flowmeter by measuring the pressure drop

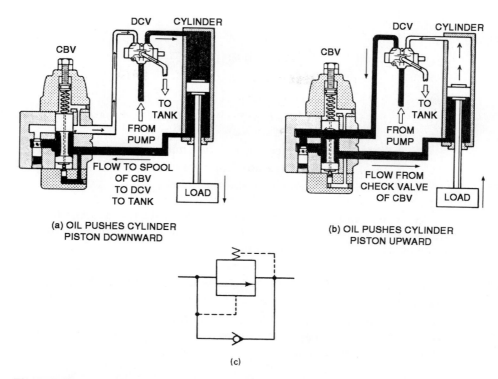

Figure 8-30. Application of counterbalance valve. *(Courtesy of Sperry Vickers, Sperry Rand Corp., Troy, Michigan.)*

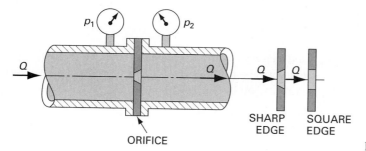

Figure 8-31. Orifice flowmeter.

(Δp) across the orifice. This is because for a given orifice, there is a unique relationship between (Δp) and Q (the flow rate through the orifice and thus the flow rate in the pipe). It can be shown that the following English-units equation relates the (Δp) vs. Q relationship for an orifice installed in a pipe to measure liquid flow rate.

$$Q = 38.1\, CA\sqrt{\frac{\Delta p}{SG}} \tag{8-1}$$

or, in metric units,

$$Q = 0.0851 \, CA\sqrt{\frac{\Delta p}{SG}} \qquad\qquad \textbf{(8-1M)}$$

where Q = flow rate (gpm, Lpm),
         C = flow coefficient (C = 0.80 for sharp-edged orifice, C = 0.60 for
            square-edged orifice),
         A = area of orifice opening (in², mm²),
         $\Delta p = p_1 - p_2$ = pressure drop across orifice (psi, kPa),
         SG = specific gravity of flowing fluid.

As seen from Eq. (8-1), the greater the flow rate, the greater will be the pressure drop and vice versa for a given orifice. Example 8-4 shows how an orifice flowmeter can be used to determine flow rate.

EXAMPLE 8-4

The pressure drop across the sharp-edged orifice of Figure 8-31 is 100 psi. The orifice has a 1-in diameter, and the fluid has a specific gravity of 0.9. Find the flow rate in units of gpm.

Solution   Substitute directly into Eq. (8-1):

$$Q = (38.1)(0.80)\left(\frac{\pi}{4} \times 1^2\right) \times \sqrt{\frac{100}{0.9}} = 252 \text{ gpm}$$

An orifice can also be used as a flow control device. As seen from Eq. (8-1), the smaller the orifice area, the smaller will be the flow rate and vice versa for a given pressure drop. This leads us to the discussion of flow control valves.

Needle Valves

Flow control valves are used to regulate the speed of hydraulic cylinders and motors by controlling the flow rate to these actuators. They may be as simple as a fixed orifice or an adjustable needle valve. Needle valves are designed to give fine control of flow in small-diameter piping. As illustrated in Figure 8-32, their name is derived from their sharp, pointed conical disk and matching seat. The graphic symbol for a needle valve (which is a variable orifice) is also given in Figure 8-32.

Figure 8-33 shows a flow control valve that is easy to read and adjust. The stem has several color rings, which, in conjunction with a numbered knob, permits reading of a given valve opening as shown. Charts are available that allow quick determination of the controlled flow rate for given valve settings and pressure drops. A locknut prevents unwanted changes in flow.

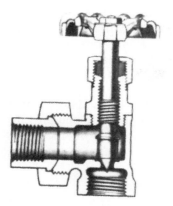

Figure 8-32. Needle valve. *(Courtesy of Crane Co., Chicago, Illinois.)*

Figure 8-33. Easy read and adjust flow control valve. *(Courtesy of Deltrol Corp., Bellwood, Illinois.)*

For a given opening position, a needle valve behaves as an orifice. However, unlike an orifice, the flow area (A) in a needle valve can be varied. Thus, Eq. (8-1) can be modified as follows to represent the pressure drop versus flow rate for a needle valve:

$$Q = C_v \sqrt{\frac{\Delta p}{SG}} \tag{8-2}$$

where $\quad Q$ = volume flow rate (gpm, Lpm),
$\quad\quad\quad C_v$ = capacity coefficient (gpm/$\sqrt{\text{psi}}$, Lpm/$\sqrt{\text{kPa}}$),
$\quad\quad\quad \Delta p$ = pressure drop across the valve (psi, kPa),
$\quad\quad\quad SG$ = specific gravity of the liquid.

In English units, the capacity coefficient is defined as the flow rate of water in gpm that will flow through the valve at a pressure drop of 1 psi. In metric units the capacity coefficient is defined as the flow rate of water in Lpm (liters per minute) that will flow through the valve at a pressure drop of 1 kPa. The value C_v is determined experimentally for each type of valve in the fully open position and is listed as the "rated C_v" in manufacturers' catalogs.

Example 8-5 shows how to calculate the capacity coefficient for a flow control valve using equivalent English and metric system values of flow rate and pressure drop. Example 8-6 illustrates how to determine the required capacity coefficient for a flow control valve that is to provide a desired hydraulic cylinder speed.

EXAMPLE 8-5

A flow control valve experiences a pressure drop of 100 psi (687 kPa) for a flow rate of 25 gpm (94.8 Lpm). The fluid is hydraulic oil with a specific gravity of 0.90. Determine the capacity coefficient.

Solution   Solving Eq. (8-2) for the capacity coefficient, we have

$$C_v = \frac{Q}{\sqrt{\Delta p / SG}}$$

Using English units yields

$$C_v = \frac{25}{\sqrt{100/0.9}} = 2.37 \text{ gpm}/\sqrt{\text{psi}}$$

For metric units the result is

$$C_v = \frac{94.8}{\sqrt{687/0.9}} = 3.43 \text{ Lpm}/\sqrt{\text{kPa}}$$

EXAMPLE 8-6

A needle valve is used to control the extending speed of a hydraulic cylinder. The needle valve is placed in the outlet line of the hydraulic cylinder, as shown in Figure 8-34. The following data are given:

1. Desired cylinder speed = 10 in/s
2. Cylinder piston diameter = 2 in (area = 3.14 in²)
3. Cylinder rod diameter = 1 in (area = 0.79 in²)
4. Cylinder load = 1000 lb

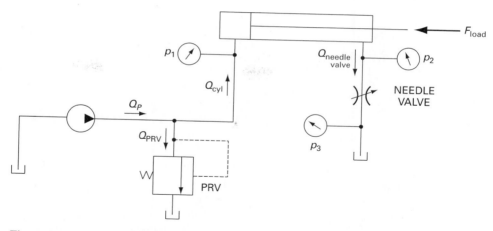

Figure 8-34. System for Example 8-6.

5. Specific gravity of oil = 0.90
6. Pressure relief valve setting (PRV setting) = 500 psi

Determine the required capacity coefficient of the needle valve.

Solution When the needle valve is fully open, all the flow from the pump goes to the hydraulic cylinder to produce maximum hydraulic cylinder speed. As the needle valve is partially closed, its pressure drop increases. This causes the back pressure p_2 to increase, which results in a greater resistance force at the rod end of the cylinder. Since this back pressure force opposes the extending motion of the cylinder, the result is an increase in cylinder blank end pressure p_1. Further closing of the needle valve ultimately results in pressure p_1 reaching and then exceeding the cracking pressure of the pressure relief valve. The result is a slower cylinder speed since part of the pump flow goes back to the oil tank through the pressure relief valve. When the cylinder speed reaches the desired value, p_1 approximately equals the PRV setting and the amount of pump flow not desired by the cylinder goes through the pressure relief valve. When this occurs, the cylinder receives the desired amount of flow rate, which equals the pump flow rate minus the flow rate through the pressure relief valve.

First, we solve for the rod end pressure p_2 that causes the blank end pressure p_1 to equal the PRV setting. This is done by summing forces on the hydraulic cylinder.

$$p_1 A_1 - F_{\text{load}} = p_2 A_2$$

where

A_1 = piston area,
A_2 = piston area minus rod area.

Substituting values yields

$$500 \text{ lb/in}^2 \times 3.14 \text{ in}^2 - 1000 \text{ lb} = p_2(3.14 - 0.79)\text{in}^2 = p_2(2.35 \text{ in}^2)$$

$$p_2 = 243 \text{ psi}$$

Next, we calculate the flow rate through the needle valve based on the desired hydraulic cylinder speed.

$$Q = A_2 v_{\text{cylinder}} = 2.35 \text{ in}^2 \times 10 \text{ in/s} = 23.5 \text{ in}^3/\text{s}$$

$$= 23.5 \text{ in}^3/\text{s} \times \frac{1 \text{ gal}}{231 \text{ in}^3} \times \frac{60 \text{ s}}{1 \text{ min}} = 6.10 \text{ gpm}$$

Since the discharge from the needle valve flows directly to the oil tank, pressure $p_3 = 0$. Thus, p_2 equals the pressure drop across the needle valve and we can solve for C_v.

$$C_v = \frac{Q}{\sqrt{\Delta p/SG}} = \frac{6.10}{\sqrt{243/0.90}} = 0.37 \text{ gpm}/\sqrt{\text{psi}}$$

Non-Pressure-Compensated Valves

There are two basic types of flow control valves: non-pressure-compensated and pressure-compensated. The non-pressure-compensated type is used where system pressures are relatively constant and motoring speeds are not too critical. They work on the principle that the flow through an orifice will be constant if the pressure drop remains constant. Figure 8-35 gives a cutaway view of a non-pressure-compensated flow control valve and its graphic symbol. The design shown also includes a check valve, which permits free flow in the direction opposite to the flow control direction.

Pressure-Compensated Valves

If the load on an actuator changes significantly, system pressure will change appreciably. Thus, the flow-rate through a non-pressure-compensated valve will change for the same flow-rate setting. Figure 8-36 illustrates the operation of a pressure-compensated valve. This design incorporates a hydrostat that maintains a constant 20-psi differential across the throttle, which is an orifice whose area can be adjusted by an external knob setting. The orifice area setting determines the flow rate to be controlled. The hydrostat is held normally open by a light spring. However, it starts to close as inlet pressure increases and overcomes the light spring force. This closes the opening through

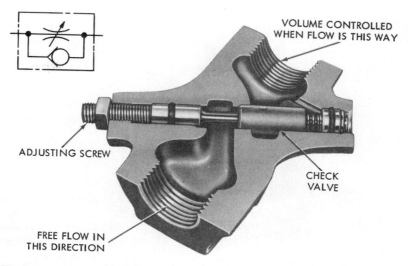

Figure 8-35. Non-pressure-compensated flow control valve. *(Courtesy of Sperry Vickers, Sperry Rand Corp., Troy, Michigan.)*

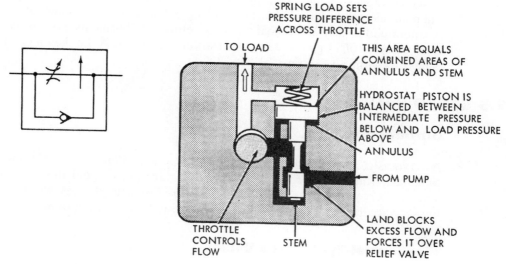

Figure 8-36. Operation of pressure-compensated flow control valve. *(Courtesy of Sperry Vickers, Sperry Rand Corp., Troy, Michigan.)*

the hydrostat and thereby blocks off all flow in excess of the throttle setting. As a result, the only oil that will pass through the valve is the amount that 20 psi can force through the throttle. Flow exceeding this amount can be used by other parts of the circuit or return to the tank via the pressure relief valve. Also included in Figure 8-36 is the graphic symbol for a pressure-compensated flow control valve.

Figure 8-37. Pressure-compensated flow control valve. *(Courtesy of Continental Hydraulics, Division of Continental Machines Inc., Savage, Minnesota.)*

In Figure 8-37 we have a see-through model of an actual pressure-compensated flow control valve, which has a pressure rating of 3000 psi. Pressure compensation will maintain preset flow within 1 to 5% depending on the basic flow rate. The dial is calibrated for easy and repeatable flow settings. Adjustments over the complete valve capacity of 12 gpm are obtained within a 270° arc. A dial key lock prevents tampering with valve settings.

8.5 SERVO VALVES

Introduction

A servo valve is a directional control valve that has infinitely variable positioning capability. Thus, it can control not only the direction of fluid flow but also the amount. Servo valves are coupled with feedback-sensing devices, which allow for the very accurate control of position, velocity, and acceleration of an actuator.

Mechanical-Type Servo Valves

Figure 8-38 shows the mechanical-type servo valve, which is essentially a force amplifier used for positioning control. In this design, a small input force shifts the spool of the servo valve to the right by a specified amount. The oil then flows through port p_1, retracting the hydraulic cylinder to the right. The action of the feedback link shifts the sliding sleeve to the right until it blocks off the flow to the hydraulic cylinder. Thus, a given input motion produces a specific and controlled amount of output motion. Such a system, where the output is fed back to modify the input is called a *closed-loop system.* One of the most common applications of this type of

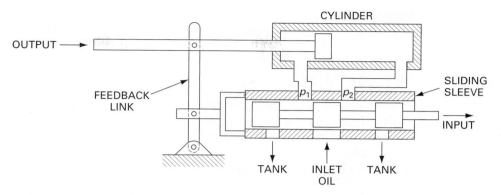

Figure 8-38. Mechanical-hydraulic servo valve.

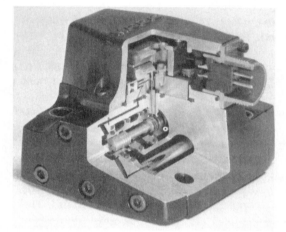

Figure 8-39. Electrohydraulic servo valve. *(Courtesy of Moog Inc., Industrial Division, East Aurora, New York.)*

mechanical-hydraulic servo valve is the hydraulic power steering system of automobiles and other transportation vehicles.

Electrohydraulic Servo Valves

Typical electrohydraulic servo valves use an electrical torque motor, a double-nozzle pilot stage, and a sliding spool second stage. Figure 8-39 gives a cutaway view of an electrohydraulic servo valve. This servo valve is an electrically controlled, proportional metering valve suitable for a variety of mobile vehicles and industrial control applications such as earth-moving vehicles, articulated arm devices, cargo-handling cranes, lift trucks, logging equipment, farm machinery, steel mill controls, utility construction, fire trucks, and servicing vehicles.

The construction and operational features of an electrohydraulic servo valve can be seen by referring to the schematic drawing of Figure 8-40. The torque motor includes coils, pole pieces, magnets, and an armature. The armature is supported for limited movement by a flexure tube. The flexure tube also provides a fluid seal between the hydraulic and electromagnetic portions of the valve. The flapper attaches to the center of the armature and extends down, inside the flexure tube. A nozzle is located on each side of the flapper so that flapper motion varies the nozzle openings. Inlet-pressurized hydraulic fluid is filtered and then supplied to each nozzle through one of the two inlet orifices located at the ends of the filter. Differential pressure between the ends of the spool is varied by flapper motion between the nozzles.

The four-way valve spool directs the flow from the supply pressure port to either of the two outlet-to-load ports in an amount proportional to spool displacement. The spool contains flow metering slots in the control lands that are uncovered by spool motion. Spool movement deflects a feedback wire that applies a torque to the armature/flapper. Electrical current in the torque motor coils causes either clockwise or counterclockwise torque on the armature. This torque displaces the flapper between the two nozzles. The differential nozzle flow moves the spool to either the right or left. The spool continues to move until the feedback torque counteracts the electromagnetic

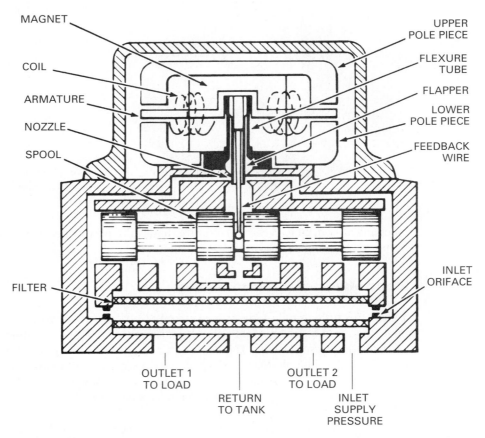

Figure 8-40. Schematic cross section of electrohydraulic servo valve. (*Courtesy of Moog Inc., Industrial Division, East Aurora, New York.*)

torque. At this point the armature/flapper is returned to center, so the spool stops and remains displaced until the electrical input changes to a new level. Therefore, valve spool position is proportional to the electrical signal.

The actual outlet flow from the valve to the external load will depend on the load pressure. Rated flow is achieved with either a +100% or −100% electrical signal, at which point the actual amount of rated flow depends on the valve pressure drop (inlet pressure minus load pressure). The operation of complete electrohydraulic servo systems is covered in Chapter 17.

8.6 PROPORTIONAL CONTROL VALVES

Proportional control valves, which are also called *electrohydraulic proportional valves*, are similar to electrohydraulic servo valves in that they both are electrically controlled. However, there are a number of differences between these two types of valves. For example, servo valves are used in closed-loop systems whereas proportional valves are used in open-loop systems. In servo valves, electrical current in a torque motor coil causes either clockwise or counterclockwise torque on an armature to control the movement of the valve spool. On the other hand, a proportional valve uses a solenoid that produces a force proportional to the current in its coils. Thus, by controlling the current in the solenoid coil, the position of the spring-loaded spool can also be controlled. This means that unlike a standard solenoid valve, a proportional valve can provide both directional and flow control capability in a single valve.

Although proportional valves are also designed to control pressure, the proportional direction control valve is the most widely used. Figure 8-41 shows an external view of a four-way proportional directional control valve along with its graphic symbol. Note that the graphic symbol contains two horizontal lines (one at the top of the symbol and one at the bottom) to indicate infinite positioning capability of the spool. Figures 8-42 and 8-43 provide a schematic cutaway view and a pictorial cutaway view, respectively, of this same valve. As seen from these three figures, a proportional directional control valve looks very similar to a conventional solenoid-actuated directional control valve. However, the spool of a proportional valve is designed specifically to provide precise metering of the oil for good speed control of cylinders and motors. To accomplish this precise control, clearances between the spool lands and mating valve bore are very small (approximately 0.0005 in). For increased precision, metering notches are machined on the spool lands to allow oil flow to begin somewhat before the lands clear the valve ports. The valve of Figure 8-41 provides a maximum flow rate of 7.9 gpm at a pressure drop of 145 psi when operating with a supply voltage of 24 VDC.

8.7 CARTRIDGE VALVES

Introduction

Market pressures and worldwide competition make the need for more efficient and economical hydraulic systems greater than ever. Integrated hydraulic circuits offer a proven way to achieve these improvements. Integrated hydraulic circuits are

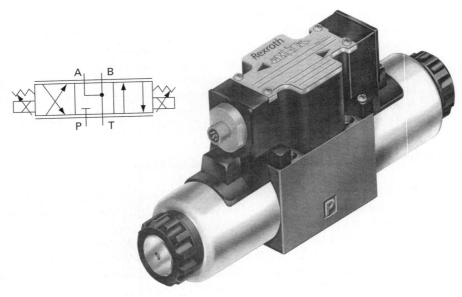

Figure 8-41. External view of four-way proportional directional control valve. *(Courtesy of Bosch Rexroth Corp., Bethlehem, Pennsylvania.)*

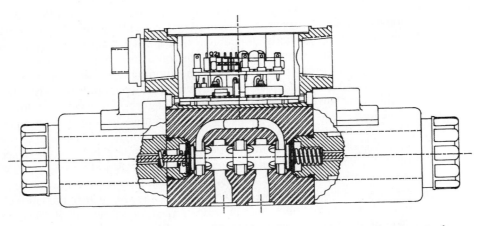

Figure 8-42. Schematic cutaway view of four-way proportional directional control valve. *(Courtesy of Bosch Rexroth Corp., Bethlehem, Pennsylvania.)*

compact hydraulic systems formed by integrating various cartridge valves and other components into a single, machined, ported manifold block.

A cartridge valve is designed to be assembled into a cavity of a ported manifold block (alone or along with other cartridge valves and hydraulic components) in order

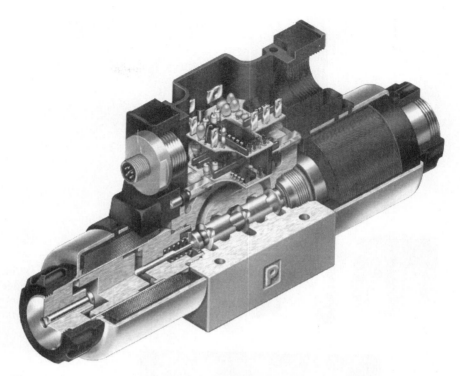

Figure 8-43. Pictorial cutaway view of four-way proportional directional control valve. *(Courtesy of Bosch Rexroth Corp., Bethlehem, Pennsylvania.)*

to perform the valve's intended function. (See Figure 8-44 for cutaway views of several threaded cartridge valves.) The cartridge valve is assembled into the manifold block either by screw threads (threaded design) or by a bolted cover (slip-in design). Figure 8-45 shows a manifold block containing a number of cartridge valves and other hydraulic components. The world map was etched on the outside surfaces of the manifold block to reflect one's entering the "world" of integrated hydraulic circuits.

Advantages and Various Valve Functions

The use of cartridge valves in ported manifold blocks provides a number of advantages over discrete, conventional, ported valves mounted at various locations in pipelines of hydraulic systems. The advantages include the following:

1. Reduced number of fittings to connect hydraulic lines between various components in a system
2. Reduced oil leakages and contamination due to fewer fittings
3. Lower system installation time and costs

Figure 8-44. Cutaway views of threaded cartridge valves. *(Courtesy of Parker Hannifin Corp., Elyria, Ohio.)*

Figure 8-45. Manifold block containing cartridge valves. *(Courtesy of Parker Hannifin Corp., Elyria, Ohio.)*

4. Reduced service time since faulty cartridge valves can be easily changed without disconnecting fittings

5. Smaller space requirements of overall system

A variety of different valve functions can be provided using cartridge valves. These include directional control, pressure relief, pressure reduction, unloading,

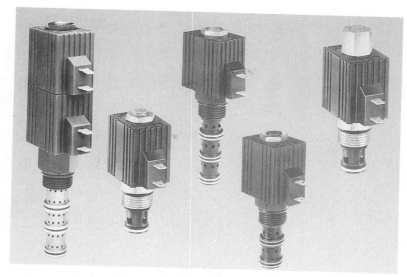

Figure 8-46. Solenoid-operated directional control cartridge valves. *(Courtesy of Parker Hannifin Corp., Elyria, Ohio.)*

counterbalance, and sequence and flow control capability. Figure 8-46 shows five different solenoid-operated directional control cartridge valves from left to right as follows:

1. Two-way, spool-type, N.C. or N.O.
2. Two-position, three-way, spool-type
3. Two position, four-way, spool-type
4. Two-way, poppet-type, N.C. or N.O.
5. Three-position, four-way, spool-type

Figures 8-47 and 8-48 show a cartridge pressure relief valve and solenoid-operated flow control (proportional) valve, respectively. Internal mechanical design and fluid flow operating features of a cartridge pressure relief valve are shown in Figure 8-49.

Integrated Hydraulics Technology

Integrated hydraulics technology can provide easier installation and servicing, greater reliability, reduced leakage, expanded design flexibility, and lighter, neater hydraulic packages for a variety of hydraulic applications. Figure 8-50 shows several manifold blocks superimposed on a symbolic hydraulic circuit diagram to represent a complete integrated hydraulic system.

Figure 8-47.   Cartridge pressure relief valve. *(Courtesy of Parker Hannifin Corp., Elyria, Ohio.)*

Figure 8-48.   Cartridge solenoid-operated flow control valve. *(Courtesy of Parker Hannifin Corp., Elyria, Ohio.)*

8.8 HYDRAULIC FUSES

Figure 8-51(a) is a schematic drawing of a hydraulic fuse, which is analogous to an electric fuse. It prevents hydraulic pressure from exceeding an allowable value in order to protect circuit components from damage. When the hydraulic pressure exceeds a design value, the thin metal disk ruptures to relieve the pressure as oil is drained back to the oil tank. After rupture, a new metal disk must be inserted before operation can be started again. Hydraulic fuses are used mainly with pressure-compensated pumps

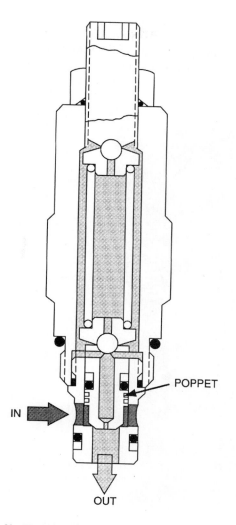
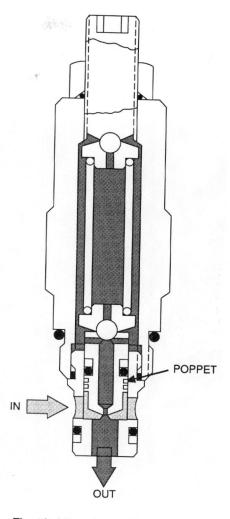

No Flow Condition

- System pressure is lower than relief valve setting.

- Poppet is seated, held in position by spring force.

- Flow is blocked at inlet.

Throttled Flow Condition

- System pressure has reached relief valve setting.

- Pressure has moved Poppet away from Seat, allowing flow to pass through valve.

- Valve is throttling flow to maintain relief pressure at inlet.

Figure 8-49. Cartridge pressure relief valve. *(Courtesy of Parker Hannifin Corp., Elyria, Ohio.)*

for fail-safe overload protection in case the compensator control on the pump fails to operate. Figure 8-51(b) is the symbolic representation of a partial circuit consisting of a pressure-compensated pump and a hydraulic fuse. A hydraulic fuse is analogous to an electrical fuse because they both are one-shot devices. On the other hand, a pressure relief valve is analogous to an electrical circuit breaker because they both are resettable devices.

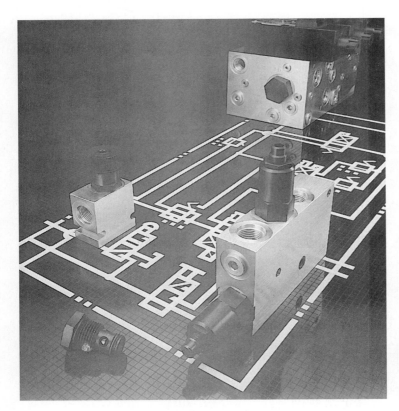

Figure 8-50. Integrated hydraulic circuit. *(Courtesy of Parker Hannifin Corp., Elyria, Ohio.)*

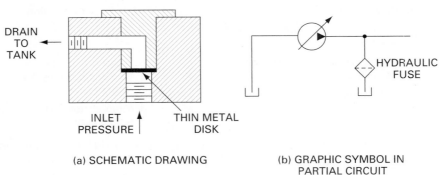

(a) SCHEMATIC DRAWING

(b) GRAPHIC SYMBOL IN PARTIAL CIRCUIT

Figure 8-51. Hydraulic fuse.

8.9 KEY EQUATIONS

Orifice flow rate versus pressure drop

Special English units:
$$Q(\text{gpm}) = 38.1 \, CA(\text{in}^2)\sqrt{\frac{\Delta p \, (\text{psi})}{SG}} \tag{8-1}$$

Metric units:
$$Q(\text{Lpm}) = 0.0851 \, CA(\text{mm}^2)\sqrt{\frac{\Delta p \, (\text{kPa})}{SG}} \tag{8-1M}$$

Flow control valve flow rate versus pressure drop

Special English units:
$$Q(\text{gpm}) = C_v\left(\frac{\text{gpm}}{\sqrt{\text{psi}}}\right)\sqrt{\frac{\Delta p \, (\text{psi})}{SG}} \tag{8-2}$$

Metric units:
$$Q(\text{Lpm}) = C_v\left(\frac{\text{Lpm}}{\sqrt{\text{kPa}}}\right)\sqrt{\frac{\Delta p \, (\text{kPa})}{SG}} \tag{8-2M}$$

EXERCISES

Questions, Concepts, and Definitions

8-1. What is the purpose of a directional control valve?

8-2. What is a check valve? What does it accomplish?

8-3. How does a pilot check valve differ from a simple check valve?

8-4. What is a four-way directional control valve?

8-5. What is a four-way, spring-centered, three-position valve?

8-6. Name three ways in which directional control valves may be actuated.

8-7. What is a solenoid? How does it work?

8-8. What is the difference between an open-center and closed-center type of directional control valve?

8-9. What is a shuttle valve? Name one application.

8-10. What is the purpose of a pressure relief valve?

8-11. What is a pressure-reducing valve? What is its purpose?

8-12. What does an unloading valve accomplish?

8-13. What is a sequence valve? What is its purpose?

8-14. Name one application of a counterbalance valve.

8-15. What is the purpose of a flow control valve?

8-16. Relative to flow control valves, define the term *capacity coefficient*.

8-17. What are the English and metric units of the term *capacity coefficient*?

8-18. What is a pressure-compensated flow control valve?

8-19. What is a servo valve? How does it work?

8-20. What is the difference between a mechanical-hydraulic and an electrohydraulic servo valve?

8-21. What is a hydraulic fuse? What electrical device is it analogous to?

8-22. What can be concluded about the pressures on the upstream and downstream sides of an orifice (and thus a valve) when oil is flowing through it?

8-23. Explain how the pilot-operated check shown in Figure 8-5 works.

8-24. Name one application for a pilot-operated check valve.

8-25. Explain how the four-way directional control valve of Figure 8-8 operates.

8-26. How do a simple pressure relief valve and a compound relief valve differ in operation?

8-27. How does an unloading valve differ from a sequence valve in mechanical construction?

8-28. Explain the operational features of the pressure-compensated flow control valve of Figure 8-36.

8-29. How many positions does a spring-offset valve have?

8-30. How many positions does a spring-centered valve have?

8-31. How are solenoids most often used in valves?

8-32. Name two ways of regulating flow to a hydraulic actuator.

8-33. What is cracking pressure?

8-34. Where are the ports of a relief valve connected?

8-35. Name the three basic functions of valves.

8-36. What is the difference in function between a pressure relief valve and a hydraulic fuse?

8-37. Relative to directional control valves, distinguish among the terms *position, way,* and *port.*

8-38. What is a cartridge valve?

8-39. What is the difference between slip-in and screw-type cartridge valves?

8-40. Name five benefits of using cartridge valves.

8-41. Name five different valve functions that can be provided using cartridge valves.

8-42. Relative to the use of cartridge valves, what are integrated hydraulic circuits?

Problems

Note: The letter *E* following an exercise number means that English units are used. Similarly, the letter *M* indicates metric units.

Pressure Relief Valve Settings

8-43E. A pressure relief valve contains a poppet with a 0.65-in^2 area on which system pressure acts. During assembly, a spring with a spring constant of 2000 lb/in is installed in the valve to hold the poppet against its seat. The adjustment mechanism is then set so that the spring initially compresses 0.15 in from its free-length condition. In order to pass full pump flow through the valve at the PRV pressure setting, the poppet must move 0.10 in from its fully closed position. Determine the

    a. Cracking pressure

    b. Full pump flow pressure (PRV pressure setting)

8-44E. What should be the initial compression of the spring in the PRV in Exercise 8-43 if the full pump flow pressure is to be 40% greater than the cracking pressure?

8-45M. A pressure relief valve contains a poppet with a 4.20-cm^2 area on which system pressure acts. During assembly a spring with a spring constant of 3200 N/cm is installed in the valve to hold the poppet against its seat. The adjustment mechanism is then set so that the spring is initially compressed 0.50 cm from its free-length condition.

In order to pass full pump flow through the valve at the PRV pressure setting, the poppet must move 0.30 cm from its fully closed position. Determine the

   a. Cracking pressure

   b. Full pump flow pressure (PRV pressure setting)

8-46M. What should be the initial compression of the spring in the PRV in Exercise 8-45 if the full pump flow pressure is to be 40% greater than the cracking pressure?

Power Losses in Valves

8-47E. A pressure relief valve has a pressure setting of 2000 psi. Compute the horsepower loss across this valve if it returns all the flow back to the tank from a 25-gpm pump.

8-48E. An unloading valve is used to unload the pump in Exercise 8-47. If the pump discharge pressure during unloading equals 30 psi, how much hydraulic horsepower is being wasted?

8-49M. A pressure relief valve has a pressure setting of 140 bars. Compute the kW power loss across this valve if it returns all the flow back to the tank from a 0.0016-m³/s pump.

8-50M. An unloading valve is used to unload the pump in Exercise 8-49. If the pump discharge pressure during unloading equals 2 bars, how much hydraulic kW power is being wasted?

Orifice Flow Rate Measurement

8-51E. A 2-in-diameter sharp-edged orifice is placed in a pipeline to measure flow rate. If the measured pressure drop is 50 psi and the fluid specific gravity is 0.90, find the flow rate in units of gpm.

8-52M. A 55-mm-diameter sharp-edged orifice is placed in a pipeline to measure flow rate. If the measured pressure drop is 300 kPa and the fluid specific gravity is 0.90, find the flow rate in units of m³/s.

8-53. For a given orifice and fluid, a graph can be generated showing the Δp-vs.-Q relationship. For the orifices and fluids in Exercises 8-51 and 8-52, plot the curves and check the answers obtained mathematically. What advantage does the graph have over the equation? What is the disadvantage of the graph?

Capacity Coefficient of Flow Control Valves

8-54E. Figure 4-20 in Chapter 4 shows the flow versus pressure drop curves for four valves when the fluid has a specific gravity of 0.9. What is the capacity coefficient of the valve identified by curve number 1 at a flow rate of 5 gpm?

8-55E. Without making any calculations, determine which of the four valves (identified by curve numbers 1, 2, 3, and 4 in Figure 4-20) has the largest capacity coefficient? Explain your answer.

8-56E. Determine the flow rate through a flow control valve that has a capacity coefficient of 1.5 gpm/$\sqrt{\text{psi}}$ and a pressure drop of 100 psi. The fluid is hydraulic oil with a specific gravity of 0.90.

8-57M. Determine the flow rate through a flow control valve that has a capacity coefficient of 2.2 Lpm/$\sqrt{\text{kPa}}$ and a pressure drop of 687 kPa. The fluid is hydraulic oil with a specific gravity of 0.90.

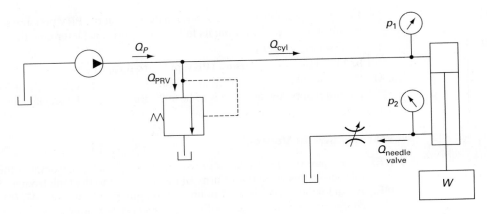

Figure 8-52. System for Exercise 8-58.

8-58E. The system of Figure 8-52 has a hydraulic cylinder with a suspended load W. The cylinder piston and rod diameters are 2 in and 1 in respectively. The pressure relief valve setting is 750 psi. Determine pressure p_2 for a constant cylinder speed if

a. $W = 2000$ lb

b. $W = 0$ (load is removed)

8-59E. For the system in Exercise 8-58 determine the cylinder speed for parts a and b if the flow control valve has a capacity coefficient of 0.5 gpm/$\sqrt{\text{psi}}$.

8-60M. Change the data in Exercise 8-58 to metric units and solve parts a and b.

8-61M. For the system in Exercise 8-60 determine the cylinder speeds for parts a and b if the flow control valve has a capacity coefficient of 0.72 Lpm/$\sqrt{\text{kPa}}$. fluid is hydraulic oil with a specific gravity of 0.90.

Hydraulic Circuit
Design and Analysis

Learning Objectives

Upon completing this chapter, you should be able to:

1. Describe the operation of complete hydraulic circuits drawn using graphic symbols for all components.
2. Troubleshoot hydraulic circuits to determine causes of malfunctions.
3. Determine the operating speeds and load-carrying capacities of regenerative cylinders.
4. Discuss the operation of air-over-oil circuits.
5. Understand the operation of mechanical-hydraulic servo systems.
6. Analyze hydraulic circuits to evaluate the safety of operation.
7. Design hydraulic circuits to perform a desired function.
8. Perform an analysis of hydraulic circuit operation, including the effects of frictional losses.
9. Analyze the speed control of hydraulic cylinders.

9.1 INTRODUCTION

Definition of Hydraulic Circuits

The material presented in previous chapters dealt with essentially fundamentals and hydraulic system components. In this chapter we discuss hydraulic circuits that primarily do not use basic electrical control devices such as push-button switches, pressure switches, limit switches, solenoids, relays, and timers. Similarly, Chapter 14 covers pneumatic circuits that primarily do not use these electrical control devices. Hydraulic and pneumatic circuits that use basic electrical control devices (shown both in the fluid power circuits and in separate electrical circuits called *ladder diagrams*) are presented in Chapter 15. Fluid power applications that use advanced electrical controls such as electrohydraulic servo systems and programmable logic controllers are presented in Chapter 17.

A hydraulic circuit is a group of components such as pumps, actuators, control valves, and conductors arranged so that they will perform a useful task. There are three important considerations when analyzing or designing hydraulic circuits:

1. Safety of operation
2. Performance of desired function
3. Efficiency of operation

It is very important for the fluid power technician or designer to have a working knowledge of components and how they operate in a circuit. Hydraulic circuits are developed using graphic symbols for all components. Before hydraulic circuits can be understood, it is necessary to know these fluid power symbols. Appendix G gives a table of symbols that conforms to the American National Standards Institute (ANSI) specifications. Many of these symbols have been presented in previous chapters, and ANSI symbols are used throughout this book. Although complete memorization of graphic symbols is not necessary, Appendix G should be studied so that the symbols become familiar. The discussions that follow will cover circuits that represent basic hydraulic technology. Convention requires that circuit diagrams be drawn with all components shown in their unactuated position.

Figure 9-1(a) provides a photograph of a hydrostatic transmission whose complete hydraulic circuit is given in Figure 9-1(b). This hydrostatic transmission, which is driven by an electric motor, is a completely integrated unit with all the controls and valving enclosed in a single, compact housing. Observe that the hydraulic circuit of Figure 9-1(b) uses graphic symbols for all the hydraulic components of the hydrostatic transmission, including the electric motor, hydraulic pump, hydraulic motor, oil reservoir, filter, and the various control valves.

Computer Analysis and Simulation of Fluid Power Systems

There are basically two types of computer software available for solving fluid power system problems. The first type performs only a mathematical analysis as it deals with the equations that define the characteristics of the components of the fluid power system. The second type recognizes the configuration of the entire fluid power system involved and thus can provide both a simulation of the system as well as a mathematical analysis.

The first type of software typically involves the programming of the equations that define the performance of the fluid power system. The operation of the system and how each component interacts within the system is defined mathematically, knowing the equations that define the characteristics of each component. The equations are programmed using computer software such as BASIC and Mathcad™. Desired input data is then fed into the computer and output results are printed out to display the

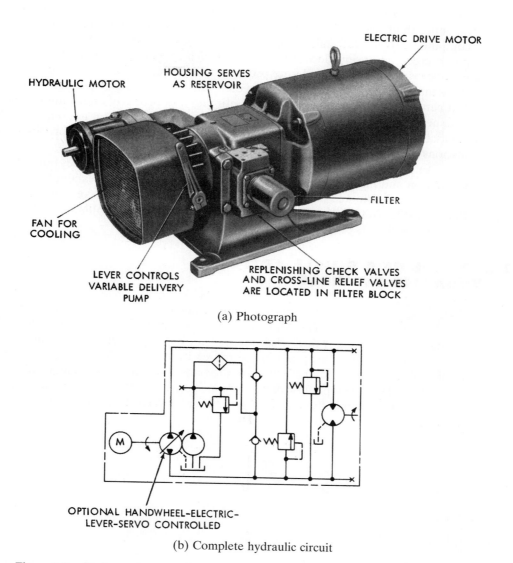

(a) Photograph

(b) Complete hydraulic circuit

Figure 9-1. Hydrostatic transmission. *(Courtesy of Sperry Vickers, Sperry Rand Corp., Troy, Michigan.)*

behavior of the fluid power system. With the use of this software, values of parameters such as pump flow rate, pump discharge pressure, and cylinder diameter can be repeatedly changed until optimum system performance is achieved. Other analysis-type software packages are available that specifically handle fluid power systems. One example is HydCalc, which performs a mathematical analysis without the user having to program equations.

The second type of computer software system recognizes the configuration of an entire fluid power circuit along with the characteristics of each component. This computer software recognizes the circuit as it is being created by the user on a computer screen. The user selects components from a computer library and hooks them together via a click, drag, and drop procedure. Mathematical analysis along with circuit simulation is graphically displayed on a computer screen. The design of the circuit can be readily modified until an optimum and properly operating system is obtained. The programming of component and system equations is not required. The computer software not only provides a mathematical analysis but also provides simulations of the operation of the fluid power circuit.

Automation Studio™ is an example of the software that provides both system simulation and mathematical analysis. Chapter 18 discusses the salient features of Automation Studio and how it is utilized. Also included with this textbook is a CD that illustrates Automation Studio in action in the design and simulation of 16 of the fluid power circuits provided throughout this book.

9.2 CONTROL OF A SINGLE-ACTING HYDRAULIC CYLINDER

Figure 9-2 shows how a two-position, three-way, manually actuated, spring-offset directional control valve (DCV) can be used to control the operation of a single-acting cylinder. In the spring-offset mode, full pump flow goes to the tank via the pressure relief valve. The spring in the rod end of the cylinder retracts the piston as oil from the blank end drains back to the tank. When the valve is manually actuated into its left envelope flow path configuration, pump flow extends the cylinder. At full extension, pump flow goes through the relief valve. Deactivation of the DCV allows the cylinder to retract as the DCV shifts into its spring-offset mode.

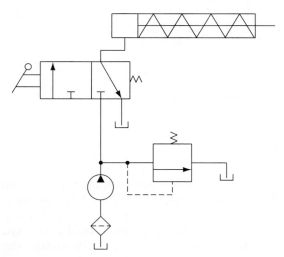

Figure 9-2. Control of single-acting hydraulic cylinder.

9.3 CONTROL OF A DOUBLE-ACTING HYDRAULIC CYLINDER

Figure 9-3 gives a circuit used to control a double-acting hydraulic cylinder. The operation is described as follows:

1. When the four-way valve is in its spring-centered position (tandem design), the cylinder is hydraulically locked. Also, the pump is unloaded back to the tank at essentially atmospheric pressure.

2. When the four-way valve is actuated into the flow path configuration of the left envelope, the cylinder is extended against its load force F_{load} as oil flows from port P through port A. Also, oil in the rod end of the cylinder is free to flow back to the tank via the four-way valve from port B through port T. Note that the cylinder could not extend if this oil were not allowed to leave the rod end of the cylinder.

3. When the four-way valve is deactivated, the spring-centered envelope prevails, and the cylinder is once again hydraulically locked.

4. When the four-way valve is actuated into the right envelope configuration, the cylinder retracts as oil flows from port P through port B. Oil in the blank end is returned to the tank via the flow path from port A to port T.

5. At the ends of the stroke, there is no system demand for oil. Thus, the pump flow goes through the relief valve at its pressure-level setting unless the four-way valve is deactivated. In any event, the system is protected from any cylinder overloads.

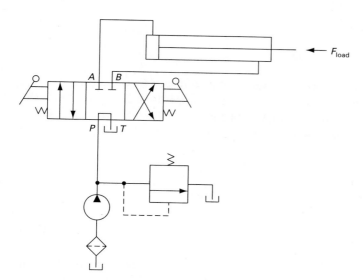

Figure 9-3. Control of a double-acting hydraulic cylinder. (This circuit is simulated on the CD included with the textbook.)

9.4 REGENERATIVE CYLINDER CIRCUIT

Operation

Figure 9-4(a) shows a regenerative circuit that is used to speed up the extending speed of a double-acting hydraulic cylinder. Note that the pipelines to both ends of the hydraulic cylinder are connected in parallel and that one of the ports of the four-way valve is blocked. A common method used to block a valve port is to simply screw a threaded plug into the port opening. The operation of the cylinder during the retraction stroke is the same as that of a regular double-acting cylinder. Fluid flows through the DCV via the right envelope during retraction. In this mode, fluid from the pump bypasses the DCV and enters the rod end of the cylinder. Fluid in the blank end drains back to the tank through the DCV as the cylinder retracts.

When the DCV is shifted into its left envelope configuration, the cylinder extends as shown in Figure 9-4(b). The speed of extension is greater than that for a regular double-acting cylinder because flow from the rod end (Q_R) regenerates with the pump flow (Q_P) to provide a total flow rate (Q_T), which is greater than the pump flow rate to the blank end of the cylinder.

Cylinder Extending Speed

The equation for the extending speed can be obtained as follows, referring to Figure 9-4(b): The total flow rate (Q_T) entering the blank end of the cylinder equals the pump flow rate (Q_P) plus the regenerative flow rate (Q_R) coming from the rod end of the cylinder:

$$Q_T = Q_P + Q_R$$

Solving for the pump flow rate, we have

$$Q_P = Q_T - Q_R$$

We know that the total flow rate equals the piston area (A_P) multiplied by the extending speed of the piston ($v_{P_{ext}}$). Similarly, the regenerative flow rate equals the difference of the piston and rod areas ($A_P - A_r$) multiplied by the extending speed of the piston. Substituting these two relationships into the preceding equation yields

$$Q_P = A_P v_{P_{ext}} - (A_P - A_r)v_{P_{ext}}$$

Solving for the extending speed of the piston, we have

$$v_{P_{ext}} = \frac{Q_P}{A_r} \tag{9-1}$$

From Eq. (9-1), we see that the extending speed equals the pump flow rate divided by the area of the rod. Thus, a small rod area (which produces a large

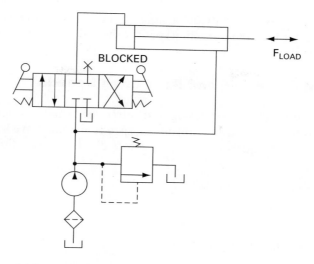

(a) Complete circuit.

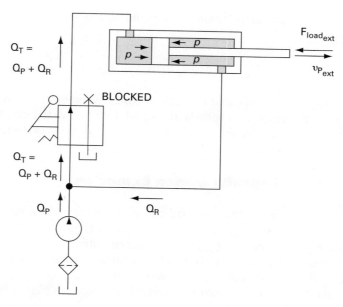

(b) Partial circuit showing flow paths during cylinder extension stroke.

Figure 9-4. Regenerative cylinder circuit.

regenerative flow rate) provides a large extending speed. In fact the extending speed can be greater than the retracting speed if the rod area is made small enough.

Ratio of Extending and Retracting Speeds

Let's find the ratio of extending and retracting speeds to determine under what conditions the extending and retracting speeds are equal. We know that the retracting speed $(v_{P_{\text{ret}}})$ equals the pump flow rate divided by the difference of the piston and rod areas:

$$v_{P_{\text{ret}}} = \frac{Q_P}{A_P - A_r} \tag{9-2}$$

Dividing Eq. (9-1) by Eq. (9-2), we have

$$\frac{v_{P_{\text{ext}}}}{v_{P_{\text{ret}}}} = \frac{Q_P/A_r}{Q_P/(A_P - A_r)} = \frac{A_P - A_r}{A_r}$$

On further simplification we obtain the desired equation:

$$\frac{v_{P_{\text{ext}}}}{v_{P_{\text{ret}}}} = \frac{A_P}{A_r} - 1 \tag{9-3}$$

From Eq. (9-3), we see that when the piston area equals two times the rod area, the extension and retraction speeds are equal. In general, the greater the ratio of piston area to rod area, the greater the ratio of extending speed to retracting speed.

Load-Carrying Capacity During Extension

It should be kept in mind that the load-carrying capacity of a regenerative cylinder during extension is less than that obtained from a regular double-acting cylinder. The load-carrying capacity $(F_{\text{load}_{\text{ext}}})$ for a regenerative cylinder during extension equals the pressure times the piston rod area rather than the pressure times piston area. This is because system pressure acts on both sides of the piston during the extending stroke of the regenerative cylinder, as shown in Figure 9-4(b). This is in accordance with Pascal's law, and thus we have

$$F_{\text{load}_{\text{ext}}} = pA_r \tag{9-4}$$

Thus, we are not obtaining more power from the regenerative cylinder during extension because the extending speed is increased at the expense of reduced load-carrying capacity.

EXAMPLE 9-1

A double-acting cylinder is hooked up in the regenerative circuit of Figure 9-4(a). The cracking pressure for the relief valve is 1000 psi. The piston area is 25 in² and the rod area is 7 in². The pump flow is 20 gpm. Find the cylinder speed, load-carrying capacity, and power delivered to the load (assuming the load equals the cylinder load-carrying capacity) during the

  a. Extending stroke
  b. Retracting stroke

Solution

  a. $v_{P_{ext}} = \dfrac{Q_P}{A_r} = \dfrac{(20 \text{ gpm})(231 \text{ in}^3/1 \text{ gal})(1 \text{ min}/60 \text{ s})}{7 \text{ in}^2} = 11.0 \text{ in/s}$

  $F_{load_{ext}} = pA_r = 1000 \text{ lb/in}^2 \times 7 \text{ in}^2 = 7000 \text{ lb}$

  $\text{Power}_{ext} = F_{load_{ext}}v_{P_{ext}} = 7000 \text{ lb} \times 11.0 \text{ in/s} = 77,000 \text{ in} \cdot \text{lb/s} = 11.7 \text{ hp}$

  b. $v_{P_{ret}} = \dfrac{Q_P}{A_p - A_r} = \dfrac{20 \times \dfrac{231}{60}}{25 - 7} = 4.28 \text{ in/s}$

  $F_{load_{ret}} = p(A_P - A_r) = 1000 \text{ lb/in}^2 \times (25 - 7) \text{ in}^2 = 18,000 \text{ lb}$

  $\text{Power}_{ret} = F_{load_{ret}}v_{P_{ret}} = 18,000 \text{ lb} \times 4.28 \text{ in/s} = 77,000 \text{ in} \cdot \text{lb/s} = 11.7 \text{ hp}$

Note that the hydraulic horsepower delivered by the pump during both the extending and retracting strokes can be found as follows:

  $HP_{pump} = \dfrac{p(\text{psi}) \times Q_{pump}(\text{gpm})}{1714} = \dfrac{1000 \text{ psi} \times 20 \text{ gpm}}{1714} = 11.7 \text{ hp}$

Thus, as expected, the hydraulic power delivered by the pump equals the power delivered to the loads during both the extending and retracting strokes.

Drilling Machine Application

Figure 9-5 shows an application using a four-way valve having a spring-centered design with a closed tank port and a pressure port open to outlet ports *A* and *B*.

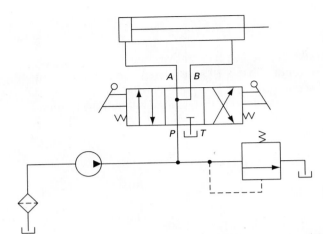

Figure 9-5. Drilling machine application. (This circuit is simulated on the CD included with textbook.)

The application is for a drilling machine, where the following operations take place:

1. The spring-centered position gives rapid spindle advance (extension).
2. The left envelope mode gives slow feed (extension) when the drill starts to cut into the workpiece.
3. The right envelope mode retracts the piston.

Why does the spring-centered position give rapid extension of the cylinder (drill spindle)? The reason is simple. Oil from the rod end regenerates with the pump flow going to the blank end. This effectively increases pump flow to the blank end of the cylinder during the spring-centered mode of operation. Once again we have a regenerative cylinder. It should be noted that the cylinder used in a regenerative circuit is actually a regular double-acting cylinder. What makes it a regenerative cylinder is the way it is hooked up in the circuit. The blank and rod ends are connected in parallel during the extending stroke of a regenerative cylinder. The retraction mode is the same as a regular double-acting cylinder.

9.5 PUMP-UNLOADING CIRCUIT

In Figure 9-6 we see a circuit using an unloading valve to unload a pump. The unloading valve opens when the cylinder reaches the end of its extension stroke because the check valve keeps high-pressure oil in the pilot line of the unloading valve. When the DCV is shifted to retract the cylinder, the motion of the piston reduces the pressure in the pilot line of the unloading valve. This resets the unloading valve until the cylinder is fully retracted, at which point the unloading valve unloads the pump. Thus, the unloading valve unloads the pump at the ends of the extending and retraction strokes as well as in the spring-centered position of the DCV.

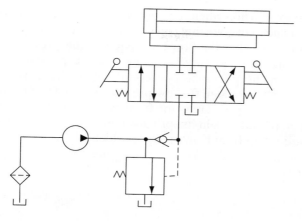

Figure 9-6. Pump-unloading circuit.

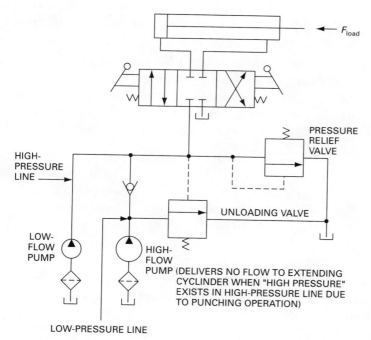

Figure 9-7. Double-pump
hydraulic system.

9.6 DOUBLE-PUMP HYDRAULIC SYSTEM

Figure 9-7 shows a circuit that uses a high-pressure, low-flow pump in conjunction with a low-pressure, high-flow pump. A typical application is a sheet metal punch press in which the hydraulic ram (cylinder) must extend rapidly over a great distance with very low pressure but high flow-rate requirements. This rapid extension of the cylinder occurs under no external load as the punching tool (connected to the end of the cylinder piston rod) approaches the sheet metal strip to be punched.

However, during the short motion portion when the punching operation occurs, the pressure requirements are high due to the punching load. During the punching operation, the cylinder travel is small and thus the flow-rate requirements are low.

The circuit shown eliminates the necessity of having a very expensive high-pressure, high-flow pump. When the punching operation begins, the increased pressure opens the unloading valve to unload the low-pressure pump. The purpose of the relief valve is to protect the high-pressure pump from overpressure at the end of the cylinder stroke and when the DCV is in its spring-centered mode. The check valve protects the low-pressure pump from high pressure, which occurs during the punching operation, at the ends of the cylinder stroke, and when the DCV is in its spring-centered mode.

EXAMPLE 9-2

For the double-pump system of Figure 9-7, what should be the pressure settings of the unloading valve and pressure relief valve under the following conditions?

a. Sheet metal punching operation requires a force of 2000 lb.

b. Hydraulic cylinder has a 1.5-in-diameter piston and 0.5-in-diameter rod.

c. During rapid extension of the cylinder, a frictional pressure loss of 100 psi occurs in the line from the high-flow pump to the blank end of the cylinder. During the same time a 50-psi pressure loss occurs in the return line from the rod end of the cylinder to the oil tank. Frictional pressure losses in these lines are negligibly small during the punching operation.

d. Assume that the unloading valve and pressure relief valve pressure settings (for their full pump flow requirements) should be 50% higher than the pressure required to overcome frictional pressure losses and the cylinder punching load, respectively.

Solution

Unloading Valve:
Back-pressure force on the cylinder equals the product of the pressure loss in the return line and the effective area of the cylinder $(A_p - A_r)$.

$$F_{\text{back pressure}} = 50 \frac{\text{lb}}{\text{in}^2} \times \frac{\pi}{4}(1.5^2 - 0.5^2)\text{in}^2 = 78.5 \text{ lb}$$

Pressure at the blank end of the cylinder required to overcome back-pressure force equals the back-pressure force divided by the area of the cylinder piston.

$$p_{\text{cyl blank end}} = \frac{78.5 \text{ lb}}{\frac{\pi}{4}(1.5 \text{ in})^2} = 44.4 \text{ psi}$$

Thus, the pressure setting of the unloading valve should be $1.50(100 + 44.4)\text{psi} = 217$ psi.

Pressure Relief Valve:

The pressure required to overcome the punching operation equals the punching load divided by the area of the cylinder piston.

$$p_{\text{punching}} = \frac{2000 \text{ lb}}{\frac{\pi}{4}(1.5 \text{ in})^2} = 1132 \text{ psi}$$

Thus, the pressure setting of the pressure relief valve should be 1.50×1132 psi $= 1698$ psi.

EXAMPLE 9-3

For the system of Example 9-2, the poppet of the pressure relief valve must move 0.10 in from its fully closed position in order to pass the full pump flow at the PRV setting (full pump flow pressure). The poppet has a 0.75-in² area on which system pressure acts. Assuming that the relief valve cracking pressure should be 10% higher than the pressure required to overcome the hydraulic cylinder punching operation, find the required

  a. Spring constant of the compression spring in the PRV
  b. Initial compression of the spring from its free-length condition as set by the spring-adjustment mechanism of the PRV (poppet held against its seat by spring)

Solution

  a. At full pump flow pressure (PRV setting), the spring force equals the hydraulic force acting on the poppet.

$$\text{Spring force} = \text{hydraulic force}$$

$$kS = p_{\text{PRV setting}}A_{\text{poppet}} = 1698\frac{\text{lb}}{\text{in}^2} \times 0.75 \text{ in}^2 = 1274 \text{ lb}$$

where

  k = spring constant (lb/in),
  S = total spring compression (in)
    = spring initial compression (l) plus full poppet stroke
    = $l + 0.10$.

Substituting the expression for S into the preceding equation yields the first of two force equations.

$$k(l + 0.10) = 1274 \text{ lb}$$

$$\text{or } kl + 0.10\,k = 1274 \text{ lb}$$

We also know that at the cracking pressure of the relief valve, the spring force equals the hydraulic force acting on the poppet. This yields the second force equation.

$$kl = p_{\text{cracking}}A_{\text{poppet}} = (1.10 \times 1132 \text{ lb/in}^2) \times 0.75 \text{ in}^2$$

or

$$kl = 934 \text{ lb}$$

Substituting the 934-lb value for kl into the first force equation yields

$$934 + 0.10k = 1274 \qquad \text{or} \qquad k = 3400 \text{ lb/in}$$

b. We can solve for the initial compression of the spring by substituting the spring constant value into the second force equation.

$$3400 \text{ lb/in} \times l(\text{in}) = 934 \text{ lb} \qquad \text{or} \qquad l = 0.275 \text{ in}$$

9.7 COUNTERBALANCE VALVE APPLICATION

Figure 9-8 illustrates the use of a counterbalance or back-pressure valve to keep a vertically mounted hydraulic cylinder in the upward position while the pump is idling. The counterbalance valve (CBV) is set to open at somewhat above the pressure required to prevent the vertical cylinder from descending due to the weight of its load. This permits the cylinder to be forced downward when pressure is applied on the top. The open-center directional control valve unloads the pump. The DCV is a solenoid-actuated, spring-centered valve with an open-center flow path configuration.

9.8 HYDRAULIC CYLINDER SEQUENCING CIRCUITS

As stated earlier, a sequence valve causes operations in a hydraulic circuit to behave sequentially. Figure 9-9 is an example where two sequence valves are used to control the sequence of operations of two double-acting cylinders. When the DCV is shifted into its left envelope mode, the left cylinder extends completely, and then the right cylinder extends. If the DCV is then shifted into its right envelope mode, the right cylinder retracts fully, and then the left cylinder retracts. This sequence of

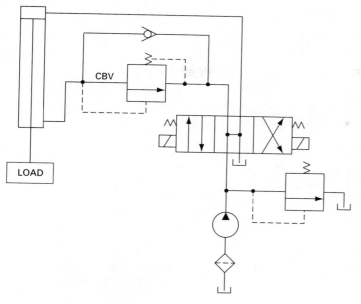

Figure 9-8. Counterbalance valve application.

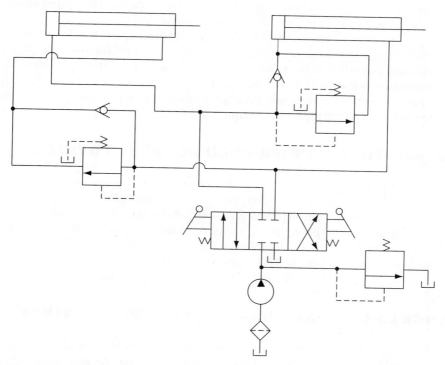

Figure 9-9. Hydraulic cylinder sequence circuit. (This circuit is simulated on the CD included with this textbook.)

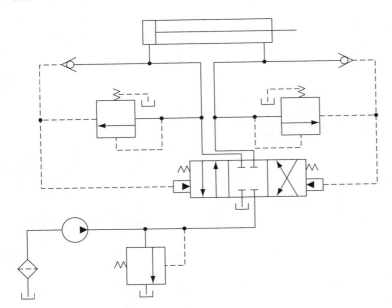

Figure 9-10. Automatic cylinder reciprocating system.

cylinder operation is controlled by the sequence valves. The spring-centered position of the DCV locks both cylinders in place.

One application of this circuit is a production operation. For example, the left cylinder could extend and clamp a workpiece via a power vise jaw. Then the right cylinder extends to drive a spindle to drill a hole in the workpiece. The right cylinder then retracts the drill spindle, and then the left cylinder retracts to release the workpiece for removal. Obviously these machining operations must occur in the proper sequence as established by the sequence valves in the circuit.

9.9 AUTOMATIC CYLINDER RECIPROCATING SYSTEM

Figure 9-10 is a circuit that produces continuous reciprocation of a hydraulic cylinder. This is accomplished by using two sequence valves, each of which senses a stroke completion by the corresponding buildup of pressure. Each check valve and corresponding pilot line prevents shifting of the four-way valve until the particular stroke of the cylinder has been completed. The check valves are needed to allow pilot oil to leave either end of the DCV while pilot pressure is applied to the opposite end. This permits the spool of the DCV to shift as required.

9.10 LOCKED CYLINDER USING PILOT CHECK VALVES

In many cylinder applications, it is necessary to lock the cylinder so that its piston cannot be moved due to an external force acting on the piston rod. One method for locking a cylinder in this fashion is by using pilot check valves, as shown in Figure 9-11. The cylinder can be extended and retracted as normally done by the

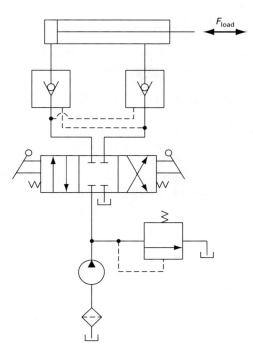

Figure 9-11. Locked cylinder using pilot check valves.

action of the directional control valve. If regular check valves were used, the cylinder could not be extended or retracted by the action of the DCV. An external force, acting on the piston rod, will not move the piston in either direction because reverse flow through either pilot check valve is not permitted under these conditions.

9.11 CYLINDER SYNCHRONIZING CIRCUITS

Cylinders Connected in Parallel

Figure 9-12 is a very interesting circuit, which seems to show how two identical cylinders can be synchronized by piping them in parallel. However, even if the two cylinders are identical, it would be necessary for the loads on the cylinders to be identical in order for them to extend in exact synchronization. If the loads are not exactly identical (as is always the case), the cylinder with the smaller load would extend first because it would move at a lower pressure level. After this cylinder has fully completed its stroke, the system pressure will increase to the higher level required to extend the cylinder with the greater load. It should be pointed out that no two cylinders are really identical. For example, differences in packing friction will vary from cylinder to cylinder. This alone would prevent cylinder synchronization for the circuit of Figure 9-12.

Cylinders Connected in Series

The circuit of Figure 9-13 shows that hooking two cylinders in series is a simple way to synchronize the two cylinders. For example, during the extending stroke of the cylinders, fluid from the pump is delivered to the blank end of cylinder 1 via the flow path shown in the upper envelope of the DCV. As cylinder 1 extends, fluid from its rod end is delivered to the blank end of cylinder 2. Note that both ends of cylinders and the entire pipeline between the cylinders is filled with fluid. Fluid returns to the oil tank from the rod end of cylinder 2, as it extends, via the DCV. For the two cylinders to be synchronized, the piston area of cylinder 2 must equal the difference between the areas of the piston and rod for cylinder 1. This can be shown by applying the continuity equation which states that the rate at which fluid leaves the rod end of cylinder 1 must equal the rate at which fluid enters cylinder 2. Thus, we have for a hydraulic fluid

$$Q_{\text{out(cyl 1)}} = Q_{\text{in(cyl 2)}}$$

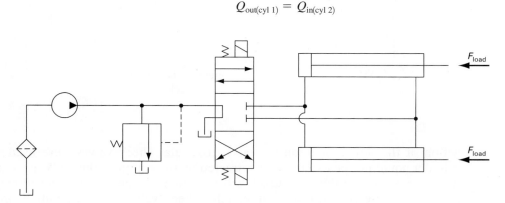

Figure 9-12. Cylinders hooked in parallel will not operate in synchronization.

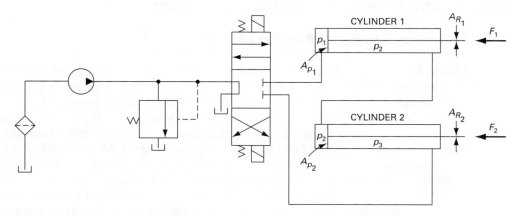

Figure 9-13. Cylinders hooked in series will operate in synchronization.

Since $Q = Av$ where A is the effective area through which fluid flows, we have

$$(A_{eff}v)_{cyl\,1} = (A_{eff}v)_{cyl\,2}$$

thus

$$(A_{P_1} - A_{R_1})v_1 = A_{P_2}v_2$$

Since for synchronization $v_1 = v_2$, we have the desired result:

$$A_{P_1} - A_{R_1} = A_{P_2}$$

It should be noted that the pump must be capable of delivering a pressure equal to that required for the piston of cylinder 1 by itself to overcome the loads acting on both extending cylinders. This is shown as follows, noting that the pressures are equal at the blank end of cylinder 2 and at the rod end of cylinder 1 per Pascal's law (refer to Figure 9-13 for area, load, and pressure identifications):

Summing forces on cylinder 1 yields

$$p_1 A_{P_1} - p_2(A_{P_1} - A_{R_1}) = F_1$$

Repeating this force summation on cylinder 2, we have

$$p_2 A_{P_2} - p_3(A_{P_2} - A_{R_2}) = F_2$$

Adding the preceding two equations together and noting that $A_{P_2} = A_{P_1} - A_{R_1}$ and that $p_3 = 0$ (due to the drain line to the tank), we obtain the desired result:

$$p_1 A_{P_1} = F_1 + F_2 \tag{9-5}$$

9.12 FAIL-SAFE CIRCUITS

Protection from Inadvertent Cylinder Extension

Fail-safe circuits are those designed to prevent injury to the operator or damage to equipment. In general they prevent the system from accidentally falling on an operator, and they also prevent overloading of the system. Figure 9-14 shows a fail-safe circuit that prevents the cylinder from accidentally falling in the event a hydraulic line ruptures or a person inadvertently operates the manual override on the pilot-actuated directional control valve when the pump is not operating. To lower the cylinder, pilot pressure from the blank end of the piston must pilot-open the check valve at the rod end to allow oil to return through the DCV to the tank. This happens when the push-button valve is actuated to permit pilot pressure actuation of the DCV or when the DCV is directly manually actuated while the pump is operating. The pilot-operated DCV allows free flow in the opposite direction to retract the cylinder when this DCV returns to its spring-offset mode.

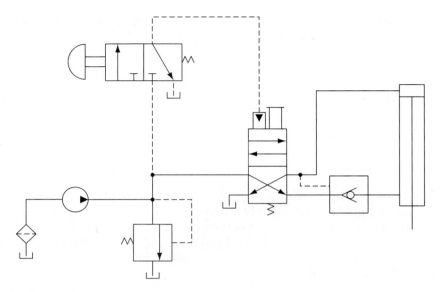

Figure 9-14. Fail-safe circuit.

Fail-Safe System with Overload Protection

Figure 9-15 shows a fail-safe circuit that provides overload protection for system components. Directional control valve 1 is controlled by push-button, three-way valve 2. When overload valve 3 is in its spring-offset mode, it drains the pilot line of valve 1. If the cylinder experiences excessive resistance during the extension stroke, sequence valve 4 pilot-actuates overload valve 3. This drains the pilot line of valve 1, causing it to return to its spring-offset mode. If a person then operates push-button valve 2, nothing will happen unless overload valve 3 is manually shifted into its blocked port configuration. Thus, the system components are protected against excessive pressure due to an excessive cylinder load during its extension stroke.

Two-Handed Safety System

The safety circuit of Figure 9-16 is designed to protect an operator from injury. For the circuit to function (extend and retract the cylinder), the operator must depress both manually actuated valves via the push buttons. Furthermore, the operator cannot circumvent this safety feature by tying down one of the buttons, because it is necessary to release both buttons to retract the cylinder. When the two buttons are depressed, the main three-position directional control valve is pilot-actuated to extend the cylinder. When both push buttons are released, the cylinder retracts.

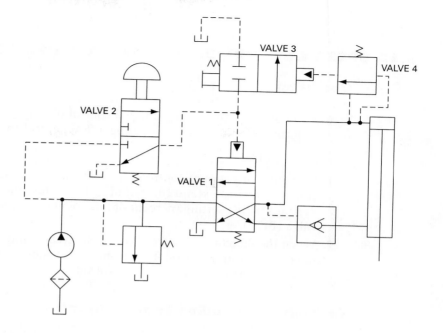

Figure 9-15. Fail-safe circuit with overload protection.

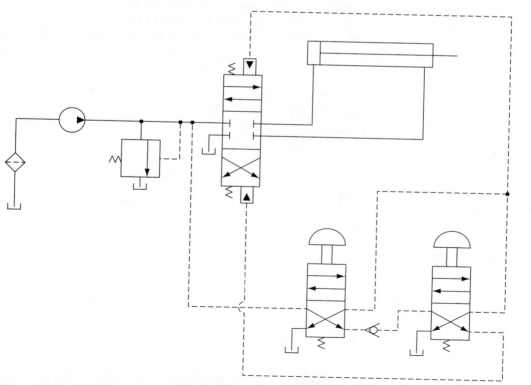

Figure 9-16. Two-handed safety circuit. (This circuit is simulated on the CD included with this textbook.)

9.13 SPEED CONTROL OF A HYDRAULIC CYLINDER

Operation

Figure 9-17 shows a circuit where speed control of a hydraulic cylinder is accomplished during the extension stroke using a flow control valve. The operation is as follows:

1. When the directional control valve is actuated, oil flows through the flow control valve to extend the cylinder. The extending speed of the cylinder depends on the setting (percent of full opening position) of the flow control valve (FCV).
2. When the directional control valve is deactuated into its spring-offset mode, the cylinder retracts as oil flows from the cylinder to the oil tank through the check valve as well as the flow control valve.

Analysis of Extending Speed Control

During the extension stroke, if the flow control valve is fully open, all the flow from the pump goes to the cylinder to produce maximum cylinder speed. As the flow control valve is partially closed its pressure drop increases. This causes an increase

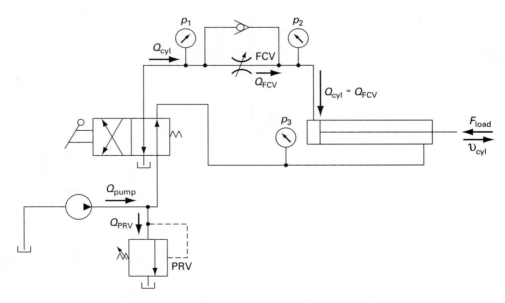

Figure 9-17.   Meter-in speed control of hydraulic cylinder during extending stroke using flow control valve. (*DCV is in manually actuated position.*)

in pressure p_1. Continued closing of the flow control valve ultimately results in pressure p_1 reaching and exceeding the cracking pressure of the pressure relief valve (PRV). The result is a slower cylinder speed since part of the pump flow goes back to the oil tank through the PRV. For the desired cylinder speed, pressure p_1 approximately equals the PRV setting, and the amount of pump flow that is not desired by the cylinder flows through the PRV. An analysis to determine the extending speed is given as follows:

The flow rate to the cylinder equals pump flow rate minus the flow rate through the PRV.

$$Q_{cyl} = Q_{pump} - Q_{PRV} \tag{9-6}$$

The flow rate through the flow control valve (FCV) is governed by Eq. (9-7).

$$Q_{FCV} = C_v\sqrt{\frac{\Delta p}{SG}} = C_v\sqrt{\frac{p_1 - p_2}{SG}} \tag{9-7}$$

where Δp = pressure drop across FCV,

        C_v = capacity coefficient of FCV,

        SG = specific gravity of oil,

Pressure $p_1 = p_{PRV}$ = PRV setting.

Also, pressure $p_3 = 0$ (ignoring small frictional pressure drop in drain line from rod end of cylinder to oil tank).

Pressure p_2 can be obtained by summing forces on the hydraulic cylinder.

$$p_2 A_{piston} = F_{load} \qquad \text{or} \qquad p_2 = F_{load}/A_{piston} \tag{9-8}$$

Also, the extending speed of the cylinder can be represented as a function of the flow rate through the flow control valve as follows:

$$v_{cyl} = Q_{cyl}/A_{piston} = Q_{FCV}/A_{piston}$$

Combining the preceding equation with Eqs. (9-7) and (9-8) yields the final result.

$$v_{cyl} = \frac{C_v}{A_{piston}}\sqrt{\frac{p_{PRV} - F_{load}/A_{piston}}{SG}} \tag{9-9}$$

As can be seen by Eq. (9-9), by varying the setting of the flow control valve, and thus the value of C_v, the desired extending speed of the cylinder can be achieved.

Meter-In Versus Meter-Out Flow Control Valve Systems

The circuit of Figure 9-17 depicts a meter-in flow control system, in which the flow control valve is placed in the line leading to the inlet port of the cylinder. Hence a meter-in flow control system controls the oil flow rate into the cylinder. Conversely, a meter-out flow control system is one in which the flow control valve is placed in the outlet line of the hydraulic cylinder. As shown in Figure 9-18, a meter-out flow control system controls the oil flow rate out of the cylinder.

Meter-in systems are used primarily when the external load opposes the direction of motion of the hydraulic cylinder. An example of the opposite situation is the case of a weight pulling downward on the piston rod of a vertical cylinder. In this case the weight would suddenly drop by pulling the piston rod down if a meter-in system is used even if the flow control valve is completely closed. Thus, the meter-out system is generally preferred over the meter-in type. One drawback of a meter-out system is the possibility of excessive pressure buildup in the rod end of the cylinder while it is extending. This is due to the magnitude of back pressure that the flow control valve can create depending on its nearness to being fully closed as well as the size of the external load and the piston-to-rod area ratio of the cylinder. In addition an excessive pressure buildup in the rod end of the cylinder results in a large pressure drop across the flow control valve. This produces the undesirable effect of a high heat generation rate with a resulting increase in oil temperature.

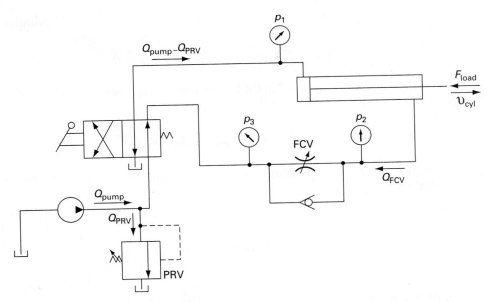

Figure 9-18. Meter-out speed control of hydraulic cylinder during extending stroke using flow control valve. (*DCV is in manually actuated position.*)

EXAMPLE 9-4

For the meter-in system of Figure 9-17 the following data are given:

$$\text{Valve capacity coefficient} = 1.0 \text{ gpm}/\sqrt{\text{psi}}$$
$$\text{Cylinder piston diameter} = 2 \text{ in (area} = 3.14 \text{ in}^2)$$
$$\text{Cylinder load} = 4000 \text{ lb}$$
$$\text{Specific gravity of oil} = 0.90$$
$$\text{Pressure relief valve setting} = 1400 \text{ psi}$$

Determine the cylinder speed.

Solution   The units for the terms in Eq. (9-9) are

$$C_v = \text{gpm}/\sqrt{\text{psi}}$$
$$Q = v_{cyl} A_{piston} = \text{gpm}$$
$$p_{PRV} = \text{psi}$$
$$F_{load}/A_{piston} = \text{psi}$$

Thus, from Eq. (9-9) we have

$$(v_{cyl}A_{piston})_{gpm} = C_v\sqrt{\frac{p_{PRV} - F_{load}/A_{piston}}{SG}} = 1.0\sqrt{\frac{1400 - 4000/3.14}{0.9}} = 11.8 \text{ gpm}$$

$$= 11.8\frac{\text{gal}}{\text{min}} \times \frac{231 \text{ in}^3}{1 \text{ gal}} \times \frac{1 \text{ min}}{60 \text{ s}} = 45.4 \text{ in}^3/\text{s}$$

Solving for the cylinder speed we have

$$v_{cyl} = \frac{45.4 \text{ in}^3/\text{s}}{A_{piston}(\text{in}^2)} = \frac{45.4}{3.14} = 14.5 \text{ in/s}$$

Note that the cylinder speed is directly proportional to the capacity coefficient of the flow control valve. Thus, for example, if the flow control valve is partially closed from its current position until its flow coefficient is cut in half (reduced from 1.0 to 0.5), the cylinder speed is cut in half (reduced from 14.5 in/s to 7.25 in/s).

EXAMPLE 9-5

For the meter-out system of Figure 9-19, which uses a suspended load, determine the pressure on each pressure gage during constant speed extension of the cylinder for

  a. No load
  b. 20,000-N load

The cylinder piston and rod diameters are 50 mm and 25 mm, respectively, and the PRV setting is 10 MPa.

Solution

  a. During extension (load is lowered) the flow rate to the cylinder is less than the pump flow rate due to the operation of the flow control valve to achieve the desired cylinder speed. The excess flow goes through the PRV. Thus, during extension, p_1 equals approximately the pressure relief valve setting of 10 MPa (assumes negligibly small pressure drops through the DCV and the pipeline connecting the pump to the blank end of the cylinder). This means that the downward hydraulic force acting on the cylinder piston equals the pressure relief valve setting of 10 MPa multiplied by the cylinder piston area.

$$F = 10 \times 10^6 \frac{N}{m^2} \times \frac{\pi}{4}(0.050 \text{ m})^2 = 19,600 \text{ N}$$

For a constant speed cylinder, the total net force acting on the piston and rod combination must equal zero. Thus, for no load, we have

$$p_1 A_p = p_2(A_p - A_r)$$

$$19,600 \text{ N} = p_2 \times \frac{\pi}{4}(0.050^2 - 0.025^2)\text{m}^2$$

$$p_2 = 13.3 \text{ MPa}$$

Finally, p_3 equals approximately zero because this location is in the return pipeline where oil drains back to the reservoir via the directional control valve (assumes negligibly small pressure drop through the DCV and pipeline returning oil back to the reservoir).

  b. For the case where the load equals 20,000 N, pressures p_1 and p_3 are the same as that found for the no-load case. However, pressure p_2 is found once again by noting that the total net force acting on the piston and rod combination must equal zero. Thus, for a load of 20,000 N, we have

$$19,600 \text{ N} + 20,000 \text{ N} = p_2 \times \frac{\pi}{4}(0.050^2 - 0.025^2)\text{m}^2$$

$$p_2 = 26.9 \text{ MPa}$$

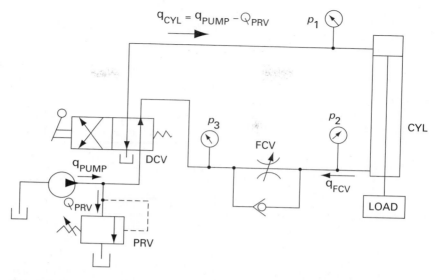

Figure 9-19. System for Example 9-5.

This example shows that the pressure in the rod end of the hydraulic cylinder is 2.69 times greater than the PRV setting. This large pressure must be taken into account in terms of its potential to damage system components. Also, the large pressure drop of 26.9 MPa across the flow control valve must be taken into account in terms of its potential to create a high heat generation rate and thus a large increase in oil temperature.

9.14 SPEED CONTROL OF A HYDRAULIC MOTOR

Figure 9-20 shows a circuit where speed control of a hydraulic motor is accomplished using a pressure-compensated flow control valve.

The operation is as follows:

1. In the spring-centered position of the tandem four-way valve, the motor is hydraulically locked.
2. When the four-way valve is actuated into the left envelope, the motor rotates in one direction. Its speed can be varied by adjusting the setting of the throttle of the flow control valve. In this way the speed can be infinitely varied as the excess oil goes through the pressure relief valve.
3. When the four-way valve is deactivated, the motor stops suddenly and becomes locked.

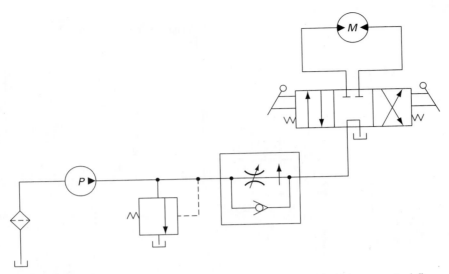

Figure 9-20. Speed control of hydraulic motor using pressure-compensated flow control valve.

4. When the right envelope of the four-way valve is in operation, the motor turns in the opposite direction. The pressure relief valve provides overload protection if, for example, the motor experiences an excessive torque load.

9.15 HYDRAULIC MOTOR BRAKING SYSTEM

When using a hydraulic motor in a fluid power system, consideration should be given to the type of loading that the motor will experience. A hydraulic motor may be driving a machine having a large inertia. This would create a flywheel effect on the motor, and stopping the flow of fluid to the motor would cause it to act as a pump. In a situation such as this, the circuit should be designed to provide fluid to the motor while it is pumping to prevent it from pulling in air. In addition, provisions should be made for the discharge fluid from the motor to be returned to the tank either unrestricted or through a relief valve. This would stop the motor rapidly but without damage to the system. Figure 9-21 shows a hydraulic motor braking circuit that possesses these desirable features for either direction of motor rotation.

9.16 HYDROSTATIC TRANSMISSION SYSTEM

Closed-Circuit Diagrams

Figures 9-20 and 9-21 are actually hydrostatic transmissions. They are called open-circuit drives because the pump draws its fluid from a reservoir. Its output is then directed to a hydraulic motor and discharged from the motor back into the

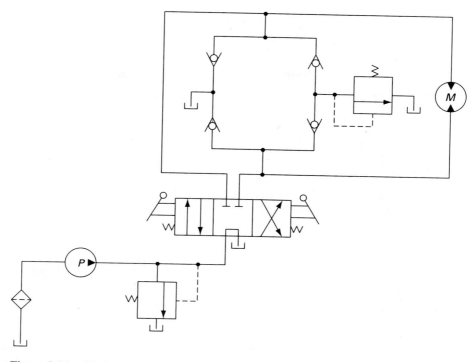

Figure 9-21. Hydraulic motor braking system.

reservoir. In a closed-circuit drive, exhaust oil from the motor is returned directly to the pump inlet. Figure 9-22 gives a circuit of a closed-circuit drive that allows for only one direction of motor rotation. The motor speed is varied by changing the pump displacement. The torque capacity of the motor can be adjusted by the pressure setting of the relief valve. Makeup oil to replenish leakage from the closed loop flows into the low-pressure side of the circuit through a line from the reservoir.

Many hydrostatic transmissions are reversible closed-circuit drives that use a variable displacement reversible pump. This allows the motor to be driven in either direction and at infinitely variable speeds depending on the position of the pump displacement control. Figure 9-23 shows a circuit of such a system using a fixed displacement hydraulic motor. Internal leakage losses are made up by a replenishing pump, which keeps a positive pressure on the low-pressure side of the system. There are two check and two relief valves to accommodate the two directions of flow and motor rotation.

Track-type Tractor Application

Figure 9-24 shows a cutaway view of a track-type tractor that is driven by two closed-circuit, reversible-direction hydrostatic transmissions. Each of the two tracks has its

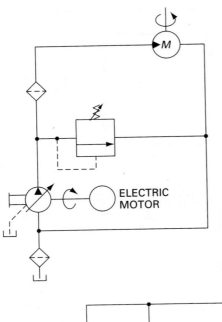

Figure 9-22. Closed-circuit, one-direction hydrostatic transmission.

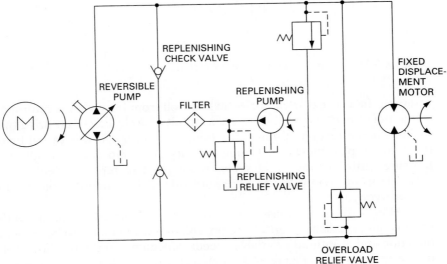

Figure 9-23. Closed-circuit, reversible-direction hydrostatic transmission. *(Courtesy of Sperry Vickers, Sperry Rand Corp., Troy, Michigan.)*

own separate hydraulic circuit. The two variable displacement piston pumps are attached directly to the diesel engine flywheel housing. Each pump delivers 17.6 gpm (66.7 Lpm) at the rated speed of 2400 rpm and 1000 psi (6900 kPa) pressure. The pressure relief valve setting is 2500 psi (17,200 kPa). The two piston motors (which power the drive sprockets for the two tracks) are mounted inboard of the main frame. These motors can run with either of two different volumetric

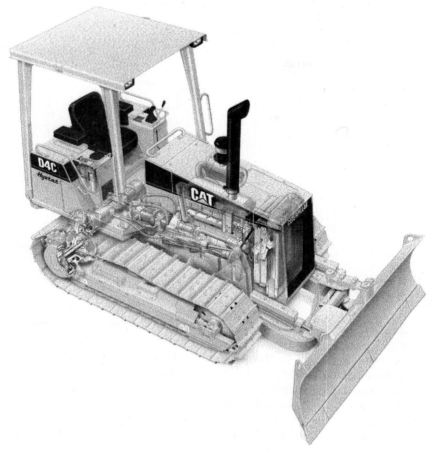

Figure 9-24. Track-type tractor driven by two hydrostatic transmissions. *(Courtesy of Caterpillar, Inc., Peoria, Illinois.)*

displacements, as controlled by the operator, to provide two speed ranges. With a single joystick, the operator can control speed, machine direction, and steering.

The front variable pitch and tilt blade is driven by hydraulic cylinders to provide the necessary down force, pry-out force, and blade control for maximum production capabilities. Applications for this tractor include finish grading, back-filling ditches, landscaping, medium land clearing, and heavy dozing.

9.17 AIR-OVER-OIL CIRCUIT

Sometimes circuits using both air and oil are used to obtain the advantages of each medium. Figure 9-25 shows a counterbalance system, which is an air-over-oil circuit. Compressed air flows through a filter, regulator, lubricator unit (FRL) and into a

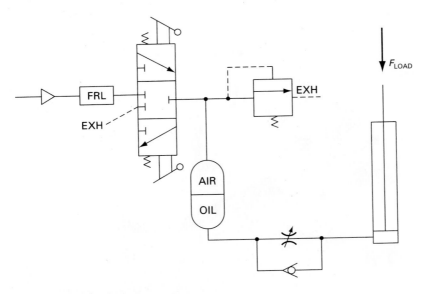

Figure 9-25. Air-over-oil circuit.

surge tank via a directional control valve (upper flow path configuration). Thus, the surge tank is pressurized by compressed air. This pushes oil out the bottom of the surge tank and to the hydraulic cylinder through a check valve and orifice hooked in parallel. This extends the cylinder to lift a load. When the directional control valve is shifted into its lower flow path mode, the cylinder retracts at a controlled rate. This happens because the variable orifice provides a controlled return flow of oil as air leaves the surge tank and exhausts into the atmosphere via the directional control valve. The load can be stopped at any intermediate position by the spring-centered position of the directional control valve. This system eliminates the need for a costly hydraulic pump and tank unit.

9.18 ANALYSIS OF HYDRAULIC SYSTEM WITH FRICTIONAL LOSSES CONSIDERED

Example 9-6 illustrates the technique for performing the analysis of a complete hydraulic system, taking frictional losses into account.

EXAMPLE 9-6

The system shown in Figure 9-26 contains a pump delivering high-pressure oil to a hydraulic motor, which drives an external load via a rotating shaft. The following data are given:

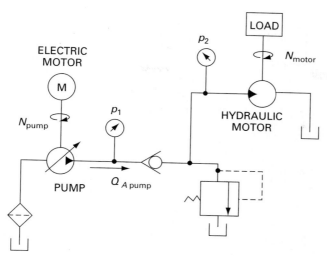

Figure 9-26. System for Example 9-6.

Pump:

$$\eta_m = 94\%$$
$$\eta_v = 92\%$$
$$V_D = 10 \text{ in}^3$$
$$N = 1000 \text{ rpm}$$
$$\text{inlet pressure} = -4 \text{ psi}$$

Hydraulic motor:

$$\eta_m = 92\%$$
$$\eta_v = 90\%$$
$$V_D = 8 \text{ in}^3$$
$$\text{inlet pressure } p_2 \text{ required to drive load} = 500 \text{ psi}$$
$$\text{motor discharge pressure} = 5 \text{ psi}$$

Pump discharge pipeline:

Pipe: 1-in schedule 40, 50 ft long (point 1 to point 2)

Fittings: two 90° elbows ($K = 0.75$ for each elbow), one check valve ($K = 4.0$)

Oil:

$$\text{viscosity} = 125 \text{ cS}$$
$$\text{specific gravity} = 0.9$$

If the hydraulic motor is 20 ft above the pump, determine the

  a. Pump flow rate
  b. Pump discharge pressure p_1
  c. Input hp required to drive the pump
  d. Motor speed
  e. Motor output hp
  f. Motor output torque
  g. Overall efficiency of system

Solution

  a. To determine the pump's actual flow rate, we first calculate the pump's theoretical flow rate.

$$(Q_T)_{\text{pump}} = \frac{(V_D)_{\text{pump}} \times N_p}{231} = \frac{10 \text{ in}^3 \times 1000 \text{ rpm}}{231} = 43.3 \text{ gpm}$$

$$(Q_A)_{\text{pump}} = (Q_T)_{\text{pump}} \times (\eta_v)_{\text{pump}} = 43.3 \times 0.94 = 40.7 \text{ gpm}$$

  b. To obtain the pump discharge pressure p_1 we need to calculate the frictional pressure loss $(p_1 - p_2)$ in the pump discharge line. Writing the energy equation between points 1 and 2, we have

$$Z_1 + \frac{p_1}{\gamma} + \frac{v_1^2}{2g} + H_p - H_m - H_L = Z_2 + \frac{p_2}{\gamma} + \frac{v_2^2}{2g}$$

Since there is no hydraulic motor or pump between stations 1 and 2, $H_m = H_p = 0$. Also, $v_1 = v_2$ and $Z_2 - Z_1 = 20$ ft. Also, per Figure 10-2 for 1-in schedule 40 pipe, the inside diameter equals 1.040 in. We next solve for the Reynolds number, friction factor, and head loss due to friction.

$$v = \frac{(Q_A)_{\text{pump}}}{A} = \frac{40.7/449 \text{ ft}^3/\text{s}}{\frac{\pi}{4}\left(\frac{1.040}{12}\text{ft}\right)^2} = \frac{0.0908 \text{ ft}^3/\text{s}}{0.00590 \text{ ft}^2} = 15.4 \text{ ft/s}$$

$$N_R = \frac{7740v(\text{ft/s}) \times D(\text{in})}{v(\text{cS})} = \frac{7740 \times 15.4 \times 1.040}{125} = 992$$

Since the flow is laminar, the friction factor can be found directly from the Reynolds number.

$$f = \frac{64}{N_R} = \frac{64}{992} = 0.0645$$

Also,

$$H_L = f\left(\frac{L_{eTOT}}{D}\right)\frac{v^2}{2g} \qquad \text{where}$$

$$L_{eTOT} = 50 + 2\left(\frac{KD}{f}\right)_{90°\text{ elbow}} + \left(\frac{KD}{f}\right)_{\text{check valve}}$$

$$= 50 + \frac{2 \times 0.75 \times 1.040/12}{0.0645} + \frac{4.0 \times 1.040/12}{0.0645} = 50 + 2.02$$

$$+ 5.37 = 57.39 \text{ ft}$$

Thus,

$$H_L = 0.0645 \times \frac{57.39}{1.040/12} \times \frac{(15.4)^2}{2 \times 32.2} = 157.3 \text{ ft}$$

Next, we substitute into the energy equation to solve for $(p_1 - p_2)/\gamma$.

$$\frac{p_1 - p_2}{\gamma} = (Z_2 - Z_1) + H_L = 20 \text{ ft} + 157.3 \text{ ft} = 177.3 \text{ ft}$$

Hence,

$$p_1 - p_2 = 177.3 \text{ ft} \times \gamma\left(\frac{\text{lb}}{\text{ft}^3}\right) = 177.3 \text{ ft} \times 0.9 \times 62.4\frac{\text{lb}}{\text{ft}^3} = 9960\frac{\text{lb}}{\text{ft}^2} = 69 \text{ psi}$$

$$p_1 = p_2 + 69 = 500 + 69 = 569 \text{ psi}$$

c. hp delivered to pump $= \dfrac{\text{pump hydraulic horsepower}}{(\eta_o)_{\text{pump}}}$

$$= \frac{(569 + 4)\text{psi} \times 40.7 \text{ gpm}}{1714 \times 0.94 \times 0.92} = 15.7 \text{ hp}$$

d. To obtain the motor speed we first need to determine the motor theoretical flow rate

$$(Q_T)_{\text{motor}} = (Q_A)_{\text{pump}} \times (\eta_v)_{\text{motor}} = 40.7 \times 0.90 = 36.6 \text{ gpm}$$

$$N_{\text{motor}} = \frac{(Q_T)_{\text{motor}} \times 231}{(V_D)_{\text{motor}}} = \frac{36.6 \times 231}{8} = 1057 \text{ rpm}$$

e. To obtain the motor output hp we first need to determine the motor input hp.

$$\text{motor input hp} = \frac{(500 - 5)\text{psi} \times 40.7 \text{ gpm}}{1714} = 11.8 \text{ hp}$$

Thus,

$$\text{motor output hp} = \text{motor input hp} \times (\eta_o)_{\text{motor}} = 11.8 \times 0.92 \times 0.90 = 9.77 \text{ hp}$$

f. motor output torque $= \dfrac{\text{motor output hp} \times 63{,}000}{N_{\text{motor}}}$

$$= \frac{9.77 \times 63{,}000}{1057} = 582 \text{ in} \cdot \text{lb}$$

g. The overall efficiency of the system is

$$(\eta_o)_{\text{overall}} = \frac{\text{motor output hp}}{\text{pump input hp}} = \frac{9.77}{15.7} = 0.622 = 62.2\%$$

Thus, 62.2% of the power delivered to the pump by the electric motor is delivered to the load by the hydraulic motor.

9.19 MECHANICAL-HYDRAULIC SERVO SYSTEM

Figure 9-27 shows an automotive power-steering example of a mechanical-hydraulic servo system (closed-loop system). Operation is as follows:

1. The input or command signal is the turning of the steering wheel.
2. This moves the valve sleeve, which ports oil to the actuator (steering cylinder).
3. The piston rod moves the wheels via the steering linkage.
4. The valve spool is attached to the linkage and thus moves with it.
5. When the valve spool has moved far enough, it cuts off oil flow to the cylinder. This stops the motion of this actuator.
6. Thus, mechanical feedback recenters (nulls) the valve (actually a servo valve) to stop motion at the desired point as determined by the position of the steering wheel. Additional motion of the steering wheel is required to cause further motion of the output wheels.

Electrohydraulic servo systems are presented in Chapter 17.

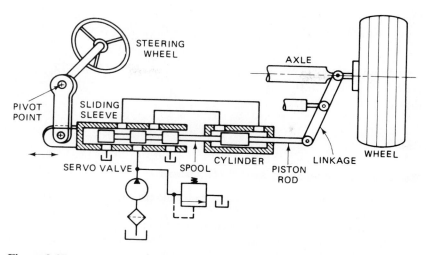

Figure 9-27. Automotive example of mechanical-hydraulic servo system.

9.20 KEY EQUATIONS

Regenerative cylinder
extending speed:

$$v_{P_{\text{ext}}} = \frac{Q_P}{A_r}$$

(9-1)

Regenerative cylinder
retracting speed:

$$v_{P_{\text{ret}}} = \frac{Q_P}{A_P - A_r}$$

(9-2)

Regenerative cylinder
extending to retracting
speed ratio:

$$\frac{v_{P_{\text{ext}}}}{v_{P_{\text{ret}}}} = \frac{A_p}{A_r} - 1$$

(9-3)

Regenerative cylinder
extending load-carrying
capacity:

$$F_{\text{load}} = pA_r$$

(9-4)

Total load-carrying
capacity of two cylinders
hooked in series and
extending in
synchronization:

$$p_1 A_{p_1} = F_1 + F_2$$

(9-5)

Flow rate through
flow control valve:

$$Q_{\text{FCV}} = C_v\sqrt{\frac{\Delta p}{SG}}$$

(9-7)

Extending velocity of
cylinder with meter-in
flow control valve:

$$v_{\text{cyl}} = \frac{C_v}{A_{\text{piston}}}\sqrt{\frac{p_{\text{PRV}} - F_{\text{load}}/A_{\text{piston}}}{SG}}$$

(9-9)

EXERCISES

Questions, Concepts, and Definitions

9-1. When analyzing or designing a hydraulic circuit, what are three important considerations?

9-2. What is the purpose of a regenerative circuit?

9-3. Why is the load-carrying capacity of a regenerative cylinder small if its piston rod area is small?

9-4. What is a fail-safe circuit?

9-5. Under what condition is a hydraulic motor braking system desirable?

9-6. What is the difference between closed-circuit and open-circuit hydrostatic transmissions?

9-7. What is meant by an air-over-oil system?

9-8. What is a mechanical-hydraulic servo system? Give one application.

9-9. What is the purpose of the check valve of Figure 9-7?

9-10. Can a hydraulic cylinder be designed so that for the same pump flow, the extending and retracting speeds will be equal? Explain your answer.

9-11. In a mechanical-hydraulic servo system, what part of the servo valve moves with the load? What part moves with the input?

9-12. For the circuit of Figure 9-13, the motions of the two cylinders are synchronized in the extension strokes. Are the motions of the two cylinders also synchronized in the retraction strokes? Explain your answer.

Problems

Note: The letter *E* following an exercise number means that English units are used. Similarly, the letter *M* indicates metric units.

Regenerative Circuits

9-13E. When the directional control valve in the system of Figure 9-5 returns to its center position, the cylinder rod moves in a given direction. Is this direction extension or retraction? During this movement determine the force and velocity. The piston and rod diameters are 3 in and 1 in, respectively. The pump flow rate is 2 gpm and the system pressure is 1000 psi.

9-14M. Repeat Exercise 9-13 with the following metric data:

Piston diameter = 75 mm

Rod diameter = 25 mm

Pump flow rate = 8 Lpm

System pressure = 7 MPa

9-15E. A double-acting cylinder is hooked up in the regenerative circuit of Figure 9-4(a). The relief valve setting is 1500 psi. The piston area is 20 in² and the rod area is 10 in². If the pump flow is 25 gpm, find the cylinder speed and load-carrying capacity for the

   a. Extending stroke

   b. Retracting stroke

9-16M. A double-acting cylinder is hooked up in the regenerative circuit of Figure 9-4(a). The relief valve setting is 105 bars. The piston area is 130 cm² and the rod area is 65 cm². If the pump flow is 0.0016 m³/s, find the cylinder speed and load-carrying capacity for the

   a. Extending stroke

   b. Retracting stroke

Troubleshooting of Circuits

9-17. Properly complete the circuit diagram of Figure 9-28. The clamp cylinder is to extend first, and then the work cylinder extends by the action of a directional control valve (DCV). By further action of the DCV, the work cylinder retracts, and then the clamp cylinder retracts.

9-18. What is wrong with the circuit of Figure 9-29?

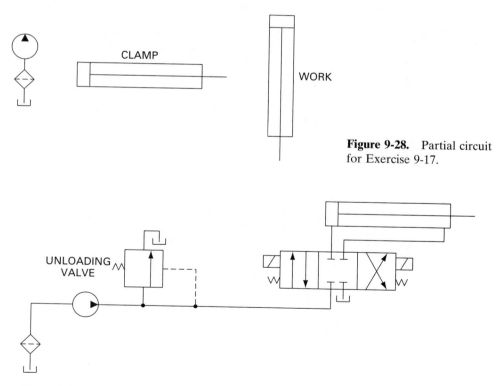

Figure 9-28. Partial circuit for Exercise 9-17.

Figure 9-29. Circuit for Exercise 9-18.

Operation of Circuits

9-19. For the circuit of Figure 9-30, which of the following is true? Explain your answer.

    a. As cylinder 1 extends, cylinder 2 extends.

    b. As cylinder 1 extends, cylinder 2 retracts.

    c. As cylinder 1 extends, cylinder 2 does not move.

9-20. What unique features does the circuit of Figure 9-31 provide in the operation of the hydraulic cylinder?

9-21. For the circuit of Figure 9-32, give the sequence of operation of cylinders 1 and 2 when the pump is turned on. Assume both cylinders are initially fully retracted.

9-22. What safety feature does the circuit of Figure 9-33 possess in addition to the pressure relief valve?

9-23. Assuming that the two double-rodded cylinders of Figure 9-34 are identical, what unique feature does this circuit possess?

9-24. For the system in Exercise 9-33, as shown in Figure 9-35, if the load on cylinder 1 is greater than the load on cylinder 2, how will the cylinders move when the DCV is shifted into the extending or retracting mode? Explain your answer.

Analysis of Synchronized Cylinders Hooked in Series

9-25E. For the system of Figure 9-13 (for the extension strokes of the cylinders), what pump pressure is required if the cylinder loads are 5000 lb each and cylinder 1 has a piston area of 10 in²?

9-26E. Repeat Exercise 9-25 for the retraction strokes of the cylinders (loads pull to right). The piston and rod areas of cylinder 2 equal 8 in² and 2 in², respectively.

9-27E. Solve Exercise 9-25 using a back pressure p_3 of 50 psi instead of zero. The piston area and rod area of cylinder 2 equal 8 in² and 2 in², respectively.

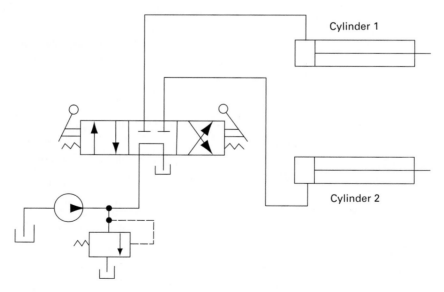

Figure 9-30. Circuit for Exercise 9-19.

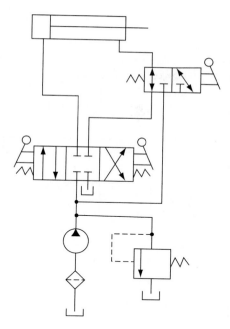

Figure 9-31.   Circuit for Exercise 9-20.

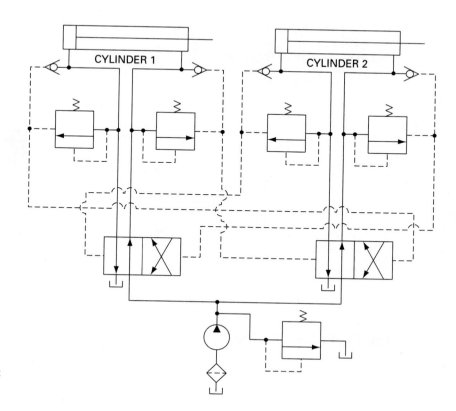

Figure 9-32.   Circuit
for Exercise 9-21.

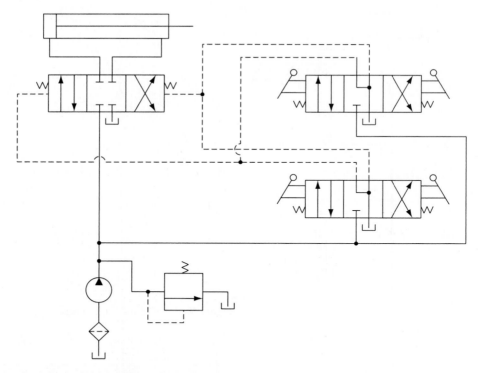

Figure 9-33.   Circuit for Exercise 9-22.

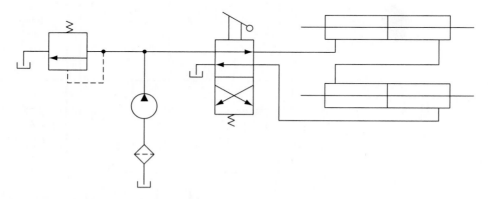

Figure 9-34.   Circuit for Exercise 9-23.

9-28M. For the system of Figure 9-13 (for the extension strokes of the cylinders), what pump pressure is required if the cylinder loads are 22,000 N each and cylinder 1 has a piston area of 65 cm²?

9-29M. Repeat Exercise 9-28 for the retraction strokes of the cylinders (loads pull to right). The piston and rod areas of cylinder 2 equal 50 cm² and 15 cm², respectively.

9-30M. Solve Exercise 9-28 using a back pressure p_3 of 300 kPa instead of zero. The piston area and rod area of cylinder 2 equal 50 cm² and 15 cm², respectively.

Analysis of Pressure Relief Valve Pressure and Spring Settings

9-31M. For the double-pump system of Figure 9-7, what should be the pressure settings of the unloading valve and pressure relief valve under the following conditions:

a. Sheet metal punching operation requires a force of 8000 N.

b. Hydraulic cylinder has a 3.75-cm-diameter piston and a 1.25-cm-diameter rod.

c. During rapid extension of the cylinder, a frictional pressure loss of 675 kPa occurs in the line from the high-flow pump to the blank end of the cylinder. During the same time, a 350-kPa pressure loss occurs in the return line from the rod end of the cylinder to the oil tank. Frictional pressure losses in these lines are negligibly small during the punching operation.

d. Assume that the unloading valve and relief valve pressure settings (for their full pump flow requirements) should be 50% higher than the pressure required to overcome frictional pressure losses and the cylinder punching load, respectively.

9-32E. A pressure relief valve contains a poppet with a 0.60-in² area on which system pressure acts. The poppet must move 0.15 in from its fully closed position in order to pass full pump flow at the PRV setting (full pump flow pressure). The pressure required to overcome the external load is 1000 psi. Assume that the PRV setting should be 50% higher than the pressure required to overcome the external load. If the valve-cracking pressure should be 10% higher than the pressure required to overcome the external load, find the required

a. Spring constant of the compression spring in the valve

b. Initial compression of the spring from its free-length condition as set by the spring-adjustment mechanism of the PRV (poppet held against its seat by spring)

Analysis of Circuits with Frictional Losses Considered

9-33E. For the fluid power system shown in Figure 9-35, determine the external load (F_1 and F_2) that each hydraulic cylinder can sustain while moving in the extending direction. Take frictional pressure losses into account. The pump produces a pressure increase of 1000 psi from the inlet port to the discharge port and a flow rate of 40 gpm. The following data are applicable.

$$\text{kinematic viscosity of oil} = 0.001 \text{ ft}^2/\text{s}$$
$$\text{specific weight of oil} = 50 \text{ lb/ft}^3$$
$$\text{cylinder piston diameter} = 8 \text{ in}$$
$$\text{cylinder rod diameter} = 4 \text{ in}$$
$$\text{all elbows are } 90° \text{ with } K \text{ factor} = 0.75$$

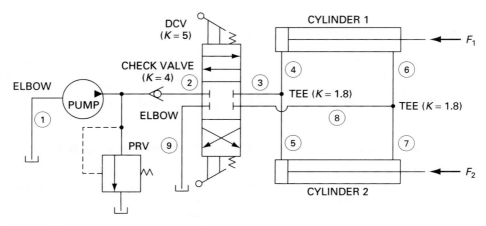

Figure 9-35. System for Exercises 9-24 and 9-33.

Pipe lengths and inside diameters are given as follows:

Pipe No.	Length (ft)	Dia (in)	Pipe No.	Length (ft)	Dia (in)
1	6	2.0	6	10	1.0
2	30	1.25	7	10	1.0
3	20	1.25	8	40	1.25
4	10	1.0	9	40	1.25
5	10	1.0			

9-34M. For the system in Exercise 9-33, as shown in Figure 9-35, convert the data to metric units and solve for the external load that each cylinder can sustain while moving in the retraction direction.

9-35E. For the system in Exercise 9-33, as shown in Figure 9-35, determine the heat-generation rate due to frictional pressure losses.

9-36E. For the system in Exercise 9-33, as shown in Figure 9-35, determine the retracting and extending speeds of both cylinders. Assume that the actual cylinder loads are equal and are less than the loads that can be sustained during motion.

9-37M. For the system in Exercise 9-34, as shown in Figure 9-35, determine the heat-generation rate due to frictional pressure losses.

9-38M. For the system in Exercise 9-34, as shown in Figure 9-35, determine the retracting and extending speeds of both cylinders. Assume that the actual cylinder loads are equal and are less than the loads that can be sustained during motion.

9-39E. Figure 9-36 shows a regenerative system in which a 25-hp electric motor drives a 90% efficient pump. Determine the external load F that the hydraulic cylinder can sustain in the regenerative mode (spring-centered position of the DCV). Pump discharge pressure is 1000 psi. Take frictional pressure losses into account. The following data are applicable:

$$\text{kinematic viscosity of oil} = 0.001 \text{ ft}^2/\text{s}$$
$$\text{specific weight of oil} = 50 \text{ lb/ft}^3$$
$$\text{cylinder piston diameter} = 8 \text{ in}$$
$$\text{cylinder rod diameter} = 4 \text{ in}$$
$$\text{all elbows are } 90° \text{ with } K \text{ factor} = 0.75$$

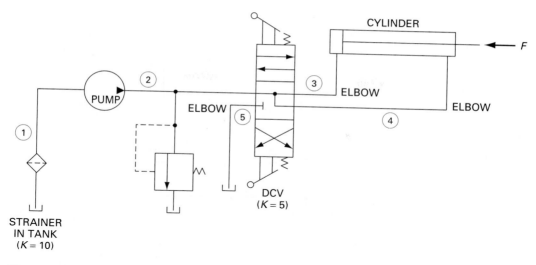

Figure 9-36. System for Exercise 9-39.

Pipe lengths and inside diameters are given as follows:

Pipe No.	Length (ft)	Dia (in)
1	2	2
2	20	1.75
3	30	1.75
4	30	1.75
5	20	1.75

9-40M. For the system in Exercise 9-39, as shown in Figure 9-36, convert the data to metric units and determine the external load F that the hydraulic cylinder can sustain in the regenerative mode.

9-41E. For the system in Exercise 9-39, as shown in Figure 9-36, determine the heat-generation rate (English units) due to frictional pressure losses in the regenerative mode.

9-42M. For the system in Exercise 9-40, as shown in Figure 9-36, determine the heat-generation rate (metric units) due to frictional pressure losses in the regenerative mode.

9-43E. For the system in Exercise 9-39, as shown in Figure 9-36, determine the cylinder speed for each position of the DCV.

9-44M. For the system in Exercise 9-40, as shown in Figure 9-36, determine the cylinder speed for each position of the DCV.

Analysis of Meter-In Flow Control Valve Systems

9-45E. For the meter-In flow control valve system of Figure 9-37, the following data are given:

$$\text{desired cylinder speed} = 10 \text{ in/s}$$
$$\text{cylinder piston diameter} = 2 \text{ in (area} = 3.14 \text{ in}^2)$$
$$\text{cylinder load} = 3000 \text{ lb}$$

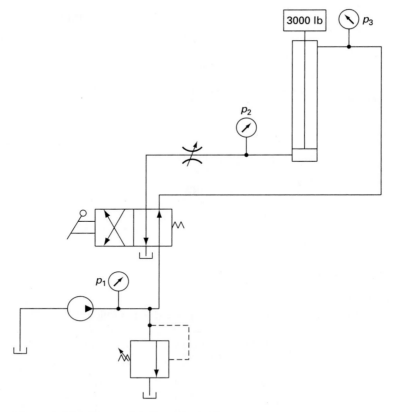

Figure 9-37. System for Exercise 9-45.

$$\text{specific gravity of oil} = 0.90$$
$$\text{pressure relief valve setting} = 1000 \text{ psi}$$

Determine the required capacity coefficient of the flow control valve.

9-46M. Change the data in Exercise 9-45 to metric units and determine the required capacity coefficient of the flow control valve.

Analysis of Meter-Out Flow Control Valve Systems

9-47E. For the system of Example 9-5, determine the pressure on each pressure gage during constant speed extension of the cylinder for

a. No load

b. 6000-lb load

The cylinder piston and rod diameters are 2.5 in and 1.5 in, respectively, and the PRV setting is 1600 psi.

9-48M. Solve the problem of Example 9-5 where the pressure drop equals 300 kPa in the pipeline from the pump outlet to the blank end of the cylinder. Also, the pressure drop equals 200 kPa in the oil return pipeline from the rod end of the cylinder to the reservoir.

Hydraulic Conductors and Fittings

10

Learning Objectives

Upon completing this chapter, you should be able to:

1. Size conductors to meet flow-rate requirements.
2. Understand the significance of the term *pressure rating of conductors.*
3. Differentiate between the terms *burst pressure* and *working pressure.*
4. Visualize the tensile stress in the wall of a conductor and understand how tensile stress varies with fluid pressure, pipe diameter, and wall thickness.
5. Determine the required wall thickness of a conductor to prevent bursting under operating fluid pressure.
6. Identify the standard commercial sizes of steel pipes and tubing.
7. Describe the various types of fittings used to connect hydraulic components to conductors.
8. Identify the construction features and function of flexible hoses.
9. Discuss the construction features and function of quick disconnect couplings.

10.1 INTRODUCTION

In a hydraulic system, the fluid flows through a distribution system consisting of conductors and fittings, which carry the fluid from the reservoir through operating components and back to the reservoir. Since power is transmitted throughout the system by means of these conducting lines (conductors and fittings used to connect system components), it follows that they must be properly designed in order for the total system to function properly.

Hydraulic systems use primarily four types of conductors:

1. Steel pipes
2. Steel tubing
3. Plastic tubing
4. Flexible hoses

The choice of which type of conductor to use depends primarily on the system's operating pressures and flow rates. In addition, the selection depends on environmental conditions such as the type of fluid, operating temperatures, vibration, and whether or not there is relative motion between connected components.

Conducting lines are available for handling working pressures up to 12,000 psi. In general, steel tubing provides greater plumbing flexibility and neater appearance and requires fewer fittings than piping. However, piping is less expensive than steel tubing. Plastic tubing is finding increased industrial usage because it is not costly and circuits can be very easily hooked up due to its flexibility. Flexible hoses are used primarily to connect components that experience relative motion. They are made from a large number of elastomeric (rubberlike) compounds and are capable of handling pressures up to 12,000 psi.

Stainless steel conductors and fittings are used if extremely corrosive environments are expected. However, they are very expensive and should be used only if necessary. Copper conductors should not be used in hydraulic systems because the copper promotes the oxidation of petroleum oils. Zinc, magnesium, and cadmium conductors should not be used either, because they are rapidly corroded by water-glycol fluids. Galvanized conductors should also be avoided because the galvanized surface has a tendency to flake off into the hydraulic fluid. When using steel pipe or steel tubing, hydraulic fittings should be made of steel except for inlet, return, and drain lines, where malleable iron may be used.

Conductors and fittings must be designed with human safety in mind. They must be strong enough not only to withstand the steady-state system pressures but also the instantaneous pressure spikes resulting from hydraulic shock. Whenever control valves are closed suddenly, this quickly stops the flowing fluid, which possesses large amounts of kinetic energy. This produces shock waves whose pressure levels can be up to four times the steady-state system design values. The sudden stopping of actuators and the rapid acceleration of heavy loads also cause pressure spikes. These high-pressure pulses are taken into account by the application of an appropriate factor of safety.

10.2 CONDUCTOR SIZING FOR FLOW-RATE REQUIREMENTS

A conductor must have a large enough cross-sectional area to handle the flow-rate requirements without producing excessive fluid velocity. Whenever we speak of fluid velocity in a conductor, we are referring to the average velocity since the actual velocity is not constant over the cross section of the pipe. As shown in Chapter 4, the velocity is zero at the pipe wall (fluid particles cling to a contacting surface due to viscosity) and reaches a maximum value at the centerline of the pipe. The average velocity is defined as the volume flow rate divided by the pipe cross-sectional area:

$$v = v_{avg} = \frac{Q}{A} \qquad \textbf{(10-1)}$$

In other words, the average velocity is that velocity which when multiplied by the pipe area equals the volume flow rate. It is also understood that the term *diameter* by itself always means inside diameter and that the pipe area is that area that corresponds to the pipe inside diameter. The maximum recommended velocity for pump suction lines is 4 ft/s (1.2 m/s) in order to prevent excessively low suction pressures and resulting pump cavitation (as discussed in Chapter 6). The maximum recommended velocity for pump discharge lines is 20 ft/s (6.1 m/s) in order to prevent turbulent flow and the corresponding excessive head losses and elevated fluid temperatures (as discussed in Chapter 5). Note that these maximum recommended values are average velocities.

EXAMPLE 10-1

A pipe handles a flow rate of 30 gpm. Find the minimum inside diameter that will provide an average fluid velocity not to exceed 20 ft/s.

Solution   First, we convert the flow rate into units of ft³/s.

$$Q(\text{ft}^3/\text{s}) = \frac{Q(\text{gpm})}{449} = \frac{30}{449} = 0.0668 \text{ ft}^3/\text{s}$$

Next, per Eq. (3-22), we solve for the minimum required pipe flow area:

$$A(\text{ft}^2) = \frac{Q(\text{ft}^3/\text{s})}{v(\text{ft}/\text{s})} = \frac{0.0668}{20} = 0.00334 \text{ ft}^2 = 0.481 \text{ in}^2$$

Finally, for a circular area we have

$$D(\text{in}) = \sqrt{\frac{4A(\text{in}^2)}{\pi}} = \sqrt{\frac{4 \times 0.481}{\pi}} = 0.783 \text{ in}$$

EXAMPLE 10-2

A pipe handles a flow rate of 0.002 m³/s. Find the minimum inside diameter that will provide an average fluid velocity not to exceed 6.1 m/s.

Solution   First, we solve for the minimum required pipe flow area:

$$A(\text{m}^2) = \frac{Q(\text{m}^3/\text{s})}{v(\text{m}/\text{s})} = \frac{0.002}{6.1} = 0.000328 \text{ m}^2$$

The minimum inside diameter can now be found.

$$D = \sqrt{\frac{4A\,(\text{m}^2)}{\pi}} = \sqrt{\frac{4 \times 0.000328}{\pi}} = 0.0204 \text{ m} = 20.4 \text{ mm}$$

10.3 PRESSURE RATING OF CONDUCTORS

Tensile Stress

A conductor must be strong enough to prevent bursting due to excessive tensile stress (also called *hoop stress*) in the wall of the conductor under operating fluid pressure. The magnitude of this tensile stress, which must be sustained by the conductor material, can be determined by referring to Figure 10-1. In Figure 10-1(a), we see the fluid pressure (p) acting normal to the inside surface of a circular pipe having a length (L). The pipe has outside diameter D_o, inside diameter D_i, and wall thickness t. Because the fluid pressure acts normal to the pipe's inside surface, a pressure force is created that attempts to separate one half of the pipe from the other half.

Figure 10-1(b) shows this pressure force F_p pushing downward on the bottom half of the pipe. To prevent the bottom half of the pipe from separating from the upper half, the upper half pulls upward with a total tensile force F. One-half of this force (or $F/2$) acts on the cross-sectional area (tL) of each wall, as shown.

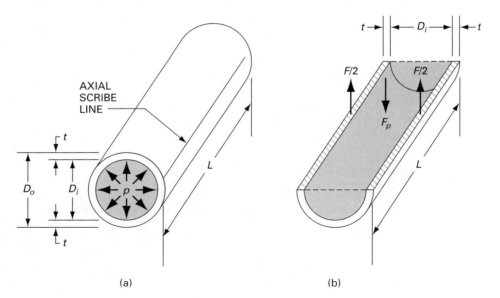

(a)   (b)

Figure 10-1. Forces in the wall of a pipe due to fluid pressure.

Since the pressure force and the total tensile force must be equal in magnitude, we have

$$F = F_p = pA$$

where A is the projected area of the lower half-pipe curved-wall surface onto a horizontal plane. Thus, A equals the area of a rectangle of width D_i and length L, as shown in Figure 10-1(b). Hence,

$$F = pA = p(LD_i)$$

The tensile stress in the pipe material equals the tensile force divided by the wall cross-sectional area withstanding the tensile force. This stress is called a tensile stress because the force (F) is a tensile force (pulls on the area over which it acts).

$$\text{tensile stress} = \frac{\text{force pulling on the pipe wall area}}{\text{pipe wall area over which force acts}}$$

Substituting variables we have

$$\sigma = \frac{F}{2tL} = \frac{pA}{2tL} = \frac{p(LD_i)}{2tL} = \frac{pD_i}{2t} \tag{10-2}$$

where σ = Greek symbol (sigma) = tensile stress.

As can be seen from Eq. (10-2), the tensile stress increases as the fluid pressure increases and also as the pipe inside diameter increases. In addition, as expected, the tensile stress increases as the wall thickness decreases, and the length of the pipe does not have any effect on the tensile stress.

Burst Pressure and Working Pressure

The burst pressure (BP) is the fluid pressure that will cause the pipe to burst. This happens when the tensile stress (σ) equals the tensile strength (S) of the pipe material. The tensile strength of a material equals the tensile stress at which the material ruptures. Notice that an axial scribe line is shown on the pipe outer wall surface in Figure 10-1(a). This scribe line shows where the pipe would start to crack and thus rupture if the tensile stress reached the tensile strength of the pipe material. This rupture will occur when the fluid pressure (p) reaches the burst pressure (BP). Thus, from Eq. (10-2) the burst pressure is

$$BP = \frac{2tS}{D_i} \tag{10-3}$$

The working pressure (WP) is the maximum safe operating fluid pressure and is defined as the burst pressure divided by an appropriate factor of safety (FS).

$$WP = \frac{BP}{FS} \tag{10-4}$$

A factor of safety ensures the integrity of the conductor by determining the maximum safe level of working pressure. Industry standards recommend the following factors of safety based on corresponding operating pressures:

$$FS = 8 \text{ for pressures from 0 to 1000 psi}$$

$$FS = 6 \text{ for pressures from 1000 to 2500 psi}$$

$$FS = 4 \text{ for pressures above 2500 psi}$$

For systems where severe pressure shocks are expected, a factor of safety of 10 is recommended.

Conductor Sizing Based on Flow-Rate and Pressure Considerations

The proper size conductor for a given application is determined as follows:

1. Calculate the minimum acceptable inside diameter (D_i) based on flow-rate requirements.
2. Select a standard-size conductor with an inside diameter equal to or greater than the value calculated based on flow-rate requirements.
3. Determine the wall thickness (t) of the selected standard-size conductor using the following equation:

$$t = \frac{D_o - D_i}{2} \qquad \text{(10-5)}$$

4. Based on the conductor material and system operating pressure (p), determine the tensile strength (S) and factor of safety (FS).
5. Calculate the burst pressure (BP) and working pressure (WP) using Eqs. (10-3) and (10-4).
6. If the calculated working pressure is greater than the operating fluid pressure, the selected conductor is acceptable. If not, a different standard-size conductor with a greater wall thickness must be selected and evaluated. An acceptable conductor is one that meets the flow-rate requirement and has a working pressure equal to or greater than the system operating fluid pressure.

The nomenclature and units for the parameters of Eqs. (10-2), (10-3), (10-4), and (10-5) are as follows:

$$BP = \text{burst pressure (psi, MPa)}$$

$$D_i = \text{conductor inside diameter (in, m)}$$

$$D_o = \text{conductor outside diameter (in, m)}$$

$$FS = \text{factor of safety (dimensionless)}$$

p = system operating fluid pressure (psi, MPa)

S = tensile strength of conductor material (psi, MPa)

t = conductor wall thickness (in, m)

WP = working pressure (psi, MPa)

σ = tensile stress (psi, MPa)

EXAMPLE 10-3

A steel tubing has a 1.250-in outside diameter and a 1.060-in inside diameter. It is made of SAE 1010 dead soft cold-drawn steel having a tensile strength of 55,000 psi. What would be the safe working pressure for this tube assuming a factor of safety of 8?

Solution First, calculate the wall thickness of the tubing using Eq. (10-5):

$$t = \frac{1.250 - 1.060}{2} = 0.095 \text{ in}$$

Next, find the burst pressure for the tubing using Eq. (10-3):

$$BP = \frac{(2)(0.095)(55,000)}{1.060} = 9860 \text{ psi}$$

Finally, using Eq. (10-4), calculate the working pressure at which the tube can safely operate:

$$WP = \frac{9860}{8} = 1,230 \text{ psi}$$

Use of Thick-Walled Conductors

Equations (10-2) and (10-3) apply only for thin-walled cylinders where the ratio D_i/t is greater than 10. This is because in thick-walled cylinders ($D_i/t \leq 10$), the tensile stress is not uniform across the wall thickness of the tube as assumed in the derivation of Eq. (10-2). For thick-walled cylinders Eq. (10-6) must be used to take into account the nonuniform tensile stress:

$$BP = \frac{2tS}{D_i + 1.2t} \qquad \textbf{(10-6)}$$

Thus, if a conductor being considered is not a thin-walled cylinder, the calculations must be done using Eq. (10-6). As would be expected, the use of Eq. (10-6) results in a smaller value of burst pressure and hence a smaller value of working pressure than that obtained from Eq. (10-3). This can be seen by comparing the two equations and noting the addition of the 1.2t term in the denominator of Eq. (10-6).

Note that the steel tubing of Example 10-3 is a thin-walled cylinder because $D_i/t = 1.060$ in/0.095 in $= 11.2 > 10$. Thus, the steel tubing of Example 10-3 can operate safely with a working pressure of 1230 psi as calculated using a factor of safety of 8. Using Eq. (10-6) for this same tubing and factor of safety yields

$$\text{WP} = \frac{\text{BP}}{8} = \frac{tS}{4(D_i + 1.2t)} = \frac{0.095 \times 55{,}000}{4(1.060 + 1.2 \times 0.095)} = 1110 \text{ psi}$$

As expected, the working pressure of 1110 psi calculated using Eq. (10-6) is less than the 1230-psi value calculated in Example 10-3 using Eq. (10-3).

10.4 STEEL PIPES

Size Designation

Pipes and pipe fittings are classified by nominal size and schedule number, as illustrated in Figure 10-2. The schedules provided are 40, 80, and 160, which are the ones most commonly used for hydraulic systems. Note that for each nominal size the outside diameter does not change. To increase wall thickness the next larger schedule number is used. Also observe that the nominal size is neither the outside nor the inside diameter. Instead, the nominal pipe size indicates the thread size for the mating connections. The pipe sizes given in Figure 10-2 are in units of inches.

NOMINAL PIPE SIZE	PIPE OUTSIDE DIAMETER	PIPE INSIDE DIAMETER		
		SCHEDULE 40	SCHEDULE 80	SCHEDULE 160
1/8	0.405	0.269	0.215	–
1/4	0.540	0.364	0.302	–
3/8	0.675	0.493	0.423	–
1/2	0.840	0.622	0.546	0.466
3/4	1.050	0.824	0.742	0.614
1	1.315	1.049	0.957	0.815
1-1/4	1.660	1.380	1.278	1.160
1-1/2	1.900	1.610	1.500	1.338
2	2.375	2.067	1.939	1.689

Figure 10-2. Common pipe sizes.

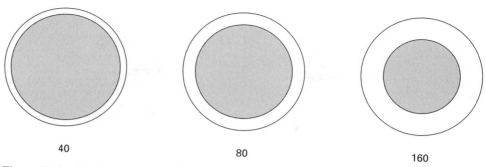

Figure 10-3. Relative size of the cross section of schedules 40, 80, and 160 pipe.

Figure 10-3 shows the relative size of the cross sections for schedules 40, 80, and 160 pipes. As shown for a given nominal pipe size, the wall thickness increases as the schedule number increases.

Thread Design

Pipes have only tapered threads whereas tube and hose fittings have straight threads and also tapered threads as required to connect to hydraulic components. As shown in Figure 10-4, pipe joints are sealed by an interference fit between the male and female threads as the pipes are tightened. This causes one of the major problems in using pipe. When a joint is taken apart, the pipe must be tightened further to reseal. This frequently requires replacing some of the pipe with slightly longer sections, although this problem has been overcome somewhat by using Teflon tape to reseal the pipe joints. Hydraulic pipe threads are the *dry-seal* type. They differ from standard pipe threads because they engage the roots and crests before the flanks. In this way, spiral clearance is avoided.

Pipes can have only male threads, and they cannot be bent around obstacles. There are, of course, various required types of fittings to make end connections and change direction, as shown in Figure 10-5. The large number of pipe fittings required in a hydraulic circuit presents many opportunities for leakage, especially as pressure increases. Threaded-type fittings are used in sizes up to $1\frac{1}{4}$ in diameter. Where larger pipes are required, flanges are welded to the pipe, as illustrated in Figure 10-6. As shown, flat gaskets or O-rings are used to seal the flanged fittings.

It is not a good idea to make connections to a pump with pipe or steel tubing. The natural vibration of the pump can, over time, damage the connection. Using pipe for connections also amplifies the pump noise. Using hose to connect to a pump at the pressure discharge port can help dampen the oil's pulsations particularly with piston pumps. All connections to pumps should be made using flexible hose.

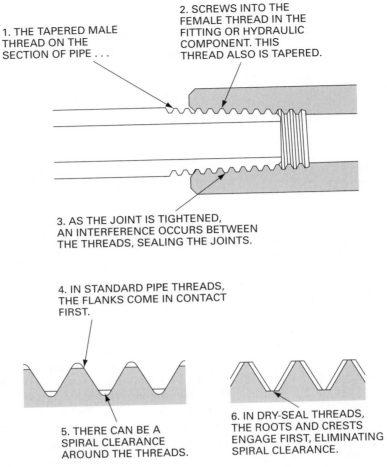

Figure 10-4. Hydraulic pipe threads are the dry-seal tapered type. *(Courtesy of Sperry Vickers, Sperry Rand Corp., Troy, Michigan.)*

10.5 STEEL TUBING

Size Designation

Seamless steel tubing is the most widely used type of conductor for hydraulic systems as it provides significant advantages over pipes. The tubing can be bent into almost any shape, thereby reducing the number of required fittings. Tubing is easier to handle and can be reused without any sealing problems. For low-volume systems, tubing can handle the pressure and flow requirements with less bulk and weight. However, tubing and its fittings are more expensive. A tubing size designation always refers to the outside diameter. Available sizes include 1/16-in

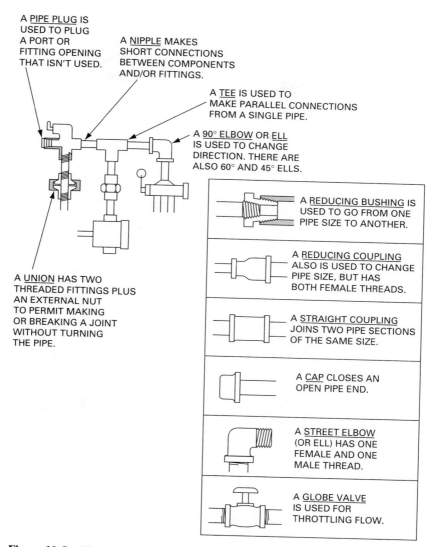

A PIPE PLUG IS USED TO PLUG A PORT OR FITTING OPENING THAT ISN'T USED.

A NIPPLE MAKES SHORT CONNECTIONS BETWEEN COMPONENTS AND/OR FITTINGS.

A TEE IS USED TO MAKE PARALLEL CONNECTIONS FROM A SINGLE PIPE.

A 90° ELBOW OR ELL IS USED TO CHANGE DIRECTION. THERE ARE ALSO 60° AND 45° ELLS.

A UNION HAS TWO THREADED FITTINGS PLUS AN EXTERNAL NUT TO PERMIT MAKING OR BREAKING A JOINT WITHOUT TURNING THE PIPE.

A REDUCING BUSHING IS USED TO GO FROM ONE PIPE SIZE TO ANOTHER.

A REDUCING COUPLING ALSO IS USED TO CHANGE PIPE SIZE, BUT HAS BOTH FEMALE THREADS.

A STRAIGHT COUPLING JOINS TWO PIPE SECTIONS OF THE SAME SIZE.

A CAP CLOSES AN OPEN PIPE END.

A STREET ELBOW (OR ELL) HAS ONE FEMALE AND ONE MALE THREAD.

A GLOBE VALVE IS USED FOR THROTTLING FLOW.

Figure 10-5. Fittings make the connections between pipes and components. *(Courtesy of Sperry Vickers, Sperry Rand Corp., Troy, Michigan.)*

increments from 1/8-in outside diameter up to 3/8-in outside diameter. For sizes from 3/8 in to 1 in the increments are 1/8 in. For sizes beyond 1 in, the increments are 1/4 in. Figure 10-7 shows some of the more common tube sizes (in units of inches) used in fluid power systems.

SAE 1010 dead soft cold-drawn steel is the most widely used material for tubing. This material is easy to work with and has a tensile strength of 55,000 psi. If greater strength is required, the tube can be made of AISI 4130 steel, which has a tensile strength of 75,000 psi.

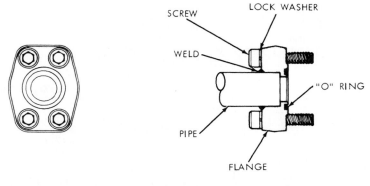

SOCKET WELD PIPE CONNECTIONS
STRAIGHT TYPE

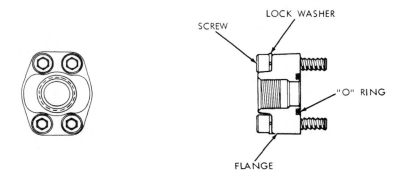

THREADED PIPE CONNECTIONS
STRAIGHT TYPE

Figure 10-6. Flanged connections for large pipes. *(Courtesy of Sperry Vickers, Sperry Rand Corp., Troy, Michigan.)*

Tube Fittings

Tubing is not sealed by threads but by special kinds of fittings, as illustrated in Figure 10-8. Some of these fittings are known as compression fittings. They seal by metal-to-metal contact and may be either the flared or flareless type. Other fittings may use O-rings for sealing purposes. The 37° flare fitting is the most widely used fitting for tubing that can be flared. The fittings shown in Figure 10-8(a) and (b) seal by squeezing the flared end of the tube against a seal as the compression nut is tightened. A sleeve inside the nut supports the tube to dampen vibrations. The standard 45° flare fitting is used for very high pressures. It is also made in an inverted design with male

TUBE OD (in)	WALL THICKNESS (in)	TUBE ID (in)	TUBE OD (in)	WALL THICKNESS (in)	TUBE ID (in)	TUBE OD (in)	WALL THICKNESS (in)	TUBE ID (in)
1/8	0.035	0.055	1/2	0.035	0.430	7/8	0.049	0.777
				0.049	0.402		0.065	0.745
3/16	0.035	0.118		0.065	0.370		0.109	0.657
				0.095	0.310			
1/4	0.035	0.180				1	0.049	0.902
	0.049	0.152	5/8	0.035	0.555		0.065	0.870
	0.065	0.120		0.049	0.527		0.120	0.760
				0.065	0.495			
5/16	0.035	0.243		0.095	0.435	1-1/4	0.065	1.120
	0.049	0.215					0.095	1.060
	0.065	0.183	3/4	0.049	0.652		0.120	1.010
				0.065	0.620			
3/8	0.035	0.305		0.109	0.532	1-1/2	0.065	1.370
	0.049	0.277					0.095	1.310
	0.065	0.245						

Figure 10-7. Common tube sizes.

threads on the compression nut. When the hydraulic component has straight thread ports, straight thread O-ring fittings can be used, as shown in Figure 10-8(c). This type is ideal for high pressures since the seal gets tighter as pressure increases.

Two assembly precautions when using flared fittings are:

1. The compression nut needs to be placed on the tubing before flaring the tube.
2. These fittings should not be overtightened. Too great a torque damages the sealing surface and thus may cause leaks.

For tubing that can't be flared, or if flaring is to be avoided, ferrule, O-ring, or sleeve compression fittings can be used [see Figure 10-8(d), (e), (f)]. The O-ring fitting permits considerable variations in the length and squareness of the tube cut.

Figure 10-9 shows a Swagelok tube fitting, which can contain any pressure up to the bursting strength of the tubing without leakage. This type of fitting can be repeatedly taken apart and reassembled and remain perfectly sealed against leakage. Assembly and disassembly can be done easily and quickly using standard tools. In the illustration, note that the tubing is supported ahead of the ferrules by the fitting body. Two ferrules grasp tightly around the tube with no damage to the tube wall. There is virtually no constriction of the inner wall, ensuring minimum flow restriction. Exhaustive tests have proven that the tubing will yield before a Swagelok tube fitting will leak. The secret of the Swagelok fitting is that all the action in the fitting moves along the tube axially instead of with a rotary motion. Since no torque is transmitted from the fitting to the tubing, there is no initial strain that might weaken the tubing. The double ferrule interaction overcomes variation in tube materials, wall thickness, and hardness.

In Figure 10-10 we see the 45° flare fitting. The flared-type fitting was developed before the compression type and for some time was the only type that could successfully seal against high pressures.

Four additional types of tube fittings are depicted in Figure 10-11: (a) union elbow, (b) union tee, (c) union, and (d) 45° male elbow. With fittings such as these, it is easy to install steel tubing as well as remove it for maintenance purposes.

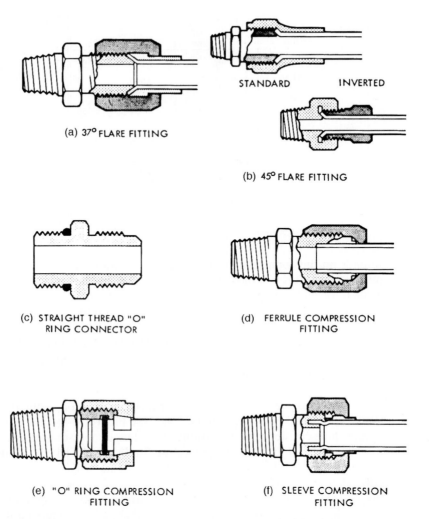

(a) 37° FLARE FITTING

STANDARD   INVERTED

(b) 45° FLARE FITTING

(c) STRAIGHT THREAD "O"
RING CONNECTOR

(d) FERRULE COMPRESSION
FITTING

(e) "O" RING COMPRESSION
FITTING

(f) SLEEVE COMPRESSION
FITTING

Figure 10-8. Threaded fittings and connectors used with tubing.
(Courtesy of Sperry Vickers, Sperry Rand Corp., Troy, Michigan.)

EXAMPLE 10-4

Select the proper size steel tube for a flow rate of 30 gpm and an operating
pressure of 1000 psi. The maximum recommended velocity is 20 ft/s, and the
tube material is SAE 1010 dead soft cold-drawn steel having a tensile strength
of 55,000 psi.

Solution   The minimum inside diameter based on the fluid velocity limitation
of 20 ft/s is the same as that found in Example 10-1 (0.783 in).

Figure 10-9. Swagelok tube fitting. *(Courtesy of Swagelok Co., Solon, Ohio.)*

Figure 10-10. The 45° flare fitting. *(Courtesy of Gould, Inc., Valve and Fittings Division, Chicago, Illinois.)*

(a)

(b)

(c)

(d)

Figure 10-11. Various steel tube fittings. (a) Union elbow, (b) union tee, (c) union, (d) 45° male elbow. *(Courtesy of Gould, Inc., Valve and Fittings Division, Chicago, Illinois.)*

From Figure 10-7, the two smallest acceptable tube sizes based on flow-rate requirements are

1-in OD, 0.049-in wall thickness, 0.902-in ID

1-in OD, 0.065-in wall thickness, 0.870-in ID

Let's check the 0.049-in wall thickness tube first since it provides the smaller velocity:

$$BP = \frac{(2)(0.049)(55,000)}{0.902} = 5980 \text{ psi}$$

$$WP = \frac{5980}{8} = 748 \text{ psi}$$

This working pressure is not adequate, so let's next examine the 0.065-in wall thickness tube:

$$BP = \frac{(2)(0.065)(55,000)}{0.870} = 8220 \text{ psi}$$

$$WP = \frac{8220}{8} = 1030 \text{ psi}$$

$$D_i/t = 0.870 \text{ in}/0.065 \text{ in} = 12.9$$

This result is acceptable, because the working pressure of 1030 psi is greater than the system-operating pressure of 1000 psi and $D_i/t > 10$.

10.6 PLASTIC TUBING

Plastic tubing has gained rapid acceptance in the fluid power industry because it is relatively inexpensive. Also, it can be readily bent to fit around obstacles, it is easy to handle, and it can be stored on reels. Another advantage is that it can be color-coded to represent different parts of the circuit because it is available in many colors. Since plastic tubing is flexible, it is less susceptible to vibration damage than steel tubing.

Fittings for plastic tubing are almost identical to those designed for steel tubing. In fact many steel tube fittings can be used on plastic tubing, as is the case for the Swagelok fitting of Figure 10-9. In another design, a sleeve is placed inside the tubing to give it resistance to crushing at the area of compression, as illustrated in Figure 10-12. In this particular design (called the Poly-Flo Flareless Tube Fitting), the sleeve is fabricated onto the fitting so it cannot be accidentally left off.

Plastic tubing is used universally in pneumatic systems because air pressures are low, normally less than 100 psi. Of course, plastic tubing is compatible with most hydraulic fluids and hence is used in low-pressure hydraulic applications.

Materials for plastic tubing include polyethylene, polyvinyl chloride, polypropylene, and nylon. Each material has special properties that are desirable for specific

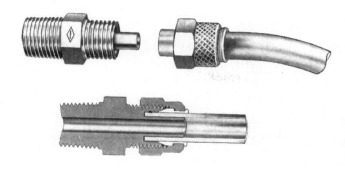

Figure 10-12. Poly-Flo Flareless Plastic Tube Fitting. *(Courtesy of Gould, Inc., Valve and Fittings Division, Chicago, Illinois.)*

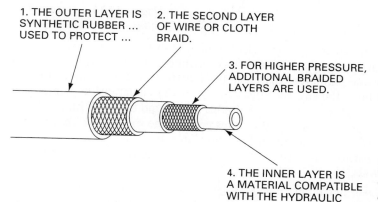

1. THE OUTER LAYER IS SYNTHETIC RUBBER ... USED TO PROTECT ...

2. THE SECOND LAYER OF WIRE OR CLOTH BRAID.

3. FOR HIGHER PRESSURE, ADDITIONAL BRAIDED LAYERS ARE USED.

4. THE INNER LAYER IS A MATERIAL COMPATIBLE WITH THE HYDRAULIC FLUID.

Figure 10-13. Flexible hose is constructed in layers. *(Courtesy of Sperry Vickers, Sperry Rand Corp., Troy, Michigan.)*

applications. Manufacturers' catalogs should be consulted to determine which material should be used for a particular application.

10.7 FLEXIBLE HOSES

Design and Size Designation

The fourth major type of hydraulic conductor is the flexible hose, which is used when hydraulic components such as actuators are subjected to movement. Examples of this are found in portable power units, mobile equipment, and hydraulically powered machine tools. Hose is fabricated in layers of elastomer (synthetic rubber) and braided fabric or braided wire, which permits operation at higher pressures.

As illustrated in Figure 10-13, the outer layer is normally synthetic rubber and serves to protect the braid layer. The hose can have as few as three layers (one being braid) or can have multiple layers to handle elevated pressures. When multiple wire layers are used, they may alternate with synthetic rubber layers, or the wire layers may be placed directly over one another.

Figure 10-14 gives some typical hose sizes and dimensions for single-wire braid and double-wire braid designs. Size specifications for a single-wire braid hose represent

HOSE SIZE	OD TUBE SIZE (in)	SINGLE-WIRE BRAID			DOUBLE-WIRE BRAID		
		HOSE ID (in)	HOSE OD (in)	MINIMUM BEND RADIUS (in)	HOSE ID (in)	HOSE OD (in)	MINIMUM BEND RADIUS (in)
4	1/4	3/16	33/64	1-15/16	1/4	11/16	4
6	3/8	5/16	43/64	2-3/4	3/8	27/32	5
8	1/2	13/32	49/64	4-5/8	1/2	31/32	7
12	3/4	5/8	1-5/64	6-9/16	3/4	1-1/4	9-1/2
16	1	7/8	1-15/64	7-3/8	1	1-9/16	11
20	1-1/4	1-1/8	1-1/2	9	1-1/4	2	16

Figure 10-14. Typical hose sizes.

the outside diameter in sixteenths of an inch of standard tubing, and the hose will have about the same inside diameter as the tubing. For example, a size 8 single-wire braid hose will have an inside diameter very close to an 8/16- or 1/2-in standard tubing. For double-braided hose, the size specification equals the actual inside diameter in sixteenths of an inch. For example, a size 8 double-wire braid hose will have a 1/2-in inside diameter. The minimum bend radii values provide the smallest values for various hose sizes to prevent undue strain or flow interference.

Figure 10-15 illustrates five different flexible hose designs whose constructions are described as follows:

a. *FC 194:* Elastomer inner tube, single-wire braid reinforcement, and elastomer cover. Working pressures vary from 375 to 2750 psi depending on the size.

b. *FC 195:* Elastomer inner tube, double-wire braid reinforcement, and elastomer cover. Working pressures vary from 1125 to 5000 psi depending on the size.

c. *FC 300:* Elastomer inner tube, polyester inner braid, single-wire braid reinforcement, and polyester braid cover. Working pressures vary from 350 to 3000 psi depending on the size.

d. *1525:* Elastomer inner tube, textile braid reinforcement, oil and mildew resistant, and textile braid cover. Working pressure is 250 psi for all sizes.

e. *2791:* Elastomer inner tube, partial textile braid, four heavy spiral wire reinforcements, and elastomer cover. Working pressure is 2500 psi for all sizes.

Hose Fittings

Hose assemblies of virtually any length and with various end fittings are available from manufacturers. See Figure 10-16 for examples of hoses with the following permanently attached end fittings: (a) straight fitting, (b) 45° elbow fitting, and (c) 90° elbow fitting.

The elbow-type fittings allow access to hard-to-get-at connections. They also permit better flexing and improve the appearance of the system.

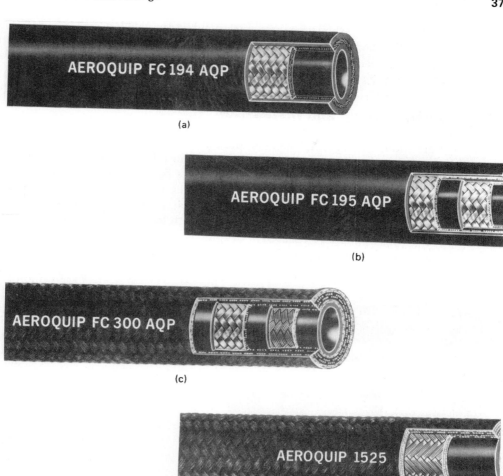

(a)

(b)

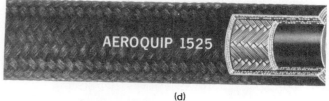

(c)

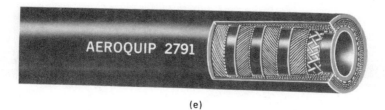

(d)

(e)

Figure 10-15. Various flexible hose designs. (a) FC 194: single-wire braid; (b) FC 195: double-wire braid; (c) FC 300: single-wire braid, polyester inner braid; (d) 1525: single-textile braid; (e) 2791: four heavy spiral wires, partial textile braid. *(Courtesy of Aeroquip Corp., Jackson, Michigan.)*

372

Figure 10-17 shows the three corresponding reusable-type end fittings. These types can be detached from a damaged hose and reused on a replacement hose. The renewable fittings idea had its beginning in 1941. With the advent of World War II, it was necessary to get aircraft with failed hydraulic lines back into operation as quickly as possible.

Hose Routing and Installation

Care should be taken in changing fluid in hoses since the hose and fluid materials must be compatible. Flexible hose should be installed so there is no kinking during operation of the system. There should always be some slack to relieve any strain and allow for the absorption of pressure surges. It is poor practice to twist the hose and use long loops in the plumbing operation. It may be necessary to use clamps to prevent chafing or tangling of the hose with moving parts. If the hose is subject to rubbing, it should be encased in a protective sleeve. Figure 10-18 gives basic information on hose routing and installation procedures.

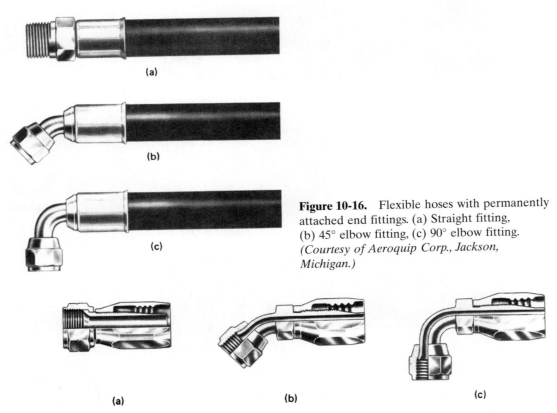

(a)

(b)

(c)

Figure 10-16. Flexible hoses with permanently attached end fittings. (a) Straight fitting, (b) 45° elbow fitting, (c) 90° elbow fitting. *(Courtesy of Aeroquip Corp., Jackson, Michigan.)*

(a) (b) (c)

Figure 10-17. Flexible hose reusable-type end fittings. (a) Straight fitting, (b) elbow fitting, (c) 90° elbow fitting. *(Courtesy of Aeroquip Corp., Jackson, Michigan.)*

Hose routing and installation

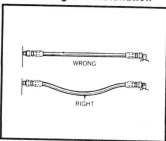

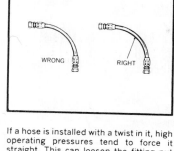

Under pressure, a hose may change in length. The range is from −4% to +2%. Always provide some slack in the hose to allow for this shrinkage or expansion. (However, excessive slack in hose lines is one of the most common causes of poor appearance.)

If a hose is installed with a twist in it, high operating pressures tend to force it straight. This can loosen the fitting nut or even burst the hose at the point of strain.

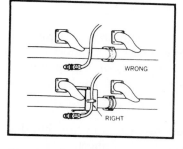

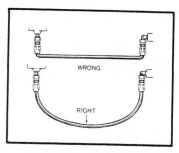

When hose lines pass near an exhaust manifold, or other heat source, they should be insulated by a heat resistant boot, firesleeve or a metal baffle. In any application, brackets and clamps keep hoses in place and reduce abrasion.

At bends, provide enough hose for a wide radius curve. Too tight a bend pinches the hose and restricts the flow. The line could even kink and close entirely. In many cases, use of the right fittings or adapters can eliminate bends or kinks.

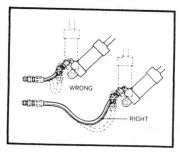

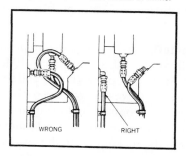

In applications where there is considerable vibration or flexing, allow additional hose length. The metal hose fittings, of course, are not flexible, and proper installation protects metal parts from undue stress, and avoids kinks in the hose.

When 90° adapters were used, this assembly became neater-looking and easier to inspect and maintain. It uses less hose, too!

Four basic adapter functions

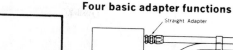

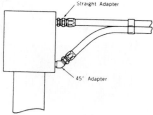

1. To join a hose to a component. Example, a valve might have a ½" female pipe thread and a hose a ¾" S.A.E. 37° swivel nut. The right Aeroquip Adapter fits both.

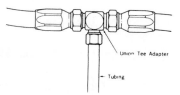

2. To connect two or more pieces of hose and tubing. Here, a T-shaped adapter connects two hoses with a length of tubing. Each end of the adapter may have a different thread.

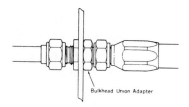

3. To provide both connection and anchor at a bulkhead. In this example, it provides an anchor in addition to connecting a hose to a tube.

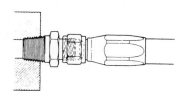

4. To eliminate the need for a bushing. Example, one end of the adapter is ¾" pipe thread, connected to the assembly, and the other is ½" S.A.E. 37° flare, which connects to an S.A.E. 37° swivel fitting. The adapter itself replaces the bushing.

Figure 10-18. Hose routing and installation information. *(Courtesy of Aeroquip Corp., Jackson, Michigan.)*

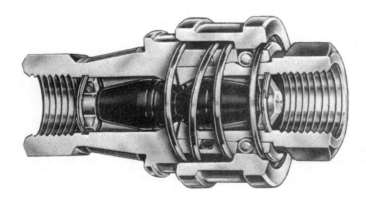

Figure 10-19. Quick disconnect coupling (cross-sectional view). *(Courtesy of Hansen Manufacturing Co., Cleveland, Ohio.)*

10.8 QUICK DISCONNECT COUPLINGS

One additional type of fitting is the quick disconnect coupling used for both plastic tubing and flexible hose. It is used mainly where a conductor must be disconnected frequently from a component. This type of fitting permits assembly and disassembly in a matter of a second or two. The three basic designs are:

1. Straight through. This design offers minimum restriction to flow but does not prevent fluid loss from the system when the coupling is disconnected.

2. One-way shutoff. This design locates the shutoff at the fluid source connection but leaves the actuator component unblocked. Leakage from the system is not excessive in short runs, but system contamination due to the entrance of dirt in the open end of the fitting can be a problem, especially with mobile equipment located at the work site.

3. Two-way shutoff. This design provides positive shutoff of both ends of pressurized lines when disconnected. See Figure 10-19 for a cutaway of this type of quick disconnect coupling. Such a coupling puts an end to the loss of fluids. As soon as you release the locking sleeve, valves in both the socket and plug close, shutting off flow. When connecting, the plug contacts an O-ring in the socket, creating a positive seal. There is no chance of premature flow or waste due to a partial connection. The plug must be fully seated in the socket before the valves will open.

10.9 METRIC STEEL TUBING

In this section we examine common metric tube sizes and show how to select the proper size tube based on flow-rate requirements and strength considerations.

Figure 10-20 shows the common tube sizes used in fluid power systems. Note that the smallest OD size is 4 mm (0.158 in), whereas the largest OD size is 42 mm

Tube OD (mm)	Wall Thickness (mm)	Tube ID (mm)	Tube OD (mm)	Wall Thickness (mm)	Tube ID (mm)	Tube OD (mm)	Wall Thickness (mm)	Tube ID (mm)
4	0.5	3	14	2.0	10	25	3.0	19
6	1.0	4	15	1.5	12	25	4.0	17
6	1.5	3	15	2.0	11	28	2.0	24
8	1.0	6	16	2.0	12	28	2.5	23
8	1.5	5	16	3.0	10	30	3.0	24
8	2.0	4	18	1.5	15	30	4.0	22
10	1.0	8	20	2.0	16	35	2.0	31
10	1.5	7	20	2.5	15	35	3.0	29
10	2.0	6	20	3.0	14	38	4.0	30
12	1.0	10	22	1.0	20	38	5.0	28
12	1.5	9	22	1.5	19	42	2.0	38
12	2.0	8	22	2.0	18	42	3.0	36

Figure 10-20. Common metric tube sizes.

(1.663 in). These values compare to 0.125 in and 1.500 in, respectively (from Figure 10-7), for common English units tube sizes.

Factors of safety based on corresponding operating pressures become

FS = 8 for pressures from 0 to 1000 psi (0 to 7 MPa or 0 to 70 bars)

FS = 6 for pressures from 1000 to 2500 psi (7 to 17.5 MPa or 70 to 175 bars)

FS = 4 for pressures above 2500 psi (17.5 MPa or 175 bars)

The corresponding tensile strengths for SAE 1010 dead soft cold-drawn steel and AISI 4130 steel are

SAE 1010	55,000 psi or 379 MPa
AISI 4130	75,000 psi or 517 MPa

EXAMPLE 10-5

Select the proper metric size steel tube for a flow rate of 0.00190 m³/s and an operating pressure of 70 bars. The maximum recommended velocity is 6.1 m/s and the tube material is SAE 1010 dead soft cold-drawn steel having a tensile strength of 379 MPa.

Solution The minimum inside diameter based on the fluid velocity limitation of 6.1 m/s is found using Eq. (3-38):

$$Q(m^3/s) = A(m^2) \times v(m/s)$$

Solving for A, we have

$$A = Q/v$$

Since $A = \frac{\pi}{4}(\text{ID})^2$, we have the final resulting equation:

$$\text{ID} = \sqrt{\frac{4Q}{\pi v}} \qquad\qquad \textbf{(10-7)}$$

Substituting values we have

$$\text{ID} = \sqrt{\frac{(4)(0.00190)}{\pi(6.1)}} = 0.0199 \text{ m} = 19.9 \text{ mm}$$

From Figure 10-20, the smallest acceptable OD tube size is

22-mm OD, 1.0-mm wall thickness, 20-mm ID

From Eq. (10-3) we obtain the burst pressure.

$$\text{BP} = \frac{2tS}{D_i} = \frac{(2)(0.001\text{m})(379 \text{ MN/m}^2)}{0.020 \text{ m}} = 37.9 \text{ MN/m}^2 = 37.9 \text{ MPa}$$

Then, we calculate the working pressure using Eq. (10-4).

$$\text{WP} = \frac{\text{BP}}{\text{FS}} = \frac{37.9 \text{ MPa}}{8} = 4.74 \text{ MPa} = 47.4 \text{ bars}$$

This pressure is not adequate (less than operating pressure of 70 bars), so let's examine the next larger size OD tube having the necessary ID.

28-mm OD, 2.0-mm wall thickness, 24-mm ID

$$\text{BP} = \frac{(2)(0.002)(379)}{0.024} = 63.2 \text{ MPa}$$

$$\text{WP} = \frac{63.2}{8} = 7.90 \text{ MPa} = 79.0 \text{ bars}$$

$$D_i/t = 24 \text{ mm}/2 \text{ mm} = 12 > 10$$

This result is acceptable.

10.10 KEY EQUATIONS

Fluid velocity:
$$v = v_{\text{avg}} = \frac{Q}{A}$$
(10-1)

Pipe tensile stress:
$$\sigma = \frac{pD_i}{2t}$$
(10-2)

Pipe burst pressure:
$$\text{BP} = \frac{2tS}{D_i}$$
(10-3)

Pipe working pressure:
$$\text{WP} = \frac{\text{BP}}{\text{FS}}$$
(10-4)

EXERCISES

Questions, Concepts, and Definitions

10-1. What is the primary purpose of the fluid distribution system?

10-2. What flow velocity is generally recommended for the discharge side of a pump?

10-3. What is the recommended flow velocity for the inlet side of a pump?

10-4. Why should conductors or fittings not be made of copper?

10-5. What metals cannot be used with water-glycol fluids?

10-6. What effect does hydraulic shock have on system pressure?

10-7. What variables determine the wall thickness and factor of safety of a conductor for a particular operating pressure?

10-8. Why should conductors have greater strength than the system working pressure requires?

10-9. Name the major disadvantages of steel pipes.

10-10. Name the four primary types of conductors.

10-11. What is meant by the term *average fluid velocity?*

10-12. Why is malleable iron sometimes used for steel pipe fittings?

10-13. Why is steel tubing more widely used than steel pipe?

10-14. What principal advantage does plastic tubing have over steel tubing?

10-15. Explain the purpose of a quick disconnect fitting.

10-16. What is the disadvantage of threaded fittings?

10-17. What is the difference between a flared fitting and a compression fitting?

10-18. Under what conditions would flexible hoses be used in hydraulic systems?

10-19. Name three factors that should be considered when installing flexible hoses.

10-20. What is the basic construction of a flexible hose?

10-21. Relative to steel pipes, for a given nominal size, does the wall thickness increase or decrease as the schedule number is increased?

10-22. How is a pipe size classified?

10-23. What is meant by the schedule number of standard pipe?

10-24. Distinguish between thin-walled and thick-walled conductors.

10-25. Name two assembly precautions when using flared fittings.

Problems

Note: The letter *E* following an exercise number means that English units are used. Similarly, the letter *M* indicates metric units.

Conductor Sizing for Flow Rate

10-26E. What size inlet line would you select for a 20-gpm pump?

10-27E. What size discharge line would you select for a 20-gpm pump?

10-28M. What metric-size inlet line would you select for a 0.002-m³/s pump?

10-29M. What metric-size discharge line would you select for a 0.002-m³/s pump?

10-30E. A pump produces a flow rate of 75 Lpm. It has been established that the fluid velocity in the discharge line should be between 6 and 7.5 m/s. Determine the minimum and maximum pipe inside diameters that should be used.

10-31. For liquid flow in a pipe, the velocity of the liquid varies inversely as the _____ of the pipe inside diameter.

10-32. For liquid flow in a pipe, doubling the pipe's inside diameter reduces the fluid velocity by a factor of _____.

10-33. For liquid flow in a pipe, derive the constants C_1 and C_2 in the following equations:

$$A(\text{in}^2) = \frac{C_1 Q(\text{gpm})}{v\left(\dfrac{\text{ft}}{\text{s}}\right)} \qquad A(\text{m}^2) = \frac{C_2 Q\left(\dfrac{\text{m}^3}{\text{s}}\right)}{v\left(\dfrac{\text{m}}{\text{s}}\right)}$$

10-34E. What minimum commercial-size tubing with a wall thickness of 0.095 in would be required at the inlet and outlet of a 30-gpm pump if the inlet and outlet velocities are limited to 5 ft/s and 20 ft/s, respectively? See Figure 10-7.

10-35M. Change the data in Exercise 10-34 to metric units and solve for the minimum commercial-size tubing at the pump inlet and outlet. See Figure 10-20.

10-36M. What minimum size commercial pipe tubing with a wall thickness of 2.0 mm would be required at the inlet and outlet of a 75-Lpm pump? The inlet and outlet velocities are limited to 1.2 m/s and 6.1 m/s, respectively. See Figure 10-20.

Pressure Rating of Conductors

10-37E. A steel tubing has a 1.250-in outside diameter and a 1.060-in inside diameter. It is made of AISI 4130 steel having a tensile strength of 75,000 psi. What would be the safe working pressure for this tube assuming a factor of safety of 8?

10-38E. For the conductor in Exercise 10-37, determine the tensile stress for an operating pressure of 1000 psi.

10-39E. Select the proper-sized steel tube for a flow rate of 20 gpm and an operating pressure of 1000 psi. The maximum recommended velocity is 20 ft/s and the factor of safety is 8.

    a. Material is SAE 1010 with a tensile strength of 55,000 psi.

    b. Material is AISI 4130 with a tensile strength of 75,000 psi.

10-40M. A steel tubing has a 30-mm outside diameter and a 24-mm inside diameter. It is made of AISI 4130 steel having a tensile strength of 517 MPa. What would be the safe working pressure in units of bars for this tube assuming a factor of safety of 8?

10-41M. For the conductor in Exercise 10-40, determine the tensile stress for an operating pressure of 10 MPa.

10-42M. Select the proper-sized metric steel tube for a flow rate of 0.001 m³/s and an operating pressure of 70 bars. The maximum recommended velocity is 6.1 m/s and the factor of safety is 8.

    a. Material is SAE 1010.

    b. Material is AISI 4130.

10-43E. A steel tube of 1-in ID has a burst pressure of 8000 psi. If the tensile strength of the tube material is 55,000 psi, find the minimum acceptable OD.

10-44M. A steel tube of 25-mm ID has a burst pressure of 50 MPa. If the tensile strength is 379 MPa, find the minimum acceptable OD.

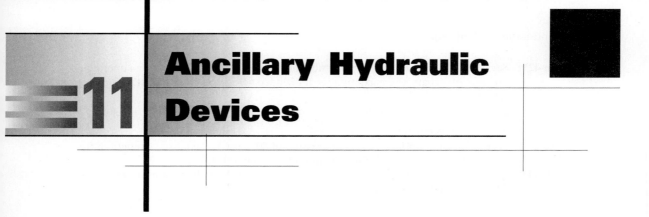

Ancillary Hydraulic Devices

Learning Objectives

Upon completing this chapter, you should be able to:

1. Describe the purpose, construction, and operation of various accumulators.
2. Explain the operation of accumulator circuits.
3. Describe the operation of pressure intensifiers and identify typical applications.
4. Explain the basic design features of reservoirs and determine the proper reservoir size for a given hydraulic system.
5. Describe the ways in which the temperature of the fluid of a hydraulic system can be controlled by heat exchangers.
6. Calculate the rise in fluid temperature as the fluid flows through restrictors such as pressure relief valves.
7. Describe the design features and operating characteristics of the most widely used types of sealing devices.
8. Describe the operation of flow measurement devices such as the rotameter and turbine flow meter.
9. Describe the operation of pressure measuring devices such as the Bourdon gage and Schrader gage.

11.1 INTRODUCTION

Ancillary hydraulic devices are those important components that do not fall under the major categories of pumps, valves, actuators, conductors, and fittings. This chapter deals with these ancillary devices, which include reservoirs, accumulators, pressure intensifiers, sealing devices, heat exchangers, pressure gages, and flow meters. Two exceptions are the components called *filters* and *strainers*, which are covered in Chapter 12, "Maintenance of Hydraulic Systems." Filters and strainers are included in Chapter 12 because these two components are specifically designed to enhance the successful maintenance of hydraulic systems.

11.2 RESERVOIRS

Design and Construction Features

The proper design of a suitable reservoir for a hydraulic system is essential to the overall performance and life of the individual components. The reservoir serves not only as a storage space for the hydraulic fluid used by the system but also as the principal location where the fluid is conditioned. The reservoir is where sludge, water, and metal chips settle and where entrained air picked up by the oil is allowed to escape. The dissipation of heat is also accomplished by a properly designed reservoir.

Figure 11-1 illustrates the construction features of a reservoir that satisfies industry's standards. This reservoir is constructed of welded steel plates. The inside surfaces

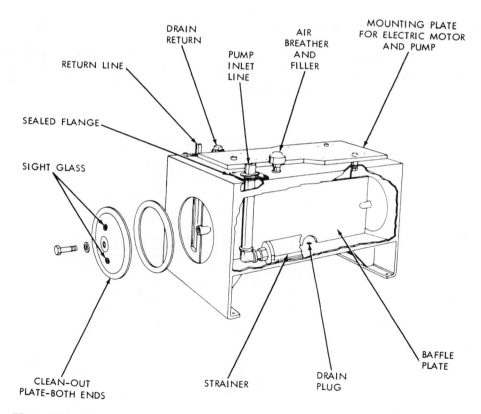

Figure 11-1. Reservoir construction. *(Courtesy of Sperry Vickers, Sperry Rand Corp., Troy, Michigan.)*

are painted with a sealer to prevent rust, which can occur due to condensed moisture. Observe that this reservoir allows for easy fluid maintenance. The bottom plate is dished and contains a drain plug at its lowest point to allow complete drainage of the tank when required. Removable covers are included to provide easy access for cleaning. A sight glass is also included to permit a visual check of the fluid level. A vented breather cap is also included and contains an air filtering screen. This allows the tank to breathe as the oil level changes due to system demand requirements. In this way, the tank is always vented to the atmosphere.

As shown in Figure 11-2, a baffle plate extends lengthwise across the center of the tank. Its height is about 70% of the height of the oil level. The purpose of the baffle plate is to separate the pump inlet line from the return line to prevent the same fluid from recirculating continuously within the tank. In this way all the fluid is uniformly used by the system.

Essentially, the baffle serves the following functions:

1. Permits foreign substances to settle to the bottom
2. Allows entrained air to escape from oil
3. Prevents localized turbulence in reservoir
4. Promotes heat dissipation through reservoir walls

As illustrated in Figure 11-1, the reservoir is constructed so that the pump and driving motor can be installed on its top surface. A smooth machined surface is provided that has adequate strength to support and hold the alignment of the two units.

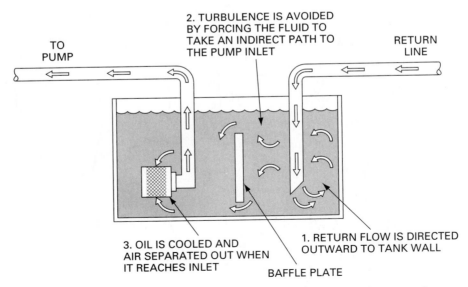

Figure 11-2. Battle plate controls direction of flow in reservoir. *(Courtesy of Sperry Vickers, Sperry Rand Corp., Troy, Michigan.)*

The return line should enter the reservoir on the side of the baffle plate that is opposite from the pump suction line. To prevent foaming of the oil, the return line should be installed within two pipe diameters from the bottom of the tank. The pump suction strainer should be well below the normal oil level in the reservoir and at least 1 in from the bottom. If the strainer is located too high, it will cause a vortex or crater to form, which will permit air to enter the suction line. In addition, suction line connections above the oil level must be tightly sealed to prevent the entrance of air into the system.

Sizing of Reservoirs

The sizing of a reservoir is based on the following criteria:

1. It must make allowance for dirt and chips to settle and for air to escape.
2. It must be able to hold all the oil that might drain into the reservoir from the system.
3. It must maintain the oil level high enough to prevent a *whirlpool* effect at the pump inlet line opening. Otherwise, air will be drawn into the pump.
4. It should have a surface area large enough to dissipate most of the heat generated by the system.
5. It should have adequate air space to allow for thermal expansion of the oil.

A reservoir having a capacity of three times the volume flow rate (in units of volume per minute) of the pump has been found to be adequate for most hydraulic systems where average demands are expected. This relationship is given by

$$\text{reservoir size (gal)} = 3 \times \text{pump flow rate (gpm)} \tag{11-1}$$

$$\text{reservoir size (m}^3\text{)} = 3 \times \text{pump flow rate (m}^3/\text{min)} \tag{11-1M}$$

Thus, a hydraulic system using a 10-gpm pump would require a 30-gal reservoir and a 0.1-m³/min pump would require a 0.3-m³ reservoir. However, the benefits of a large reservoir are usually sacrificed for mobile and aerospace applications due to weight and space limitations.

11.3 ACCUMULATORS

Definition of Accumulator

An accumulator is a device that stores potential energy by means of either gravity, mechanical springs, or compressed gases. The stored potential energy in the accumulator is a quick secondary source of fluid power capable of doing useful work as required by the system.

There are three basic types of accumulators used in hydraulic systems. They are identified as follows:

1. Weight-loaded, or gravity, type
2. Spring-loaded type
3. Gas-loaded type

Weight-Loaded Accumulator

The weight-loaded accumulator is historically the oldest. This type consists of a vertical, heavy-wall steel cylinder, which incorporates a piston with packings to prevent leakage. A deadweight is attached to the top of the piston (see Figure 11-3). The force of gravity of the deadweight provides the potential energy in the accumulator. This type of accumulator creates a constant fluid pressure throughout the full volume output of the unit regardless of the rate and quantity of output. In the other types of accumulators, the fluid output pressure decreases as a function of the volume output of the accumulator. The main disadvantage of this type of accumulator is its extremely large size and heavy weight, which makes it unsuitable for mobile equipment.

Spring-Loaded Accumulator

A spring-loaded accumulator is similar to the weight-loaded type except that the piston is preloaded with a spring, as illustrated in Figure 11-4. The compressed spring

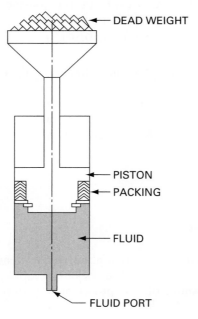

Figure 11-3. Weight-loaded accumulator. *(Courtesy of Greer Olaer Products Division/Greer Hydraulics Inc., Los Angeles, California.)*

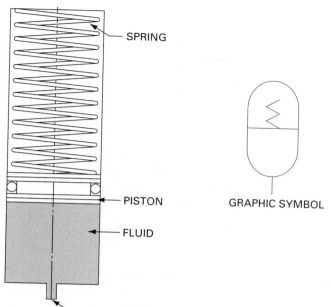

SPRING

PISTON

FLUID

FLUID PORT

GRAPHIC SYMBOL

Figure 11-4. Spring-loaded accumulator. *(Courtesy of Greer Olaer Products Division/Greer Hydraulics Inc., Los Angeles, California.)*

is the source of energy that acts against the piston, forcing the fluid into the hydraulic system to drive an actuator. The pressure generated by this type of accumulator depends on the size and preloading of the spring. In addition, the pressure exerted on the fluid is not a constant. The spring-loaded accumulator typically delivers a relatively small volume of oil at low pressures. Thus, they tend to be heavy and large for high-pressure, large-volume systems. This type of accumulator should not be used for applications requiring high cycle rates because the spring will fatigue, resulting in an inoperative accumulator.

Gas-Loaded Accumulators

Gas-loaded accumulators (frequently called *hydropneumatic accumulators*) have been found to be more practical than the weight- and spring-loaded types. The gas-loaded type operates in accordance with Boyle's law of gases, which states that for a constant temperature process, the pressure of a gas varies inversely with its volume. Thus, for example, the gas volume of the accumulator would be cut in half if the pressure were doubled. The compressibility of gases accounts for the storage of potential energy. This energy forces the oil out of the accumulator when the gas expands due to the reduction of system pressure when, for example, an actuator rapidly moves a load. Nitrogen is the gas used in accumulators because (unlike air) it contains no moisture. In addition, nitrogen is an inert gas and thus will not support combustion. The sizing of gas-loaded accumulators for given applications is covered in Chapter 14, after Boyle's law of gases is discussed.

Gas-loaded accumulators fall into two main categories:

1. Nonseparator type
2. Separator type

Nonseparator-Type Accumulator.   The nonseparator type of accumulator consists of a fully enclosed shell containing an oil port on the bottom and a gas charging valve on the top (see Figure 11-5). The gas is confined in the top and the oil at the bottom of the shell. There is no physical separator between the gas and oil, and thus the gas pushes directly on the oil. The main advantage of this type is its ability to handle large volumes of oil. The main disadvantage is absorption of the gas in the oil due to the lack of a separator. This type must be installed vertically to keep the gas confined at the top of the shell. This type is not recommended for use with high-speed pumps because the entrapped gas in the oil could cause cavitation and damage to the pump. Absorption of gas in the oil also makes the oil compressible, resulting in spongy operation of the hydraulic actuators.

Separator-Type Accumulator.   The commonly accepted design of gas-loaded accumulators is the separator type. In this type there is a physical barrier between the gas and the oil. This barrier effectively uses the compressibility of the gas. The three major classifications of the separator accumulator are

1. Piston type
2. Diaphragm type
3. Bladder type

Piston Accumulator.   The piston type of accumulator consists of a cylinder containing a freely floating piston with proper seals, as illustrated in Figure 11-6. The piston serves as the barrier between the gas and oil. A threaded lock ring provides a safety feature, which prevents the operator from disassembling the unit while it is

GAS VALVE

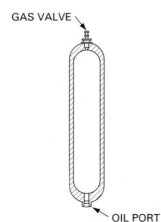

OIL PORT

Figure 11-5.   Nonseparator-type accumulator. *(Courtesy of Greer Olaer Products Division/Greer Hydraulics Inc., Los Angeles, California.)*

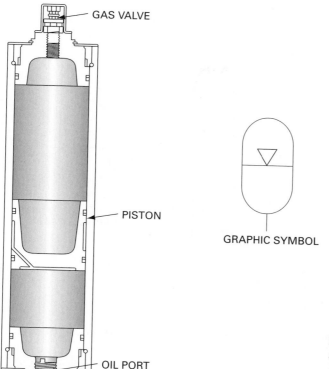

GAS VALVE

PISTON

OIL PORT

GRAPHIC SYMBOL

Figure 11-6. Piston-type gas-loaded accumulator. *(Courtesy of Greer Olaer Products Division/Greer Hydraulics Inc., Los Angeles, California.)*

precharged. The main disadvantages of the piston types of accumulator are that they are expensive to manufacture and have practical size limitations. Piston and seal friction may also be a problem in low-pressure systems. Also, appreciable leakage tends to occur over a long period, requiring frequent precharging. Piston accumulators should not be used as pressure pulsation dampeners or shock absorbers because of the inertia of the piston and the friction of the seals. The principal advantage of the piston accumulator is its ability to handle very high or low temperature system fluids through the use of compatible O-ring seals.

Diaphragm Accumulator. The diaphragm-type accumulator consists of a diaphragm, secured in the shell, which serves as an elastic barrier between the oil and gas (see Figure 11-7). A shutoff button, which is secured at the base of the diaphragm, covers the inlet of the line connection when the diaphragm is fully stretched. This prevents the diaphragm from being pressed into the opening during the precharge period. On the gas side, the screw plug allows control of the charge pressure and charging of the accumulator by means of a charging device.

Figure 11-8 illustrates the operation of a diaphragm-type accumulator. The hydraulic pump delivers oil into the accumulator and deforms the diaphragm. As the pressure increases, the volume of gas decreases, thus storing energy. In the reverse case, where additional oil is required in the circuit, it comes from the accumulator as

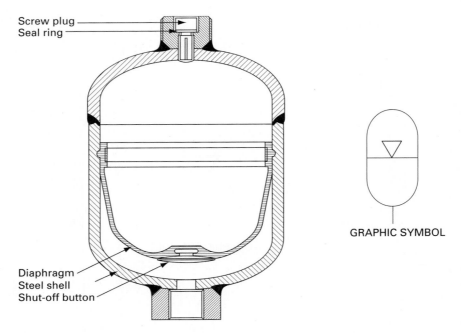

Figure 11-7. Diaphragm-type accumulator. (*Courtesy of Robert Bosch Corp., Broadview, Illinois.*)

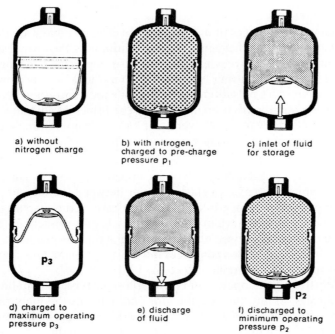

Figure 11-8. Operation of a diaphragm-type accumulator. (*Courtesy of Robert Bosch Corp., Broadview, Illinois.*)

the pressure drops in the system by a corresponding amount. The primary advantage of this type of accumulator is its small weight-to-volume ratio, which makes it suitable almost exclusively for airborne applications.

Bladder Accumulator. A bladder-type accumulator contains an elastic barrier (bladder) between the oil and gas, as illustrated in Figure 11-9. The bladder is fitted in the accumulator by means of a vulcanized gas-valve element and can be installed or removed through the shell opening at the poppet valve. The poppet valve closes the inlet when the accumulator bladder is fully expanded. This prevents the bladder from being pressed into the opening. The greatest advantage of this type of accumulator is the positive sealing between the gas and oil chambers. The lightweight bladder provides quick response for pressure regulating, pump pulsation, and shock-dampening applications.

Figure 11-10 illustrates the operation of a bladder-type accumulator. The hydraulic pump delivers oil into the accumulator and deforms the bladder. As the

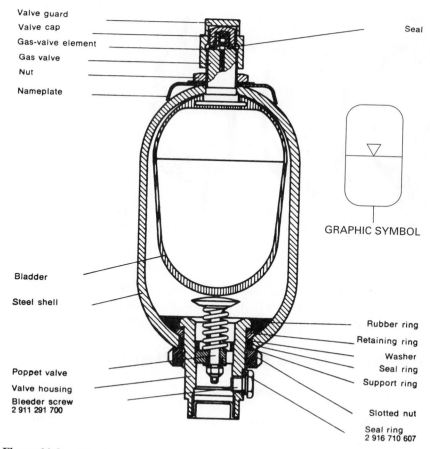

Figure 11-9. Bladder-type accumulator. *(Courtesy of Robert Bosch Corp., Broadview, Illinois.)*

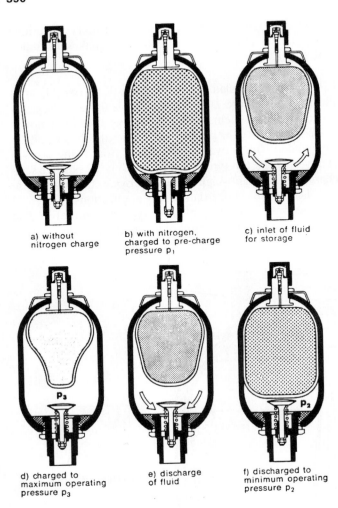

a) without
nitrogen charge

b) with nitrogen,
charged to pre-charge
pressure p_1

c) inlet of fluid
for storage

d) charged to
maximum operating
pressure p_3

e) discharge
of fluid

f) discharged to
minimum operating
pressure p_2

Figure 11-10. Operation of a bladder-type accumulator. *(Courtesy of Robert Bosch Corp., Broadview, Illinois.)*

pressure increases, the volume of gas decreases, thus storing energy. In the reverse case, where additional oil is required in the circuit, it comes from the accumulator as pressure drops in the system by a corresponding amount.

11.4 APPLICATIONS OF ACCUMULATORS

Basic Applications

There are four basic applications where accumulators are used in hydraulic systems.

1. An auxiliary power source
2. A leakage compensator

3. An emergency power source

4. A hydraulic shock absorber

The following is a description and the accompanying circuit diagram of each of these four applications.

Accumulator as an Auxiliary Power Source

One of the most common applications of accumulators is as an auxiliary power source. The purpose of the accumulator in this application is to store oil delivered by the pump during a portion of the work cycle. The accumulator then releases this stored oil on demand to complete the cycle, thereby serving as a secondary power source to assist the pump. In such a system where intermittent operations are performed, the use of an accumulator results in being able to use a smaller-sized pump.

This application is depicted in Figure 11-11 in which a four-way valve is used in conjunction with an accumulator. When the four-way valve is manually actuated, oil flows from the accumulator to the blank end of the cylinder. This extends the piston until it reaches the end of its stroke. While the desired operation is occurring (the cylinder is in the fully extended position), the accumulator is being charged by the pump. The four-way valve is then deactivated for the retraction of the cylinder. Oil flows from the pump and accumulator to retract the cylinder rapidly. The accumulator size is selected to supply adequate oil during the retraction stroke. The sizing of gas-loaded accumulators as an auxiliary power source is presented in Chapter 14.

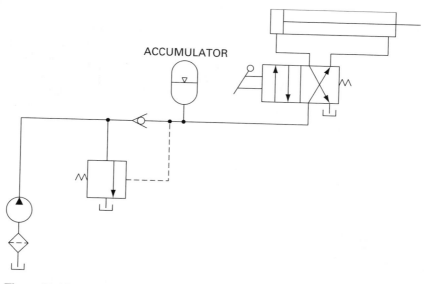

ACCUMULATOR

Figure 11-11. Accumulator as an auxiliary power source.

Accumulator as a Leakage Compensator

A second application for accumulators is as a compensator for internal or external leakage during an extended period of time during which the system is pressurized but not in operation. As shown in Figure 11-12, for this application the pump charges the accumulator and system until the maximum pressure setting on the pressure switch is obtained. The contacts on the pressure switch then open to automatically stop the electric motor that drives the pump. The accumulator then supplies leakage oil to the system during a long period. Finally, when system pressure drops to the minimum pressure setting of the pressure switch, it closes the electrical circuit of the pump motor (not shown) until the system has been recharged. The use of an accumulator as a leakage compensator saves electrical power and reduces heat in the system. Electrical circuit diagrams such as those used in this application to control the pump motor are presented in Chapter 15.

Accumulator as an Emergency Power Source

In some hydraulic systems, safety dictates that a cylinder be retracted even though the normal supply of oil pressure is lost due to a pump or electrical power failure. Such an application requires the use of an accumulator as an emergency power source, as depicted in Figure 11-13. In this circuit, a solenoid-actuated, three-way valve is used in conjunction with the accumulator. When the three-way valve is energized, oil flows to the blank end of the cylinder and also through the check valve into the accumulator and rod end of the cylinder. The accumulator charges as the cylinder extends. If the pump fails due to an electrical failure, the solenoid will

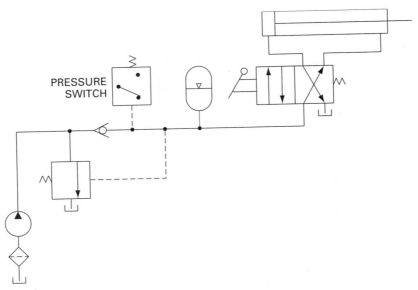

Figure 11-12. Accumulator as a leakage compensator.

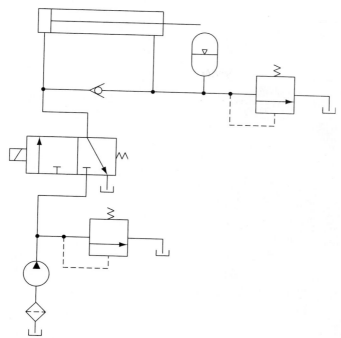

Figure 11-13. Accumulator as an emergency power source.

de-energize, shifting the valve to its spring-offset mode. Then the oil stored under pressure is forced from the accumulator to the rod end of the cylinder. This retracts the cylinder to its starting position.

Figure 11-14 shows an accumulator application involving a machine that transports and handles huge logs. The circuit for the hydraulic breaking system of this machine is given in Figure 11-15. This circuit shows that in case of low oil pressure, as sensed by a low-pressure warning switch, to gas-charged accumulators ensure that adequate pressurized oil can be sent to the hydraulic brake valves. This would allow for adequate hydraulic braking action to take place on the wheels to stop any travel motion of the machine. Braking occurs if the operator pushes on the pedal of the pedal-actuated hydraulic power brake valve.

Accumulator as a Hydraulic Shock Absorber

One of the most important industrial applications of accumulators is the elimination or reduction of high-pressure pulsations or hydraulic shock. Hydraulic shock (or *water hammer*, as it is frequently called) is caused by the sudden stoppage or deceleration of a hydraulic fluid flowing at relatively high velocity in a pipeline. One example where this occurs is in the case of a rapidly closing valve. This creates a compression wave where the rapidly closing valve is located. This compression wave travels at the speed of sound upstream to the end of the pipe and back again to the closed valve, causing an increase in the line pressure. This wave travels

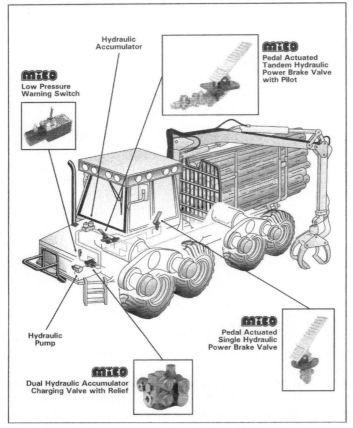

MICO Hydraulic Components and Brake Systems
Logging Forwarder example

Figure 11-14. Log transport/ handling machine. *(Courtesy of MICO, Incorporated, North Mankato, Minnesota.)*

back and forth along the entire pipe length until its energy is finally dissipated by friction. The resulting rapid pressure pulsations or high-pressure surges may cause damage to the hydraulic system components. If an accumulator is installed near the rapidly closing valve, as shown in Figure 11-16, the pressure pulsations or high-pressure surges are suppressed.

11.5 PRESSURE INTENSIFIERS

Introduction

Although a pump is the primary power source for a hydraulic system, auxiliary units are frequently employed for special purposes. One such auxiliary unit is the pressure intensifier or booster.

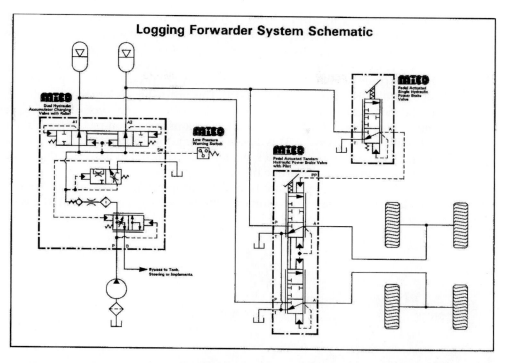

Figure 11-15. Hydraulic braking system for log transport/handling machine of Figure 11-14. *(Courtesy of MICO, Incorporated, North Mankato, Minnesota.)*

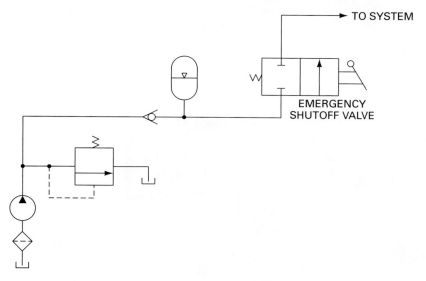

Figure 11-16. Accumulator as a hydraulic shock absorber.

A pressure intensifier is used to increase the pressure in a hydraulic system to a value above the pump discharge pressure. It accepts a high-volume flow at relatively low pump pressure and converts a portion of this flow to high pressure.

Figure 11-17 shows a cutaway view of a Racine pressure intensifier. The internal construction consists of an automatically reciprocating large piston that has two small rod ends (also see Figure 11-18). This piston has its large area (total area of piston) exposed to pressure from a low-pressure pump. The force of the low-pressure oil moves the piston and causes the small area of the piston rod to force the oil out at intensified high pressure. This device is symmetrical about a vertical centerline. Thus, as the large piston reciprocates, the left- and right-hand halves of the unit duplicate each other during each stroke of the large piston.

The increase in pressure is in direct proportion to the ratio of the large piston area and the rod area. The volume output is inversely proportional to this same ratio.

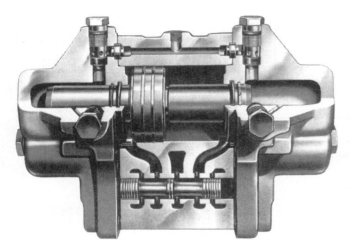

Figure 11-17. Cutaway view of pressure intensifier. *(Courtesy of Rexnord Inc., Hydraulic Components Division, Racine, Wisconsin.)*

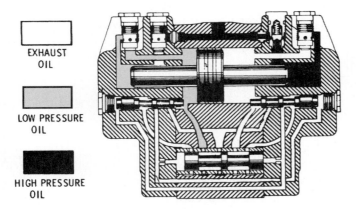

EXHAUST OIL

LOW PRESSURE OIL

HIGH PRESSURE OIL

Figure 11-18. Oil flow paths of pressure intensifier. *(Courtesy of Rexnord Inc., Hydraulic Components Division, Racine, Wisconsin.)*

$$\frac{\text{high discharge pressure}}{\text{low inlet pressure}} = \frac{\text{area of piston}}{\text{area of rod}} = \frac{\text{high inlet flow rate}}{\text{low discharge flow rate}} \quad \textbf{(11-2)}$$

Racine pressure intensifiers are available with area ratios of 3:1, 5:1, and 7:1, developing pressures to 5000 psi and flows to 7 gpm. There are many applications for pressure intensifiers, such as the elimination of a high-pressure/low-flow pump used in conjunction with a low-pressure/high-flow pump. In an application such as a punch press, it is necessary to extend a hydraulic cylinder rapidly using little pressure to get the ram near the sheet metal strip as quickly as possible. Then the cylinder must exert a large force using only a small flow rate. The large force is needed to punch the workpiece from the sheet metal strip. Since the strip is thin, only a small flow rate is required to perform the punching operation in a short time. The use of the pressure intensifier results in a significant cost savings in this application, because it replaces the expensive high-pressure pump that would normally be required.

EXAMPLE 11-1

Oil at 20 gpm and 500 psi enters the low-pressure inlet of a 5:1 Racine pressure intensifier. Find the discharge flow and pressure.

Solution   Substitute directly into Eq. (11-2):

$$\frac{\text{high discharge pressure}}{500 \text{ psi}} = \frac{5}{1} = \frac{20 \text{ gpm}}{\text{low discharge flow rate}}$$

Solving for the unknown quantities, we have

$$\text{high discharge pressure} = 5(500 \text{ psi}) = 2500 \text{ psi}$$

$$\text{low discharge flow rate} = \frac{20}{5} = 4 \text{ gpm}$$

Pressure Intensifier Circuit

Figure 11-19 gives the circuit for a punch press application where a pressure intensifier is used to eliminate the need for a high-pressure/low-flow pump. This circuit also includes a pilot check valve and sequence valve. The operation is as follows: When the pressure in the cylinder reaches the sequence valve pressure setting, the intensifier starts to operate. The high-pressure output of the intensifier closes the pilot check valve and pressurizes the blank end of the cylinder to perform the punching operation. A pilot check valve is used instead of a regular check valve to permit retraction of the cylinder. Very high pressures can be supplied by a pressure intensifier operating on a low-pressure pump. The intensifier should be installed near the cylinder to keep the high-pressure lines as short as possible.

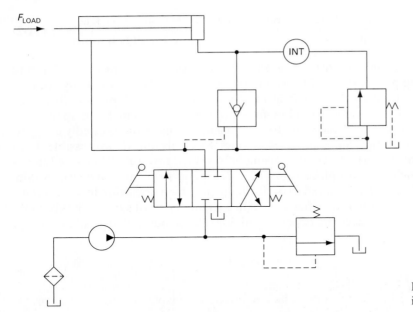

Figure 11-19. Pressure intensifier circuit.

Air-over-Oil Intensifier System

In Figure 11-20 we see an air-over-oil intensifier circuit, which drives a cylinder over a large distance at low pressure and then over a small distance at high pressure. Shop air can be used to extend and retract the cylinder during the low-pressure portion of the cycle. The system operates as follows: Valve 1 extends and retracts the cylinder using shop air at approximately 80 psi. Valve 2 applies air pressure to the top end of the hydraulic intensifier. This produces high hydraulic pressure at the bottom end of the intensifier. Actuation of valve 1 directs air to the approach tank. This forces oil at 80 psi through the bottom of the intensifier to the blank end of the cylinder. When the cylinder experiences its load (such as the punching operation in a punch press), valve 2 is actuated, which sends shop air to the top end of the intensifier. The high-pressure oil cannot return to the approach tank because this port is blocked off by the downward motion of the intensifier piston. Thus, the cylinder receives high-pressure oil at the blank end to overcome the load. When valve 2 is released, the shop air is blocked, and the top end of the intensifier is vented to the atmosphere. This terminates the high-pressure portion of the cycle. When valve 1 is released, the air in the approach tank is vented, and shop air is directed to the return tank. This delivers oil at shop pressure to the rod end of the cylinder, causing it to retract. Oil enters the bottom end of the intensifier and flows back to the approach tank. This completes the entire cycle. Figure 11-21 shows an air-oil intensifier and its graphic symbol. This type of intensifier is capable of producing output hydraulic pressures up to 3000 psi.

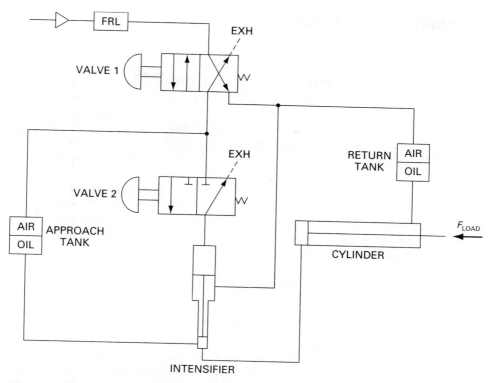

Figure 11-20. Air-over-oil intensifier circuit.

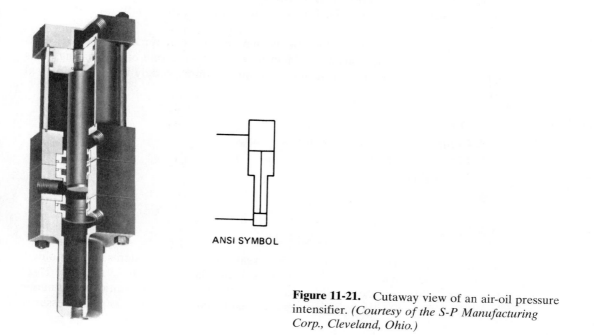

Figure 11-21. Cutaway view of an air-oil pressure intensifier. *(Courtesy of the S-P Manufacturing Corp., Cleveland, Ohio.)*

11.6 SEALING DEVICES

Introduction

Oil leakage, located anywhere in a hydraulic system, reduces efficiency and increases power losses. Internal leakage does not result in loss of fluid from the system because the fluid returns to the reservoir. Most hydraulic components possess clearances that permit a small amount of internal leakage. This leakage increases as component clearances between mating parts increase due to wear. If the entire system leakage becomes large enough, most of the pump's output is bypassed, and the actuators will not operate properly. External leakage represents a loss of fluid from the system. In addition, it is unsightly and represents a safety hazard. Improper assembly of pipe fittings is the most common cause of external leakage. Overtightened fittings may become damaged, or vibration can cause properly tightened fittings to become loose. Shaft seals on pumps and cylinders may become damaged due to misalignment or excessive pressure.

Seals are used in hydraulic systems to prevent excessive internal and external leakage and to keep out contamination. Seals can be of the positive or nonpositive type and can be designed for static or dynamic applications. Positive seals do not allow any leakage whatsoever (external or internal). Nonpositive seals (such as the clearance used to provide a lubricating film between a valve spool and its housing bore) permit a small amount of internal leakage.

Static seals are used between mating parts that do not move relative to each other. Figure 11-22 shows some typical examples, which include flange gaskets and seals. Note that these seals are compressed between two rigidly connected parts. They represent a relatively simple and nonwearing joint, which should be trouble-free if properly assembled. Figure 11-23 shows a number of die-cut gaskets used for flange-type joints.

Dynamic seals are assembled between mating parts that move relative to each other. Hence, dynamic seals are subject to wear because one of the mating parts rubs against the seal. The following represent the most widely used types of seal configurations:

1. O-rings
2. Compression packings (V- and U-shapes)
3. Piston cup packings
4. Piston rings
5. Wiper rings

O-Rings

The O-ring is one of the most widely used seals for hydraulic systems. It is a molded, synthetic rubber seal that has a round cross section in its free state. See Figure 11-24 for several different-sized O-rings, which can be used for most static and dynamic conditions. These O-ring seals give effective sealing through a wide range of pressures,

BASIC FLANGE JOINTS

GASKET

METAL-TO-METAL JOINTS

Figure 11-22. Static seal flange joint applications. *(Courtesy of Sperry Vickers, Sperry Rand Corp., Troy, Michigan.)*

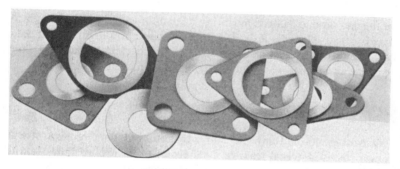

Figure 11-23. Die-cut gaskets used for flanged joints. *(Courtesy of Crane Packing Co., Morton Grove, Illinois.)*

Figure 11-24. Several different-sized O-rings. *(Courtesy of Crane Packing Co., Morton Grove, Illinois.)*

temperatures, and movements with the added advantages of sealing pressure in both directions and providing low running friction on moving parts.

As illustrated in Figure 11-25, an O-ring is installed in an annular groove machined into one of the mating parts. When it is initially installed, it is compressed at both its inside and outside diameters. When pressure is applied, the O-ring is forced against a third surface to create a positive seal. The applied pressure also forces the O-ring to push even harder against the surfaces in contact with its inside and outside diameters. As a result, the O-ring is capable of sealing against high pressures. However,

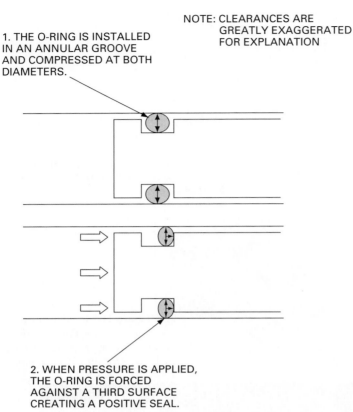

1. THE O-RING IS INSTALLED IN AN ANNULAR GROOVE AND COMPRESSED AT BOTH DIAMETERS.

NOTE: CLEARANCES ARE GREATLY EXAGGERATED FOR EXPLANATION

2. WHEN PRESSURE IS APPLIED, THE O-RING IS FORCED AGAINST A THIRD SURFACE CREATING A POSITIVE SEAL.

Figure 11-25. O-ring operation. *(Courtesy of Sperry Vickers, Sperry Rand Corp., Troy, Michigan.)*

O-rings are not generally suited for sealing rotating shafts or where vibration is a problem.

At very high pressures, the O-ring may extrude into the clearance space between mating parts, as illustrated in Figure 11-26. This is unacceptable in a dynamic application because of the rapid resulting seal wear. This extrusion is prevented by installing a backup ring, as shown in Figure 11-26. If the pressure is applied in both directions, a backup ring must be installed on both sides of the O-ring.

Compression Packings

V-ring packings are compression-type seals that are used in virtually all types of reciprocating motion applications. These include rod and piston seals in hydraulic and pneumatic cylinders, press rams, jacks, and seals on plungers and pistons in reciprocating pumps. They are also readily suited to certain slow rotary applications such as valve stems. These packings (which can be molded into U-shapes as well as V-shapes) are frequently installed in multiple quantities for more effective sealing. As illustrated in Figure 11-27, these packings are compressed by tightening a flanged

NOTE: CLEARANCES ARE
GREATLY EXAGGERATED
FOR EXPLANATION

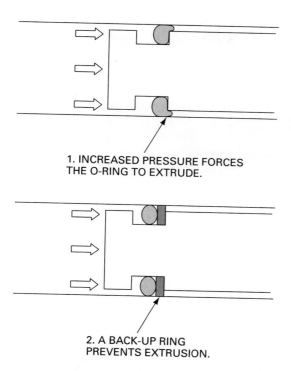

1. INCREASED PRESSURE FORCES
THE O-RING TO EXTRUDE.

2. A BACK-UP RING
PREVENTS EXTRUSION.

Figure 11-26. Backup ring prevents extrusion of O-ring. *(Courtesy of Sperry Vickers, Sperry Rand Corp., Troy, Michigan.)*

follower ring against them. Proper adjustment is essential since excessive tightening will hasten wear.

In many applications these packings are spring-loaded to control the correct force as wear takes place. However, springs are not recommended for high-speed or quick reverse motion on reciprocating applications. Figure 11-28(a) shows several different-sized V-ring packings, whereas Figure 11-28(b) shows two different-sized sets of V-ring packings stacked together.

Piston Cup Packings

Piston cup packings are designed specifically for pistons in reciprocating pumps and pneumatic and hydraulic cylinders. They offer the best service life for this type of application, require a minimum recess space and minimum recess machining, and are simply and quickly installed. Figure 11-29 shows the typical installation for single-acting and double-acting operations. Sealing is accomplished when pressure pushes the cup lip outward against the cylinder barrel. The backing plate and retainers clamp the cup packing tightly in place, allowing it to handle very high pressures. Figure 11-30 shows several different-sized piston cup packings.

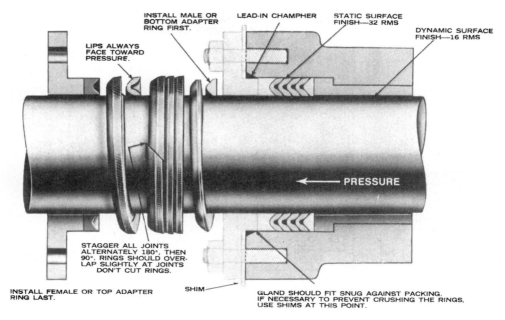

Figure 11-27. Application of V-ring packings. *(Courtesy of Crane Packing Co., Morton Grove, Illinois.)*

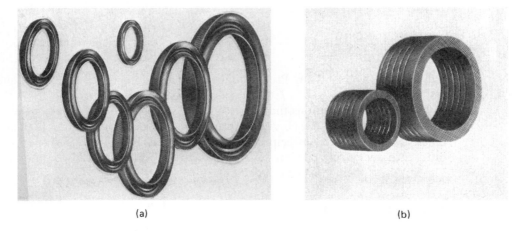

(a)   (b)

Figure 11-28. V-ring packings. (a) Several different sizes. (b) Two sizes in stacked arrangement. *(Courtesy of Crane Packing Co., Morton Grove, Illinois.)*

Piston Rings

Piston rings are seals that are universally used for cylinder pistons, as shown in Figure 11-31. Metallic piston rings are made of cast iron or steel and are usually plated or given an outer coating of materials such as zinc phosphate or manganese phosphate to prevent rusting and corrosion. Piston rings offer substantially less opposition to

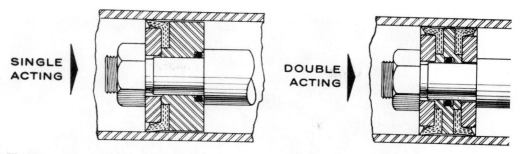

Figure 11-29. Typical applications of piston cup packings. *(Courtesy of Crane Packing Co., Morton Grove, Illinois.)*

Figure 11-30. Several different-sized piston cup packings. *(Courtesy of Crane Packing Co., Morton Grove, Illinois.)*

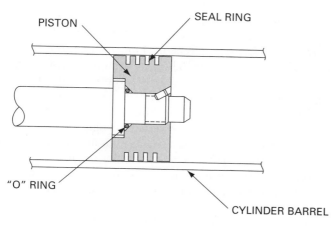

Figure 11-31. Use of piston rings for cylinder pistons. *(Courtesy of Sperry Vickers, Sperry Rand Corp., Troy, Michigan.)*

motion than do synthetic rubber (elastomer) seals. Sealing against high pressures is readily handled if several rings are used, as illustrated in Figure 11-31.

Figure 11-32 shows a number of nonmetallic piston rings made out of tetrafluoroethylene (TFE), a chemically inert, tough, waxy solid. Their extremely low coefficient of friction (0.04) permits them to be run completely dry and at the same time prevents scoring of the cylinder walls. This type of piston ring is an ideal solution to many

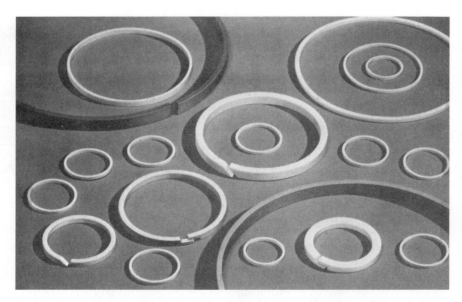

Figure 11-32.   TFE nonmetallic piston rings. *(Courtesy of Crane Packing Co., Troy, Michigan.)*

applications where the presence of lubrication can be detrimental or even dangerous. For instance, in an oxygen compressor, just a trace of oil is a fire or explosion hazard.

Wiper Rings

Wiper rings are seals designed to prevent foreign abrasive or corrosive materials from entering a cylinder. They are not designed to seal against pressure. They provide insurance against rod scoring and add materially to packing life. Figure 11-33(a) shows several different-sized wiper rings, and Figure 11-33(b) shows a typical installation arrangement. The wiper ring is molded from a synthetic rubber, which is stiff enough to wipe all dust or dirt from the rod yet pliable enough to maintain a snug fit. The rings are easily installed with a snap fit into a machined groove in the gland. This eliminates the need for and expense of a separate retainer ring.

Natural rubber is rarely used as a seal material because it swells and deteriorates with time in the presence of oil. In contrast, synthetic rubber materials are compatible with most oils. The most common types of materials used for seals are leather, Buna-N, silicone, neoprene, tetrafluoroethylene, viton, and, of course, metals.

1. Leather. This material is rugged and inexpensive. However, it tends to squeal when dry and cannot operate above 200°F, which is inadequate for many hydraulic systems. Leather does operate well at cold temperatures to about −60°F.

2. Buna-N. This material is rugged and inexpensive and wears well. It has a rather wide operating temperature range (−50°F to 230°F) during which it maintains its good sealing characteristics.

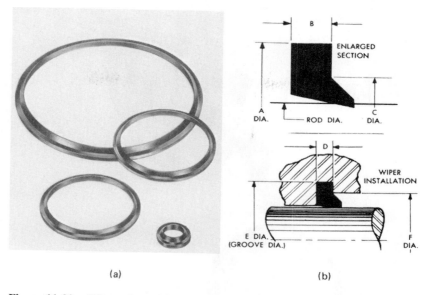

Figure 11-33. Wiper rings. (a) Various sizes. (b) Installation arrangement. *(Courtesy of Crane Packing Co., Morton Grove, Illinois.)*

3. Silicone. This elastomer has an extremely wide temperature range ($-90°F$ to $450°F$). Hence, it is widely used for rotating shaft seals and static seals where a wide operating temperature is expected. Silicone is not used for reciprocating seal applications because it has low tear resistance.

4. Neoprene. This material has a temperature range of $-65°F$ to $250°F$. It is unsuitable above $250°F$ because it has a tendency to vulcanize.

5. Tetrafluoroethylene. This material is the most widely used plastic for seals of hydraulic systems. It is a tough, chemically inert, waxy solid, which can be processed only by compacting and sintering. It has excellent resistance to chemical breakdown up to temperatures of $700°F$. It also has an extremely low coefficient of friction. One major drawback is its tendency to flow under pressure, forming thin, feathery films. This tendency to flow can be greatly reduced by the use of filler materials such as graphite, metal wires, glass fibers, and asbestos.

6. Viton. This material contains about 65% fluorine. It has become almost a standard material for elastomer-type seals for use at elevated temperatures up to $500°F$. Its minimum operating temperature is $-20°F$.

Durometer Hardness Tester

Physical properties frequently used to describe the behavior of elastomers are as follows: hardness, coefficient of friction, volume change, compression set, tensile strength, elongation modulus, tear strength, squeeze stretch, coefficient of thermal

Figure 11-34. Durometer.
(Courtesy of Shore Instruments &
Mfg. Co., Jamaica, New York.)

expansion, and permeability. Among these physical properties, hardness is among
the most important since it has a direct relationship to service performance. A durometer (see Figure 11-34) is an instrument used to measure the indentation hardness
of rubber and rubberlike materials. As shown, the hardness scale has a range from
0 to 100. The durometer measures 100 when pressed firmly on flat glass. High durometer readings indicate a great resistance to denting and thus a hard material. A durometer hardness of 70 is the most common value for seal materials. A hardness of 80
is usually specified for rotating motion to eliminate the tendency toward side motion
and bunching in the groove. Values between 50 and 60 are used for static seals on
rough surfaces. Hard seal materials (values between 80 and 90) have less breakaway
friction than softer materials, which have a greater tendency to deform and flow into
surface irregularities. As a result, harder materials are used for dynamic seals.

11.7 HEAT EXCHANGERS

Introduction

Heat is generated in hydraulic systems because no component can operate at 100%
efficiency. Significant sources of heat include the pump, pressure relief valves, and

Figure 11-35. Air-cooled heat exchanger. *(Courtesy of American Standard, Heat Transfer Division, Buffalo, New York.)*

flow control valves. Heat can cause the hydraulic fluid temperature to exceed its normal operating range of 110°F to 150°F. Excessive temperature hastens oxidation of the hydraulic oil and causes it to become too thin. This promotes deterioration of seals and packings and accelerates wear between closely fitting parts of hydraulic components of valves, pumps, and actuators.

The steady-state temperature of the fluid of a hydraulic system depends on the heat-generation rate and the heat-dissipation rate of the system. If the fluid operating temperature in a hydraulic system becomes excessive, this means that the heat-generation rate is too large relative to the heat-dissipation rate. Assuming that the system is reasonably efficient, the solution is to increase the heat-dissipation rate. This is accomplished by the use of *coolers*, which are commonly called "heat exchangers." In some applications, the fluid must be heated to produce a satisfactory value of viscosity. This is typical when, for example, mobile hydraulic equipment is to operate in below-0°F temperatures. In these cases, the heat exchangers are called "heaters." However, for most hydraulic systems, the natural heat-generation rate is sufficient to produce high enough temperatures after an initial warm-up period. Hence, the problem usually becomes one of using a heat exchanger to provide adequate cooling.

There are two main types of heat-dissipation heat exchangers: air coolers and water coolers. Figure 11-35 shows an air cooler in which the hydraulic fluid is pumped through tubes banded to fins. It can handle oil flow rates up to 200 gpm and employs a fan to increase the heat-transfer rate.

The air cooler shown uses tubes that contain special devices—turbulators to mix the warmer and cooler oil for better heat transfer—because the oil near the center of the tube is warmer than that near the wall. Light, hollow, metal spheres are randomly inserted inside the tubes. These spheres cause the oil to tumble over itself to provide thorough mixing to produce a lighter and better cooler.

Figure 11-36. Water-cooled shell and tube heat exchanger. *(Courtesy of American Standard, Heat Transfer Division, Buffalo, New York.)*

In Figure 11-36 we see a water cooler. In this type of heat exchanger, water is circulated through the unit by flowing around the tubes, which contain the hydraulic fluid. The design shown has a tough ductile, red-brass shell; unique flanged, yellow-brass baffles; seamless nonferrous tubes; and cast-iron bonnets. It provides heat-transfer surface areas up to 124 ft² (11.5 m²).

Fluid Temperature Rise Across Pressure Relief Valves

The following equations permit the calculation of the fluid temperature rise as it flows through a restriction such as a pressure relief valve:

$$\frac{\text{temperature}}{\text{increase (°F)}} = \frac{\text{heat-generation rate (Btu/min)}}{\text{oil specific heat (Btu/lb/°F)} \times \text{oil flow rate (lb/min)}} \qquad \textbf{(11-3)}$$

$$\frac{\text{temperature}}{\text{increase (°C)}} = \frac{\text{heat-generation rate (kW)}}{\text{oil specific heat (kJ/kg/°C)} \times \text{oil flow rate (kg/s)}} \qquad \textbf{(11-3M)}$$

$$\text{specific heat of oil} = 0.42 \text{ Btu/lb/°F} \qquad \textbf{(11-4)}$$

$$\text{specific heat of oil} = 1.8 \text{ kJ/kg/°C} \qquad \textbf{(11-4M)}$$

$$\text{oil flow-rate (lb/min)} = 7.42 \times \text{oil flow-rate (gpm)} \qquad \textbf{(11-5)}$$

$$\text{oil flow-rate (kg/s)} = 895 \times \text{oil flow-rate (m}^3\text{/s)} \qquad \textbf{(11-5M)}$$

Examples 11-2 and 11-3 show how to determine the fluid temperature rise across pressure relief valves.

EXAMPLE 11-2

Oil at 120°F and 1000 psi is flowing through a pressure relief valve at 10 gpm. What is the downstream oil temperature?

Solution First, calculate the horsepower lost and convert to the heat-generation rate in units of Btu/min:

$$HP = \frac{p \text{ (psi)} \times Q \text{ (gpm)}}{1714} = \frac{(1000)(10)}{1714} = 5.83$$

Since 1 HP = 42.4 Btu/min, we have

$$Btu/min = 5.83 \times 42.4 = 247$$

Next, calculate the oil flow rate in units of lb/min and the temperature increase:

$$\text{oil flow rate} = 7.42(10) = 74.2 \text{ lb/min}$$

$$\text{temperature increase} = \frac{247}{0.42 \times 74.2} = 7.9°F$$

$$\text{downstream oil temperature} = 120 + 7.9 = 127.9°F$$

EXAMPLE 11-3

Oil at 50°C and 70 bars is flowing through a pressure relief valve at 0.000632 m³/s. What is the downstream oil temperature?

Solution First, calculate the heat-generation rate in units of kW:

$$kW = \frac{p \text{ (Pa)} \times Q \text{(m}^3/\text{s)}}{1000} = \frac{(7 \times 10^6)(632 \times 10^{-6})}{1000} = 4.42 \text{ kW}$$

Next, calculate the oil flow rate in units of kg/s and the temperature increase:

$$\text{oil flow rate} = (895)(0.000632) = 0.566 \text{ kg/s}$$

$$\text{temperature increase} = \frac{4.42}{(1.8)(0.566)} = 4.3°C$$

$$\text{downstream oil temperature} = 50 + 4.3 = 54.3°C$$

Sizing of Heat Exchangers

When sizing heat exchangers in English units, a heat load value is calculated for the entire system in units of Btu/hr. The calculation of the system heat load can be readily calculated by noting that 1 hp equals 2544 Btu/hr. Similar calculations are made when sizing heat exchangers in metric units.

Examples 11-4 and 11-5 show how to determine the heat load value for a hydraulic system and thereby establish the required heat-exchanger rating to dissipate all the generated heat.

EXAMPLE 11-4

A hydraulic pump operates at 1000 psi and delivers oil at 20 gpm to a hydraulic actuator. Oil discharges through the pressure relief valve (PRV) during 50% of the cycle time. The pump has an overall efficiency of 85%, and 10% of the power is lost due to frictional pressure losses in the hydraulic lines. What rating heat exchanger is required to dissipate all the generated heat?

Solution

$$\text{pump HP loss} = \text{HP}_{\text{input}} - \text{HP}_{\text{output}} = \frac{\text{HP}_{\text{output}}}{\eta_o} - \text{HP}_{\text{output}}$$

$$= \left(\frac{1}{\eta_o} - 1\right)\text{HP}_{\text{output}} = \left(\frac{1}{\eta_o} - 1\right)\frac{p\,(\text{psi}) \times Q\,(\text{gpm})}{1714}$$

$$= \left(\frac{1}{0.85} - 1\right) \times \frac{1000 \times 20}{1714} = 0.1765 \times \frac{1000 \times 20}{1714} = 2.06 \text{ hp}$$

$$\text{PRV average HP loss} = 0.50 \times \frac{1000 \times 20}{1714} = 5.83 \text{ hp}$$

$$\text{line average HP loss} = 0.50 \times 0.10 \times \frac{1000 \times 20}{1714} = 0.58 \text{ hp}$$

$$\text{total average HP loss} = 8.47$$

$$\text{heat-exchanger rating} = 8.47 \times 2544 = 21{,}500 \text{ Btu/hr}$$

EXAMPLE 11-5

A hydraulic pump operates at 70 bars and delivers oil at 0.00126 m³/s to a hydraulic actuator. Oil discharges through the pressure relief valve (PRV) during 50% of the cycle time. The pump has an overall efficiency of 85%, and 10% of the power is lost due to frictional pressure losses in the hydraulic lines. What heat-exchanger rating is required to dissipate all the generated heat?

Solution   Per the solution to Example 11-4, we have

$$\text{pump kW loss} = \left(\frac{1}{\eta_o} - 1\right) \times \text{pump kW output}$$

$$= \left(\frac{1}{\eta_o} - 1\right) \times \frac{p(\text{Pa}) \times Q(\text{m}^3/\text{s})}{1000}$$

$$= 0.1765 \times \frac{(7 \times 10^6) \times (1260 \times 10^{-6})}{1000} = 1.56\,\text{kW}$$

$$\text{PRV average kW loss} = 0.50 \times \frac{7 \times 10^6 \times 1260 \times 10^{-6}}{1000} = 4.41\,\text{kW}$$

$$\text{line average kW loss} = 0.50 \times 0.10 \times \frac{7 \times 10^6 \times 1260 \times 10^{-6}}{1000} = 0.44\,\text{kW}$$

$$\text{total kW loss} = 6.41\,\text{kW}$$

$$\text{heat-exchanger rating} = 6.41\,\text{kW}$$

11.8 PRESSURE GAGES

Pressure-measuring devices are needed in hydraulic circuits for a number of reasons. In addition to testing and troubleshooting, they are used to adjust pressure settings of pressure control valves and to determine forces exerted by hydraulic cylinders and torques delivered by hydraulic motors.

One of the most widely used pressure-measuring devices is the Bourdon gage (see Figure 11-37, which shows an assortment of Bourdon gages, each having different pressure ranges). The Bourdon gage contains a sealed tube formed in the shape of an arc (refer to Figure 11-38). When pressure is applied at the port opening, the tube starts to straighten somewhat. This activates a linkage-gear system, which moves the pointer to indicate the pressure on the dial. The scale of most Bourdon gages reads zero when the gage is open to the atmosphere, because the gages are calibrated to read pressure above atmospheric pressure or gage pressure. Some Bourdon gages are capable of reading pressures below atmospheric or vacuum (suction) pressures, such as those existing in pump inlet lines. The range for vacuum gages is from 0 to 30 in of mercury, which represents a perfect vacuum.

A second common type of pressure-measuring device is the Schrader gage. As illustrated in Figure 11-39, pressure is applied to a spring-loaded sleeve and piston.

Figure 11-37. Bourdon gages with different pressure ranges. *(Courtesy of Span Instruments, Inc., Plano, Texas.)*

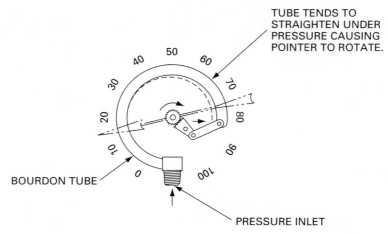

TUBE TENDS TO STRAIGHTEN UNDER PRESSURE CAUSING POINTER TO ROTATE.

BOURDON TUBE

PRESSURE INLET

Figure 11-38. Operation of Bourdon gage. *(Courtesy of Sperry Vickers, Sperry Rand Corp., Troy, Michigan.)*

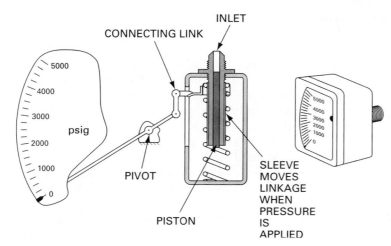

CONNECTING LINK

INLET

psig

PIVOT

PISTON

SLEEVE MOVES LINKAGE WHEN PRESSURE IS APPLIED

Figure 11-39. Operation of a Schrader gage. *(Courtesy of Sperry Vickers, Sperry Rand Corp., Troy, Michigan.)*

As the pressure moves the sleeve, it actuates the indicating pointer through mechanical linkages.

11.9 FLOWMETERS

Introduction

Flow-rate measurements are frequently required to evaluate the performance of hydraulic components as well as to troubleshoot a hydraulic system. They can be used to check the volumetric efficiency of pumps and also to determine leakage paths within a hydraulic circuit.

Rotameter

Probably the most common type of flowmeter is the rotameter, which consists of a metering float in a calibrated vertical tube, as shown in Figure 11-40. The operation of the rotameter is as follows (refer to Figure 11-41):

The metering float is free to move vertically in the tapered glass tube. The fluid flows through the tube from bottom to top. When no fluid is flowing, the float rests at the bottom of the tapered tube, and its maximum diameter is usually so selected that it blocks the small end of the tube almost completely. When flow begins in the pipeline, the fluid enters the bottom of the meter and raises the float. This increases the flow area between the float and tube until an equilibrium position is reached. At this position, the weight of the float is balanced by the upward force of the fluid on the float. The greater the flow rate, the higher the float rises in the tube. The tube is graduated to allow a direct reading of the flow rate.

Sight Flow Indicator

Sometimes it is desirable to determine whether or not fluid is flowing in a pipeline and to observe the flowing fluid visually. Such a device for accomplishing this is called a *sight flow indicator*. It does not measure the rate of flow but instead indicates only whether or not there is flow. The sight flow indicator shown in Figure 11-42 has two windows located on opposite sides of the body fittings to give the best possible visibility.

Disk Piston

Figure 11-43 shows another type of flowmeter, which incorporates a *disk piston*. When the fluid passes through the measuring chamber, the disk piston develops a rotary motion, which is transmitted through gearing to a pointer on a dial.

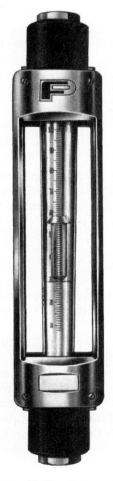

Figure 11-40. Rotameter. *(Courtesy of Fischer & Porter Co., Worminster, Pennsylvania.)*

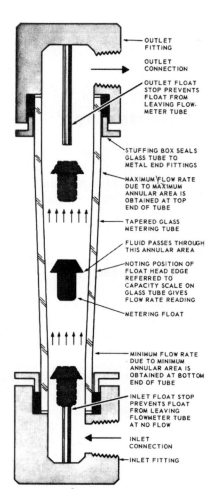

OUTLET FITTING

OUTLET CONNECTION

OUTLET FLOAT STOP PREVENTS FLOAT FROM LEAVING FLOW-METER TUBE

STUFFING BOX SEALS GLASS TUBE TO METAL END FITTINGS

MAXIMUM FLOW RATE DUE TO MAXIMUM ANNULAR AREA IS OBTAINED AT TOP END OF TUBE

TAPERED GLASS METERING TUBE

FLUID PASSES THROUGH THIS ANNULAR AREA

NOTING POSITION OF FLOAT HEAD EDGE REFERRED TO CAPACITY SCALE ON GLASS TUBE GIVES FLOW RATE READING

METERING FLOAT

MINIMUM FLOW RATE DUE TO MINIMUM ANNULAR AREA IS OBTAINED AT BOTTOM END OF TUBE

INLET FLOAT STOP PREVENTS FLOAT FROM LEAVING FLOWMETER TUBE AT NO FLOW

INLET CONNECTION

INLET FITTING

Figure 11-41. Operation of rotameter. *(Courtesy of Fischer & Porter Co., Worminster, Pennsylvania.)*

Figure 11-42. Sight flow indicator. *(Courtesy of Fischer & Porter Co., Worminster, Pennsylvania.)*

416

Turbine Flowmeter

A schematic drawing of a turbine-type flowmeter is given in Figure 11-44. This design incorporates a turbine rotor mounted in a housing connected in a pipeline whose fluid flow rate is to be measured. The fluid causes the turbine to rotate at a speed proportional to the flow rate. The rotation of the turbine generates an electrical impulse every time a turbine blade passes a sensing device. An electronic device connected to the sensor converts the pulses to flow-rate information.

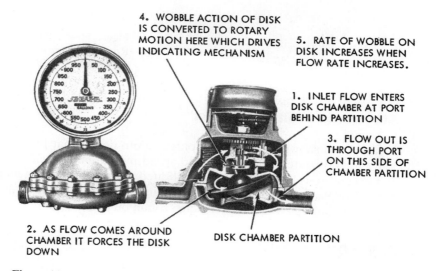

4. WOBBLE ACTION OF DISK IS CONVERTED TO ROTARY MOTION HERE WHICH DRIVES INDICATING MECHANISM

5. RATE OF WOBBLE ON DISK INCREASES WHEN FLOW RATE INCREASES.

1. INLET FLOW ENTERS DISK CHAMBER AT PORT BEHIND PARTITION

3. FLOW OUT IS THROUGH PORT ON THIS SIDE OF CHAMBER PARTITION

2. AS FLOW COMES AROUND CHAMBER IT FORCES THE DISK DOWN

DISK CHAMBER PARTITION

Figure 11-43. Flowmeter with disk piston. *(Courtesy of Sperry Vickers, Sperry Rand Corp., Troy, Michigan.)*

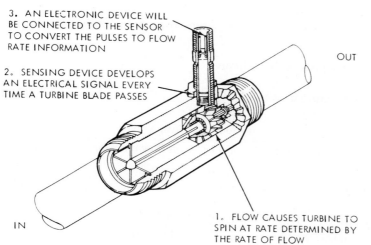

3. AN ELECTRONIC DEVICE WILL BE CONNECTED TO THE SENSOR TO CONVERT THE PULSES TO FLOW RATE INFORMATION

2. SENSING DEVICE DEVELOPS AN ELECTRICAL SIGNAL EVERY TIME A TURBINE BLADE PASSES

OUT

IN

1. FLOW CAUSES TURBINE TO SPIN AT RATE DETERMINED BY THE RATE OF FLOW

Figure 11-44. Turbine flowmeter. *(Courtesy of Sperry Vickers, Sperry Rand Corp., Troy, Michigan.)*

Figure 11-45. Digital Electronic Readout. *(Courtesy of Flo-Tech, Inc., Mundelein, Illinois.)*

Electronic Digital Readout

Figure 11-45 shows a digital electronic readout device that provides five-digit displays for flow-rate and pressure and speed measurements accurate to ±0.15% of full scale. The scale can be factory calibrated to display values in units such as gpm, L/min, psi, Pascals, bars, rpm, in/min, m/min, and so on. Data are updated two times per second and provide an over-range condition indication.

11.10 KEY EQUATIONS

Adequate reservoir size

English units: Reservoir size (gal) = 3 × pump flow rate (gpm)   **(11-1)**

Metric units: Reservoir size (m³) = 3 × pump flow rate (m³/min)   **(11-1M)**

Temperature rise across a pressure relief valve

English units: Temperature increase (°F)   **(11-3)**

$$= \frac{\text{heat-generation rate (Btu/min)}}{\text{oil specific heat (Btu/lb/°F)} \times \text{oil flow rate (lb/min)}}$$

Metric units: Temperature increase (°C)   **(11-3M)**

$$= \frac{\text{heat-generation rate (kW)}}{\text{oil specific heat (kJ/kg/°C)} \times \text{oil flow rate (kg/s)}}$$

EXERCISES

Questions, Concepts, and Definitions

11-1. Name four criteria by which the size of a reservoir is determined.

11-2. What are the three most common reservoir designs?

11-3. What is the purpose of the reservoir breather?

11-4. For what is a reservoir baffle plate used?

11-5. Name the three basics types of accumulators.

11-6. Name the three major classifications of gas-loaded accumulators. Give one advantage of each classification.

11-7. Describe four applications of accumulators.

11-8. What is a pressure intensifier? List one application.

11-9. What is the difference between a positive and a nonpositive seal?

11-10. Explain, by example, the difference between internal and external leaks.

11-11. What is the difference between a static and dynamic seal?

11-12. Why are backup rings sometimes used with O-rings?

11-13. Name three types of seals in addition to an O-ring.

11-14. Are wiper seals designed to seal against pressure? Explain your answer.

11-15. Name four types of materials used for seals.

11-16. What is the purpose of a durometer?

11-17. What is the purpose of a heat exchanger?

11-18. Name the important factors to consider when selecting a heat exchanger.

11-19. Name two types of flow-measuring devices.

11-20. Name two types of pressure-measuring devices.

11-21. Why is it desirable to measure flow rates and pressures in a hydraulic system?

11-22. What advantage does a digital readout fluid parameter-measuring device have over an analog device?

11-23. What is a sight flow indicator and what is its purpose?

Problems

Note: The letter *E* following an exercise number means that English units are used. Similarly, the letter *M* indicates metric units.

Troubleshooting of Circuits

11-24. What is wrong with the circuit in Figure 11-46?

11-25. What is wrong with the circuit in Figure 11-47?

Reservoir Sizing

11-26E. What size reservoir should be used for a hydraulic system using a 15-gpm pump?

11-27M. What would be an adequate size reservoir for a hydraulic system using a 0.001-m³/s pump?

Pressure Intensifiers

11-28E. Oil at 21 gpm and 1000 psi enters the low-pressure inlet of a 3:1 Racine pressure intensifier. Find the discharge flow rate and pressure.

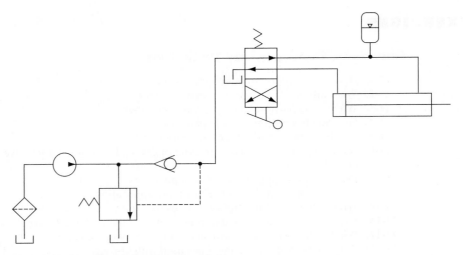

Figure 11-46. Circuit for Exercise 11-24.

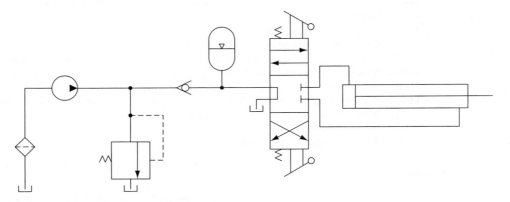

Figure 11-47. Circuit for Exercise 11-25.

11-29M. Oil at 0.001 m³/s and 70 bars enters the low-pressure inlet of a 3:1 Racine pressure intensifier. Find the discharge flow rate and pressure.

Oil Temperature Rise across Pressure Relief Valves

11-30E. Oil at 130°F and 2000 psi is flowing through a pressure relief valve at 15 gpm. What is the downstream oil temperature?

11-31M. Oil at 60°C and 140 bars is flowing through a pressure relief valve at 0.001 m³/s. What is the downstream temperature?

Heat Exchanger Ratings

11-32E. A hydraulic pump operates at 2000 psi and delivers oil at 15 gpm to a hydraulic actuator. Oil discharges through the pressure relief valve (PRV) during 60% of the cycle time. The pump has an overall efficiency of 82%, and 15% of the power is lost due to frictional pressure losses in the hydraulic lines. What heat-exchanger rating is required to dissipate all the generated heat?

11-33M. A hydraulic pump operates at 140 bars and delivers oil at 0.001 m³/s to a hydraulic actuator. Oil discharges through the pressure relief valve (PRV) during 60% of the cycle time. The pump has an overall efficiency of 82%, and 15% of the power is lost due to frictional pressure losses in the hydraulic lines. What heat exchanger rating is required to dissipate all the generated heat?

Heat-Generation Rate

11-34. A BTU is the amount of heat required to raise 1 lb of water 1°F. Derive the conversion factor between BTUs and joules.

11-35E. A hydrostatic transmission that is driven by a 12-hp electric motor delivers 10 hp to the output shaft. Assuming that 75% of the power loss is due to heat loss, calculate the total BTU heat loss over a 5-hr period.

11-36E. A hydraulic press with a 12-gpm pump cycles every 6 min. During the 2-min high-pressure portion of the cycle, oil is dumped over a high-pressure relief valve at 3000 psi. During the 4-min low-pressure portion of the cycle, oil is dumped over a low-pressure relief valve at 600 psi. If the travel time of the cylinder is negligibly small, calculate the total BTU heat loss per hour generated by dumping oil through both relief valves.

11-37M. A pump delivers oil to a hydraulic motor at 20 L/min at a pressure of 15 MPa. If the motor delivers 4 kW, and 80% of the power loss is due to internal leakage, which heats the oil, calculate the heat-generation rate in kJ/min.

12 Maintenance of Hydraulic Systems

Learning Objectives

Upon completing this chapter, you should be able to:

1. Identify the most common causes of hydraulic system breakdown.
2. Understand the significance of oxidation and corrosion prevention of hydraulic fluids.
3. Discuss the various types of fire-resistant fluids.
4. Recognize the significance of foam-resistant fluids.
5. Understand the significance of the neutralization number of a hydraulic fluid.
6. Explain the environmental significance of properly maintaining and disposing of hydraulic fluids.
7. Describe the operation of filters and strainers and specify the locations where filters and strainers should be located in hydraulic circuits.
8. Understand the significance of the parameter *Beta ratio* relative to how well a filter traps particles.
9. Calculate the Beta ratio and Beta efficiency of filters.
10. Understand the concept of specifying fluid cleanliness levels required for various hydraulic components.
11. Discuss the mechanism of the wear of moving parts of hydraulic components due to solid-particle contamination of the fluid.
12. Describe the problems caused by gases in hydraulic fluids.
13. Describe how to troubleshoot fluid power circuits effectively, depending on the symptoms of the problem involved.
14. Understand the importance of safety and that there should be no compromise in safety when fluid power systems are designed, installed, operated, and maintained.
15. Describe the key environmental issues dealing with hydraulic systems.

12.1 INTRODUCTION

Causes of Hydraulic System Problems

In the early years of fluid power systems, maintenance was frequently performed on a hit-or-miss basis. The prevailing attitude was to fix the problem when the system broke down. However, with today's highly sophisticated machinery and the advent of mass production, industry can no longer afford to operate on this basis. The cost of downtime is prohibitive.

The following is a list of the most common causes of hydraulic system breakdown:

1. Clogged or dirty oil filters
2. Inadequate supply of oil in the reservoir
3. Leaking seals
4. Loose inlet lines that cause the pump to take in air
5. Incorrect type of oil
6. Excessive oil temperature
7. Excessive oil pressure

Preventive Maintenance

Most of these kinds of problems can be eliminated if a planned preventive maintenance program is undertaken. This starts with the fluid power designer in the selection of high-quality, properly sized components. The next step is the proper assembly of the various components. This includes applying the correct amount of torque to the various tube fittings to prevent leaks and, at the same time, not distort the fitting. Parts should be cleaned as they are assembled, and the system should be completely flushed with clean oil prior to putting it into service. It is important for the total system to provide easy access to components requiring periodic inspection such as filters, strainers, sight gages, drain and fill plugs, flowmeters, and pressure and temperature gages.

Over half of all hydraulic system problems have been traced directly to the oil. This is why the sampling and testing of the fluid is one of the most important preventive maintenance measures that can be undertaken. Figure 12-1 shows a hydraulic fluid test kit, which provides a quick, easy method to test for hydraulic system contamination. Even small hydraulic systems may be checked. The test kit may be used on the spot to determine whether fluid quality permits continued use. Tests that can be performed include the determination of viscosity, water content, and particulate contamination level. Viscosity is measured using a Visgage viscosity comparator. Water content is determined by the hot plate method. Contamination is evaluated by filtering a measured amount of hydraulic fluid, examining the particles caught on the filter under a microscope, and comparing what is seen with a series of photos indicating contamination levels. The complete test requires only approximately 10 min.

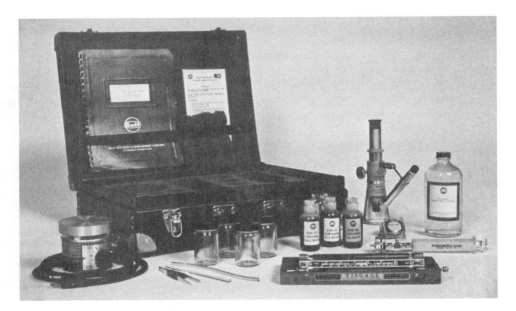

Figure 12-1.   Hydraulic fluid test kit. *(Courtesy of Gulf Oil Corp., Houston, Texas.)*

Training of Maintenance Personnel and Record Keeping

It is vitally important for maintenance personnel and machine operators to be trained to recognize early symptoms of potential hydraulic problems. For example, a noisy pump may be due to cavitation caused by a clogged inlet filter. This may also be due to a loose intake fitting, which allows air to be taken into the pump. If the cavitation noise is due to such an air leak, the oil in the reservoir will be covered with foam. When air becomes entrained in the oil, it causes spongy operation of hydraulic actuators. A sluggish actuator may be due to fluid having too high a viscosity. However, it can also be due to excessive internal leakage through the actuator or one of its control valves.

For preventive maintenance techniques to be truly effective, it is necessary to have a good report and records system. These reports should include the following:

1. The types of symptoms encountered, how they were detected, and the date.
2. A description of the maintenance repairs performed. This should include the replacement of parts, the amount of downtime, and the date.
3. Records of dates when oil was tested, added, or changed. Dates of filter changes should also be recorded.

Safety and Environmental Issues

Proper maintenance procedures for external oil leaks are also essential. Safety hazards due to oil leaking on the floor and around machinery must be prevented. In some process industries, external oil leakage is prohibitive because of contamination of the end product. Loose mounting bolts or brackets should be tightened as soon as they are detected because they can cause misalignment of the shafts of actuators and pumps, which can result in shaft seal or packing damage. A premature external oil leak can occur that will require costly downtime for repair.

Environmental rules and regulations have been established concerning the operation of fluid power systems. The fluid power industry is responding by developing efficient, cost-effective ways to meet these regulations, which deal with the following four issues:

1. Developing biodegradable fluids
2. Maintaining and disposing of hydraulic fluids
3. Reducing oil leakage
4. Reducing noise levels

12.2 OXIDATION AND CORROSION OF HYDRAULIC FLUIDS

Oxidation, which is caused by the chemical reaction of oxygen from the air with particles of oil, can seriously reduce the service life of a hydraulic fluid. Petroleum oils are especially susceptible to oxidation because oxygen readily unites with both carbon and hydrogen molecules. Most products of oxidation are soluble in oil and are acidic in nature, which can cause corrosion of parts throughout the system. The products of oxidation include insoluble gums, sludge, and varnish, which tend to increase the viscosity of the oil.

There are a number of parameters that hasten the rate of oxidation once it begins. Included among these are heat, pressure, contaminants, water, and metal surfaces. However, oxidation is most dramatically affected by temperature. The rate of oxidation is very slow below 140°F but doubles for every 20°F temperature rise. Additives are incorporated in many hydraulic oils to inhibit oxidation. Since this increases the costs, they should be specified only if necessary, depending on temperature and other environmental conditions.

Rust and corrosion are two different phenomena, although they both contaminate the system and promote wear. Rust is the chemical reaction between iron or steel and oxygen. The presence of moisture in the hydraulic system provides the necessary oxygen. One primary source of moisture is from atmospheric air, which enters the reservoir through the breather cap. Figure 12-2 shows a steel part that has experienced rusting due to moisture in the oil.

Corrosion, on the other hand, is the chemical reaction between a metal and acid. The result of rusting or corrosion is the "eating away" of the metal surfaces of hydraulic components. This can cause excessive leakage past the sealing surfaces of the affected parts. Figure 12-3 shows a new valve spool and a used one which has

Figure 12-2. Rust caused by moisture in the oil. *(Courtesy of Sperry Vickers, Sperry Rand Corp., Troy, Michigan.)*

Figure 12-3. Corrosion caused by acid formation in the hydraulic oil. *(Courtesy of Sperry Vickers, Sperry Rand Corp., Troy, Michigan.)*

areas of corrosion caused by acid formation in the hydraulic oil. Rust and corrosion can be resisted by incorporating additives that plate on the metal surfaces to prevent chemical reaction.

12.3 FIRE-RESISTANT FLUIDS

It is imperative that a hydraulic fluid not initiate or support a fire. Most hydraulic fluids will, however, burn under certain conditions. There are many hazardous applications where human safety requires the use of a fire-resistant fluid. Examples include coal mines, hot-metal processing equipment, aircraft, and marine fluid power systems.

A fire-resistant fluid is one that can be ignited but will not support combustion when the ignition source is removed. Flammability is defined as the ease of ignition and ability to propagate a flame. The following are the usual characteristics tested for in order to determine the flammability of a hydraulic fluid:

1. **Flash point**—the temperature at which the oil surface gives off sufficient vapors to ignite when a flame is passed over the surface
2. **Fire point**—the temperature at which the oil will release sufficient vapor to support combustion continuously for five seconds when a flame is passed over the surface
3. **Autogenous ignition temperature (AIT)**—the temperature at which ignition occurs spontaneously

Fire-resistant fluids have been developed to reduce fire hazards. There are basically four different types of fire-resistant hydraulic fluids in common use:

1. **Water-glycol solutions.** This type consists of an actual solution of about 40% water and 60% glycol. These solutions have high viscosity index values, but the viscosity rises as the water evaporates. The operating temperature range runs from −10°F to about 180°F. Most of the newer synthetic seal materials are compatible with water-glycol solutions. However, metals such as zinc, cadmium, and magnesium react with water-glycol solutions and therefore should not be used. In addition, special paints must be used.

2. **Water-in-oil emulsions.** This type consists of about 40% water completely dispersed in a special oil base. It is characterized by the small droplets of water completely surrounded by oil. The water provides a good coolant property but tends to make the fluid more corrosive. Thus, greater amounts of corrosion inhibitor additives are necessary. The operating temperature range runs from −20°F to about 175°F. As is the case with water-glycol solutions, it is necessary to replenish evaporated water to maintain proper viscosity. Water-in-oil emulsions are compatible with most rubber seal materials found in petroleum-based hydraulic systems.

3. **Straight synthetics.** This type is chemically formulated to inhibit combustion and in general has the highest fire-resistant temperature. Typical fluids in this category are the phosphate esters or chlorinated hydrocarbons. Disadvantages of straight synthetics include low viscosity index, incompatibility with most natural or synthetic rubber seals, and high costs. In particular, the phosphate esters readily dissolve paints, pipe thread compounds, and electrical insulation.

4. **High-water-content fluids.** This type consists of about 90% water and 10% concentrate. The concentrate consists of fluid additives that improve viscosity, lubricity, rust protection, and protection against bacteria growth. Advantages of high-water-content fluids include high fire resistance, outstanding cooling characteristics, and low cost, which is about 20% of the cost of petroleum-based hydraulic fluids. Maximum operating temperatures should be held to 120°F to minimize evaporation. Due to a somewhat higher density and lower viscosity compared to petroleum-based

fluids, pump inlet conductors should be sized to keep fluid velocities low enough to prevent the formation of vapor bubbles, which causes cavitation. High-water-content fluids are compatible with most rubber seal materials, but leather, paper, or cork materials should not be used since they tend to deteriorate in water.

12.4 FOAM-RESISTANT FLUIDS

Air can become dissolved or entrained in hydraulic fluids. For example, if the return line to the reservoir is not submerged, the jet of oil entering the liquid surface will carry air with it. This causes air bubbles to form in the oil. If these bubbles rise to the surface too slowly, they will be drawn into the pump intake. This can cause pump damage due to cavitation, as discussed in Chapter 5.

In a similar fashion, a small leak in the suction line can cause the entrainment of large quantities of air from the atmosphere. This type of leak is difficult to find since air leaks in rather than oil leaking out. Another adverse effect of entrained and dissolved air is the great reduction in the bulk modulus of the hydraulic fluid. This can have serious consequences in terms of stiffness and accuracy of hydraulic actuators. The amount of dissolved air can be greatly reduced by properly designing the reservoir since this is where the vast majority of the air is picked up.

Another method is to use premium-grade hydraulic fluids that contain foam-resistant additives. These additives are chemical compounds, which break out entrained air to separate quickly the air from the oil while it is in the reservoir.

12.5 FLUID LUBRICATING ABILITY

Hydraulic fluids must have good lubricity to prevent wear between the closely fitted working parts. Direct metal-to-metal contact is avoided by the film strength of fluids having adequate viscosity, as shown in Figure 12-4. Hydraulic parts that are affected include pump vanes, valve spools, piston rings, and rod bearings.

Wear is the actual removal of surface material due to the frictional force between two mating surfaces. This can result in a change in component dimension, which can lead to looseness and subsequent improper operation.

The friction force F is the force parallel to the two mating surfaces that are sliding relative to each other. This friction force actually opposes the sliding movement between the two surfaces. The greater the frictional force, the greater the wear and heat generated. This, in turn, results in power losses and reduced life, which, in turn, increase maintenance costs.

It has been determined that the friction force F is proportional to the normal force N that forces the two surfaces together. The proportionality constant is called the *coefficient of friction (CF)*:

$$F = (CF) \times N \qquad \textbf{(12-1)}$$

Thus, the greater the value of coefficient of friction and normal force, the greater the frictional force and hence wear. The magnitude of the normal force depends on

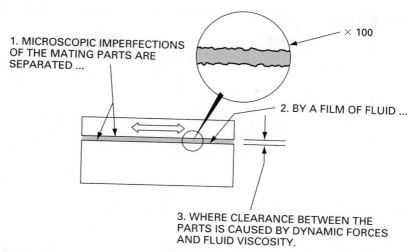

1. MICROSCOPIC IMPERFECTIONS OF THE MATING PARTS ARE SEPARATED ...

× 100

2. BY A FILM OF FLUID ...

3. WHERE CLEARANCE BETWEEN THE PARTS IS CAUSED BY DYNAMIC FORCES AND FLUID VISCOSITY.

Figure 12-4. Lubricating film prevents metal-to-metal contact. *(Courtesy Sperry Vickers, Sperry Rand Corp., Troy, Michigan.)*

the amount of power and forces being transmitted and thus is independent of the hydraulic fluid properties. However, the coefficient of friction depends on the ability of the fluid to prevent metal-to-metal contact of the closely fitting mating parts.

Equation (12-1) can be rewritten to solve for the coefficient of friction, which is a dimensionless parameter:

$$CF = \frac{F}{N} \tag{12-2}$$

It can be seen now that CF can be experimentally determined to give an indication of the antiwear properties of a fluid if F and N can be measured.

12.6 FLUID NEUTRALIZATION NUMBER

The neutralization number is a measure of the relative acidity or alkalinity of a hydraulic fluid and is specified by a pH factor. A fluid having a small neutralization number is recommended because high acidity or alkalinity causes corrosion of metal parts as well as deterioration of seals and packing glands.

For an acidic fluid, the neutralization number equals the number of milligrams (mg) of potassium hydroxide necessary to neutralize the acid in a 1-g sample of the fluid. In the case of an alkaline (basic) fluid, the neutralization number equals the amount of alcoholic hydrochloric acid (expressed as an equivalent number of milligrams of potassium) that is necessary to neutralize the alkaline in a 1-g sample of the hydraulic fluid. Hydraulic fluids normally become acidic rather than basic with use. Hydraulic fluids that have been treated with additives to inhibit the formation of acids are usually able to keep this number at a low value between 0 and 0.1.

12.7 PETROLEUM-BASE VERSUS FIRE-RESISTANT FLUIDS

The petroleum-base fluid, which is the most widely used type, is refined from selected crude oil. During the refining process, additives are included to meet the requirements of good lubricity, high viscosity index, and oxidation/foam resistance. Petroleum-based fluids dissipate heat well, are compatible with most seal materials, and resist oxidation well for operating temperatures below 150°F.

The primary disadvantage of a petroleum-based fluid is that it will burn. As a result, the fire-resistant fluid has been developed. This greatly reduces the danger of a fire. However, fire-resistant fluids generally have a higher specific gravity than do petroleum-based fluids. This may cause cavitation problems in the pump due to excessive vacuum pressure in the pump inlet line unless proper design steps are implemented. In addition, fire-resistant fluids generally have significantly lower lubricity than do petroleum-based fluids. Also, most fire-resistant fluids are more expensive and have more compatibility problems with seal materials. Therefore, fire-resistant fluids should be used only if hazardous operating conditions exist. Manufacturer's recommendations should be followed very carefully when changing from a petroleum-based fluid to a fire-resistant fluid, and vice versa. Normally, thorough draining, cleaning, and flushing are required. It may even be necessary to change seals and gaskets on the various hydraulic components.

12.8 MAINTAINING AND DISPOSING OF FLUIDS

Controlling pollution and conserving natural resources are important goals to achieve for the benefit of society. Thus, it is important to minimize the generation of waste hydraulic fluids and to dispose of them in an environmentally sound manner. These results can be accomplished by implementing fluid-control and preventive maintenance programs along with following proper fluid-disposal procedures. The following recommendations should be adhered to for properly maintaining and disposing of hydraulic fluids:

1. Select the optimum fluid for the application involved. This includes consideration of the system operating pressures and temperatures, as well as the desired fluid properties of specific gravity, viscosity, lubricity, oxidation resistance, and bulk modulus.

2. Use a well-designed filtration system to reduce contamination and increase the useful life of the fluid. Filtration should be continuous, and filters should be changed at regular intervals.

3. Follow proper storage procedures for the unused fluid supply. For example, outdoor storage is not recommended, especially if the fluid is stored in drums. Drum markings and labels may become illegible due to inclement weather. Also, drum seams may weaken due to expansion and contraction, leading to fluid leakage and contamination. Indoor storage facilities should include racks and shelves to provide adequate protection of drums from accidental damage. Tanks used for bulk storage should be well constructed of sheet steel using riveted or welded seams.

4. Transporting the fluids from the storage containers to the hydraulic systems should be done carefully, since the chances for contamination increase greatly with handling. Any transfer container used to deliver fluid from the storage drums or tanks to the hydraulic systems should be clearly labeled. These transfer containers should be covered when not in use and returned to the fluid-storage area after each use to prevent contamination with other products.

5. Operating fluids should be checked regularly for viscosity, acidity, bulk modulus, specific gravity, water content, color, additive levels, concentration of metals, and particle contamination.

6. The entire hydraulic system, including pumps, piping, fittings, valves, solenoids, filters, actuators, and the reservoir, should be maintained according to manufacturer's specifications.

7. Corrective action should be taken to reduce or eliminate leakage from operating hydraulic systems. Typically, leakage occurs due to worn seals or loose fittings. A preventive maintenance program should be implemented to check seals, fittings, and other equipment-operating conditions that may affect leakage, at regular intervals.

8. Disposal of fluids must be done properly, because a hydraulic fluid is considered to be a waste material when it has deteriorated to the point where it is no longer suitable for use in hydraulic systems. The Environmental Protection Agency (EPA) has instituted regulations that do not permit the practice of mixing hazardous wastes (such as solvents) with waste hydraulic fluids being disposed. Also not permitted is the burning of these waste fluids in nonindustrial boilers. An acceptable way to dispose of fluid is to use a disposal company that is under contract to pick up waste hydraulic fluids.

Pollution control and conservation of natural resources are critical environmental issues for society. Properly maintaining and disposing of hydraulic fluids represent a cost-effective way to achieve a cleaner environment while conserving natural resources.

12.9 FILTERS AND STRAINERS

Introduction

Modern hydraulic systems must be dependable and provide high accuracy. This requires highly precision-machined components. The worst enemy of a precision-made hydraulic component is contamination of the fluid. Essentially, contamination is any foreign material in the fluid that results in detrimental operation of any component of the system. Contamination may be in the form of a liquid, gas, or solid and can be caused by any of the following:

1. Built into system during component maintenance and assembly. Contaminants here include metal chips, bits of pipe threads, tubing burrs, pipe dope, shreds of plastic tape, bits of seal material, welding beads, bits of hose, and dirt.

2. Generated within system during operation. During the operation of a hydraulic system, many sources of contamination exist. They include moisture due to water condensation inside the reservoir, entrained gases, scale caused by rust, bits of worn seal materials, particles of metal due to wear, and sludges and varnishes due to oxidation of the oil.

3. Introduced into system from external environment. The main source of contamination here is due to the use of dirty maintenance equipment such as funnels, rags, and tools. Disassembled components should be washed using clean hydraulic fluid before assembly. Any oil added to the system should be free of contaminants and poured from clean containers.

Strainers

As indicated in Section 11.2, reservoirs help to keep the hydraulic fluid clean. In fact, some reservoirs contain magnetic plugs at their bottom to trap iron and steel particles carried by the fluid (see Figure 12-5). However, this is not adequate, and in reality the main job of keeping the fluid clean is performed by filters and strainers.

Filters and strainers are devices for trapping contaminants. Specifically, a filter is a device whose primary function is to retain, by some porous medium, insoluble contaminants from a fluid. Basically, a strainer (see Figure 12-6) is a coarse filter. Strainers are constructed of a wire screen that rarely contains openings less than 100 mesh (U.S. Sieve No.). The screen is wrapped around a metal frame. As shown in Figure 12-7, a 100-mesh screen has openings of 0.0059 in, and thus a strainer removes only the larger particles. Observe that the lower the mesh number, the coarser the screen. Because strainers have low-pressure drops, they are usually installed in the pump suction line to remove contaminants large enough to damage the pump. A pressure gage is normally installed in the suction line between the pump and strainer to indicate the condition of the strainer. A drop in pressure indicates that the strainer is becoming clogged. This can starve the pump, resulting in cavitation and increased pump noise.

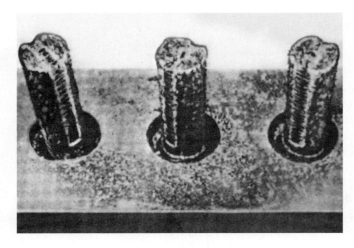

Figure 12-5. Magnetic plugs trap iron and steel particles. *(Courtesy of Sperry Vickers, Sperry Rand Corp., Troy, Michigan.)*

Filters

A filter can consist of materials in addition to a screen. Particle sizes removed by filters are measured in micrometers (or microns). As illustrated in Figure 12-7, 1 μm is one-millionth of a meter, or 0.000039 in. The smallest-sized particle that can normally be removed by a strainer is 0.0059 in or approximately 150 μm. On the other hand, filters can remove particles as small as 1 μm. Studies have shown that particles as small as 1 μm can have a damping effect on hydraulic systems (especially servo systems) and can also accelerate oil deterioration. Figure 12-7 also gives the relative sizes of micronic particles magnified 500 times. Another way to visualize the size of a micrometer is to note the following comparisons:

A grain of salt has a diameter of about 100 μm.

A human hair has a diameter of about 70 μm.

The lower limit of visibility is about 40 μm.

One-thousandth of an inch equals about 25 μm.

There are three basic types of filtering methods used in hydraulic systems: mechanical, absorbent, and adsorbent.

1. Mechanical. This type normally contains a metal or cloth screen or a series of metal disks separated by thin spacers. Mechanical-type filters are capable of removing only relatively coarse particles from the fluid.

2. Absorbent. These filters are porous and permeable materials such as paper, wood pulp, diatomaceous earth, cloth, cellulose, and asbestos. Paper filters are normally impregnated with a resin to provide added strength. In this type of filter, the

Figure 12-6. Inlet strainer. *(Courtesy of Sperry Vickers, Sperry Rand Corp., Troy, Michigan.)*

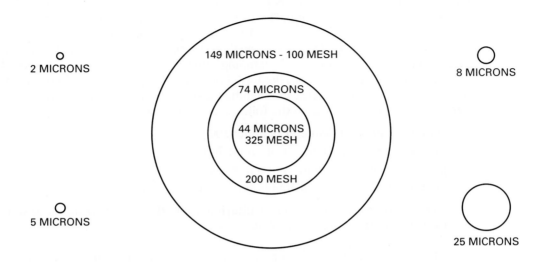

RELATIVE SIZES

LOWER LIMIT OF VISIBILITY (NAKED EYE)... 40 MICRONS
WHITE BLOOD CELLS .... 25 MICRONS
RED BLOOD CELLS .... 8 MICRONS
BACTERIA (COCCI) .... 2 MICRONS

LINEAR EQUIVALENTS

1 INCH _____ 25.4 MILLIMETERS _____ 25,400 MICRONS
1 MILLIMETER _____ .0394 INCHES _____ 1,000 MICRONS
1 MICRON _____ 1/25,400 OF AN INCH _____ .001 MILLIMETERS
1 MICRON _____ 3.94×10^{-5} INCHES _____ .000039 INCHES

SCREEN SIZES

MESHES PER LINEAR INCH	U.S. SIEVE NO.	OPENING IN INCHES	OPENING IN MICRONS
52.36	50	.0117	297
72.45	70	.0083	210
101.01	100	.0059	149
142.86	140	.0041	105
200.00	200	.0029	74
270.26	270	.0021	53
323.00	325	.0017	44

Figure 12-7. Relative and absolute sizes of micronic particles. *(Courtesy of Sperry Vickers, Sperry Rand Corp., Troy, Michigan.)*

particles are actually absorbed as the fluid permeates the material. As a result, these filters are used for extremely small particle filtration.

3. Adsorbent. Adsorption is a surface phenomenon and refers to the tendency of particles to cling to the surface of the filter. Thus, the capacity of such a filter depends on the amount of surface area available. Adsorbent materials used include activated clay and chemically treated paper. Charcoal and Fuller's earth should not be used because they remove some of the essential additives from the hydraulic fluid.

Some filters are designed to be installed in the pressure line and normally are used in systems where high-pressure components such as valves are more dirt sensitive than the pump. Pressure line filters are accordingly designed to sustain system operating pressures. Return line filters are used in systems that do not have a very large reservoir to permit contaminants to settle to the bottom. A return line filter is needed in systems containing close-tolerance, high-performance pumps, because inlet line filters, which have limited pressure drop allowance, cannot filter out extremely small particles in the 1- to 5-μm range.

Figure 12-8 shows a versatile filter that can be directly welded into reservoirs for suction or return line installations or mounted into the piping for pressure line applications. This filter can remove particles as small as 3 μm. It also contains an indicating element that signals the operator when cleaning is required. The operation of this Tell-Tale filter is dependent on fluid passing through a porous filter media, which traps contamination. The Tell-Tale indicator monitors the pressure differential buildup due to dirt, reporting the condition of the filter element. It can handle flow rates to 700 gpm and pressures from suction to 300 psi. This filter is available with or without

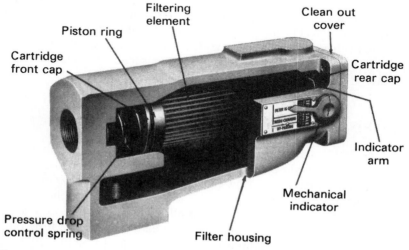

Figure 12-8. Cutaway view of a Tell-Tale filter. *(Courtesy of Parker Hannifin Corp., Hazel Park, Michigan.)*

bypass relief valves. The optional bypass ensures continued flow no matter how dirt-clogged the filter might become. The bypass valves can be set to provide the maximum allowable pressure drop to match system requirements.

Location of Filters in Hydraulic Circuits

Figure 12-9 shows the four typical locations where filters are installed in hydraulic circuits. The considerations for using these four filter locations are described as follows:

1. Single location for proportional flow filters. Figure 12-9(a) shows the location for a proportional flow filter. As the name implies, proportional flow filters (also called bypass filters) are exposed to only a percentage of the total system flow during operation. It is assumed that on a recirculating basis the probability of mixture of the fluid within the system will force all the fluid through the proportional flow filter. The primary disadvantages of proportional flow filtration are that there is no positive protection of any specific components within the system, and there is no way to know when the filter is dirty.

2. Three locations for full flow filters. Figure 12-9(b), (c), and (d) show the three locations for full flow filtration filters, which accept all the flow of the pump. Figure 12-9(b) shows the location on the suction side of the pump, whereas Figure 12-9(c) and (d) show the filter on the pressure side of the pump and in the return line, respectively.

In general, there is no best single place to put a filter. The basic rule of thumb is the following: Consider where the dirt enters the system, and put the filter/filters where they do the most good. Good hydraulic systems have multiple filters. There should always be a filter in the pump inlet line and a high-pressure filter in the pump discharge line. Placing the pump discharge filter between the pump and the pressure relief valve can provide very good filtration because oil is flowing through the filter even when the working part of the circuit is inactive and the pump discharge is going directly to the reservoir.

Flow Capacity of Filters

One of the parameters involved in the selection of a filter for a given application is the maximum flow rate that a filter is designed to handle. This parameter, which is called the flow capacity, is specified by filter manufacturers. To assure proper filter operation, it is necessary to determine the maximum flow rate in the line containing the filter. This allows for the selection of a filter having a flow capacity equal to or greater than the actual maximum flow rate through the filter. As shown in Example 12-1, the flow rate through a filter can exceed pump flow rate.

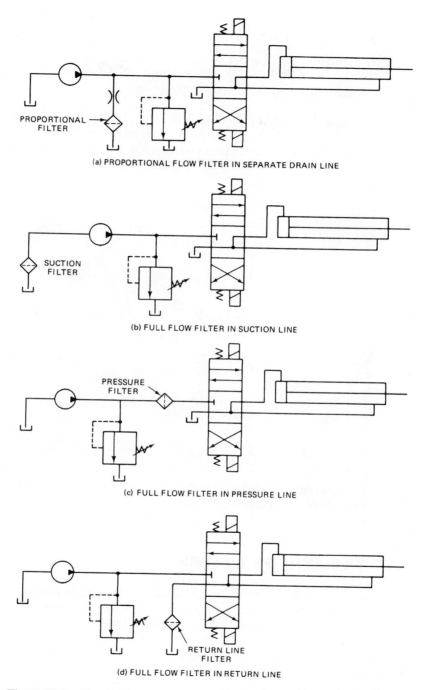

(a) PROPORTIONAL FLOW FILTER IN SEPARATE DRAIN LINE

PROPORTIONAL FILTER

(b) FULL FLOW FILTER IN SUCTION LINE

SUCTION FILTER

(c) FULL FLOW FILTER IN PRESSURE LINE

PRESSURE FILTER

(d) FULL FLOW FILTER IN RETURN LINE

RETURN LINE FILTER

Figure 12-9. Four common circuit locations for filters.

EXAMPLE 12-1

Determine the minimum flow capacity of the return line filter of Figure 12-9(d). The pump flow rate is 20 gpm and the cylinder piston and rod diameters are 5 in and 3 in, respectively.

Solution   Note that during the cylinder retraction stroke, the flow rate through the return line filter exceeds the pump flow rate. Thus, the flow capacity of the filter must equal or exceed the flow rate it receives during the cylinder retraction stroke. The cylinder retraction speed is found first:

$$v_{ret} = \frac{Q_{pump}}{A_p - A_R}$$

where

$$Q_{pump} = 20 \frac{gal}{min} \times \frac{231 \text{ in}^3}{1 \text{ gal}} = 4620 \text{ in}^3/min$$

$$A_P = \frac{\pi}{4}(5 \text{ in})^2 = 19.63 \text{ in}^2 \text{ and } A_R = \frac{\pi}{4}(3 \text{ in})^2 = 7.07 \text{ in}^2$$

Thus, we have $v_{ret} = \dfrac{4620 \text{ in}^3/min}{19.63 \text{ in}^2 - 7.07 \text{ in}^2} = 368 \text{ in}/min$

The flow rate through the filter during the cylinder retraction stroke can now be found.

$$Q_{filter} = A_p v_{ret} = 19.63 \text{ in}^2 \times 368 \frac{in}{min} \times \frac{1 \text{ gal}}{231 \text{ in}^3} = 31.3 \text{ gpm}$$

Thus, the filter must have a flow capacity of at least 31.3 gpm rather than the pump flow-rate value of 20 gpm.

12.10 BETA RATIO OF FILTERS

Filters are rated according to the smallest size of particles they can trap. Filter ratings used to be identified by nominal and absolute values in micrometers. A filter with a nominal rating of 10 μm is supposed to trap 95% of the entering particles greater than 10 μm in size. The absolute rating represents the size of the largest opening or pore in the filter and thus indicates the largest-size particle that could pass through the filter. Hence, the absolute rating of a 10-μm nominal size filter would be larger than 10 μm.

A better parameter for establishing how well a filter traps particles is called the Beta ratio, or Beta rating. The Beta ratio is determined during laboratory testing of a filter receiving a specified steady-state flow containing a fine dust of selected particle size. The test begins with a clean filter and ends when the pressure drop across the filter reaches a specified value indicating that the filter has reached the saturation point. This is when the contaminant capacity has been reached, which is a measure of the service life or acceptable time interval between filter element changes in an actual operating system.

By mathematical definition, the Beta ratio equals the number of upstream particles of greater size than N μm divided by the number of downstream particles greater in size than N μm (as counted during the test), where N is the selected particle size for the given filter. This ratio is represented by the following equation:

$$\text{Beta ratio} = \frac{\text{no. upstream particles of size} > N\mu m}{\text{no. downstream particles of size} > N\mu m} \qquad \text{(12-3)}$$

A Beta ratio of 1 would mean that no particles above the specified size N are trapped by the filter. A Beta ratio of 50 means that 50 particles are trapped for every one that gets through the filter. Most filters have Beta ratings greater than 75 when N equals the absolute rating. A filter efficiency value can be calculated using the following equation:

$$\text{Beta efficiency} = \frac{\text{no. upstream particles} - \text{no. downstream particles}}{\text{no. upstream particles}} \qquad \text{(12-4)}$$

where the particle size is greater than a specified value of N μm.

Thus, we have the following relationship between Beta efficiency and Beta ratio:

$$\text{Beta efficiency} = 1 - \frac{1}{\text{Beta ratio}} \qquad \text{(12-5)}$$

Hence, a filter with a Beta ratio of 50 would have an efficiency of $1 - \frac{1}{50}$, or 98%. Note from Eq. (12-5), that the higher the Beta ratio the higher the Beta efficiency. The designation $B_{20} = 50$ identifies a particle size of 20 μm and a Beta ratio of 50 for a particular filter. Thus, a designation of $B_{20} = 50$ means that 98% of the particles larger than 20 μm would be trapped by the filter during the time a clean filter becomes saturated.

12.11 FLUID CLEANLINESS LEVELS

The basis for controlling the particle contamination of a hydraulic fluid is to measure the fluid's cleanliness level. This means counting the particles per unit volume for specific particle sizes and comparing the results to a required cleanliness level. This allows for the selection of the proper filtration system for a given hydraulic application. Sensitive optical instruments are used to count the number of particles in the specified size ranges. The result of the counting is a report of the number of particles greater than a certain size found per milliliter of fluid.

Figure 12-10 provides a table showing a cleanliness level standard accepted by the ISO (International Standards Organization). This table shows the ISO code

Code No.	No. of Particles per Milliliter	Code No.	No. of Particles per Milliliter
30	10,000,000	14	160
29	5,000,000	13	80
28	2,500,000	12	40
27	1,300,000	11	20
26	640,000	10	10
25	320,000	9	5
24	160,000	8	2.5
23	80,000	7	1.3
22	40,000	6	0.64
21	20,000	5	0.32
20	10,000	4	0.16
19	5,000	3	0.08
18	2,500	2	0.04
17	1,300	1	0.02
16	640	0.9	0.01
15	320	0.8	0.005

Figure 12-10. ISO code numbers for fluid cleanliness levels.

number that is used to represent either the number of particles per milliliter of fluid of size greater than 5 micrometers or greater than 15 micrometers. In using the code, two numbers are used, separated by a slash. The left-most number corresponds to particle sizes greater than 5 micrometers and the right-most number corresponds to particle sizes greater than 15 micrometers. For example, a code designation of ISO 18/15 indicates that per milliliter of fluid, there are 2500 particles of size greater than 5 micrometers and 320 particles of size greater than 15 micrometers. The ISO code is widely used because it can easily represent most particle distributions. For example, a fluid with a very high silt content and a low large-particle distribution can be represented by 25/10.

The 5 micrometers number gives an indication of the silting condition of the fluid where very fine particles collect in the space between moving parts of a component. This leads to sticking or sluggish action of the component such as a solenoid-actuated valve. The 15 micrometers number indicates the quantity of larger particles that contribute to wear problems in components such as a hydraulic cylinder or pump.

Many manufacturers of hydraulic equipment specify the fluid cleanliness level required for providing the expected life of their components. Thus, using a fluid with a higher-than-required contamination level will result in a shorter component life. Figure 12-11 gives typical ISO code cleanliness levels required for various hydraulic components. These required cleanliness levels can be used to select the proper filtration system for a given hydraulic application.

Component	ISO Code
Servo Valves	14/11
Vane and Piston Pumps/Motors	16/13
Directional and Pressure Control Valves	16/13
Gear Pumps/Motors	17/14
Flow Control Valves and Cylinders	18/15

Figure 12-11. Typical fluid cleanliness levels required for hydraulic components.

12.12 WEAR OF MOVING PARTS DUE TO SOLID-PARTICLE CONTAMINATION OF THE FLUID

All hydraulic fluids contain solid contaminants (dirt) to one degree or another. However, the necessity of having a proper filter in a given location in a hydraulic system cannot be overstated. In fact, excessive solid contaminants in the hydraulic fluid will cause premature failure of even excellently designed hydraulic systems.

One of the major problems caused by solid contaminants is that they prevent the hydraulic fluid from providing proper lubrication of moving internal parts of hydraulic components such as pumps, hydraulic motors, valves, and actuators. As an example, Figure 12-12 shows a hydraulic cylinder having a radial clearance between the bore of the cylinder and the piston's outer cylindrical surface. This figure shows the cylinder bore surface to be worn over a given axial length due to excessive solid particle contamination of the fluid. Such a wear problem often includes a scored piston seal and cylinder bore. This problem typically means that channels are cut through the outer surface of the seal and tiny grooves are cut into the cylinder bore surface. This wear causes excessive internal leakage, prevents the cylinder from positioning accurately, and results in premature cylinder failure.

Solid contaminants can be classified by their size relative to the clearance between the moving parts of a hydraulic component, such as the radial clearance between the piston and bore of the hydraulic cylinder of Figure 12-12. There are three relative sizes: smaller than, equal to, and larger than the clearance. All three contaminant sizes can contribute to wear problems. Contaminants that are smaller than the clearance can collect inside the clearance when the hydraulic cylinder is not operating. These contaminants block lubricant flow through the clearance when cylinder actuation is initiated. Contaminants of the same size as the clearance rub against the mating surfaces, causing a breakdown in the fluid lubricating film. Large contaminants interfere with lubrication by collecting at the entrance to the clearance and blocking fluid flow between the mating surfaces.

In addition to the internal leakage between the piston and cylinder bore, a similar wear and leakage problem can occur around the rod seal of a hydraulic cylinder. This wear produces an external leakage that can create a messy leak as well as become a safety hazard to personnel in the area.

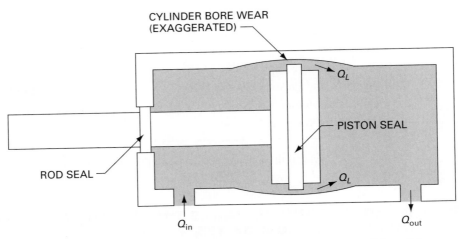

Figure 12-12. Hydraulic cylinder with a bore that is worn due to solid-particle contamination of the fluid.

The majority of hydraulic system breakdowns are due to excessive contamination of the hydraulic fluid. Wear of moving parts due to this contamination is one of the major reasons for these failures.

12.13 PROBLEMS CAUSED BY GASES IN HYDRAULIC FLUIDS

Gases can be present in a hydraulic fluid (or any other liquid) in three ways: free air, entrained gas, and dissolved air.

Free Air

Air can exist in a free pocket located at some high point of a hydraulic system (such as the highest elevation of a given pipeline). This free air either existed in the system when it was initially filled or was formed due to air bubbles in the hydraulic fluid rising into the free pocket. Free air can cause the hydraulic fluid to possess a much lower stiffness (bulk modulus), resulting in spongy and unstable operation of hydraulic actuators.

Entrained Gas

Entrained gas (gas bubbles within the hydraulic fluid) is created in two ways. Air bubbles can be created when the flowing hydraulic fluid sweeps air out of a free pocket and carries it along the fluid stream. Entrained gas can also occur when the pressure drops below the vapor pressure of the hydraulic fluid. When this happens, bubbles of hydraulic fluid vapor are created within the fluid stream. Entrained gases (either in the form of air bubbles or fluid vapor bubbles) can cause cavitation problems in pumps and valves. Entrained gases can also greatly reduce the hydraulic fluid's effective bulk modulus, resulting in spongy and unstable operation of hydraulic actuators. Relative to entrained gas, there are two important considerations that need to be understood: vapor pressure and cavitation.

1. Vapor Pressure. Vapor pressure is defined as the pressure at which a liquid starts to boil (vaporize) and thus begins changing into a vapor (gas). The vapor pressure of a hydraulic fluid (or any other liquid) increases with an increase in temperature. Petroleum-based hydraulic fluids and phosphate ester fire-resistant fluids have very low vapor pressures even at the maximum operating temperatures of typical hydraulic systems (150°F, or 65°C). However, this statement is not true for water-based fire-resistant fluids such as water-glycol solutions and water-in-oil emulsions. Because water-glycol solutions and water-in-oil emulsions contain a high percentage of water, they possess vapor pressures of several inches of Hg abs at an operating temperature of 150°F. On the other hand, petroleum-based fluids and phosphate ester possess vapor pressures of less than 0.1 in of Hg abs at 150°F. As a result, water-glycol solutions and water-in-oil emulsions have a much greater tendency to vaporize in the suction line of a pump and cause pump cavitation.

Figure 12-13 gives the vapor pressure–vs.-temperature curves for pure water, water-glycol solutions, and water-in-oil emulsions. Note that this relationship for pure

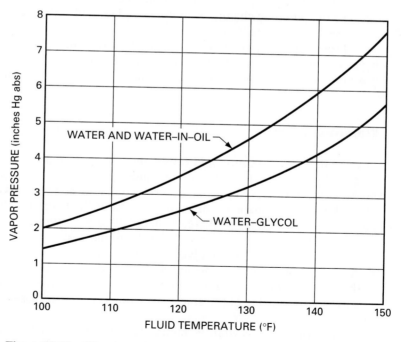

Figure 12-13. Vapor pressure–vs.-temperature curves for water, water-in-oil emulsions, and water-glycol solutions.

water and water-in-oil emulsions is essentially the same and thus is represented by a single curve. We know from experience that water starts to boil at 212°F (100°C) when the pressure is atmospheric (30 in Hg abs). However, as shown in Figure 12-13, water will also start to boil at 150°F when the pressure is reduced to about 7.7 in Hg abs. Similarly, at 150°F water-glycol boils at about 5.5 in Hg abs and water-in-oil boils at about 7.7 in Hg abs. Thus, for water-in-oil emulsions at 150°F, if the suction pressure at the inlet to a pump is reduced to 7.7 in Hg abs (3.77 psia), boiling will occur and vapor bubbles will form at the inlet port of the pump. This boiling causes cavitation, which is the formation and collapse of vapor bubbles.

2. Cavitation.   Cavitation occurs because the vapor bubbles collapse rapidly as they are exposed to the high pressure at the outlet port of the pump, creating extremely high local fluid velocities. This high-velocity fluid impacts on internal metal surfaces of the pump. The resulting high-impact forces cause flaking or pitting of the surfaces of internal components, such as gear teeth, vanes, and pistons. This damage results in premature pump failure. In addition the tiny flakes or particles of metal move downstream of the pump and enter other parts of the hydraulic system, causing damage to other components. Cavitation can also interfere with lubrication of mating moving surfaces and thus produce increased wear.

One indication of pump cavitation is a loud noise emanating from the pump. The rapid collapsing of gas bubbles produces vibrations of pump components, which are transmitted into pump noise. Cavitation also causes a decrease in pump flow rate because the pumping chambers do not completely fill with the hydraulic fluid. As a result, system pressure becomes erratic.

Frequently, entrained air is present due to a leak in the suction line or a leaking pump shaft seal. In addition, any entrained air that did not escape while the fluid was in the reservoir will enter the pump suction line and cause cavitation.

Dissolved Air

Dissolved air is in solution and thus cannot be seen and does not add to the volume of the hydraulic fluid. Hydraulic fluids can hold an amazingly large amount of air in solution. A hydraulic fluid, as received at atmospheric pressure, typically contains about 6% of dissolved air by volume. After the hydraulic fluid is pumped, the amount of dissolved air increases to about 10% by volume.

Dissolved air creates no problem in hydraulic systems as long as the air remains dissolved. However, if the dissolved air comes out of solution, it forms bubbles in the hydraulic fluid and thus becomes entrained air. The amount of air that can be dissolved in the hydraulic fluid increases with pressure and decreases with temperature. Thus, dissolved air will come out of solution as the pressure decreases or the temperature increases.

To avoid pump cavitation, pump manufacturers specify a minimum allowable vacuum pressure at the pump inlet port based on the type of fluid being pumped, the maximum operating temperature, and the rated pump speed. The following rules will control or eliminate pump cavitation by keeping the suction pressure above the vapor pressure of the fluid:

1. Keep suction line velocities below 4 ft/s (1.2 m/s).
2. Keep pump inlet lines as short as possible.
3. Minimize the number of fittings in the pump inlet line.
4. Mount the pump as close as possible to the reservoir.
5. Use low-pressure drop-pump inlet filters or strainers.
6. Use a properly designed reservoir that will remove the entrained air from the fluid before it enters the pump inlet line.
7. Use a proper oil, as recommended by the pump manufacturer.
8. Keep the oil temperature from exceeding the recommended maximum temperature level (usually 150°F, or 65°C).

12.14 TROUBLESHOOTING HYDRAULIC SYSTEMS

Introduction

Hydraulic systems depend on proper flow and pressure from the pump to provide the necessary actuator motion for producing useful work. Therefore, flow and pressure measurements are two important means of troubleshooting faulty operating hydraulic circuits. Temperature is a third parameter, which is frequently monitored when troubleshooting hydraulic systems because it affects the viscosity of the oil. Viscosity, in turn, affects leakage, pressure drops, and lubrication.

The use of flowmeters can tell whether or not the pump is producing proper flow. Flowmeters can also indicate whether or not a particular actuator is receiving the expected flow rate. Figure 12-14 shows a flowmeter that can be installed permanently or used for hydraulic system checkout or troubleshooting in pressure lines to 3000 psi. This direct-reading flowmeter monitors fluid flow rates to determine pump performance, flow regulator settings, or hydraulic system performance. It is intended for use in mobile and industrial oil hydraulic circuits. It can be applied to pressure lines, return lines, or drain lines. A moving indicator on the meter provides direct readings and eliminates any need for electrical connections on readout devices. These flowmeters are available for capacities up to 100 gpm (380 Lpm) and also can be calibrated to read pump speed or fluid velocity when connected in a pipe of known diameter.

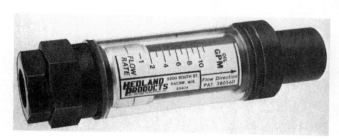

Figure 12-14. In-line flowmeter. (*Courtesy of Heland Products, Racine, Wisconsin.*)

Pressure measurements can provide a good indication of leakage problems and faulty components such as pumps, flow control valves, pressure relief valves, strainers, and actuators. Excessive pressure drops in pipelines can also be detected by the use of pressure measurements. The Bourdon gage is the most commonly used type of pressure-measuring device. This type of gage can be used to measure vacuum pressures in suction lines as well as absolute pressures anywhere in a hydraulic circuit.

Figure 12-15 shows a combination flow-pressure test kit. This unit measures both pressure (using a Bourdon gage) as well as flow rate and can be quickly installed in a hydraulic line because it uses quick couplers at each end.

A portable hydraulic circuit tester (all components are built into a convenient-to-carry container) is shown in Figure 12-16. This unit not only measures pressure

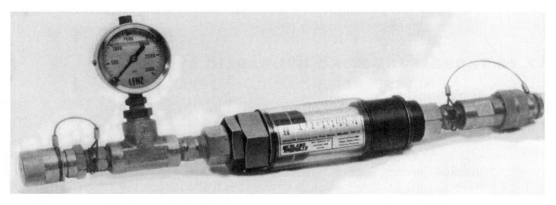

Figure 12-15. Flow-pressure test kit. *(Courtesy of Heland Products, Racine, Wisconsin.)*

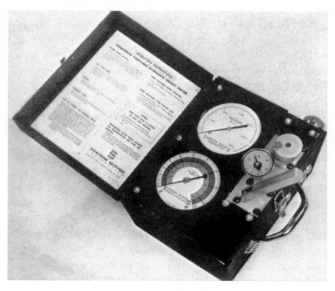

Figure 12-16. Portable hydraulic circuit tester. *(Courtesy of Schroeder Brothers Corp., McKees Rock, Pennsylvania.)*

and flow rate but also temperature. By connecting this tester to the hydraulic circuit, a visual means is provided to determine the efficiency of the system and to determine which component in the system, if any, is not working properly. Testing a hydraulic system with this tester consists of the following:

1. Measure pump flow at no-load conditions.
2. Apply desired pressure with the tester load valve on each component to find out how much of the fluid is not available for power because it may be
   a. Flowing at a lower rate because of slippage inside the pump due to worn parts.
   b. Flowing over pressure relief valves due to worn seats or weak or improperly set springs.
   c. Leaking past valve spools and seats back into the fluid supply reservoir without having reached the working cylinder or motor.
   d. Leaking past the cylinder packing or motor parts directly into the return line without having produced any useful work.

Probable Causes of Hydraulic System Problems

When troubleshooting hydraulic circuits, it should be kept in mind that a pump produces the flow of a fluid. However, there must be resistance to flow in order to have pressure. The following is a list of hydraulic system operating problems and the corresponding probable causes that should be investigated during troubleshooting:

1. **Noisy pump**
   a. Air entering pump inlet
   b. Misalignment of pump and drive unit
   c. Excessive oil viscosity
   d. Dirty inlet strainer
   e. Chattering relief valve
   f. Damaged pump
   g. Excessive pump speed
   h. Loose or damaged inlet line

2. **Low or erratic pressure**
   a. Air in the fluid
   b. Pressure relief valve set too low
   c. Pressure relief valve not properly seated
   d. Leak in hydraulic line
   e. Defective or worn pump
   f. Defective or worn actuator

3. No pressure

a. Pump turning in wrong direction

b. Ruptured hydraulic line

c. Low oil level in reservoir

d. Pressure relief valve stuck open

e. Broken pump shaft

f. Full pump flow bypassed to tank due to faulty valve or actuator

4. Actuator fails to move

a. Faulty pump

b. Directional control valve fails to shift

c. System pressure too low

d. Defective actuator

e. Pressure relief valve stuck open

f. Actuator load is excessive

g. Check valve in backwards

5. Slow or erratic motion of actuator

a. Air in system

b. Viscosity of fluid too high

c. Worn or damaged pump

d. Pump speed too low

e. Excessive leakage through actuators or valves

f. Faulty or dirty flow control valves

g. Blocked air breather in reservoir

h. Low fluid level in reservoir

i. Faulty check valve

j. Defective pressure relief valve

6. Overheating of hydraulic fluid

a. Heat exchanger turned off or faulty

b. Undersized components or piping

c. Incorrect fluid

d. Continuous operation of pressure relief valve

e. Overloaded system

f. Dirty fluid

g. Reservoir too small

h. Inadequate supply of oil in reservoir

i. Excessive pump speed

j. Clogged or inadequate-sized air breather

12.15 SAFETY CONSIDERATIONS

There should be no compromise in safety when hydraulic circuits are designed, operated, and maintained. However, human errors are unavoidable, and accidents can occur, resulting in injury to operating and maintenance personnel. This can be greatly reduced by eliminating all unsafe conditions dealing with the operation and maintenance of the system. Many safe practices have been proven effective in preventing safety hazards, which can be harmful to the health and safety of personnel.

The Occupational Safety and Health Administration (OSHA) of the Department of Labor describes and enforces safety standards at the industry location where hydraulic equipment is operated. For detailed information on OSHA standards and requirements, the reader should request a copy of OSHA publication 2072, General Industry Guide for Applying Safety and Health Standards, 29 CFR 1910. These standards and requirements deal with the following categories:

1. **Workplace standards.** In this category are included the safety of floors, entrance and exit areas, sanitation, and fire protection.
2. **Machines and equipment standards.** Important items are machine guards; inspection and maintenance techniques; safety devices; and the mounting, anchoring, and grounding of fluid power equipment. Of big concern are noise levels produced by operating equipment.
3. **Materials standards.** These standards cover items such as toxic fumes, explosive dust particles, and excessive atmospheric contamination.
4. **Employee standards.** Concerns here include employee training, personnel protective equipment, and medical and first-aid services.
5. **Power source standards.** Standards are applied to power sources such as electrohydraulic, pneumatic, and steam supply systems.
6. **Process standards.** Many industrial processes are included such as welding, spraying, abrasive blasting, part dipping, and machining.
7. **Administrative regulations.** Industry has many administrative responsibilities which it must meet. These include the displaying of OSHA posters stating the rights and responsibilities of both the employer and employee. Industry is also required to keep safety records on accidents, illnesses, and other exposure-type occurrences. An annual summary must also be posted.

It is important that safety be incorporated into hydraulic systems to ensure compliance with OSHA regulations. The basic rule to follow is that there should be no compromise when it comes to the health and safety of people at their place of employment.

12.16 ENVIRONMENTAL ISSUES

Environmental rules and regulations have been established concerning the operation of fluid power systems. The fluid power industry is responding by developing efficient, cost-effective ways to meet these regulations, which deal with four issues: developing biodegradable fluids, maintaining and disposing of hydraulic fluids, reducing oil leakage, and reducing noise levels.

1. Developing biodegradable fluids. This issue deals with preventing environmental damage caused by potentially harmful material leaking from fluid power systems. Fluids commonly used in hydraulic systems are mineral based and hence are not biodegradable. Oil companies are developing vegetable-based fluids that are biodegradable and compatible with fluid power equipment. Fluid power–equipment manufacturers are testing their products to ensure compatibility with the new biodegradable fluids.

2. Maintaining and disposing of hydraulic fluids. It is important to minimize the generation of waste hydraulic fluids and to dispose of them in an environmentally sound manner. These results can be accomplished by implementing fluid control and preventive maintenance programs along with proper fluid-disposal programs. Proper maintenance and disposal of hydraulic fluids represent cost-effective ways of achieving a cleaner environment while conserving natural resources.

3. Reducing oil leakage. Hydraulic fluid leakage can occur at pipe fittings in hydraulic systems and at mist-lubricators in pneumatic systems. This leakage represents an environmental issue because the federal Environmental Protection Agency (EPA) has identified oil as a hazardous air pollutant. To resolve this issue, the fluid power industry is striving to produce zero-leakage systems. New seals and fittings are being designed that can essentially eliminate oil leakage. In addition, prelube and nonlube pneumatic components are being developed to eliminate the need for pipeline-installed lubricators and thus prevent oil-mist leaks.

4. Reducing noise levels. Hydraulic power units such as pumps and motors can operate at noise levels exceeding the limits established by OSHA (Occupational Safety and Health Administration). New standards for indoor systems require that pumps and motors operate at reduced noise levels without reducing power or efficiency. Fluid power manufacturers are offering power units that produce lower noise levels. In addition, noise-reduction methods such as modifying hydraulic hose designs, adding sound filters, baffles, or coatings, and providing equipment vibration-absorbing mounts are being developed.

Meeting stricter environmental requirements represents a challenge to which the fluid power industry is responding. These environmental issues make careers in the fluid power industry both challenging and exciting.

12.17 WATER HYDRAULICS

Definition

Webster's definition of hydraulics specifically refers to water as follows: "Hydraulics is the science dealing with water and other liquids in motion, the laws of their actions and their engineering applications." The name *hydraulics* was used to identify liquid fluid power systems when hydraulics was first introduced in the eighteenth century because the working fluid was water. In the early twentieth century, oil replaced water as the working fluid because oil allowed for more compact, high-pressure, high-power systems.

Challenges of Water Hydraulics

Using water instead of oil increases problems with cavitation because water has a much higher vapor pressure (about 7.7 in Hg abs versus 0.1 in Hg abs at 150°F). Thus, for example, water has a much greater tendency to vaporize in the suction line of a pump and cause pump cavitation. Cavitation causes erosion of metallic components as well as noise, vibration, and reduced efficiency. Water's lower viscosity and higher specific gravity also contribute to a greater tendency for corroding metal components. This is due to the resulting higher fluid velocity and increased turbulent flow. To reduce erosion, components on water-based systems are often made of expensive stainless steels and ceramics. In addition, water needs to be mixed with additives (about 5% of total fluid volume) to improve lubricity, reduce corrosion, and control the growth of bacteria. Another concern when using water is the creation of high-pressure spikes caused when fluid velocities change rapidly such as in the case of a fast-closing valve. These pressure spikes are greater when using water because the bulk modulus of water is about 40% greater than that of oil. Accumulators can be used to solve this problem along with using valves designed for controlled shifting to control fluid acceleration and deceleration. For water hydraulic systems the operating temperature must not fall below about 35°F because the freezing temperature of water is 32°F. In applications subject to freezing, the appropriate amount of antifreeze is added to the water.

Advantages of Water Hydraulics

Water hydraulics does have a number of significant advantages over oil hydraulics. Water has a lower viscosity than oil, resulting in less heat generation and thus lower energy losses, although at the expense of greater leakage especially in values. Also, water costs less, is nonflammable, does not contaminate the environment, and is more compatible with a number of applications such as in the food and paper processing industries. In addition, water is readily available and does not deteriorate in the same manner as oil. There is also less cost associated with the disposal and storage of water.

Promising Applications for Water Hydraulics

Where water hydraulics appears to have the greatest potential to replace oil hydraulics is in low-pressure systems (150 to 750 psi). For pressure above 1000 psi, water-based systems are much more expensive than oil-based systems due to the greater cost of materials such as stainless steel and ceramics needed to prevent erosion. For the low-pressure systems, less expensive materials such as plastics can be used. This means there is a greater promise of benefiting from the use of water-based systems for applications that fall between the high pressures (greater than 750 psi) of compact oil-based hydraulic systems and the low pressures (less than 150 psi) of pneumatic systems.

12.18 KEY EQUATIONS

Beta ratio of filters:

$$\text{Beta ratio} = \frac{\text{no. upstream particles of size} > N\mu\text{m}}{\text{no. downstream particles of size} > N\mu\text{m}} \tag{12-3}$$

Beta efficiency of filters:

$$\text{Beta efficiency} = \frac{\text{no. upstream particles} - \text{no. downstream particles}}{\text{no. upstream particles}} \tag{12-4}$$

$$\text{Beta efficiency} = 1 - \frac{1}{\text{Beta ratio}} \tag{12-5}$$

EXERCISES

Questions, Concepts, and Definitions

12-1. Name five of the most common causes of hydraulic system breakdown.

12-2. To what source has over half of all hydraulic system problems been traced?

12-3. What is the difference between oxidation and corrosion?

12-4. Under what conditions should a fire-resistant fluid be used?

12-5. Define the terms *flash point, fire point*, and *autogenous ignition temperature*.

12-6. Name the four different types of fire-resistant fluids.

12-7. Name three disadvantages of fire-resistant fluids.

12-8. What is a foam-resistant fluid, and why would it be used?

12-9. Why must a hydraulic fluid have good lubricating ability?

12-10. Define the term *coefficient of friction*.

12-11. What is the significance of the neutralization number?

12-12. Why should normal operating temperatures be controlled below 140°F in most hydraulic systems?

12-13. What effect does a higher specific gravity have on the inlet of a pump?

12-14. Why is it necessary to use precautions when changing from a petroleum-based fluid to a fire-resistant fluid, and vice versa?

12-15. Explain the environmental significance of properly maintaining and disposing of hydraulic fluids.

12-16. Identify eight recommendations that should be followed for properly maintaining and disposing of hydraulic fluids.

12-17. Name two items that should be included in reports dealing with maintenance procedures.

12-18. What is the difference between a filter and strainer?

12-19. Name the three ways in which hydraulic fluid becomes contaminated.

12-20. What is a 10-μm filter?

12-21. Name the three basic types of filtering methods.

12-22. What is the purpose of an indicating filter?

12-23. Name the four locations where filters are typically installed in hydraulic circuits.

12-24. What three devices are commonly used in the troubleshooting of hydraulic circuits?

12-25. What single most important concept should be understood when troubleshooting hydraulic systems?

12-26. Name five things that can cause a noisy pump.

12-27. Name four causes of low or erratic pressure.

12-28. Name four causes of no pressure.

12-29. If an actuator fails to move, name five possible causes.

12-30. If an actuator has slow or erratic motion, name five possible causes.

12-31. Give six reasons for the overheating of the fluid in a hydraulic system.

12-32. What does OSHA stand for? What is OSHA attempting to accomplish?

12-33. Name and give a brief explanation of the seven categories of safety for which OSHA has established standards.

12-34. Why is loss of pressure in a hydraulic system not a symptom of pump malfunction?

12-35. What happens when a filter becomes filled with contaminants?

12-36. What factors influence cylinder friction?

12-37. What is the difference between nominal and absolute ratings of filters?

12-38. What is the significance of specifying required levels of the cleanliness of hydraulic fluids for various hydraulic components?

12-39. Describe the ISO cleanliness level standard.

12-40. Name one of the major hydraulic system problems caused by solid contaminants in the hydraulic fluid.

12-41. How do solid contaminants in the hydraulic fluid cause wear of the moving parts of hydraulic components?

12-42. In what three ways can gases be present in hydraulic fluids?

12-43. What is meant by the term *vapor pressure?*

12-44. Explain how cavitation causes damage to hydraulic pumps.

12-45. What do pump manufacturers recommend to users to avoid pump cavitation?

12-46. Name six rules that will control or eliminate pump cavitation.

12-47. Describe the environmental issues dealing with developing biodegradable fluids, reducing oil leakage, maintaining and disposal of hydraulic fluids, and reducing noise levels.

Problems

Flow Capacity of Filters

12-48M. Determine the minimum flow capacity of the return line filter in Figure 12-9(d). The pump flow rate is 75 Lpm and the cylinder piston and rod diameters are 125 mm and 75 mm, respectively.

12-49. The piping of the circuit in Figure 12-9(d) is modified as follows. First, the pipeline from the upper discharge port of the directional control valve is connected to the rod end of the cylinder. Then the pipeline from the lower discharge port of the directional control valve is connected to the blank end of the cylinder. What effect does this have on the minimum flow capacity of the return line filter?

Beta Ratio of Filters

12-50. Determine the Beta ratio of a filter when, during test operation, 30,000 particles greater than 20 μm enter the filter and 1050 of these particles pass through the filter.

12-51. For the filter in Exercise 12-50, what is the Beta efficiency?

12-52. What is the relationship between Beta ratio and Beta efficiency?

12-53. What is meant by the Beta ratio designation $B_{10} = 75$?

12-54. A Beta ratio of 75 means that 75 particles are trapped for every _____ that get through the filter.

ISO Cleanliness Level Standard

12-55. What is meant by an ISO code designation of 26/9?

12-56. For the ISO code designation in Exercise 12-55, what is the significance of each of the two numbers that are separated by the slash?

Pneumatics: Air Preparation and Components

13

Learning Objectives

Upon completing this chapter, you should be able to:

1. Apply the perfect gas laws to determine the interactions of pressure, volume, and temperature of a gas.
2. Describe the purpose, construction, and operation of compressors.
3. Calculate the power required to drive compressors to satisfy system requirements.
4. Determine the size of compressor air receivers for meeting system pressure and flow-rate requirements.
5. Explain the purpose and operation of fluid conditioners, including filters, regulators, lubricators, mufflers, and air dryers.
6. Calculate pressure losses in pneumatic pipelines.
7. Perform an analysis of moisture removal from air using aftercoolers and air dryers.
8. Determine how the flow rate of air can be controlled by valves.
9. Describe the purpose, construction, and operation of pneumatic pressure control valves, flow control valves, and directional control valves.
10. Discuss the construction and operation of pneumatic cylinders and motors.
11. Determine the air-consumption rate of pneumatically driven equipment.

13.1 INTRODUCTION

Pneumatics Versus Hydraulics

Pneumatic systems use pressurized gases to transmit and control power. As the name implies, pneumatic systems typically use air (rather than some other gas) as the fluid medium, because air is a safe, low-cost, and readily available fluid. It is particularly safe in environments where an electrical spark could ignite leaks from system components.

There are several reasons for considering the use of pneumatic systems instead of hydraulic systems. Liquids exhibit greater inertia than do gases. Therefore, in hydraulic systems the weight of oil is a potential problem when accelerating and decelerating actuators and when suddenly opening and closing valves. Liquids also exhibit greater viscosity than do gases. This results in larger frictional pressure and power losses. Also, since hydraulic systems use a fluid foreign to the atmosphere, they require special reservoirs and no-leak system designs. Pneumatic systems use air that is exhausted directly back into the surrounding environment. Generally speaking, pneumatic systems are less expensive than hydraulic systems.

However, because of the compressibility of air, it is impossible to obtain precise, controlled actuator velocities with pneumatic systems. Also, precise positioning control is not obtainable. In applications where actuator travel is to be smooth and steady against a variable load, the air exhaust from the actuator is normally metered. Whereas pneumatic pressures are quite low due to explosion dangers involved if components such as air tanks should rupture (less than 250 psi), hydraulic pressures can be as high as 12,000 psi. Thus, hydraulics can be high-power systems, whereas pneumatics are confined to low-power applications.

Use of Compressed Air

In pneumatic systems, compressors are used to compress and supply the necessary quantities of air. Compressors are typically of the piston, vane, or screw type. Basically, a compressor increases the pressure of a gas by reducing its volume as described by the perfect gas laws. Pneumatic systems normally use a large centralized air compressor, which is considered to be an infinite air source similar to an electrical system where you merely plug into an electrical outlet for electricity. In this way, pressurized air can be piped from one source to various locations throughout an entire industrial plant. The compressed air is piped to each circuit through an air filter to remove contaminants, which might harm the closely fitting parts of pneumatic components such as valves and cylinders. The air then flows through a pressure regulator, which reduces the pressure to the desired level for the particular circuit application. Because air is not a good lubricant, pneumatic systems require a lubricator to inject a very fine mist of oil into the air discharging from the pressure regulator. This prevents wear of the closely fitting moving parts of pneumatic components.

Free air from the atmosphere contains varying amounts of moisture. This moisture can be harmful in that it can wash away lubricants and thus cause excessive wear and corrosion. Hence, in some applications, air dryers are needed to remove this undesirable moisture. Since pneumatic systems exhaust directly into the atmosphere, they are capable of generating excessive noise. Therefore, mufflers are mounted on exhaust ports of air valves and actuators to reduce noise and prevent operating personnel from possible injury, resulting not only from exposure to noise but also from high-speed airborne particles.

Figure 13-1. Pneumatically powered hoist. *(Courtesy of Ingersoll-Rand Corp., Southern Pines, North Carolina.)*

Industrial Applications

Industrial applications of pneumatic systems are growing at a rapid pace. Typical examples include stamping, drilling, hoisting, punching, clamping, assembling, riveting, materials handling, and logic controlling operations. Figure 13-1 is a photograph of a pneumatically powered hoist that has a 1600-lb capacity. This hoist is driven by a pneumatic motor that operates with 90-psi air and has a maximum air consumption rate at a rated load of 65 standard cubic feet per min. Loads can be lifted and lowered at rates of 12 to 25 ft/min and 14 to 51 ft/min respectively. Figure 13-2 shows a pneumatically controlled dextrous hand designed to study machine dexterity and human manipulation in applications such as robotics and tactile sensing. Pneumatic actuators give the hand humanlike grasping and manipulating capability. Key operating characteristics include high speed in performing manipulation tasks, strength to easily grasp hand-sized objects that have varying densities, and force grasping control. The hand possesses three fingers and an opposing thumb. Each joint is positioned by two pneumatic actuators (located in an actuator pack) driving a high-strength tendon.

13.2 PROPERTIES OF AIR

Introduction

Air is actually a mixture of gases containing about 21% oxygen, 78% nitrogen, and 1% other gases such as argon and carbon dioxide. The preceding percentage values

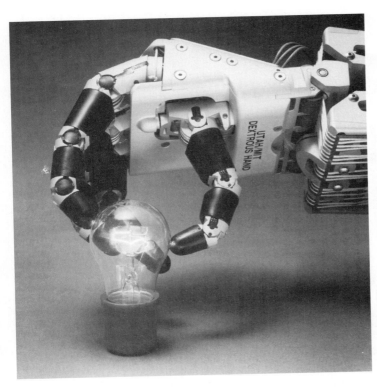

Figure 13-2. Pneumatically controlled dextrous hand. *(Courtesy of Sarcos, Inc., Salt Lake City, Utah.)*

are based on volume. Air also contains up to 4% water vapor depending on the humidity. The percent of water vapor in atmospheric air can vary constantly from hour to hour even at the same location.

Earth is surrounded by a blanket of air—the atmosphere. Because air has weight, the atmosphere exerts a pressure at any point due to the column of air above that point. The reference point is sea level, where the atmosphere exerts a pressure of 14.7 psia (101 kPa abs). Figure 13-3 shows how the atmospheric pressure decreases with altitude. For the region up to an altitude of 20,000 ft (6.1 km), the relationship is nearly linear, with a drop in pressure of about 0.5 psi per 1000-ft change in altitude (11 kPa per km).

When making pneumatic circuit calculations, atmospheric pressure of 14.7 psia is used as a standard. The corresponding standard specific weight value for air is 0.0752 lb/ft³ at 14.7 psia and 68°F (11.8 N/m³ at 101 kPa abs and 20°C). As shown in Section 13.3 in a discussion of perfect gas laws, the density of a gas depends not only on its pressure but also on its temperature.

Air is not only readily compressible, but its volume will vary to fill the vessel containing it because the air molecules have substantial internal energy and are at a considerable distance from each other. This accounts for the sensitivity of density changes with respect to changes in pressure and temperature.

Free air is considered to be air at actual atmospheric conditions. Since atmospheric pressure and temperature vary from day to day, the characteristics of free air

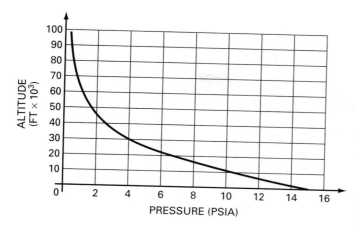

Figure 13-3. Pressure variation in the atmosphere.

vary accordingly. Thus, when making pneumatic circuit calculations, the term *standard air* is used. Standard air is sea-level air having a temperature of 68°F, a pressure of 14.7 psia (20°C and 101 kPa abs), and a relative humidity of 36%.

Absolute Pressures and Temperatures

Circuit calculations dealing with volume and pressure changes of air must be performed using absolute pressure and absolute temperature values. Eqs. (13-1) and (13-2) permit the calculation of absolute pressures and temperatures, respectively:

$$\text{absolute pressure (psia)} = \text{gage pressure (psig)} + 14.7 \qquad \textbf{(13-1)}$$

$$\text{absolute pressure (Pa abs)} = \text{gage pressure (Pa gage)} + 101{,}000 \qquad \textbf{(13-1M)}$$

$$\text{absolute temperature } (^{\circ}\text{R}) = \text{temperature} (^{\circ}\text{F}) + 460 \qquad \textbf{(13-2)}$$

$$\text{absolute temperature } (\text{K}) = \text{temperature } (^{\circ}\text{C}) + 273 \qquad \textbf{(13-2M)}$$

The units of absolute temperature in the English system are degrees Rankine, abbreviated °R. A temperature of 0°R (−460°F) is the temperature at which all molecular motion ceases to exist and the volume and pressure of a gas theoretically become zero. The units of absolute temperature in the metric system are degrees Kelvin, abbreviated K. From Eq. (13-2M), we note that a temperature of 0 K (absolute zero) equals −273°C.

Figure 13-4 gives the graphical representation of these four temperature scales using a mercury thermometer reading a room temperature of 68°F (20°C, 528°R, 293 K). As shown, absolute zero is 0°R = −460°F = −273°C = 0 K.

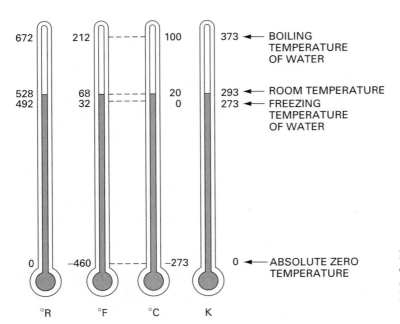

672 ← BOILING TEMPERATURE OF WATER
212 / 100 ← BOILING TEMPERATURE OF WATER
373 ← BOILING TEMPERATURE OF WATER

528 / 68 / 20 / 293 ← ROOM TEMPERATURE
492 / 32 / 0 / 273 ← FREEZING TEMPERATURE OF WATER

0 / −460 / −273 / 0 ← ABSOLUTE ZERO TEMPERATURE

°R   °F   °C   K

Figure 13-4. A comparison of the Fahrenheit (°F), Celsius (°C), Rankine (°R), and Kelvin (K) temperature scales.

13.3 THE PERFECT GAS LAWS

Introduction

During the sixteenth century, scientists discovered the laws that determine the interactions of pressure, volume, and temperature of a gas. These laws are called the "perfect gas laws" because they were derived on the basis of a perfect gas. Even though perfect gases do not exist, air behaves very closely to that predicted by Boyle's law, Charles' law, Gay-Lussac's law, and the general gas law for the pressure and temperature ranges experienced by pneumatic systems. Each of these laws (for which absolute pressure and temperature values must be used) is defined and applied to a particular problem as follows.

Boyle's Law

Boyle's law states that if the temperature of a given amount of gas is held constant, the volume of the gas will change inversely with the absolute pressure of the gas:

$$\frac{V_1}{V_2} = \frac{p_2}{p_1} \qquad \textbf{(13-3)}$$

Boyle's law is demonstrated by the cylinder piston system of Figure 13-5. As shown, the air in the cylinder is compressed at constant temperature from volume

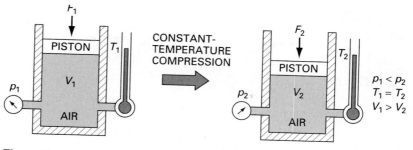

Figure 13-5. Air undergoing a constant-temperature process.

V_1 to V_2 by increasing the force applied to the piston from F_1 to F_2. Since the volume decreases, the pressure increases, as depicted by the pressure gage.

EXAMPLE 13-1

The 2-in-diameter piston of the pneumatic cylinder of Figure 13-6 retracts 4 in from its present position ($p_1 = 20$ psig; $V_1 = 20$ in³) due to the external load on the rod. If the port at the blank end of the cylinder is blocked, find the new pressure, assuming the temperature does not change.

Solution

$$V_1 = 20 \text{ in}^3$$

$$V_2 = 20 \text{ in}^3 - \frac{\pi}{4}(2)^2(4) = 7.43 \text{ in}^3$$

$$p_1 = 20 + 14.7 = 34.7 \text{ psia}$$

Substituting into Eq. (13-3), which defines Boyle's law, we have

$$\frac{20}{7.43} = \frac{p_2}{34.7}$$

$$p_2 = 93.4 \text{ psia} = 78.7 \text{ psig}$$

Charles' Law

Charles' law states that if the pressure on a given amount of gas is held constant, the volume of the gas will change in direct proportion to the absolute temperature:

$$\frac{V_1}{V_2} = \frac{T_1}{T_2} \tag{13-4}$$

Charles' law is demonstrated by the cylinder-piston system of Figure 13-7. As shown, the air in the cylinder is heated while the piston rod is supporting a weight W.

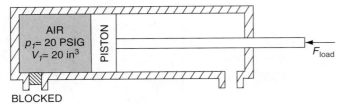

AIR
p_1= 20 PSIG
V_1= 20 in^3

PISTON

F_{load}

BLOCKED

Figure 13-6. System for Example 13-1.

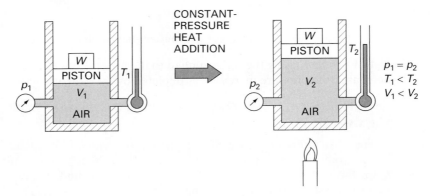

Figure 13-7. Air undergoing a constant-pressure process.

Since the weight maintains a constant force on the piston, the pressure remains constant and the volume increases.

EXAMPLE 13-2

The cylinder of Figure 13-6 has an initial position where p_1 = 20 psig and V_1 = 20 in^3 as controlled by the load on the rod. The air temperature is 60°F. The load on the rod is held constant to maintain constant air pressure, but the air temperature is increased to 120°F. Find the new volume of air at the blank end of the cylinder.

Solution

$$T_1 = 60 + 460 = 520°R$$

$$T_2 = 120 + 460 = 580°R$$

Substituting into Eq. (13-4), which defines Charles' law, yields the answer:

$$\frac{20}{V_2} = \frac{520}{580}$$

$$V_2 = 22.3 \text{ in}^3$$

Gay-Lussac's Law

Gay-Lussac's law states that if the volume of a given gas is held constant, the pressure exerted by the gas is directly proportional to its absolute temperature:

$$\frac{p_1}{p_2} = \frac{T_1}{T_2} \qquad \text{(13-5)}$$

Gay-Lussac's law is demonstrated by the closed cylinder in Figure 13-8. As shown, heat is added to the air in the constant-volume cylinder, which causes an increase in temperature and pressure.

EXAMPLE 13-3

The cylinder in Figure 13-6 has a locked position (V_1 = constant). p_1 = 20 psig, and T_1 = 60°F. If the temperature increases to 160°F, what is the new pressure in the blank end?

Solution

$$p_1 = 20 + 14.7 = 34.7 \text{ psia}$$

$$T_1 = 60 + 460 = 520°\text{R and } T_2 = 160 + 460 = 620°\text{R}$$

Substituting into Eq. (13-5), which defines Gay-Lussac's law, we obtain

$$\frac{34.7}{p_2} = \frac{520}{620}$$

or

$$p_2 = 41.4 \text{ psia} = 26.7 \text{ psig}$$

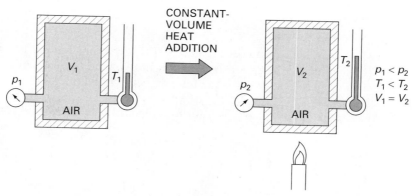

Figure 13-8. Air undergoing a constant-volume process.

Chapter 13

General Gas Law

Boyle's, Charles', and Gay-Lussac's laws can be combined into a single general gas law, as defined by

$$\frac{p_1 V_1}{T_1} = \frac{p_2 V_2}{T_2} \tag{13-6}$$

The general gas law contains all three gas parameters (pressure, temperature, and volume), since none are held constant during a process from state 1 to state 2. By holding T, p, or V constant, the general gas law reduces to Boyle's, Charles', or Gay-Lussac's law, respectively. For example, if $T_1 = T_2$, the general gas law reduces to $p_1 V_1 = p_2 V_2$, which is Boyle's law.

The general gas law is used in Chapter 14 to size gas-loaded accumulators. In addition, Examples 13-4 and 13-5 illustrate its use.

EXAMPLE 13-4

Gas at 1000 psig and 100°F is contained in the 2000-in³ cylinder of Figure 13-9. A piston compresses the volume to 1500 in³ while the gas is heated to 200°F. What is the final pressure in the cylinder?

Solution   Solve Eq. (13-6) for p_2 and substitute known values:

$$p_2 = \frac{p_1 V_1 T_2}{V_2 T_1}$$

$$= \frac{(1000 + 14.7)(2000)(200 + 460)}{(1500)(100 + 460)} = \frac{(1014.7)(2000)(660)}{1500(560)}$$

$$= 1594.5 \text{ psia} = 1579.8 \text{ psig}$$

EXAMPLE 13-5

Gas at 70 bars gage pressure and 37.8°C is contained in the 12,900-cm³ cylinder of Figure 13-9. A piston compresses the volume to 9680 cm³ while the gas is heated to 93.3°C. What is the final pressure in the cylinder?

Solution   Solve Eq. (13-6) for p_2 and substitute known values:

$$p_2 = \frac{p_1 V_1 T_2}{V_2 T_1} = \frac{(70 \times 10^5 + 1 \times 10^5)(12,900)(93.3 + 273)}{(9680)(37.8 + 273)}$$

$$= 111.5 \times 10^5 \text{ Pa absolute} = 111.5 \text{ bars absolute}$$

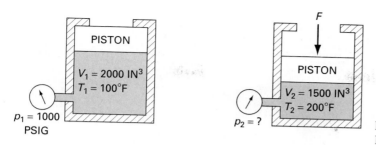

Figure 13-9. System for Example 13-4.

13.4 COMPRESSORS

Introduction

A compressor is a machine that compresses air or another type of gas from a low inlet pressure (usually atmospheric) to a higher desired pressure level. This is accomplished by reducing the volume of the gas. Air compressors are generally positive displacement units and are either of the reciprocating piston type or the rotary screw or rotary vane types.

Piston Compressors

Figure 13-10 illustrates many of the design features of a piston-type compressor. Such a design contains pistons sealed with piston rings operating in precision-bored, close-fitting cylinders. Note that the cylinders have air fins to help dissipate heat. Cooling is necessary with compressors to dissipate the heat generated during compression. When air is compressed, it picks up heat as the molecules of air come closer together and bounce off each other at faster and faster rates. Excessive temperature can damage the metal components as well as increase input power requirements. Portable and small industrial compressors are normally air-cooled, whereas larger units must be water-cooled. Figure 13-11 shows a typical small-sized, two-stage compressor unit. Observe that it is a complete system containing not only a compressor but also the compressed air tank (receiver), electric motor and pulley drive, pressure controls, and instrumentation for quick hookup and use. It is driven by a 10-hp motor, has a 120-gal receiver, and is designed to operate in the 145- to 175-psi range with a capacity of 46.3 cfm (cubic ft per min).

Figure 13-12 gives a cutaway view of a direct-drive, two-cylinder, piston-type compressor. The fan in the forefront accelerates the air cooling of the compressor by providing forced airflow.

A single-piston compressor can provide pressure up to about 150 psi. Above 150 psi, the compression chamber size and heat of compression prevent efficient pumping action. For compressors having more than one cylinder, staging can be used to improve pumping efficiency. Staging means dividing the total pressure among two or more cylinders by feeding the exhaust from one cylinder into the

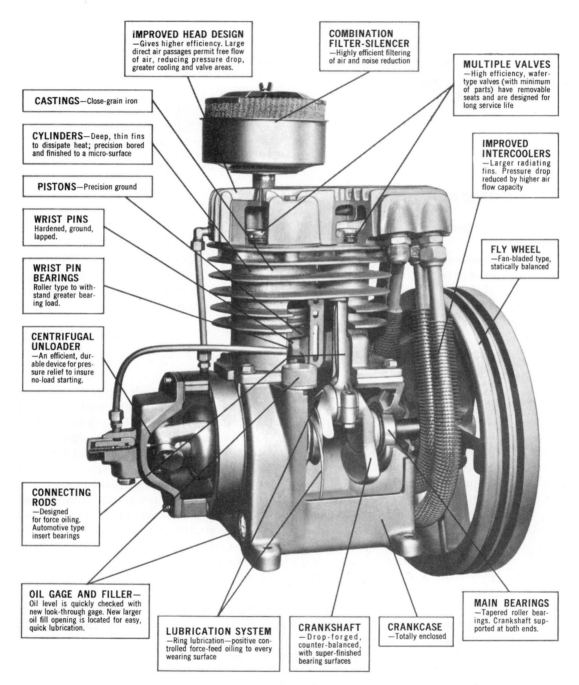

iMPROVED HEAD DESIGN
—Gives higher efficiency. Large direct air passages permit free flow of air, reducing pressure drop, greater cooling and valve areas.

COMBINATION FILTER-SILENCER
—Highly efficient filtering of air and noise reduction

MULTIPLE VALVES
—High efficiency, wafer-type valves (with minimum of parts) have removable seats and are designed for long service life

CASTINGS—Close-grain iron

CYLINDERS—Deep, thin fins to dissipate heat; precision bored and finished to a micro-surface

IMPROVED INTERCOOLERS
—Larger radiating fins. Pressure drop reduced by higher air flow capacity

PISTONS—Precision ground

WRIST PINS
Hardened, ground, lapped.

FLY WHEEL
—Fan-bladed type, statically balanced

WRIST PIN BEARINGS
Roller type to withstand greater bearing load.

CENTRIFUGAL UNLOADER
—An efficient, durable device for pressure relief to insure no-load starting.

CONNECTING RODS
—Designed for force oiling. Automotive type insert bearings

OIL GAGE AND FILLER—
Oil level is quickly checked with new look-through gage. New larger oil fill opening is located for easy, quick lubrication.

LUBRICATION SYSTEM
—Ring lubrication—positive controlled force-feed oiling to every wearing surface

CRANKSHAFT
—Drop-forged, counter-balanced, with super-finished bearing surfaces

CRANKCASE
—Totally enclosed

MAIN BEARINGS
—Tapered roller bearings. Crankshaft supported at both ends.

Figure 13-10.   Design features of a piston-type compressor. *(Courtesy of Kellogg-American, Inc., Oakmont, Pennsylvania.)*

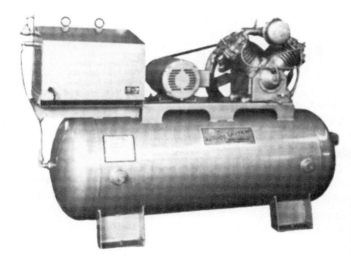

Figure 13-11. Complete piston-type, two-stage compressor unit. *(Courtesy of Kellogg-American, Inc., Oakmont, Pennsylvania.)*

Figure 13-12. Direct-drive, fan-cooled, piston-type compressor. *(Courtesy of Gast Manufacturing Corp., Benton Harbor, Michigan.)*

inlet of the next. Because effective cooling can be implemented between stages, multistage compressors can dramatically increase the efficiency and reduce input power requirements. In multistage piston compressors, successive cylinder sizes decrease, and the intercooling removes a significant portion of the heat of compression. This increases air density and the volumetric efficiency of the compressor. This is shown in Figure 13-13 by the given pressure capacities for the various number of stages of a piston-type compressor.

NUMBER OF STAGES	PRESSURE CAPACITY (PSI)
1	150
2	500
3	2500
4	5000

Figure 13-13. Effect of number of stages on pressure capacity.

Compressor Starting Unloader Controls

An air compressor must start, run, deliver air to the system as needed, stop, and be ready to start again without the attention of an operator. Since these functions usually take place after a compressed air system has been brought up to pressure, automatic controls are required to work against the air pressure already established by the compressor.

If an air compressor is started for the very first time, there is no need for a starting unloader control since there is not yet an established pressure against which the compressor must start. However, once a pressure has been established in the compressed air piping, a starting unloader is needed to prevent the established air pressure from pushing back against the compressor, preventing it from coming up to speed. Figure 13-14 shows a pressure-switch-type unloader control. When the pressure switch shuts the electric motor off, pressure between the compressor head and the check valve is bled off to the atmosphere through the release valve. The compressor is then free to start again whenever needed.

Figure 13-15 illustrates the operation of the centrifugal-type unloader control. To provide a greater degree of protection for motors and drives, an unloader valve operates by the air compressor itself rather than by the switch. This type is preferred on larger compressors. A totally enclosed centrifugal unloader operated by and installed on the compressor crankshaft is best for this purpose.

Once an air compressor is equipped with a starting unloader, it may be operated automatically by the pressure switch, as depicted in Figure 13-16. This is the normal method, using an adjustable start-stop control switch. Normal air compressor operation calls for 50 to 80% running time when using pressure switch controls. An air compressor that cycles too often (more than once each 6 min) or one that runs more than 80% of the time delivering air to the tank should be regulated by a constant-speed control.

Screw Compressors

There is a current trend toward increased use of the rotary-type compressor due to technological advances, which have produced stronger materials and better manufacturing processes. Figure 13-17 shows a cutaway view of a single-stage, screw-type compressor, which is very similar to a screw pump that was previously discussed. Compression is accomplished by rolling the trapped air into a progressively smaller volume as the screws rotate. Figure 13-18 illustrates the unsymmetrical profile of

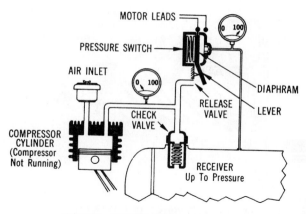

MOTOR LEADS

PRESSURE SWITCH

AIR INLET

DIAPHRAM

LEVER

RELEASE VALVE

CHECK VALVE

COMPRESSOR CYLINDER (Compressor Not Running)

RECEIVER Up To Pressure

DIAGRAM — PRESSURE SWITCH CONTROL

Figure 13-14. Pressure-switch-type unloader control. *(Courtesy of Kellogg-American, Inc., Oakmont, Pennsylvania.)*

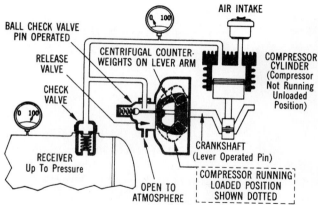

BALL CHECK VALVE PIN OPERATED

AIR INTAKE

RELEASE VALVE

CENTRIFUGAL COUNTER-WEIGHTS ON LEVER ARM

CHECK VALVE

COMPRESSOR CYLINDER (Compressor Not Running Unloaded Position)

RECEIVER Up To Pressure

CRANKSHAFT (Lever Operated Pin)

OPEN TO ATMOSPHERE

COMPRESSOR RUNNING LOADED POSITION SHOWN DOTTED

DIAGRAM — CENTRIFUGAL UNLOADER

Figure 13-15. Centrifugal-type unloader control. *(Courtesy of Kellogg-American, Inc., Oakmont, Pennsylvania.)*

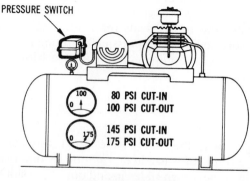

PRESSURE SWITCH

80 PSI CUT-IN
100 PSI CUT-OUT

145 PSI CUT-IN
175 PSI CUT-OUT

TYPICAL PRESSURE SETTINGS

Figure 13-16. Typical pressure settings for pressure switch. *(Courtesy of Kellogg-American, Inc., Oakmont, Pennsylvania.)*

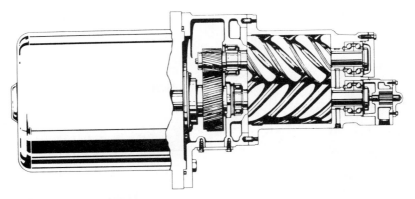

Figure 13-17. Single-stage screw compressor. *(Courtesy of Ingersoll-Rand Co., Washington, New Jersey.)*

Figure 13-18. Unsymmetrical profile of screw rotors. *(Courtesy of Ingersoll-Rand Co., Washington, New Jersey.)*

the two rotors. The rotors turn freely, with a carefully controlled clearance between both rotors and the housing, protected by a film of oil. Rotor wear will not occur, since metal-to-metal contact is eliminated. A precisely measured amount of filtered and cooled air is injected into the compression chamber, mixing with the air as it is compressed. The oil lubricates the rotors, seals the rotor clearances for high-compression efficiency, and absorbs heat of compression, resulting in low discharge air temperatures. Single-stage screw compressors are available with capacities up to 1450 cfm and pressures of 120 psi.

Vane Compressors

Figure 13-19 shows a cutaway view of the sliding-vane-type rotary compressor. In this design, a cylindrical slotted rotor turns inside of a stationary outer casing. Each rotor slot contains a rectangular vane, which slides in and out of the slot due to centrifugal force. As the rotor turns, air is trapped and compressed between the vanes

Figure 13-19. Sliding-vane-type rotary compressor. *(Courtesy of Gast Manufacturing Corp., Benton Harbor, Michigan.)*

and then discharged through a port to the receiver. Rotary sliding vane compressors can operate up to approximately 50 psi in a single stage and up to 150 psi in a two-stage design. This low-pressure, low-volume type of compressor is normally used for instrument and other laboratory-type air needs.

Air Capacity Rating of Compressors

Air compressors are generally rated in terms of cfm of free air, defined as air at actual atmospheric conditions. Cfm of free air is called scfm when the compressor inlet air is at the standard atmospheric conditions of 14.7 psia and 68°F. The abbreviation scfm means standard cubic feet per minute. Therefore, a calculation is necessary to determine the compressor capacity in terms of cfm of free air or scfm for a given application. In metric units a similar calculation is made using m^3/min of free air or standard m^3/min where standard atmospheric conditions are 101,000 Pa abs and 20°C.

The equation that allows for this calculation is derived by solving the general gas law Eq. (13-6) for V_1 as follows:

$$V_1 = V_2 \left(\frac{p_2}{p_1}\right) \left(\frac{T_1}{T_2}\right)$$

In the above equation, subscript 1 represents compressor inlet atmospheric conditions (standard or actual) and subscript 2 represents compressor discharge conditions. Dividing both sides of this equation by time (t) converts volumes V_1 and V_2 into volume flow rates Q_1 and Q_2, respectively. Thus, we have the desired equation:

$$Q_1 = Q_2 \left(\frac{p_2}{p_1}\right) \left(\frac{T_1}{T_2}\right) \qquad \textbf{(13-7)}$$

Note that absolute pressure and temperature values must be used in Eq. (13-7).

EXAMPLE 13-6

Air is used at a rate of 30 cfm from a receiver at 90°F and 125 psi. If the atmospheric pressure is 14.7 psia and the atmospheric temperature is 70°F, how many cfm of free air must the compressor provide?

Solution   Substituting known values into Eq. (13-7) yields

$$Q_1 = Q_2 \left(\frac{p_2}{p_1}\right)\left(\frac{T_1}{T_2}\right) = 30 \times \frac{125 + 14.7}{14.7} \times \frac{70 + 460}{90 + 460}$$

$$= 275 \text{ cfm of free air}$$

In other words, the compressor must receive atmospheric air (14.7 psia and 70°F) at a rate of 275 cfm in order to deliver air (125 psi and 90°F) at 30 cfm.

Sizing of Air Receivers

The sizing of air receivers requires taking into account parameters such as system pressure and flow-rate requirements, compressor output capability, and the type of duty of operation. Basically, a receiver is an air reservoir. Its function is to supply air at essentially constant pressure. It also serves to dampen pressure pulses either coming from the compressor or the pneumatic system during valve shifting and component operation. Frequently a pneumatic system demands air at a flow rate that exceeds the compressor capability. The receiver must be capable of handling this transient demand.

Equations (13-8) and (13-8M) can be used to determine the proper size of the receiver in English units and metric units, respectively.

$$V_r = \frac{14.7t(Q_r - Q_c)}{p_{max} - p_{min}} \tag{13-8}$$

$$V_r = \frac{101t(Q_r - Q_c)}{p_{max} - p_{min}} \tag{13-8M}$$

where t = time that receiver can supply required amount of air (min),

Q_r = consumption rate of pneumatic system (scfm, standard m³/min),

Q_c = output flow rate of compressor (scfm, standard m³/min),

p_{max} = maximum pressure level in receiver (psi, kPa),

p_{min} = minimum pressure level in receiver (psi, kPa),

V_r = receiver size (ft³, m³).

EXAMPLE 13-7

a. Calculate the required size of a receiver that must supply air to a pneumatic system consuming 20 scfm for 6 min between 100 and 80 psi before the compressor resumes operation.

b. What size is required if the compressor is running and delivering air at 5 scfm?

Solution

a.
$$V_r = \frac{14.7 \times 6 \times (20 - 0)}{100 - 80} = 88.2 \text{ ft}^3 = 660 \text{ gal}$$

b.
$$V_r = \frac{14.7 \times 6 \times (20 - 5)}{100 - 80} = 66.2 \text{ ft}^3 = 495 \text{ gal}$$

It is common practice to increase the calculated size of the receiver by 25% for unexpected overloads and by another 25% for possible future expansion needs.

Power Required to Drive Compressors

Another important design consideration is to determine the power required to drive an air compressor to meet system pressure and flow-rate requirements. Equations (13-9) and (13-9M) can be used to determine the theoretical power required to drive an air compressor.

$$\text{theoretical horsepower (HP)} = \frac{p_{\text{in}}Q}{65.4}\left[\left(\frac{p_{\text{out}}}{p_{\text{in}}}\right)^{0.286} - 1\right] \tag{13-9}$$

$$\text{theoretical power (kW)} = \frac{p_{\text{in}}Q}{17.1}\left[\left(\frac{p_{\text{out}}}{p_{\text{in}}}\right)^{0.286} - 1\right] \tag{13-9M}$$

where p_{in} = inlet atmospheric pressure (psia, kPa abs),
p_{out} = outlet pressure (psia, kPa abs),
Q = flow rate (scfm, standard m³/min).

To determine the actual power, the theoretical power from Eq. (13-9) is divided by the overall compressor efficiency η_o.

EXAMPLE 13-8

Determine the actual power required to drive a compressor that delivers 100 scfm of air at 100 psig. The overall efficiency of the compressor is 75%.

Solution Since absolute pressures must be used in Eq. (13-9), we have $p_{in} = 14.7$ psia and $p_{out} = 114.7$ psia. Substituting directly into Eq. (13-9) yields the theoretical horsepower required.

$$\text{HP}_{\text{theor}} = \frac{14.7 \times 100}{65.4}\left[\left(\frac{114.7}{14.7}\right)^{0.286} - 1\right] = 18.0 \text{ hp}$$

The actual horsepower required is

$$\text{HP}_{\text{act}} = \frac{\text{HP}_{\text{theor}}}{h_o} = \frac{18.0}{0.75} = 24.0 \text{ hp}$$

13.5 FLUID CONDITIONERS

Introduction

The purpose of fluid conditioners is to make air a more acceptable fluid medium for the pneumatic system as well as operating personnel. Fluid conditioners include filters, regulators, lubricators, mufflers, and air dryers.

Air Filters

The function of a filter is to remove contaminants from the air before it reaches pneumatic components such as valves and actuators. Generally speaking, in-line filters contain filter elements that remove contaminants in the 5- to 50-μm range. Figure 13-20 shows a cutaway view of a filter that uses 5-μm cellulose felt, reusable, surface-type elements. These elements have gaskets molded permanently to each end to prevent air bypass and make element servicing foolproof. These elements have a large ratio of air to filter media and thus can hold an astonishing amount of contamination on the surface without suffering significant pressure loss. The baffling system used in these filters mechanically separates most of the contaminants before they reach the filter element. In addition, a quiet zone prevents contaminants collected in the bowl from reentering the airstream. Also shown in Figure 13-20 is the ANSI symbol for an air filter.

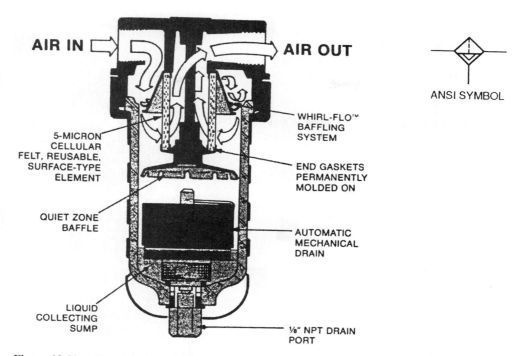

AIR IN

AIR OUT

ANSI SYMBOL

5-MICRON
CELLULAR
FELT, REUSABLE,
SURFACE-TYPE
ELEMENT

QUIET ZONE
BAFFLE

LIQUID
COLLECTING
SUMP

WHIRL-FLO™
BAFFLING
SYSTEM

END GASKETS
PERMANENTLY
MOLDED ON

AUTOMATIC
MECHANICAL
DRAIN

⅛" NPT DRAIN
PORT

Figure 13-20.   Operation of air filter. *(Courtesy of Wilkerson Corp., Englewood, Colorado.)*

Air Pressure Regulators

So that a constant pressure is available for a given pneumatic system, a pressure regulator is used. Figure 13-21 illustrates the design features of a pressure regulator whose operation is as follows:

Airflow enters the regulator at A. Turning adjusting knob B clockwise (viewed from knob end) compresses spring C, causing diaphragm D and main valve E to move, allowing flow across the valve seat area. Pressure in the downstream area is sensed through aspirator tube F to the area H above diaphragm D. As downstream pressure rises, it offsets the load of spring C. Diaphragm D and valve E move to close the valve against its seat, stopping airflow through the regulator. The holding pressure of spring C and downstream pressure H are in balance, at reduced outlet pressure. Any airflow demand downstream, such as opening a valve, will cause the downstream pressure to drop. Spring C will again push open valve E, repeating the sequence in a modulating fashion to maintain the downstream pressure setting. A rise in downstream pressure above the set pressure will cause diaphragm D to lift off the top of valve stem J, thus relieving the excess pressure to the atmosphere under knob B. When the downstream pressure returns to the set pressure, the diaphragm reseats on the valve stem, and the system is again in equilibrium. The ANSI symbol of an air pressure regulator is also shown in Figure 13-21.

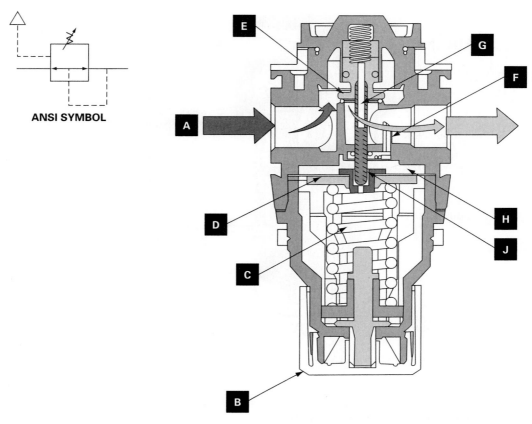

ANSI SYMBOL

Figure 13-21. Air pressure regulator. *(Courtesy of Wilkerson Corp., Englewood, Colorado.)*

Air Lubricators

A lubricator ensures proper lubrication of internal moving parts of pneumatic components. Figure 13-22 illustrates the operation of a lubricator, which inserts every drop of oil leaving the drip tube, as seen through the sight dome, directly into the airstream. These drops of oil are transformed into an oil mist prior to their being transported downstream. This oil mist consists of both coarse and fine particles. The coarse particles may travel distances of 20 ft or more, while the fine particles often reach distances as great as 300 ft from the lubricator source. These oil mist particles are created when a portion of the incoming air passes through the center of the variable orifice and enters the mist generator, mixing with the oil delivered by the drip tube. This air-oil mixture then rejoins any air that has bypassed the center of the variable orifice and continues with that air toward its final destination. Oil reaching the mist generator is first pushed up the siphon tube, past the adjustment screw to the drip tube located within the sight dome. This is accomplished by diverting a small amount of air from the mainstream through the bowl pressure control valve, into the bowl or reservoir. This valve is located so that it will close, shutting off the air supply to the bowl when the fill plug is loosened or removed,

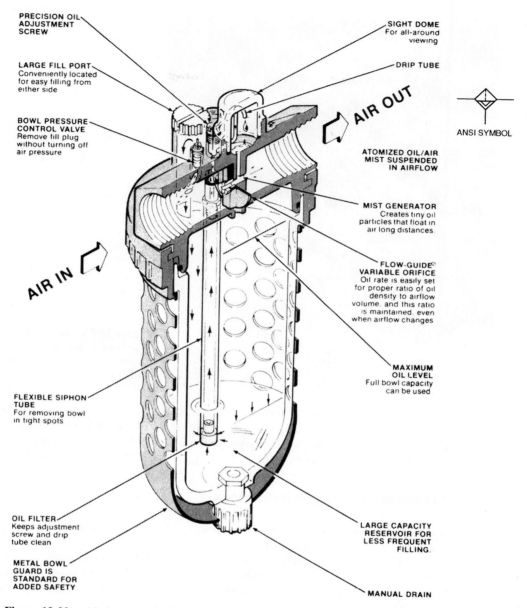

PRECISION OIL ADJUSTMENT SCREW

LARGE FILL PORT
Conveniently located for easy filling from either side

BOWL PRESSURE CONTROL VALVE
Remove fill plug without turning off air pressure

AIR IN

FLEXIBLE SIPHON TUBE
For removing bowl in tight spots

OIL FILTER
Keeps adjustment screw and drip tube clean

METAL BOWL GUARD IS STANDARD FOR ADDED SAFETY

SIGHT DOME
For all-around viewing

DRIP TUBE

AIR OUT

ANSI SYMBOL

ATOMIZED OIL/AIR MIST SUSPENDED IN AIRFLOW

MIST GENERATOR
Creates tiny oil particles that float in air long distances.

FLOW-GUIDE® VARIABLE ORIFICE
Oil rate is easily set for proper ratio of oil density to airflow volume, and this ratio is maintained, even when airflow changes

MAXIMUM OIL LEVEL
Full bowl capacity can be used

LARGE CAPACITY RESERVOIR FOR LESS FREQUENT FILLING.

MANUAL DRAIN

Figure 13-22. Air lubricator. *(Courtesy of Wilkerson Corp., Englewood, Colorado.)*

permitting refilling of the bowl or reservoir without shutting off the air supply line. On replacement of the fill plug, the bowl pressure control valve will open automatically, causing the bowl to be pressurized once again and ready to supply lubrication where it is needed.

Also shown in Figure 13-22 is the ANSI symbol for an air lubricator.

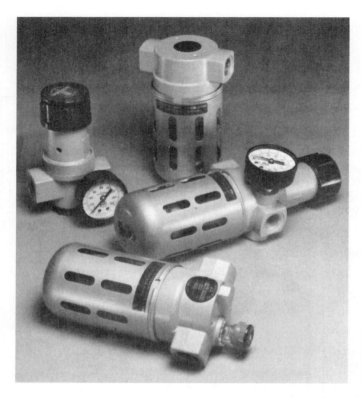

Figure 13-23.   Individual filter, regulator, lubricator units. *(Courtesy of C. A. Norgren Co., Littleton, Colorado.)*

Figure 13-23 shows an individual filter, two individual pressure regulators, and an individual lubricator. In contrast, in Figure 13-24 we see a combination filter-regulator-lubricator unit (FRL). Also shown is its ANSI symbol. In both Figures 13-23 and 13-24, the units with the pressure gages are the pressure regulators.

Pneumatic Pressure Indicators

Figure 13-25(a) shows a pneumatic pressure indicator that provides a two-color, two-position visual indication of air pressure. The rounded lens configuration provides a 180° view of the indicator status, which is a fluorescent signal visible from the front and side. This indicator is easily panel-mounted using the same holes as standard electrical pilot lights. However, they are completely pneumatic, requiring no electrical power.

These pneumatic pressure indicators are field adjustable for either one input with spring return or two inputs with memory. This memory does not require continuous pressure to maintain its last signal input. Field conversion may be made to select either single-input, spring return, or two-input maintained modes of operation. Figure 13-25(b) shows the adjustment on the rear of the indicator housing. By using the same adjustment, either of the two display colors and its individual input may be selected for single-input operation. In the center position, this adjustment allows the

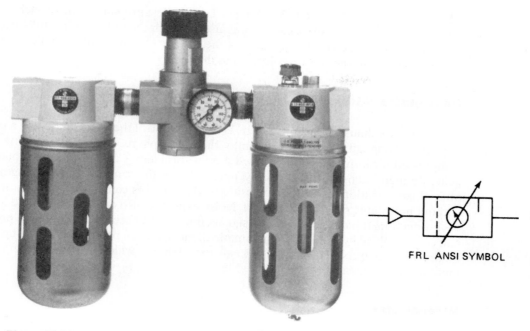

FRL ANSI SYMBOL

Figure 13-24. Combination filter, regulator, lubricator unit. *(Courtesy of C. A. Norgren Co., Littleton, Colorado.)*

(a)

(b)

Figure 13-25. Pneumatic pressure indicator. (a) Front view. (b) Rear view. *(Courtesy of Numatics Inc., Highland, Michigan.)*

indicator to accept two inputs for a maintained (memory) mode of operation. If both inputs are on simultaneously, the indicator will assume an intermediate position and show parts of both colors.

These indicators come in a variety of color combinations and are completely compatible with pneumatic systems. They are available with pressure ranges of

0.5 to 30 psi, 25 to 150 psi, and 45 to 150 psi. The smallest pressure value of each pressure range (0.5, 25, and 45 psi) is the pressure at which the indicator has fully transferred to the second color. The actuation time, or time elapsed until the indicator has fully transferred to the second color, is less than 1 s.

Pneumatic Silencers

A pneumatic exhaust silencer (muffler) is used to control the noise caused by a rapidly exhausting airstream flowing into the atmosphere. The increased use of compressed air in industry has created a noise problem. Compressed air exhausts generate high-intensity sound energy, much of it in the same frequency ranges as normal conversation. Excessive exposure to these noises can cause loss of hearing without noticeable pain or discomfort. Noise exposure also causes fatigue and lowers production. It blocks out warning signals, thus causing accidents. This noise problem can be solved by installing a pneumatic silencer at each pneumatic exhaust port. Figure 13-26 depicts several types of exhaust silencers, which are designed not to build up back pressure with continued use.

Aftercoolers

Air from the atmosphere contains varying amounts of moisture in the form of water vapor. Compressors do not remove this moisture. Cooling of compressed air in piping causes condensation of moisture, much of which is eventually carried along into air-operated tools and machines. Water washes away lubricants causing excessive wear in components containing moving parts such as cylinders, valves, and motors. Water also causes rusting of metallic surfaces and damage to plumbing components such as conductors and fittings.

An aftercooler is a heat exchanger that has two functions. First, it serves to cool the hot air discharged from the compressor to a desirable level (about 80 to 100°F) before it enters the receiver. Second, it removes most of the moisture from the air

ANSI SYMBOL

Figure 13-26. Pneumatic silencers. *(Courtesy of C. A. Norgren Co., Littleton, Colorado.)*

discharged from the compressor by virtue of cooling the air to a lower temperature. Figure 13-27 shows an aftercooler that is installed in the air line between the compressor and the air receiver. In this aftercooler, the moist air from the compressor flows on the outside of tubes inside of which flows cool water. The water flows in an opposite direction to the airflow. The tubes contain internal baffles to provide proper water velocity and turbulence for high heat transfer rates. After passing around the tubes, the cooled air enters the moisture-separating chamber, which effectively traps out condensed moisture.

Air Dryers

Aftercoolers remove only about 85% of the moisture from the air leaving the compressor. Air dryers are installed downstream of aftercoolers when it is important to remove enough moisture from the air so that the air will not become saturated as it flows through the pneumatic system. There are three basic types of air dryers: chemical, adsorption, and refrigeration.

In chemical air dryers, moisture is absorbed by pellets made of dryer agent materials, such as dehydrated chalk or calcium chloride. A chemical process turns the pellets into a liquid that is drained from the system. The pellets are replaced on a planned maintenance schedule.

Adsorption dryers remove moisture, using beds made of materials such as activated alumina or silica gel. This is a mechanical process that involves the capturing

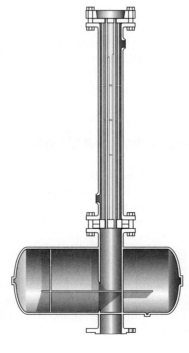

Figure 13-27.   Aftercooler.
(Courtesy of Ingersoll-Rand Co., Washington, New Jersey.)

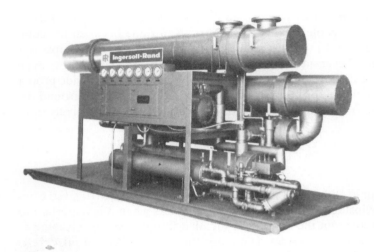

Figure 13-28. Chiller air dryer.
*(Courtesy of Ingersoll-Rand Co.,
Washington, New Jersey.)*

of moisture in the pores of the bed material. On a planned maintenance schedule, the beds are replenished or reactivated by the application of heat and a dryer gas.

Refrigeration dryers are basically refrigerators that use commercial refrigerants. In these dryers, the moist air passes through a heat exchanger where it is cooled as it flows around coils containing a liquid refrigerant. Refrigeration dryers can achieve lower dew points and thus lower moisture contents than can chemical or adsorption type dryers. Small to medium-size refrigeration dryers typically pass the moist air directly across refrigerant coils. The large-size units are called *chiller dryers* and operate by first cooling water and running the cool water through coils over which flows the moist air.

Figure 13-28 shows a chiller air dryer, which removes virtually all moisture by lowering the temperature of the pressurized air to a dew point of 40°F. It is shipped completely assembled, piped, and wired. All that is needed are the connections to the air line, the electric power system, the cooling water circuit, and the condensate discharge line.

13.6 ANALYSIS OF MOISTURE REMOVAL FROM AIR

The amount of moisture in air is identified by the term *humidity*. When air contains the maximum amount of moisture that it can hold at a given temperature, the air is said to be *saturated*. The term *relative humidity* is defined as the ratio of the actual amount of moisture contained in the air to the amount of moisture it would contain if it were saturated. The relative humidity depends on the air temperature. For example, for air containing a given amount of moisture, if the air temperature goes down, the relative humidity goes up, and vice versa.

This brings us to the term *dew point*, which is defined as the temperature at which the air is saturated and thus the relative humidity is 100%. When air is saturated (air is at the dew point temperature), any decrease in temperature will cause water to condense out of the air. This condensation process takes place in aftercoolers to remove moisture from the air. High-pressure air discharged from compressors contains much more moisture per unit volume than does the atmospheric air entering the compressor. The actual amount of moisture per cubic foot in the compressed air depends on the amount of moisture in the entering atmospheric air and the compressor discharge pressure.

Figure 13-29 provides a graph that shows the amount of moisture contained in saturated air at various temperatures and pressures. For example, Figure 13-29 shows that saturated free air entering a compressor at 80°F and 14.7 psia (zero gage pressure) contains 1.58 lb of moisture per 1000 ft³. This 1.58 value is obtained at the upper left-hand portion of the graph where the 80°F temperature curve (if extended to a pressure value of zero gage pressure) would intersect the vertical axis containing the pounds of moisture content values. If the compressor increases the air pressure of this 1000 ft³ of saturated free air to 100 psig and the temperature is cooled (in an aftercooler) back to 80°F, the maximum amount of moisture this air can hold is reduced to 0.20 lb. The 0.20 value is obtained as follows: Find the point of intersection of the vertical axis at 100 psig and the 80°F temperature curve. Then, from this point, follow a horizontal line to the left until it intersects the vertical axis containing the pounds of moisture content values.

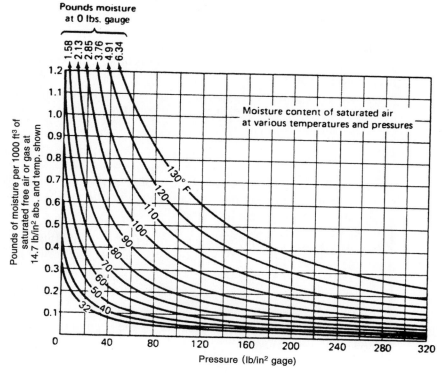

Figure 13-29. Moisture content of saturated air at various temperatures and pressures. *(Courtesy of the Compressed Air and Gas Institute,* Compressed Air and Gas Handbook, *5e.)*

The value of 0.20 means that 1.38 lb out of the 1.58 lb of moisture per 1000 ft³ of saturated free air (or 87.3% of the moisture) would condense out of the air in the form of water, to be drained away by the aftercooler. This shows that compressing air and then cooling it back to its original temperature is an effective way to remove moisture. Note that as a compressor compresses air, it also increases the air temperature due to the heat of compression.

Air dryers remove so much of the moisture from the air that the air does not become saturated as it flows through pneumatic systems that are located inside factory buildings. This is because the relative humidity of the air is so low that the dew point is not reached and thus condensation does not occur.

EXAMPLE 13-9

A compressor delivers 100 scfm of air at 100 psig to a pneumatic system. Saturated atmospheric air enters the compressor at 80°F.

a. If the compressor operates 8 hours per day, determine the number of gallons of moisture delivered to the pneumatic system by the compressor per day.

b. How much moisture per day would be received by the pneumatic system if an aftercooler is installed to cool the compressed air temperature back to 80°F?

c. How much moisture per day would be received by the pneumatic system if an air dryer is installed to cool the compressed air temperature to 40°F?

Solution

a. Per Figure 13-29, the atmospheric air entering the compressor contains 1.58 lb of moisture per 1000 ft³. Thus, the rate at which moisture enters the compressor can be found.

$$\text{moisture rate (lb/min)} = \text{entering moisture content (lb/ft}^3) \\ \times \text{entering scfm flow rate ft}^3/\text{min)}$$

$$= \frac{1.58}{1000} \text{ lb/ft}^3 \times 100 \text{ ft}^3/\text{min} = 0.158 \text{ lb/min}$$

The number of gallons per day received by the pneumatic system can be found knowing that water weighs 8.34 lb/gal.

$$\text{gal/day} = 0.158 \frac{\text{lb}}{\text{min}} \times \frac{60 \text{ min}}{1 \text{ hr}} \times \frac{8 \text{ hr}}{\text{day}} \times \frac{1 \text{ gal}}{8.34 \text{ lb}} = 9.09 \text{ gal/day}$$

b. Per Figure 13-29, if the compressed air is cooled back to 80°F, the maximum amount of moisture the air (leaving the aftercooler) can hold per 1000 ft³ of free air is 0.20 lb. Since $(1.58 - 0.20)/1.58 = 0.873$, this means that 87.3% of the moisture would condense out of the air and be drained away by the aftercooler. The gallons of moisture per day received by the pneumatic system are

$$\text{gal/day} = (1 - 0.873)(9.09 \text{ gal/day}) = 1.15 \text{ gal/day}$$

c. Per Figure 13-29, the maximum amount of moisture the air (leaving the air dryer) can hold per 1000 ft³ of free air is 0.05 lb. This represents a 96.8% moisture removal rate. Thus, the moisture received per day by the pneumatic system is

$$\text{gal/day} = (1 - 0.968)(9.09 \text{ gal/day}) = 0.29 \text{ gal/day}$$

Air leaves the dryer and enters the pneumatic system at 40°F. As long as the air temperature in the pneumatic system stays above the 40°F value reached in the air dryer (which is typically the case for indoor systems), none of the moisture will condense into water. Thus, the 0.29 gal/day of moisture entering the pneumatic system would remain in the form of water vapor. Moisture in the air in the form of water vapor does not cause harm to components because water vapor is a gas whereas water is a liquid. Refrigeration dryers are capable of lowering the temperature of the compressed air to as low as 35°F.

13.7 AIR FLOW-RATE CONTROL WITH ORIFICES

Flow Rate Through an Orifice

Since a valve is a variable orifice, it is important to evaluate the flow rate of air through an orifice. Such a relationship is discussed for liquid flow in Chapter 8. However, because of the compressibility of air, the relationship describing the flow rate of air is more complex.

Equations (13-10) and (13-10M) provide for the calculation of air volume flow rates through orifices using English and metric units, respectively.

$$Q = 22.7 C_v \sqrt{\frac{(p_1 - p_2)(p_2)}{T_1}} \tag{13-10}$$

$$Q = 0.0698 C_v \sqrt{\frac{(p_1 - p_2)(p_2)}{T_1}} \tag{13-10M}$$

where Q = volume flow rate (scfm, std m^3/min),
         C_v = flow capacity constant,
         p_1 = upstream pressure (psia, kPa abs),
         p_2 = downstream pressure (psia, kPa abs),
         T_1 = upstream temperature (°R, K).

The preceding equations are valid when p_2 is more than $0.53p_1$ or when p_2 is more than 53% of p_1. Beyond this region, the flow through the orifice is said to be choked. Thus, the volume flow rate through the orifice increases as the pressure drop $p_1 - p_2$ increases until p_2 becomes equal to $0.53p_1$. Any lowering of p_2 to values below $0.53p_1$ does not produce any increase in volume flow rate, as would be predicted by Eqs. (13-10) or (13-10M), because the downstream fluid velocity reaches the speed of sound. Thus, the pressure ratio p_2/p_1 must be calculated to determine if the flow is choked before using Eqs. (13-10) and (13-10M).

From a practical point of view, this means that a downstream pressure of 53% of the upstream pressure is the limiting factor for passing air through a valve to an actuator. Thus, for example, with 100-psia line pressure, if the pressure at the inlet of an actuator drops to 53 psia, the fluid velocity is at its maximum. No higher fluid velocity can be attained even if the pressure at the inlet of the actuator drops below 53 psia. Assuming an upstream pressure of 100 psia, the volume flow rate must be calculated for a downstream pressure of 53 psia using Eqs. (13-10) or (13-10M) even though the downstream pressure may be less than 53 psia.

By the same token, if p_2 is less than $0.53p_1$, increasing the value of p_1 will result in a greater pressure drop across the valve but will not produce an increase in fluid velocity, because the orifice is already choked. Thus, increasing the ratio of p_1/p_2 beyond 1/0.53 (or 1.89) does not produce any increase in volume flow rate. However, it should be noted that raising the value of p_1 beyond 1.89 p_2 will increase the mass flow rate because the density of air increases as the pressure rises. Thus raising the value of p_1 beyond 1.89 p_2 increases the mass flow rate even though the volume flow rate remains at the choked value.

Sizing of Valves Based on Flow Rates

Values of the flow capacity constant C_v are determined experimentally by valve manufacturers and are usually given in table form for various sizes of valves. The proper-size valve can be selected from manufacturers' catalogs for a given application. Knowing the system flow rate (Q), the upstream air temperature (T_1), the maximum acceptable pressure drop across the valve ($p_1 - p_2$), and the required pressure downstream of the valve for driving an actuator, the corresponding flow capacity constant (C_v) can be calculated using Eqs. (13-10) or (13-10M). Selecting a valve with a C_v greater than or equal to that calculated from Eqs. (13-10) or (13-10M) will provide a valve of adequate size for the application involved. A large C_v indicates a large-size valve, because for the same valve pressure drop and valve downstream pressure, the volume flow rate increases directly with the flow capacity constant, C_v.

EXAMPLE 13-10

Air at 80°F passes through a ½-in-diameter orifice having a flow capacity constant of 7.4. If the upstream pressure is 80 psi, what is the maximum flow rate in units of scfm of air?

Solution

$$T_1 = 80 + 460 = 540°R$$

$$p_1 = 80 + 14.7 = 94.7 \text{ psia}$$

The maximum flow rate occurs when the orifice is choked ($p_2 = 0.53p_1$). Thus,

$$p_2 = 0.53 \times 94.7 = 50.2 \text{ psia}$$

Substituting directly into Eq. (13-10) yields

$$Q = 22.7 \times 7.4\sqrt{\frac{(94.7 - 50.2)(50.2)}{540}} = 22.7 \times 7.4 \times 2.03$$

$$= 341 \text{ scfm of air}$$

13.8 AIR CONTROL VALVES

Pressure Regulators

Air control valves are used to control the pressure, flow rate, and direction of air in pneumatic circuits. Pneumatic pressure control valves are air line regulators that are installed at the inlet of each separate pneumatic circuit. As such, they establish the working pressure of the particular circuit. Sometimes air line regulators are installed within a circuit to provide two or more different pressure levels for separate portions of the circuit. A cutaway view of an actual pressure regulator (whose operation is discussed in Section 13.5) is given in Figure 13-30. The desired pressure level is established by the T-handle, which exerts a compressive force on the spring. The spring transmits a force to the diaphragm, which regulates the opening and closing of the control valve. This regulates the airflow rate to establish the desired downstream pressure.

Check Valves

In Figure 13-31 we see a check valve that shuts off instantaneously against reverse flow and opens at low cracking pressures in the forward direction. As shown in the schematic views, the disk seals before reverse flow is established, thus avoiding fluid

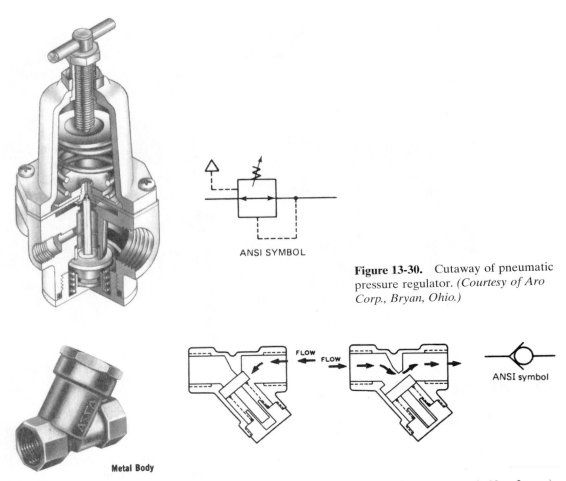

ANSI SYMBOL

Figure 13-30. Cutaway of pneumatic pressure regulator. *(Courtesy of Aro Corp., Bryan, Ohio.)*

Metal Body

Figure 13-31. Pneumatic check valve. *(Courtesy of Automatic Switch Co., Florham Park, New Jersey.)*

shock on reversal of pressure differential. Although the design shown has a metal body, lightweight plastic body designs with fittings suitable for plastic or metal tubing are also available.

Shuttle Valves

Figure 13-32(a) is a photograph of a pneumatic shuttle valve that automatically selects the higher of two input pressures and connects that pressure to the output port while blocking the lower pressure. This valve has two input ports and one output port and employs a free-floating spool with an open-center action. At one end of the spool's travel, it connects one input with the output port. At the other end of its travel, it connects the second input with the output port.

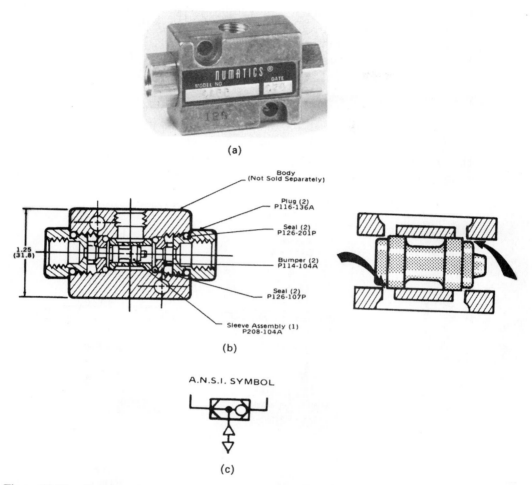

Figure 13-32. Shuttle valve. (a) External view. (b) Internal view and spool-port configuration. (c) ANSI symbol. *(Courtesy of Numatics Incorporated, Highland, Michigan.)*

When a pressure is applied to an input port, the air shifts the spool and then moves through the sleeve ports and out the output port. When the pressure is removed from the input port, the air in the output port exhausts back through the shuttle valve and out one of the input ports. It normally exhausts out the input port through which it entered, but there is no guarantee, and it may exhaust out the other. If a signal is applied to the second input port, a similar action takes place.

If, while one input is pressurized, the second input port receives a pressure that is 1.5 psig greater than the first, the higher pressure will appear at the output. If the second input is the same as the first, no change will take place until the first signal is exhausted. Then, as it drops in pressure, the second input will predominate.

The open-center action of the shuttle valve is shown in Figure 13-32(b), where the arrows indicate the clearance in the center position. As the spool shifts through the center position, it lacks a few thousandths of an inch of blocking the ports. Thus, it is not possible for the spool to find a center position that will block all exhaust action. It will always move to one position or the other.

Figure 13-32(c) gives the ANSI symbol for a pneumatic shuttle valve.

Two-Way Directional Control Valves

In Figure 13-33 we see an air-operated (air-piloted), two-way, pneumatic valve. As shown, this valve is available to operate either normally open or normally closed. The poppet-type construction provides a tight shutoff, and variations in the pilot air pressure or main line pressure do not affect the operation of these valves. The

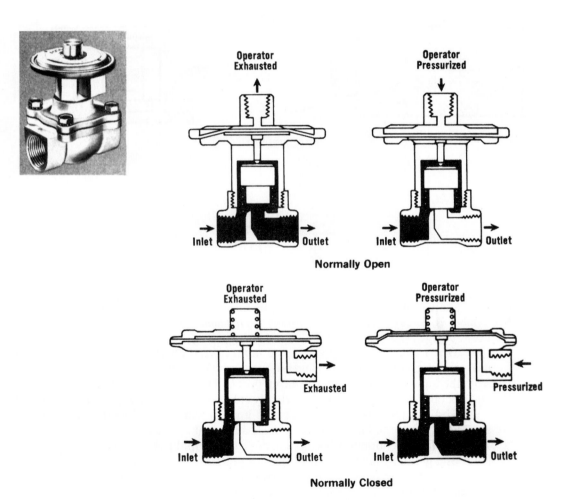

Figure 13-33. Two-way, air-piloted valve. *(Courtesy of Automatic Switch Co., Florham Park, New Jersey.)*

pilot pressure need not be constant. These valves will handle dry or lubricated air and provide long life.

Three-Way and Four-Way Directional Control Valves

Figure 13-34 shows a multipurpose three-way or open-exhaust, four-way, push-button directional control valve. The three-way valves are three-port, multipurpose valves, and any port can be pressurized. The four-way valves may also be used as normally open or closed three-ways by plugging the appropriate cylinder port. Exhaust is through two screened ports. The force required to operate these valves is 2.5 lb.

Figure 13-35 shows a palm-button directional control valve. The large mushroom heads are extra heavy-duty operators especially designed to survive the day-after-day pounding of heavy, gloved hands in stamping press, foundry, and other similar applications. The large, rounded button is padded with a soft synthetic rubber cover, which favors the operator's hand.

In Figure 13-36, we see a limit valve that uses a roller-level actuator. These directional control valves are available as multipurpose three-ways or open-exhaust four-ways. This type of valve is normally actuated by a cylinder piston rod at the ends or limits of its extension or retraction strokes.

In Figure 13-37, we see a hand-lever-operated, four-way directional control valve. The hand lever is used with two- or three-position valves. Hand movement of the lever causes the spool to move. The lever is directly connected to the spool. Detents, which provide a definite "feel" when the spool is in a specific position, are available.

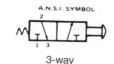

3-way

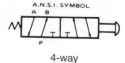

4-way

Figure 13-34. Push-button directional control valve. *(Courtesy of Numatics Incorp., Highland, Michigan.)*

3-WAY ANSI SYMBOL

4-WAY ANSI SYMBOL

Figure 13-35. Palm-button valve. *(Courtesy of Numatics Incorp., Highland, Michigan.)*

3-WAY ANSI SYMBOL

4-WAY ANSI SYMBOL

Figure 13-36. Limit valve. *(Courtesy of Numatics Incorp., Highland, Michigan.)*

2 POSITION ANSI SYMBOL

3 POSITION ANSI SYMBOL
(ALL PORTS BLOCKED IN CENTER POSITION)

Figure 13-37. Hand-lever-operated, four-way valve. *(Courtesy of Skinner Precision Industries, Inc., New Britain, Connecticut.)*

Figure 13-38 illustrates the internal construction features of a four-way, two-position, solenoid-actuated directional control valve. The single-solenoid operator shown will move the spool when energized, and a spring will return the spool when the solenoid is de-energized.

Using two solenoids, a two-position valve can be shifted by energizing one solenoid momentarily. The valve will remain in the shifted position until the opposite solenoid is energized momentarily.

Three-position valves will remain in the spring-centered position until one of the solenoids is energized. Energizing the solenoid causes the spool to shift and stay shifted until the solenoid is de-energized. When the solenoid is de-energized, the spool will return to the center position.

Flow Control Valves

A flow control valve is illustrated in Figure 13-39. As shown, a spring-loaded disk allows free flow in one direction and an adjustable or controlled flow in the opposite direction. Flow adjustment is performed by a tapered brass stem that controls the flow through the cross hole in the disk.

The adjustable knob contains a unique locking device that consists of a plastic metering knob and thumb latch pawl. The valve bonnet is scribed with graduations to serve as a position indicator for the stem. When the pawl is in the up position, it

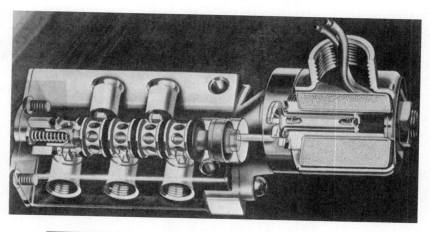

ANSI SYMBOL
2 POSITION-SINGLE SOLENOID

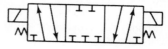

ANSI SYMBOL
3 POSITION-DOUBLE SOLENOID

Figure 13-38. Solenoid-actuated directional control valve. (*Courtesy of Skinner Precision Industries, Inc., New Britain, Connecticut.*)

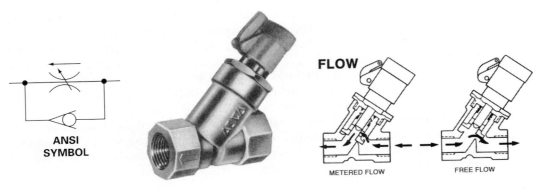

ANSI SYMBOL

FLOW

METERED FLOW   FREE FLOW

Figure 13-39. Flow control valve. *(Courtesy of Automatic Switch Co., Florham Park, New Jersey.)*

creates a friction lock on the knurled bonnet, and the knob cannot rotate. When the pawl is at 90° to the knob, the knob is free to rotate. Mounting in any position will not affect operation.

Sizing of Valves

As in the case of hydraulic systems, it is important that valves be properly sized in pneumatic systems. If valves are too small, excessive pressure drops will occur, leading to high operating cost and system malfunction. Similarly, oversized valves are undesirable due to high component costs and space requirements. Per Eq. (13-10), the flow capacity constant is a direct indication of the size of a valve. Selecting a valve from manufacturers' catalogs with a C_v value equal to or greater than that calculated from Eq. (13-10) will provide an adequately sized valve. Example 13-11 shows how this calculation is done for a particular application.

EXAMPLE 13-11

A pneumatically powered impact tool requires 50 scfm of air at 100 psig. What size valve (C_v) should be selected for this application if the valve pressure drop should not exceed 12 psi, and the upstream air temperature is 80°F?

Solution Convert the upstream temperature and downstream pressure into absolute units.

$$T_1 = 80 + 460 = 540°R$$

$$p_2 = 100 + 14.7 = 114.7 \text{ psia}$$

Next, solve Eq. (13-10) for C_v and substitute known values:

$$C_v = \frac{Q}{22.7}\sqrt{\frac{T_1}{(p_1 - p_2)(p_2)}} = \frac{50}{22.7}\sqrt{\frac{540}{12 \times 114.7}} = 1.38$$

Thus, any valve with a C_v of 1.38 or greater can be selected. If a C_v less than 1.38 is selected, excessive pressure drops will occur, leading to system malfunction. However, selecting a C_v that is much greater than 1.38 will result in a greatly oversized valve, which increases the costs of the pneumatic system.

13.9 PNEUMATIC ACTUATORS

Introduction

Pneumatic systems make use of actuators in a fashion similar to that of hydraulic systems. However, because air is the fluid medium rather than hydraulic oil, pressures are lower, and hence pneumatic actuators are of lighter construction. For example, air cylinders make extensive use of aluminum and other nonferrous alloys to reduce weight, improve heat transfer characteristics, and minimize the corrosive action of air.

Pneumatic Cylinders

Figure 13-40 illustrates the internal construction features of a typical double-acting pneumatic cylinder. The piston uses wear-compensating, pressure-energized U-cup seals to provide low-friction sealing and smooth chatter-free movement of this 200-psi, pressure-rated cylinder. The end plates use ribbed aluminum alloy to provide strength while minimizing weight. Self-aligning Buna-N seals provide a positive leakproof cushion with check valve action, which reverts to free flow on cylinder reversal. The cushion adjustment, which uses a tapered self-locking needle at each end, provides positive control over the stroke, which can be as large as 20 in.

ANSI SYMBOL

Figure 13-40. Construction of pneumatic cylinder. *(Courtesy of Aro Corp., Bryan, Ohio.)*

Figure 13-41 depicts a rotary index table driven by a double-acting pneumatic cylinder. The inlet pressure can be adjusted to provide exact force for moving the load and to prevent damage in case of accidental obstructions. A rack and gear drive transmits the straight-line motion of the air cylinder to the rotary motion with full power throughout its cycle. Through the use of different cams, the table can be indexed in 90°, 60°, 45°, 30°, or 15° increments.

Pneumatic Rotary Actuators

In Figure 13-42 we see a pneumatic rotary actuator, which is available in five basic models to provide a range of torque outputs from 100 to 10,000 in · lb using 100-psi air. Standard rotations are 94°, 184°, and 364°. The cylinder heads at each end serve as positive internal stops for the enclosed floating pistons. The linear motion of the piston is modified into rotary motion by a rack and pinion made of hardened steel for durability.

Figure 13-41. Air cylinder-drive rotary index table. *(Courtesy of Allenair Corp., Mineola, New York.)*

Figure 13-42. Pneumatic rotary actuator. *(Courtesy of Flo-Tork, Inc., Orrville, Ohio.)*

Rotary Air Motors

Rotary air motors can be used to provide a smooth source of power. They are not susceptible to overload damage and can be stalled for long periods without any heat problems. They can be started and stopped very quickly and with pressure regulation and metering of flow can provide infinitely variable torque and speed.

Figure 13-43 shows a vane air motor that contains four vanes and can deliver up to 1.7 hp using 100-psi air. The vanes are self-sealing since they take up their own wear, ensuring consistent output over the life of the motor. This self-sealing feature exists because compressed air, entering the motor, forces the vanes to slide outward in the radial slots of the eccentric-mounted rotor. In this way, the outer tips of the vanes maintain contact with the housing cam ring throughout the entire 3000-rpm speed range of the motor. Because of their cool running operation, these air motors can be used in ambient temperatures up to 250°F. Typical applications for this air motor include mixing equipment, conveyor drives, food packaging, hoists, tension devices, and turn tables.

Figure 13-44 gives the following performance curves in both English and metric units for the vane motor of Figure 13-43: output power vs. speed, torque vs. speed, and air consumption vs. speed. These performance curves, which are given at five different pressure levels, are determined by actual test data. Observe that output power increases with either speed or pressure. Note that torque also increases with pressure. However, the torque vs. speed curves exhibit a different characteristic. At a given pressure level, torque increases with speed from zero speed to about 250 rpm. From 250 rpm to the maximum speed of 3000 rpm, torque decreases with speed. Also observe that the starting torque (torque produced under load at zero speed) is lower than the running torque over most of the speed range. As a result, higher inlet pressure (as controlled by a pressure regulator) may be required to start driving a large load torque. Finally, note that air consumption increases with either speed or pressure. With 100-psi air and a maximum operating speed of 3000 rpm, the motor consumption rate equals 72 cfm of free air.

Figure 13-43.   Vane air motor. *(Courtesy of Gast Manufacturing, Inc., Benton Harbor, Michigan.)*

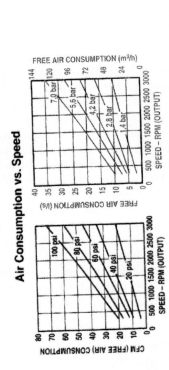

Note: Performance data represents a 4-vane model with no exhaust restriction.

Torque vs. Speed

Output Power vs. Speed

Air Consumption vs. Speed

Figure 13-44. Performance curves for the vane air motor of Figure 13-43. *(Courtesy of Gast Manufacturing, Inc., Benton Harbor, Michigan.)*

Figure 13-45 shows a radial piston air motor. The five-cylinder piston design provides even torque at all speeds due to overlap of the five power impulses occurring during each revolution of the motor. At least two pistons are on the power stroke at all times. The smooth overlapping power flow and accurate balancing make these motors vibrationless at all speeds. This smooth operation is especially noticeable at low speeds when the flywheel action is negligible. This air motor has relatively little exhaust noise, and this can be further reduced by use of an exhaust muffler. It is suitable for continuous operation using 100-psi air pressure and can deliver up to 15 HP.

In Figure 13-46 we see an axial piston air motor, which can deliver up to 3 HP using 100-psi air. The power pulses for these five-piston axial design motors are the same as those for the radial piston design. At least two pistons are on the power stroke at all times, providing even torque at all speeds.

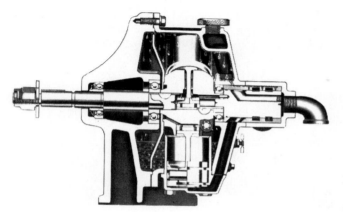

Figure 13-45. Radial piston air motor. *(Courtesy of Gardner-Denver Co., Quincy, Illinois.)*

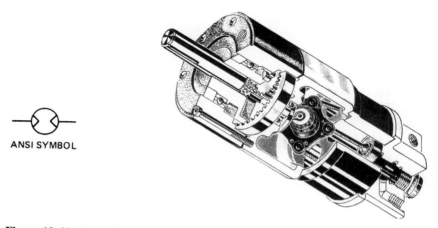

ANSI SYMBOL

Figure 13-46. Axial piston air motor. *(Courtesy of Gardner-Denver Co., Quincy, Illinois.)*

Air Requirements of Pneumatic Actuators

Pneumatic actuators are used to drive a variety of power tools for performing useful work. The air requirements of these tools in terms of flow rate and pressure depend on the application involved. Figure 13-47 gives the airflow requirements in scfm and standard m³/min for a number of average-size pneumatic tools designed to operate at a nominal pressure of 100 psig (687 kPa gage).

EXAMPLE 13-12

A single-acting pneumatic cylinder with a 1.75-in piston diameter and 6-in stroke drives a power tool using 100-psig air at 80°F. If the cylinder reciprocates at 30 cycles/min, determine the air-consumption rate in scfm (cfm of air at standard atmospheric conditions of 14.7 psia and 68°F).

Solution   The volume per minute (Q_2) of 100-psig, 80°F air consumed by the cylinder is found first.

$$Q_2(\text{ft}^3/\text{min}) = \text{displacement volume (ft)}^3 \times \text{reciprocation rate (cycles/min)}$$

$$= \text{piston area (ft)}^2 \times \text{piston stroke (ft)} \times \text{recip. rate (cycles/min)}$$

$$= \frac{p}{4}\left(\frac{1.75}{12}\right)^2 \times \frac{6}{12} \times 30 = 0.251 \text{ ft}^3/\text{min}$$

To obtain the volume per minute (Q_1) of air (scfm) consumed by the cylinder, we use Eq. (13-7):

$$Q_1 = Q_2\left(\frac{p_2}{p_1}\right)\left(\frac{T_1}{T_2}\right)$$

PNEUMATIC TOOL	scfm	STANDARD m³/min
HOISTS	5	0.14
PAINT SPRAYERS	10	0.28
IMPACT WRENCHES	10	0.28
HAMMERS	20	0.57
GRINDERS	30	0.85
SANDERS	40	1.13
ROTARY DRILLS	60	1.70
PISTON DRILLS	80	2.36

Figure 13-47.   Air requirements of various average-sized pneumatic tools designed for operations at 100 psig (687 kPa gage).

where $p_2 = 100 + 14.7 = 114.7$ psia,
$p_1 = p_{atm} = 14.7$ psia,
$T_2 = 80 + 460 = 540°R$,
$T_1 = 68 + 460 = 528°R$.

Substituting values yields

$$Q_1 = 0.251\left(\frac{114.7}{14.7}\right)\left(\frac{528}{540}\right) = 1.91 \text{ scfm}$$

If we ignore the temperature increase from standard atmospheric temperature (68°F) to the air temperature at the cylinder (80°F), the value of Q_1 becomes

$$Q_1 = 0.251\left(\frac{114.7}{14.7}\right)\left(\frac{528}{528}\right) = 1.96 \text{ scfm}$$

Thus, ignoring the increase in air temperature results in only a 2% error $\left(\frac{1.96 - 1.91}{1.91} \times 100\%\right)$. However, if the air temperature had increased to a value of 180°F, for instance, the percent error would equal 21%.

EXAMPLE 13-13

For the pneumatic cylinder-driven power tool of Exercise 13-12, at what rate can reciprocation take place? The following metric data apply:

a. Piston diameter = 44.5 mm
b. Piston stroke = 152 mm
c. Air pressure and temperature (at the pnuematic cylinder) = 687 kPa gage and 27°C
d. Available flow rate = 0.0555 standard m³/min (cfm of air at standard atmospheric conditions of 101 kPa abs and 20°C)

Solution   First, solve for Q_2, which equals the volume per minute of air at 687 kPa gage and 27°C consumed by the cylinder. Using Eq. (13-7) yields

$$Q_2 = Q_1\left(\frac{p_1}{p_2}\right)\left(\frac{T_2}{T_1}\right)$$

where $p_2 = 687 + 101 = 788$ kPa abs,
$p_1 = p_{atm} = 101$ kPa abs,
$T_2 = 27 + 273 = 300$ K,
$T_1 = 20 + 273 = 293$ K,
$Q_1 = 0.0555$ standard m³/min of air.

Substituting values yields an answer for Q_2:

$$Q_2 = 0.0555\left(\frac{101}{788}\right)\left(\frac{300}{293}\right) = 0.00728 \ \text{m}^3/\text{min}$$

Next, solve for the corresponding reciprocation rate:

$$Q_2(\text{m}^3/\text{min}) = \text{area}(\text{m}^2) \times \text{stroke}(\text{m}) \times \text{recip. rate (cycles/min)}$$

$$0.00728 = \frac{p}{4}(0.0445)^2 \times 0.152 \times \text{recip. rate (cycles/min)}$$

$$\text{recip. rate} = 30 \ \text{cycles/min}$$

13.10 KEY EQUATIONS

Absolute pressure

English units: absolute pressure (psia) = gage pressure (psig) + 14.7   **(13-1)**

Metric units: absolute pressure (Pa abs)

$\qquad\qquad\qquad\qquad$ = gage pressure (Pa gage) + 101,000   **(13-1M)**

Absolute temperature

English units: absolute temperature (°R) = temperature (°F) + 460   **(13-2)**

Metric units: absolute temperature (K) = temperature (°C) + 273   **(13-2M)**

General gas law: $\dfrac{p_1 V_1}{T_1} = \dfrac{p_2 V_2}{T_2}$ **(13-6)**

Air capacity ratings of compressors: $Q_1 = Q_2\left(\dfrac{p_2}{p_1}\right)\left(\dfrac{T_1}{T_2}\right)$ **(13-7)**

Air receiver size

English units: $V_r(\text{ft})^3 = \dfrac{14.7t(\text{min}) \times (Q_r - Q_c)\text{scfm}}{(p_{\text{max}} - p_{\text{min}})\text{psi}}$ **(13-8)**

Metric units:
$$V_r(m)^3 = \frac{101t(\text{min}) \times (Q_r - Q_c)\text{std m}^3/\text{min}}{(p_{\max} - p_{\min})\text{kPa}}$$
(13-8M)

Power required to drive a compressor

English units:
$$\text{theoretical horsepower (hp)} = \frac{p_{\text{in}}Q}{65.4}\left[\left(\frac{p_{\text{out}}}{p_{\text{in}}}\right)^{0.286} - 1\right]$$
(13-9)

Metric units:
$$\text{theoretical power (kW)} = \frac{p_{\text{in}}Q}{17.1}\left[\left(\frac{p_{\text{out}}}{p_{\text{in}}}\right)^{0.286} - 1\right]$$
(13-9M)

where p_{in} = inlet atmospheric pressure (psia, kPa abs),
p_{out} = outlet pressure (psia, kPa abs),
Q = flow rate (scfm, standard m³/min).

Flow rate of air through an orifice

English units:
$$Q(\text{scfm}) = 22.7\, C_v \sqrt{\frac{(p_1 - p_2)\text{psi} \times p_2(\text{psia})}{T_1(°R)}}$$
(13-10)

Metric units: $Q(\text{standard m}^3/\text{min})$
$$= 0.0698 C_v \sqrt{\frac{(p_1 - p_2)\text{kPa} \times p_2(\text{kPa abs})}{T_1(K)}}$$
(13-10M)

EXERCISES

Questions, Concepts, and Definitions

13-1. Name three reasons for considering the use of pneumatics instead of hydraulics.
13-2. What is the difference between free air and standard air?
13-3. State Boyle's, Charles', and Gay-Lussac's laws.
13-4. Name three types of air compressors.
13-5. What is a multistage compressor?
13-6. Describe the function of an air filter.
13-7. Describe the function of an air pressure regulator.
13-8. Why would a lubricator be used in a pneumatic system?
13-9. What is a pneumatic pressure indicator?
13-10. Why are exhaust silencers used in pneumatic systems?
13-11. What is the difference between an aftercooler and a chiller air dryer?
13-12. What is meant by the term *dew point*?

13-13. What is the relative humidity when the actual air temperature and the dew point temperature are equal?

13-14. How do pneumatic actuators differ from hydraulic actuators?

13-15. Name the steps required to size an air compressor.

13-16. Name four functions of an air receiver.

13-17. Relative to air motors, define the term *starting torque*.

13-18. Define the term *flow capacity constant* when dealing with valve flow rates.

Problems

Note: The letter *E* following an exercise means that English units are used. Similarly, the letter *M* indicates metric units.

The Perfect Gas Laws

13-19E. The 2-in-diameter piston of the pneumatic cylinder of Figure 13-6 retracts 5 in from its present position ($p_1 = 30$ psig, $V_1 = 20$ in^3) due to the external load on the rod. If the port at the blank end of the cylinder is blocked, find the new pressure, assuming the temperature does not change.

13-20E. The cylinder of Figure 13-6 has an initial position in which $p_1 = 30$ psig and $V_1 = 20$ in^3, as controlled by the load on the rod. The air temperature is 80°F. The load on the rod is held constant to maintain constant air pressure, but the air temperature is increased to 150°F. Find the new volume of air at the blank end of the cylinder.

13-21E. The cylinder of Figure 13-6 has a locked position ($V_1 = $ constant), $p_1 = 30$ psig, and $T_1 = 80$°F. If the temperature increases to 160°F, what is the new pressure in the blank end?

13-22E. Gas at 1200 psig and 120°F is contained in the 2000-in^3 cylinder of Figure 13-9. A piston compresses the volume to 1500 in^3 while the gas is heated to 250°F. What is the final pressure in the cylinder?

13-23M. The 50-mm-diameter piston of the pneumatic cylinder of Figure 13-6 retracts 130 mm from its present position ($p_1 = 2$ bars gage, $V_1 = 130$ cm^3) due to the external load on the rod. If the port at the blank end of the cylinders is blocked, find the new pressure, assuming the temperature does not change.

13-24M. The cylinder of Figure 13-6 has an initial position where $p_1 = 2$ bars gage and $V_1 = 130$ cm^3, as controlled by the load on the rod. The air temperature is 30°C. The load on the rod is held constant to maintain constant air pressure, but the air temperature is increased to 65°C. Find the new volume of air at the blank end of the cylinder.

13-25M. The cylinder of Figure 13-6 has a locked position ($V_1 = $ constant), $p_1 = 2$ bars gage, and $T_1 = 25$°C. If the temperature increases to 70°C, what is the new pressure in the blank end?

13-26M. Gas at 80 bars gage and 50°C is contained in the 1290-cm^3 cylinder of Figure 13-9. A piston compresses the volume to 1000 cm^3 while the gas is heated to 120°C. What is the final pressure in the cylinder?

13-27. Convert a temperature of 160°F to degrees Celsius, Rankine, and Kelvin.

Compressor Flow and Receiver Size

13-28E. Air is used at a rate of 30 cfm from a receiver at 100°F and 150 psi. If the atmospheric pressure is 14.7 psia and the atmospheric temperature is 80°F, how many cfm of free air must the compressor provide?

13-29E. **a.** Calculate the required size of a receiver that must supply air to a pneumatic system consuming 30 scfm for 10 min between 120 psi and 100 psi before the compressor resumes operation.

b. What size is required if the compressor is running and delivering air at 6 scfm?

13-30M. Air is used at a rate of 1.0 m³/min from a receiver at 40°C and 1000 kPa gage. If the atmospheric pressure is 101 kPa abs and the atmospheric temperature is 20°C, how many m³/min of free air (standard m³/min in this case) must the compressor provide?

13-31M. Change the data of Exercise 13-29 to metric units and solve parts a and b.

Compressor Power Requirements

13-32E. Determine the actual power required to drive a compressor that delivers air at 200 scfm at 120 psig. The overall efficiency of the compressor is 72%.

13-33M. Determine the output pressure of a compressor operating with the following data:

a. Actual power required to drive the compressor is 20 kW.

b. Overall efficiency of the compressor is 75%.

c. Compressor delivers four standard m³/min.

d. Compressor inlet pressure is 100 kPa abs.

Moisture Removal from Air

13-34E. A compressor delivers 50 scfm of air at 125 psig to a pneumatic system. Saturated atmospheric air enters the compressor at 80°F. If the compressor operates 16 hr per day

a. Determine the number of gallons of moisture delivered by the compressor to the pneumatic system per day

b. How much moisture per day would be received by the pneumatic system if an aftercooler were installed to cool the compressed air temperature back to 80°F

c. How much moisture per day would be received by the pneumatic system if an air dryer were installed to cool the compressed air temperature to 40°F

13-35E. What compressor discharge pressure is required for 85% of the moisture to be removed by an aftercooler? The compressor receives saturated atmospheric air at 90°F and the aftercooler returns the compressed air temperature back to 90°F.

Air Control Valve Flow Rate

13-36E. Air at 100°F passes through a $\frac{1}{2}$-in-diameter orifice having a flow capacity constant of 7. If the upstream pressure is 125 psi, what is the maximum flow rate in units of scfm of air?

13-37M. A flow control valve in a pneumatic system is providing a flow rate of 10 standard m³/min of air. The valve upstream pressure is 400 kPa abs and the downstream pressure is 180 kPa abs. Is the valve choked?

Pneumatic Actuator Flow Rate Requirements

13-38E. A single-acting air cylinder with a $2\frac{1}{2}$-in-diameter piston and 12-in stroke operates at 100 psig and reciprocates at 30 cycles per min. Compute the air consumption in scfm of air.

13-39M. A single-acting air cylinder with a 6-cm-diameter piston and 30-cm stroke operates at 700 kPa gage pressure and reciprocates at 30 cycles per min. Compute the air consumption in standard m³/min.

13-40M. A double-acting air cylinder has a 50-mm-diameter piston and a 2.5-cm stroke, operates at 600 kPa gage pressure, and reciprocates at 80 cycles/min. Determine the time it takes to consume 100 m³ of standard atmospheric air.

13-41E. A double-acting air cylinder with a 2-in-diameter piston and 12-in stroke reciprocates at 200 cycles/min using 100-psig air. What is the scfm flow rate of air to the cylinder? Ignore the piston rod cross-sectional area and assume the temperature remains constant.

13-42E. A rotary vane air motor has a displacement volume of 4 in³/rev and operates at 1750 rpm using 100-psig air. Calculate the scfm rate of consumption and HP output of the motor. Assume the temperature remains constant.

13-43M. A rotary vane air motor has a displacement volume of 80 cm³/rev and operates at 1750 rpm using 700 kPa gage pressure air. Calculate the standard m³/min rate of consumption and kW power output of the motor. Assume the temperature remains constant.

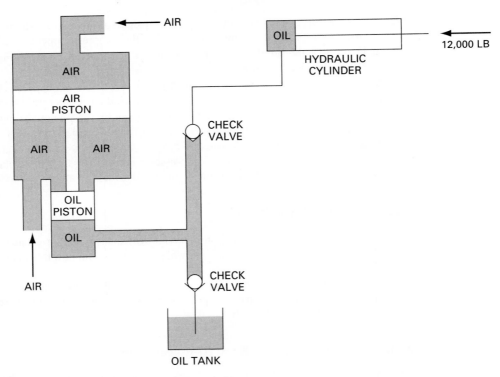

Figure 13-48. System for Exercise 13-44.

Air-Hydraulic Intensifier

13-44E. An air-hydraulic intensifier is connected to a hydraulic cylinder driving a 12,000-lb load, as shown in Figure 13-48. The diameter of the hydraulic cylinder piston is 1.5 in. The following data applies to the intensifier:

air piston diameter = 8 in
oil piston diameter = 1 in
intensifier stroke = 2 in
intensifier cycle frequency = 1 stroke/s

Determine the

    a. Volume displacement of the intensifier oil piston
    b. Volume displacement of the hydraulic cylinder piston per intensifier stroke
    c. Movement of the hydraulic cylinder piston per intensifier stroke
    d. Volume displacement of the blank end of the intensifier air cylinder piston
    e. Flow rate of oil from the intensifier
    f. Air-consumption rate of intensifier in scfm

13-45M. Change the data of the air-hydraulic intensifier system of Exercise 13-44 to metric units and solve parts a–f.

 **Pneumatics: Circuits and Applications**

Learning Objectives

Upon completing this chapter, you should be able to:

1. Explain the important considerations that must be taken into account when analyzing or designing a pneumatic circuit.
2. Determine pressure losses in pipelines of pneumatic circuits.
3. Evaluate the economic costs of energy losses due to friction in pneumatic systems.
4. Determine the economic cost of air leakage in pneumatic systems.
5. Troubleshoot pneumatic circuits for determining causes of system malfunction.
6. Read pneumatic circuit diagrams and describe the corresponding system operation.
7. Analyze the operation of pneumatic vacuum systems.
8. Size gas-loaded accumulators.

14.1 INTRODUCTION

The material presented in Chapter 13 discussed the fundamentals of pneumatics in regard to air preparation and component operation. This chapter discusses pneumatic circuits and applications. A pneumatic circuit consists of a variety of components, such as compressors, receivers, filters, pressure regulators, lubricators, mufflers, air dryers, actuators, control valves, and conductors, arranged so that a useful task can be performed.

In a pneumatic circuit the force delivered by a cylinder and the torque delivered by a motor are determined by the pressure levels established by pressure regulators placed at the desired locations in the circuit. Similarly, the linear speed of a pneumatic cylinder and the rotational speed of an air motor are determined by flow control valves placed at desired locations in the circuit. The direction of flow in various flow paths is established by the proper location of directional control valves. After the pressurized air is spent driving actuators, it is then exhausted back into the atmosphere.

Figure 14-1 shows a riveting assembly machine, which performs continuous, high-speed, repetitive production of riveted components. The control system contains many pneumatic components such as regulators, filters, lubricators, solenoid valves, and cylinders. These machines were designed to operate under tough production-line conditions with a minimum of downtime for maintenance and adjustment.

Figure 14-2 shows an application of pneumatics to solve a nail polish brush assembly problem in industry. The problem and its solution are described as follows:

A cosmetic manufacturer had been manually inserting a brush into nail polish bottles. The process required a worker to pick up each brush and insert it into the open nail polish bottle. The bottles were transported to and from the assembly area via a

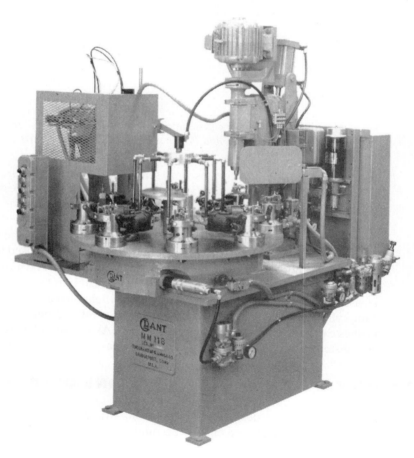

Figure 14-1. Pneumatically controlled riveting assembly machine. *(Courtesy of C. A. Norgren Co., Littleton, Colorado.)*

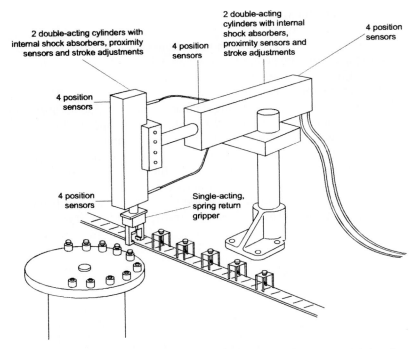

Figure 14-2.   Pneumatic system for inserting a brush into nail polish bottles.
(Courtesy of National Fluid Power Association, Milwaukee, Wisconsin.)

conveyor system. The bottle caps were seated and twisted onto the bottles at a later assembly station. In addition to being labor intensive, the process required a high degree of individual accuracy in order to fit the brush tips into the small hole in the nail polish bottle. This greatly limited production line speeds.

It was decided to automate this brush assembly process, as shown in Figure 14-2. The operation is as follows: A pneumatically actuated gripper is used to grasp the brush. Pneumatic cylinders then raise the gripper holding the brush and move it over the open nail polish bottle. The cylinders then lower the brush into the bottle and the gripper releases the brush. The caps are again put on and tightened at a later station. The pneumatic cylinders employ adjustable internal stops so that different sizes of bottles and brushes can be used on the same assembly line. As shown in Figure 14-2, an indexing table instead of a conveyor system is used to transport the bottle to the next station.

14.2 PNEUMATIC CIRCUIT DESIGN CONSIDERATIONS

When analyzing or designing a pneumatic circuit, the following four important considerations must be taken into account:

1. Safety of operation
2. Performance of desired function

3. Efficiency of operation
4. Costs

Safety of operation means that an operator must be protected by the use of built-in emergency stop features as well as safety interlock provisions that prevent unsafe, improper operation. Although compressed air is often quiet, it can cause sudden movements of machine components. These movements could injure a technician who, while troubleshooting a circuit, inadvertently opens a flow control valve that controls the movement of the actuator.

Performance of the desired function must be accomplished on a repeatable basis. Thus, the system must be relatively insensitive to adverse conditions such as high ambient temperatures, humidity, and dust. Shutting down a pneumatic system due to failure or misoperation can result in the stoppage of a production line. Stoppage can result in very large costs, especially if the downtime is long because of difficulty in repairing the pneumatic system involved.

Efficiency of operation and costs are related design parameters. A low-efficiency compressor requires more electrical power to operate, which increases the system operating costs. Although atmospheric air is "free," compressed air is not. Yet if a pneumatic system leaks air into the atmosphere without making significant noise, it is often ignored, because the air is clean. On the other hand, a hydraulic leak would be fixed immediately, because it is messy and represents a safety hazard to personnel in the vicinity of the leak.

Pneumatic circuit air losses through various leakage areas with a combined area of a 0.25-in-diameter hole would equal about 70 scfm for an operating pressure of 100 psig. Examples of such leakage areas include the imperfect sealing surfaces of improperly installed pipe fittings. A typical cost of compressing air to 100 psig is about $0.35 per 1000 ft^3 of standard air. Therefore, it costs about $0.35 to compress 1000 ft^3 of air from 14.7 psig to 100 psig. Thus, the yearly cost of such a leaking pneumatic system operating without any downtime is

$$\text{yearly cost} = \frac{\$0.35}{1000 \text{ ft}^3} \times 70 \frac{\text{ft}^3}{\text{min}} \times \frac{60 \text{ min}}{1 \text{ hr}} \times \frac{24 \text{ hr}}{1 \text{ day}} \times \frac{365 \text{ days}}{1 \text{ yr}}$$

$$= \$12,900/\text{yr}$$

Another cause of increased operating costs is significantly undersized components such as pipes and valves. Such components cause excessive pressure losses due to friction. As a result the compressor must operate at much higher output pressure, which requires greater input power. Of course, greatly oversized components result in excessive initial installation costs along with improved operating efficiencies. Thus, a compromise must be made between higher initial costs with lower operating energy costs and lower initial costs with higher operating energy costs based on the expected life of the pneumatic system.

14.3 AIR PRESSURE LOSSES IN PIPELINES

As in the case for liquids, when air flows through a pipe, it loses energy due to friction. The energy loss shows up as a pressure loss, which can be calculated using the Harris formula:

$$p_f = \frac{cLQ^2}{3600\,(\text{CR}) \times d^5} \tag{14-1}$$

where p_f = pressure loss (psi),
c = experimentally determined coefficient,
L = length of pipe (ft),
Q = flow rate (scfm),
CR = compression ratio = pressure in pipe/atmospheric pressure,
d = inside diameter of pipe (in).

For schedule 40 commercial pipe, the experimentally determined coefficient can be represented as a function of the pipe inside diameter:

$$c = \frac{0.1025}{d^{0.31}} \tag{14-2}$$

Substituting Eq. (14-2) into the Harris formula yields a single usable equation for calculating pressure drops in air pipelines:

$$p_f = \frac{0.1025LQ^2}{3600\,(\text{CR}) \times d^{5.31}} \tag{14-3}$$

Tabulated values of d and $d^{5.31}$ are given in Figure 14-3 for schedule 40 common pipe sizes.

Nominal Pipe Size (in)	Inside Diameter d (in)	$d^{5.31}$	Nominal Pipe Size (in)	Inside Diameter d (in)	$d^{5.31}$
$\frac{3}{8}$	0.493	0.0234	$1\frac{1}{2}$	1.610	12.538
$\frac{1}{2}$	0.622	0.0804	2	2.067	47.256
$\frac{3}{4}$	0.824	0.3577	$2\frac{1}{2}$	2.469	121.419
1	1.049	1.2892	3	3.068	384.771
$1\frac{1}{4}$	1.380	5.5304	$3\frac{1}{2}$	3.548	832.550

Figure 14-3.   Tabulated values of d and $d^{5.31}$ for schedule 40 common pipe sizes.

EXAMPLE 14-1

A compressor delivers 100 scfm of air through a 1-in schedule 40 pipe at a receiver pressure of 150 psi. Find the pressure loss for a 250-ft length of pipe.

Solution   First, solve for the compression ratio:

$$CR = \frac{150 + 14.7}{14.7} = 11.2$$

Next, find the value of $d^{5.31}$ from Figure 14-3:

$$d^{5.31} = 1.2892$$

Finally, using the Harris formula, the pressure loss is found:

$$p_f = \frac{0.1025 \times 250 \times (100)^2}{3600 \times 11.2 \times 1.2892} = 4.93 \text{ psi}$$

If a 3/4-in pipe is used instead of the 1-in size, the pressure loss equals 17.8 psi. This value represents a 260% increase in pressure loss, which points out the need to size pneumatic pipes adequately.

EXAMPLE 14-2

A compressor delivers 150 scfm of air through a pipe at a receiver pressure of 120 psig. What minimum size schedule 40 pipe should be used if the pressure loss is to be limited to 0.05 psi per foot of pipe length?

Solution   First, solve for the compression ratio:

$$CR = \frac{120 + 14.7}{14.7} = 9.16$$

Next, solve for $d^{5.31}$ from Eq. 14-3:

$$d^{5.31} = \frac{0.0125 \, Q^2}{3600 \times (CR) \times P_f/L} = \frac{0.1025 \times 150^2}{3600 \times 9.16 \times 0.05} = 1.398$$

Thus, the minimum required inside diameter can now be found:

$$d = (1.398)^{1/5.31} = (1.398)^{0.188} = 1.065 \text{ in}$$

From Figure 14-3 the minimum size pipe that can be used is 1-1/4 in schedule 40 which has a 1.380 inside diameter. The next smaller size is a 1 in schedule 40 which has only a 1.049-in inside diameter.

FITTING	NOMINAL PIPE SIZE (in)						
	$\frac{3}{8}$	$\frac{1}{2}$	$\frac{3}{4}$	1	$1\frac{1}{4}$	$1\frac{1}{2}$	2
GATE VALVE (FULLY OPEN)	0.30	0.35	0.44	0.56	0.74	0.86	1.10
GLOBE VALVE (FULLY OPEN)	14.0	18.6	23.1	29.4	38.6	45.2	58.0
TEE (THROUGH RUN)	0.50	0.70	1.10	1.50	1.80	2.20	3.30
TEE (THROUGH BRANCH)	2.50	3.30	4.20	5.30	7.00	8.10	10.44
90° ELBOW	1.40	1.70	2.10	2.60	3.50	4.10	5.20
45° ELBOW	0.50	0.78	0.97	1.23	1.60	1.90	2.40

Figure 14-4. Equivalent length of various fittings (ft).

The frictional losses in pneumatic fittings can be computed using the Harris formula if the equivalent lengths of the fittings are known. The term L in the Harris formula would then represent the total equivalent length of the pipeline including its fittings. Figure 14-4 gives equivalent length values in feet for various types of fittings.

EXAMPLE 14-3

If the pipe in Example 14-1 has two gate valves, three globe valves, five tees (through run), four 90° elbows, and six 45° elbows, find the total pressure loss.

Solution   The total equivalent length of the pipe is

$$L = 250 + 2(0.56) + 3(29.4) + 5(1.50) + 4(2.60) + 6(1.23)$$

$$= 250 + 1.12 + 88.2 + 7.5 + 10.4 + 7.38 = 364.6 \text{ ft}$$

Substituting into the Harris formula yields the answer:

$$p_f = \frac{0.1025 \times 364.6 \times (100)^2}{3600 \times 11.2 \times 1.2892} = 7.19 \text{ psi}$$

Thus, the total pressure loss in the pipes, valves, and fittings is 46% greater than the pressure loss in only the pipes. This shows that valves and fittings must also be adequately sized to avoid excessive pressure losses.

14.4 ECONOMIC COST OF ENERGY LOSSES IN PNEUMATIC SYSTEMS

As mentioned in Section 14.2, although atmospheric air is free, compressed air is not. Not all the energy required to drive a compressor is actually received by pneumatic actuators to drive their respective loads. In between the compressor prime mover and the pneumatic actuators are components in which frictional energy losses occur. Because these energy losses are due to friction, they are lost forever in the form of heat. In addition to the cost of the wasted energy, there is, of course, frictional wear of mating sliding components (such as between the piston and wall of a pneumatic cylinder), which contributes to maintenance costs as well as downtime costs. In this section we determine the dollar costs of energy losses due to friction and air leakage in pneumatic systems. Examples 14-4 and 14-5 show how this cost analysis is accomplished.

EXAMPLE 14-4

A compressor delivers air at 100 psig and 270 scfm.

a. Determine the actual hp required to drive the compressor if the overall efficiency of the compressor is 75%.

b. Repeat part a assuming the compressor is required to provide air at 115 psig to offset a 15-psi pressure loss in the pipeline due to friction.

c. Calculate the cost of electricity per year for parts a and b. Assume the efficiency of the electric motor driving the compressor is 92% and that the compressor operates 3000 hr per year. The cost of electricity is $0.11/kWh.

Solution

a. Using Eq. (13-9) and dividing by η_o, the actual horsepower (hp) required to drive the compressor at 100 psig is

$$\text{actual hp (at 100 psig)} = \frac{p_{in}Q}{65.4\eta_o}\left[\left(\frac{p_{out}}{p_{in}}\right)^{0.286} - 1\right]$$

$$= \frac{14.7 \times 270}{65.4 \times 0.75}\left[\left(\frac{114.7}{14.7}\right)^{0.286} - 1\right]$$

$$= 64.7 \text{ hp}$$

b. The actual hp required to drive the compressor at 115 psig is

$$\text{actual hp (at 115 psig)} = \frac{14.7 \times 270}{65.4 \times 0.75}\left[\left(\frac{129.7}{14.7}\right)^{0.286} - 1\right]$$

$$= 69.9 \text{ hp}$$

c. Since 0.746 kW = 1 hp, the actual power required to drive the compressor at 100 psig equals

$$64.7 \text{ hp} \times \frac{0.746 \text{ kW}}{1 \text{ hp}} = 48.3 \text{ kW}$$

Thus, the electric power required to drive the electric motor at 100-psig air delivery is

$$\frac{48.3 \text{ kW}}{0.92} = 52.5 \text{ kW}$$

The cost of electricity per year at a pressure of 100 psig is now found as follows:

$$\text{yearly cost} = \text{power rate} \times \text{time per year} \times \text{unit cost of elec.}$$

$$= 52.5 \text{ kW} \times 3000 \text{ hr/yr} \times \$0.11/\text{kWh} = \$17,300/\text{yr}$$

The electric power required to drive the electric motor at 115-psig air delivery is

$$\frac{69.9}{0.92} \times 0.746 = 56.7 \text{ kW}$$

The cost of electricity per year at a pressure of 115 psig is now found as follows:

$$\text{yearly cost} = 56.7 \text{ kW} \times 3000 \text{ hr/yr} \times \$0.11/\text{kWh} = \$18,700/\text{yr}$$

Hence, the cost of the compressor having to provide the additional 15 psi to offset the pressure loss in the pipeline is $1400/yr.

EXAMPLE 14-5

The compressor in Example 14-4 delivers air at 100 psi. If the compressor is required to provide an additional 70 scfm to compensate for air leakage from the pneumatic circuit into the atmosphere, what is the yearly cost of the leakage?

Solution From Example 14-4 the actual power required to drive the compressor at 100 psig and 270 scfm is 64.7 hp. The corresponding electric power required to drive the electric motor is 52.5 kW. Thus, the additional electric power required to drive the electric motor of the compressor to compensate for the air leakage of 70 scfm is

$$\frac{70}{270} \times 52.3 \text{ kW} = 13.6 \text{ kW}$$

As a result, if the pneumatic system operates with no downtime, the yearly cost of the 70-scfm leakage is

$$\text{yearly cost} = \text{power rate} \times \text{time per year} \times \text{unit cost of elec.}$$
$$= 13.6\,\text{kW} \times 24\,\text{hr/day} \times 365\,\text{days/yr} \times \$0.11/\text{kWh}$$
$$= \$13,100/\text{yr}$$

As stated in Section 14.2, 70 scfm is the amount of 100-psi air that would pass through the area of a 0.25-in-diameter hole. Thus, if all the various leakage areas of a pneumatic circuit (such as those existing in imperfectly sealed pipe fittings) add up to this hole area, the total leakage into the atmosphere would equal 70 scfm. The leakage area of a 0.25-in-diameter hole can easily occur even if only a few pipe fittings experience small leaks. The total cost of the air leakage plus the 15-psi pressure loss in the pipeline (from Example 14-4) equals the very significant amount of \$14,500/yr.

14.5 BASIC PNEUMATIC CIRCUITS

Introduction

In this section we present a number of basic pneumatic circuits using pneumatic components that have been previously discussed. Pneumatic circuits are similar to their hydraulic counterparts. One difference is that no return lines are used in pneumatic circuits because the exhausted air is released directly into the atmosphere. This is depicted by a short dashed line leading from the exhaust port of each valve. Also, no input device (such as a pump in a hydraulic circuit) is shown, because most pneumatic circuits use a centralized compressor as their source of energy. The input to the circuit is located at some conveniently located manifold, which leads directly into the filter-regulator-lubricator (FRL) unit.

Operation of Single-Acting Cylinder

Figure 14-5 shows a simple pneumatic circuit, which consists of a three-way valve controlling a single-acting cylinder. The return stroke is accomplished by a compression spring located at the rod end of the cylinder. When the push-button valve

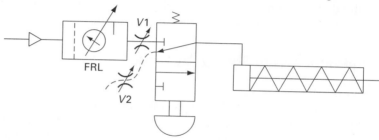

Figure 14-5. Operation of a single-acting cylinder.

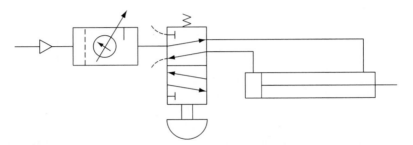

Figure 14-6. Operation of a double-acting cylinder.

is actuated, the cylinder extends. It retracts when the valve is deactivated. Needle valves $V1$ and $V2$ permit speed control of the cylinder extension and retraction strokes, respectively.

Operation of Double-Acting Cylinder

In Figure 14-6 we see the directional control of a double-acting cylinder using a four-way valve. Note that control of a double-acting cylinder requires a DCV with four different functioning ports (each of the two exhaust ports perform the same function). Thus, a four-way valve has four different functioning ports. In contrast, the control of a single-acting, spring-return cylinder requires a DCV with only three ports. Hence a three-way valve has only three ports, as shown in Figure 14-5. Actuation of the push-button valve extends the cylinder. The spring-offset mode causes the cylinder to retract under air power.

Air Pilot Control of Double-Acting Cylinder

In Figure 14-7 we see a circuit in which a double-acting cylinder can be remotely operated through the use of an air-pilot-actuated DCV. Push-button valves $V1$ and $V2$ are used to direct airflow (at low pressure such as 10 psi) to actuate the air-piloted DCV, which directs air at high pressure such as 100 psi to the cylinder. Thus, operating personnel can use low-pressure push-button valves to remotely control the operation of a cylinder that requires high-pressure air for performing its intended function. When $V1$ is actuated and $V2$ is in its spring-offset mode, the cylinder extends. Deactivating $V1$ and then actuating $V2$ retracts the cylinder.

Cylinder Cycle Timing System

Figure 14-8 shows a circuit that employs a limit valve to provide a timed cylinder extend and retract cycle. When push-button valve $V3$ is momentarily actuated, valve $V2$ shifts to extend the cylinder. When the piston rod cam actuates limit valve $V4$, it shifts $V2$ into its opposite mode to retract the cylinder. Flow control valve $V1$ controls the flow rate and thus the cylinder speed.

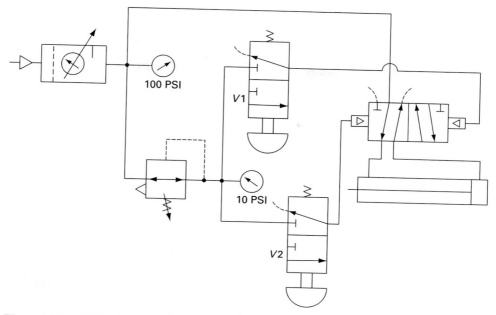

Figure 14-7. Air pilot control of a double-acting cylinder. *(This circuit is simulated on the CD included with this textbook.)*

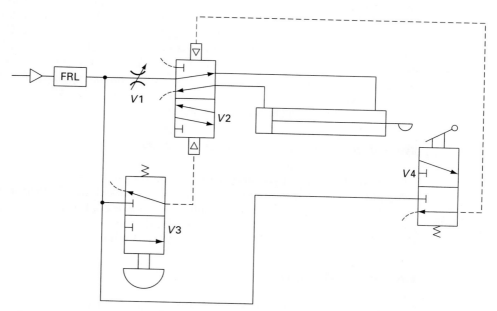

Figure 14-8. Cylinder cycle timing system.

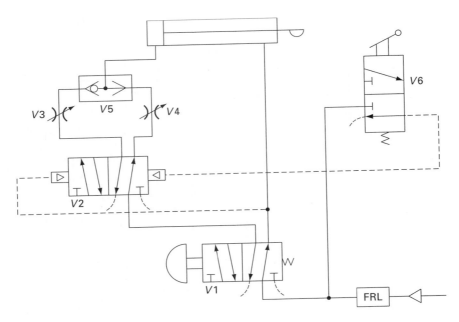

Figure 14-9. Two-step speed control system.

Two-Step Speed Control System

A two-step speed control system is shown in Figure 14-9. The operation is as follows, assuming that flow control valve $V3$ is adjusted to allow a greater flow rate than valve $V4$. Initially, the cylinder is fully retracted. When push-button valve $V1$ is actuated, airflow goes through valves $V2$, $V3$, and the shuttle valve $V5$ to extend the cylinder at high speed. When the piston rod cam actuates valve $V6$, valve $V2$ shifts. The flow is therefore diverted to valve $V4$ and through the shuttle valve. However, due to the low flow setting of valve $V4$, the extension speed of the cylinder is reduced. After the cylinder has fully extended, valve $V1$ is released by the operator to cause retraction of the cylinder.

Two-Handed Safety Control System

Figure 14-10 shows a two-handed safety control circuit. Both palm-button valves ($V1$ and $V2$) must be actuated to cause the cylinder to extend. Retraction of the cylinder will not occur unless both palm buttons are released. If both palm-button valves are not operated together, the pilot air to the three-position valve is vented. Hence, this three-way valve goes into its spring-centered mode, and the cylinder is locked.

Control of Air Motor

In Figure 14-11 we see a circuit used to control an air motor. The operation is as follows. When the START push-button valve is actuated momentarily, the air pilot valve shifts to supply air to the motor. When the STOP push-button valve is actuated momentarily, the air pilot valve shifts into its opposite mode to shut off the supply of air to the motor. The flow control valve is used to adjust the speed of the motor.

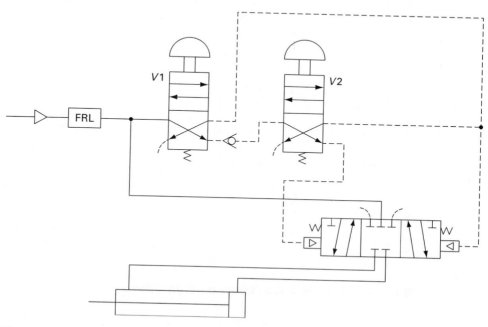

Figure 14-10. Two-handed safety control circuit.

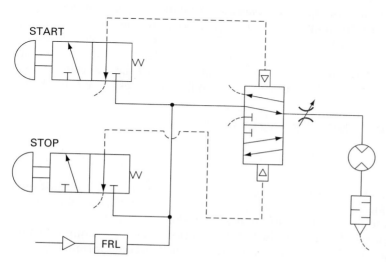

Figure 14-11. Control of an air motor. *(This circuit is simulated on the CD included with this textbook.)*

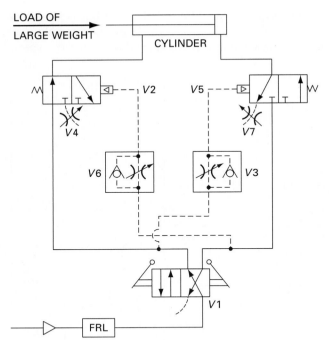

Figure 14-12. Deceleration air cushion of a pneumatic cylinder.

Deceleration Air Cushion of Cylinder

Figure 14-12 shows a circuit that provides an adjustable deceleration air cushion at both ends of the stroke of a cylinder when it drives a load of large weight. The operation is as follows:

Valve $V1$ supplies air to the rod end of the cylinder and to the pilot of valve $V5$ through flow control valve $V3$. Free air exhausting from the blank end of the cylinder permits a fast cylinder-retraction stroke until valve $V5$ operates due to increased pressure at its pilot. When valve $V5$ is actuated, the cylinder blank end exhaust is restricted by valve $V7$. The resulting pressure buildup in the blank end of the cylinder acts as an air cushion to gradually slow down the large weight load. For the extension stroke, valves $V2$, $V4$, and $V6$ behave in a fashion similar to that of valves $V5$, $V7$, and $V3$.

14.6 PNEUMATIC VACUUM SYSTEMS

Introduction

When we think of the force caused by a fluid pressure acting on the surface area of an object, we typically envision the pressure to be greater than atmospheric pressure. However, there are a number of applications where a vacuum air pressure is used to perform a useful function. Industrial applications where a vacuum pressure is used include materials handling, clamping, sealing, and vacuum forming.

In terms of materials-handling applications, a pneumatic vacuum can be used to lift smoothly objects that have a flat surface and are not more than several hundred pounds in weight. Examples of such objects include glass plates, sheet metal, sheets of paper, and floor-covering materials, such as ceramic tile and sheets of linoleum. The weight limitation is due to the fact that the maximum suction pressure equals 1 atm of pressure in magnitude.

Materials-Handling Application

Figure 14-13 shows a materials-handling application where a vacuum cup (sometimes called a "suction cup") is used to establish the force capability to lift a flat

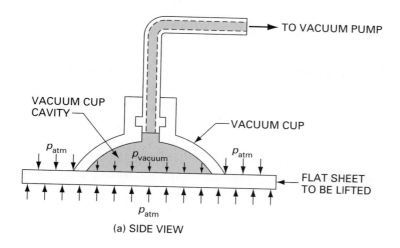

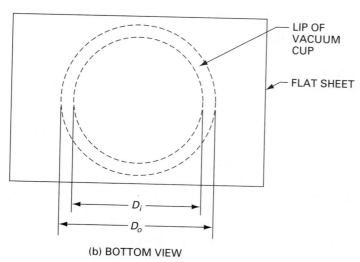

Figure 14-13. Vacuum cup used to lift a flat sheet.

sheet. The cup is typically made of a flexible material such as rubber so that a seal can be made where its lip contacts the surface of the flat sheet.

A vacuum pump (not shown) is turned on to remove air from the cavity between the inside of the cup and top surface of the flat sheet. As the pressure in the cavity falls below atmospheric pressure, the atmospheric pressure acting on the bottom of the flat sheet pushes the flat sheet up against the lip of the cup. This action results in vacuum pressure in the cavity between the cup and the flat sheet that causes an upward force to be exerted on the flat sheet.

Analysis of Suction Lift Force

The magnitude of this force can be determined by algebraically summing the pressure forces on the top and bottom surfaces of the flat sheet as follows:

$$F = p_{atm}A_o - p_{suction}A_i \tag{14-4}$$

where F = the upward force the suction cup exerts on the flat sheet (lb, N),

p_{atm} = the atmospheric pressure in absolute units (psia, Pa abs),

A_o = the area of the outer circle of the suction cup lip

$$= \frac{\pi}{4}D_o^2 (in^2, m^2),$$

D_o = the diameter of the suction cup lip outer circle (in, m),

$p_{suction}$ = the suction pressure inside the cup cavity in absolute units (psia, Pa abs),

A_i = the area of the inner circle of the suction cup lip

$$= \frac{\pi}{4}D_i^2 (in^2, m^2),$$

D_i = the diameter of the suction cup inner lip circle (in, m).

Note in Figure 14-13(a) that the atmospheric pressure on the top and bottom surfaces of the flat sheet cancel out away from the outer circle area of the cup lip. If all the air were removed from the cup cavity, we would have a perfect vacuum, and thus the suction pressure would be equal to zero in absolute pressure units. Thus, for a perfect vacuum, Eq. (14-4) becomes

$$F = p_{atm}A_o \tag{14-5}$$

When a perfect vacuum exists, the maximum lift force is produced, as can be seen by comparing Eqs. (14-4) and (14-5). Examination of Eq. (14-5) shows that the lift force is limited by the value of atmospheric pressure and the size of the suction cup (area of the outer circle of the cup lip). The exact amount of suction pressure developed cannot be guaranteed, and the resulting suction force must be at least as large as the weight of the object to be lifted. Thus, a factor of safety is applied with a value of between 2 and 4, depending on the application. When large, flat sheets are to be lifted, four to eight suction cups are used. In this way the sheet can be lifted uniformly. In addition, the load-lifting capacity is multiplied by the number of suction cups used.

Figure 14-14 is a photograph that shows a pneumatic vacuum system that uses vacuum cups to lift and transport large sheet metal panels in a factory. In this application the vacuum cups are attached to a frame that is positioned over the sheet metal panels by an overhead crane. The frame is raised and lowered via a pulley-cable system. A variety of actual, different-sized vacuum cups is displayed in the photograph of Figure 14-15.

EXAMPLE 14-6

How heavy an object can be lifted with a suction cup having a 6-in lip outside diameter and a 5-in lip inside diameter for each suction pressure?

a. -10 psig = 10-psi vacuum = 10-psi suction
b. Zero absolute (a perfect vacuum)

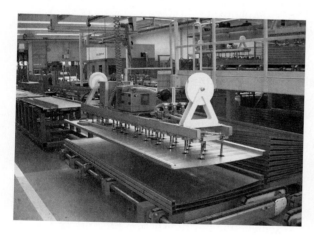

Figure 14-14. Factory application of the lifting and transporting of large sheet metal panels using pneumatic vacuum system technology. *(Courtesy of Schmalz, Inc., Raleigh, North Carolina.)*

Figure 14-15. A variety of actual, different-sized vacuum cups. *(Courtesy of Schmalz, Inc., Raleigh, North Carolina.)*

Solution

a. The suction pressure (which must be in absolute units) equals

$$p_{suction} (\text{abs}) = p_{suction} (\text{gage}) + p_{atm}$$

$$= -10 + 14.7 = 4.7 \text{ psia}$$

The maximum weight that can be lifted can now be found using Eq. (14-4):

$$F = \text{maximum weight } W = p_{atm} A_o - p_{suction} A_i$$

$$= 14.7 \times \frac{\pi}{4}(6)^2 - 4.7 \times \frac{\pi}{4}(5)^2 = 416 - 92 = 324 \text{ lb}$$

b. Substituting directly into Eq. (14-5), we have

$$F = W = 14.7 \times \frac{\pi}{4}(6)^2 = 416 \text{ lb}$$

If we use a factor of safety of 2, the answers to parts a and b become 162 lb and 208 lb, respectively.

Time to Achieve Desired Vacuum Pressure

When a suction cup is placed on the top of a flat sheet and the vacuum pump is turned on, a certain amount of time must pass before the desired vacuum pressure is achieved. The time it takes to produce the desired vacuum pressure can be determined from Eq. (14-6):

$$t = \frac{V}{Q} \ln \left(\frac{p_{atm}}{p_{vaccum}} \right) \qquad \textbf{(14-6)}$$

where t = the time required to achieve the desired suction pressure (min),

V = the total volume of the space in the suction cup cavity and connecting pipeline up to the location of the vacuum pump (ft^3, m^3),

$\ln$ = the natural logarithm to the base e, where e is approximately 2.718,

Q = the flow rate produced by the vacuum pump (scfm, standard m^3/min),

p_{atm} = atmospheric pressure in absolute units (psia, Pa abs),

p_{vacuum} = the desired vacuum pressure in absolute units (psia, Pa abs).

Because p_{atm}/p_{vacuum} is a ratio, it is dimensionless. Thus, any desired units can be used for p_{atm} and p_{vacuum} as long as the units are the same and are absolute. For example, inches of mercury absolute could also be used for both pressures instead of using psia or pascals absolute. Thus, for example, if p_{atm} is five times as large as p_{vacuum}, the pressure ratio will equal 5 no matter what units are used, as long as they are the same units and are absolute.

EXAMPLE 14-7

A pneumatic vacuum lift system has a total volume of 6 ft³ inside the suction cup and associated pipeline leading to the vacuum pump. The vacuum pump produces a flow rate of 4 scfm when turned on. The desired suction pressure is 6 in Hg abs and atmospheric pressure is 30 in Hg abs. Determine the time required to achieve the desired vacuum pressure.

Solution Substituting into Eq. (14-6) yields the solution:

$$t = \frac{V}{Q} \ln \left(\frac{p_{atm}}{p_{vacuum}} \right) = \frac{6 \text{ ft}^3}{4 \text{ ft}^3/\text{min}} \ln \left(\frac{30 \text{ in Hg abs}}{6 \text{ in Hg abs}} \right) = 2.41 \text{ min}$$

Because division by zero cannot be done, a perfect suction pressure of zero absolute cannot be substituted into Eq. (14-6). To find the approximate time required to come close to obtaining a perfect vacuum, a very nearly perfect suction pressure of 0.50 in Hg abs (or the equivalent pressure of 0.245 psia) can be used. Substituting these values into Eq. (14-6) yields the time it takes to achieve an almost perfect vacuum:

$$t = \frac{6 \text{ ft}^3}{4 \text{ ft}^3/\text{min}} \ln \left(\frac{30 \text{ in Hg abs}}{0.5 \text{ in Hg abs}} \right) = \frac{6 \text{ ft}^3}{4 \text{ ft}^3/\text{min}} \ln \left(\frac{14.7 \text{ psia}}{0.245 \text{ psia}} \right) = 6.14 \text{ min}$$

Thus, it takes 2.55 times (6.14 min/2.41 min) as long to achieve a very nearly perfect vacuum as it does to achieve a vacuum of 1/5 atm (6 in Hg/30 in Hg).

14.7 SIZING OF GAS-LOADED ACCUMULATORS

Example 14-8 illustrates how to size gas-loaded accumulators as an auxiliary power source in hydraulic systems. Boyle's law is used, assuming that the gas temperature change inside the accumulator is negligibly small.

EXAMPLE 14-8

The circuit of Figure 14-16 has been designed to crush a car body into bale size using a gas-loaded accumulator and 6-in-diameter hydraulic cylinder. The hydraulic cylinder is to extend 100 in during a period of 10 s. The time between

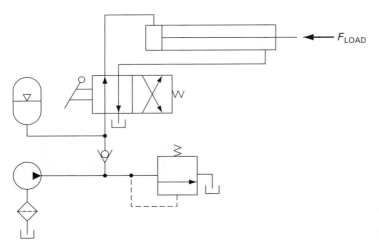

Figure 14-16. Use of gas-loaded accumulator in hydraulic system for crushing car bodies.

crushing strokes is 5 min. The following accumulator gas absolute pressures are given:

$$p_1 = \text{gas precharge pressure} = 1200 \text{ psia}$$

$$p_2 = \text{gas charge pressure when pump is turned on}$$

$$= 3000 \text{ psia} = \text{pressure relief valve setting}$$

$$p_3 = \text{minimum pressure required to actuate load} = 1800 \text{ psia}$$

a. Calculate the required size of the accumulator.
b. What are the pump hydraulic horsepower and flow requirements with and without an accumulator?

Solution   Figure 14-17 shows the three significant accumulator operating conditions:
1. **Preload** [Figure 14-17(a)]. This is the condition just after the gas has been introduced into the top of the accumulator. Note that the piston (assuming a piston design) is all the way down to the bottom of the accumulator.
2. **Charge** [Figure 14-17(b)]. The pump has been turned on, and hydraulic oil is pumped into the accumulator since p_2 is greater than p_1. During this phase, the four-way valve is in its spring-offset position. Thus, system pressure builds up to the 3000-psia level of the pressure relief valve setting.
3. **Final position of accumulator piston at end of cylinder stroke** [Figure 14-17(c)]. The four-way valve is actuated to extend the cylinder against its load. When the system pressure drops below 3000 psia, the accumulator 3000-psia gas pressure forces oil out of the accumulator into the system to

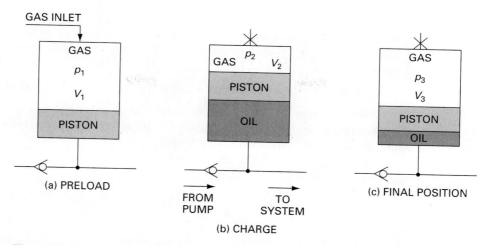

Figure 14-17. Operation of gas-loaded piston accumulator.

assist the pump during the rapid extension of the cylinder. The accumulator gas pressure reduces to a minimum value of p_3, which must not be less than the minimum value of 1800 psia required to drive the load.

a. Use Eq. (13-3):

$$p_1V_1 = p_2V_2 = p_3V_3$$

where V_1 = required accumulator size, and also use $V_{\text{hydraulic cylinder}} = V_3 - V_2$ (assuming negligible assistance from the pump). Thus, we have

$$V_3 = \frac{p_2V_2}{p_3} = \frac{3000V_2}{1800} = 1.67\,V_2$$

$$V_{\text{hydraulic cylinder}} = \frac{\pi}{4}(6)^2 \times 100 = 2830 \text{ in}^3 = V_3 - V_2$$

Solving the preceding equations yields

$$V_2 = 4230 \text{ in}^3 \quad V_3 = 7060 \text{ in}^3$$

Therefore, we have a solution:

$$V_1 = \frac{p_2V_2}{p_1} = \frac{(3000)(4230)}{1200} = 10{,}550 \text{ in}^3 = 45.8\text{-gal accumulator}$$

530 Chapter 14

Note that this is a very large accumulator because it is required to do a big job. For example, if the cylinder stroke were only 10 in, a 4.58-gal accumulator would suffice.

b. *With accumulator* (pump charges accumulator twice in 5 min):

Note that in 5 min the pump must recharge the accumulator only to the extent of the volume displaced in the cylinder during extension and retraction. Ignoring the diameter of the hydraulic cylinder rod, this volume equals $2(V_3 - V_2) = 2(2830/231)$ gal = 24.5 gal. Thus, we have

$$Q_{pump} = \frac{24.5 \text{ gal}}{5 \text{ min}} = 4.90 \text{ gpm (a small size pump)}$$

$$hp_{pump} = \frac{(3000)(4.90)}{1714} = 8.58 \text{ hp (a small horse power requirement)}$$

Without accumulator (pump extends cylinder in 10 s):

$$Q_{pump} = \frac{\left(\frac{2830}{231}\right) \text{ gal}}{\left(\frac{1}{6}\right) \text{ min}} = 73.4 \text{ gpm (a very large pump)}$$

$$hp_{pump} = \frac{(1800)(73.4)}{1714} = 76.6 \text{ hp (a very large horse power requirement)}$$

The results show that an accumulator, by handling large transient demands, can dramatically reduce the size and power requirements of the pump.

14.8 PNEUMATIC CIRCUIT ANALYSIS USING METRIC UNITS

Examples 14-9 through 14-11 illustrate the analysis of pneumatic circuits using metric units.

EXAMPLE 14-9

A 75% efficient compressor delivers air at 687 kPa gage and 7.65 standard m³/min. Calculate the cost of electricity per year if the efficiency of the electric motor driving the compressor is 92% and the compressor operates 3000 hr per year. The cost of electricity is $0.11/kWh.

Solution Using Eq. (13-8M) and dividing by η_o, the actual power needed to drive the compressor is

$$\text{actual power (kW)} = \frac{p_{in}Q}{17.1\eta_o}\left[\left(\frac{p_{out}}{p_{in}}\right)^{0.286} - 1\right]$$

$$= \frac{101 \times 7.65}{17.1 \times 0.75}\left[\left(\frac{788}{101}\right)^{0.286} - 1\right] = 48.2 \text{ kW}$$

Thus, the electric power required to drive the electric motor is $48.2/0.92$ kW = 52.4 kW. The cost of electricity per year can now be found:

$$\text{yearly cost} = \text{power rate} \times \text{time per year} \times \text{unit cost of elec.}$$

$$= 52.4 \text{ kW} \times 3000 \text{ hr/yr} \times \frac{\$0.11}{\text{kWh}} = \$17,300/\text{yr}$$

EXAMPLE 14-10

A pneumatic vacuum lift system uses four suction cups, each having a 100-mm lip outside diameter and a 80-mm lip inside diameter. The vacuum system is to lift large steel sheets weighing 1000 N. The total volume inside the cup cavities and associated pipelines up to the vacuum pump is 0.15 m³. If a factor of safety of 2 is used, what flow rate must the vacuum pump deliver if the time required to produce the desired vacuum pressure is 1.0 min?

Solution First, solve for the required vacuum pressure using Eq. (14-4):

$$F = p_{atm} A_o - p_{vacuum} A_i$$

$$\frac{1000 \times 2}{4} = 101,000 \times \frac{\pi}{4}(0.100)^2 - p_{vacuum} \times \frac{\pi}{4}(0.080)^2$$

$$500 = 793 - 0.00503 p_{vacuum}$$

$$p_{vacuum} = 58,300 \text{ Pa abs}$$

Now we can solve for the required vacuum pump flow rate using Eq. (14-6):

$$Q = \frac{V}{t}\ln\left(\frac{p_{atm}}{p_{vacuum}}\right) = \frac{0.15}{1.0}\ln\left(\frac{101,000}{58,300}\right)$$

$$= 0.0824 \text{ standard m}^3/\text{min of air}$$

EXAMPLE 14-11

The circuit of Figure 14-16 has been designed to crush a car body into bale size using a 152-mm-diameter hydraulic cylinder. The hydraulic cylinder is to extend 2.54 m during a period of 10 s. The time between crushing strokes is 5 min. The following accumulator gas absolute pressures are given:

p_1 = gas precharge pressure = 84 bars abs

p_2 = gas charge pressure when pump is turned on

= 210 bars abs = pressure relief valve setting

p_3 = minimum pressure required to actuate load = 126 bars abs

a. Calculate the required size of the accumulator.
b. What are the pump hydraulic kW power and the flow requirements with and without an accumulator?

Solution Figure 14-17 shows the three significant accumulator operating conditions (preload, charge, and final position of accumulator piston at end of cylinder stroke).

a. Use Eq. (13-3).

$$p_1 V_1 = p_2 V_2 = p_3 V_3$$

where V_1 = required accumulator size, and also use

$$V_{\text{hydraulic cylinder}} = V_3 - V_2$$

Thus, we have

$$V_3 = \frac{p_2 V_2}{p_3} = \frac{210}{126} V_2 = 1.67 V_2$$

$$V_{\text{hydraulic cylinder}} = \frac{\pi}{4}(0.152)^2 \times 2.54 = 0.0461 \text{ m}^3 = V_3 - V_2$$

Solving the preceding equations yields

$$V_2 = 0.0688 \text{ m}^3 \quad V_3 = 0.115 \text{ m}^3$$

Therefore, we have a solution

$$V_1 = \frac{p_2 V_2}{p_1} = \frac{(210)(0.0688)}{84} = 0.172 \text{ m}^3 = 172 \text{ L}$$

b. *With accumulator* (pump charges accumulator twice in 5 min):

Ignoring the diameter of the hydraulic cylinder rod yields (see solution to Example 14-8):

$$Q_{pump} = \frac{2(V_3 - V_2)}{300 \text{ s}} = \frac{2(46.1 \text{ L})}{300 \text{ s}} = 0.307 \text{ L/s (a small size pump)}$$

$$kW_{pump} = \frac{(210 \times 10^5)(30.7 \times 10^{-5})}{1000} = 6.45 \text{ kW (a small power requirement)}$$

Without accumulator (pump extends cylinder in 10 s):

$$Q_{pump} = \frac{46.1 \text{ L}}{10 \text{ s}} = 4.61 \text{ L/s (a very large pump)}$$

$$kW_{pump} = \frac{(126 \times 10^5)(461 \times 10^{-5})}{1000} = 58.1 \text{ kW (a very large power requirement)}$$

14.9 KEY EQUATIONS

Air pressure loss in a pipeline:

$$p_f(\text{psi}) = \frac{0.1025 L(\text{ft}) \times [Q(\text{scfm})]^2}{3600(\text{CR}) \times [d(\text{in})]^{5.31}} \tag{14-3}$$

Pneumatic suction lift force:

$$F = p_{atm} A_o - p_{suction} A_i \tag{14-4}$$

Time to achieve vacuum pressure

English units:

$$t(\text{min}) = \frac{V(\text{ft}^3)}{Q(\text{scfm})} \ln\left[\frac{p_{atm}(\text{psia})}{p_{vaccum}(\text{psia})}\right] \tag{14-6}$$

Metric units:

$$t(\text{min}) = \frac{V(\text{m}^3)}{Q(\text{standard m}^3/\text{min})} \ln\left[\frac{p_{atm}(\text{Pa abs})}{p_{vaccum}(\text{Pa abs})}\right] \tag{14-6M}$$

EXERCISES

Questions, Concepts, and Definitions

14-1. Name the four important considerations that must be taken into account when analyzing or designing a pneumatic circuit.

14-2. Why are air leaks into the atmosphere from a pneumatic system often ignored?

14-3. What undesirable consequence occurs when components of a pneumatic system, such as pipes and valves, are undersized?

14-4. What undesirable consequence occurs when components of a pneumatic system, such as pipes and valves, are oversized?

14-5. What effect do air pressure losses in pipelines have on the operation of the compressor?

14-6. What effect do air leaks into the atmosphere from a pneumatic system have on the operation of the compressor?

14-7. Explain what is meant by the following expression: Atmospheric air is free, but compressed air is not.

14-8. What is a pneumatic vacuum system?

14-9. Name three applications of pneumatic vacuum systems.

14-10. Why is a factor of safety used when applying a vacuum lift system?

14-11. Name one disadvantage of using a vacuum lift system.

14-12. What benefit is achieved in using an accumulator as an auxiliary power source?

14-13. What are the three significant accumulator operating conditions?

Problems

Note: The letter *E* following an exercise number means that English units are used. Similarly, the letter *M* indicates metric units.

Air Pressure Losses in Pipelines

14-14E. A compressor delivers 150 scfm of air through a 1-in schedule 40 pipe at a receiver pressure of 125 psig. Find the pressure loss for a 150-ft length of pipe.

14-15E. If the pipe in Exercise 14-14 has three gate valves, two globe valves, four tees (through run), and five 90° elbows, find the pressure loss.

14-16M. A compressor delivers air at 3 standard m³/min through a 25-mm inside diameter pipe at a receiver pressure of 1000 kPa gage. Find the pressure loss for a 100-m length of pipe.

14-17M. If the pipe in Exercise 14-16 has two gate valves, three globe valves, five tees (through run), four 90° elbows, and six 45° elbows, find the pressure loss.

14-18E. A compressor delivers 200 scfm of air through a pipe at a receiver pressure of 140 psig. What minimum size schedule 40 pipe should be used if the pressure loss is to be limited to 0.10 psi per foot of pipe length?

14-19M. A compressor delivers air at 4 standard $\frac{m^3}{min}$ through a pipe at a receiver pressure of 800 kpa gage. What minimum size inside diameter pipe should be used if the pressure loss is to be limited to 2 kPa per meter of pipe length?

Economic Cost of Energy Losses

14-20E. A 70% efficient compressor delivers air to a pneumatic system at 100 psig and 200 scfm. The efficiency of the electric motor driving the compressor is 90%, and the compressor operates 4000 hours per year. If the cost of electricity is $0.10/kWh, determine the cost of electricity per year.

14-21E. For the system in Exercise 14-20, the compressor is required to provide air at 112 psig to offset a 12-psi pressure loss in the pipelines due to friction. In addition, the compressor is required to provide an additional 50 scfm of air to compensate for air leakage from the pneumatic system into the atmosphere. What is the additional cost of electricity per year due to these two types of energy losses?

14-22M. A 70% efficient compressor delivers air to a pneumatic system at 690 kPa gage and 6 standard m³/min. The efficiency of the electric motor driving the compressor is 90%, and the compressor operates 4000 hours per year. If the cost of electricity is $0.10/kWh, determine the cost of electricity per year.

14-23M. For the system in Exercise 14-22, the compressor is required to provide air at 790 kPa gage to offset a 100-kPa pressure loss in the pipelines due to friction. In addition, the compressor is required to provide an additional 1.5 standard m³/min of air to compensate for air leakage from the pneumatic system into the atmosphere. What is the additional cost of electricity per year due to these two types of energy losses?

Operation of Pneumatic Circuits

14-24. What does the circuit of Figure 14-18 accomplish when the manual shutoff valve $V1$ is opened?

14-25. Consider the circuit of Figure 14-19.
   a. What happens to the cylinder when valve $V4$ is depressed?
   b. What happens to the cylinder when valve $V5$ is depressed?

14-26. For the circuit of Figure 14-20, what happens to the cylinders in each case?
   a. Valve $V1$ is actuated and held.
   b. Valve $V1$ is released and valve $V2$ is actuated and held. Valves $V3$ and $V4$ are sequence valves.

14-27. For the circuit in Exercise 14-26, as shown in Figure 14-20, cylinder 1 will not hold against a load while cylinder 2 is retracting. Modify this circuit by adding a pilot check valve and appropriate piping so that cylinder 1 will hold while cylinder 2 is retracting.

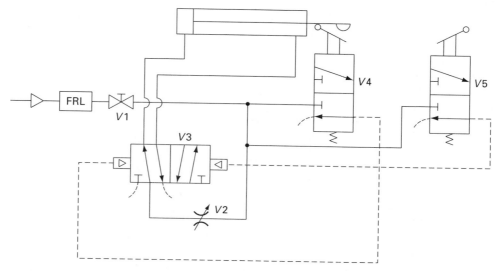

Figure 14-18. Circuit for Exercise 14-24. (*This circuit is simulated on the CD included with this textbook.*)

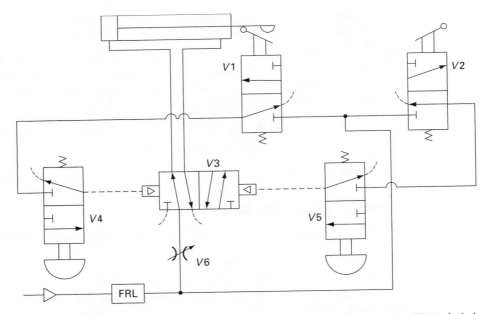

Figure 14-19. Circuit for Exercise 14-25. (*This circuit is simulated on the CD included with this textbook.*)

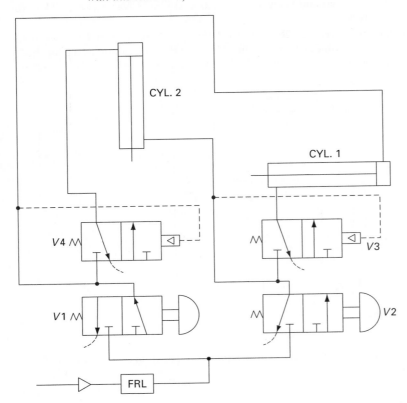

Figure 14-20. Circuit for Exercise 14-26.

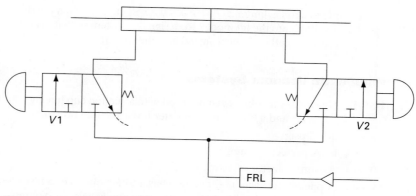

Figure 14-21. System for Exercise 14-28.

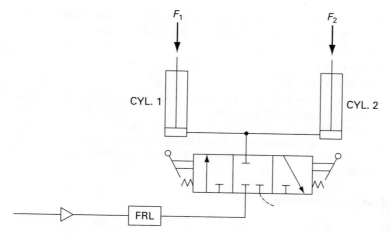

Figure 14-22. System for Exercise 14-30.

14-28. The pneumatic system of Figure 14-21 contains a double-rod cylinder controlled by two three-way, two-position directional control valves. Describe the four operating conditions for this system.

14-29. For the system of Exercise 14-28, as shown in Figure 14-21, the cylinder is free (both ends vented to the atmosphere) in the unactuated (spring offset) position of the directional control valves. Redesign the system using the same components to accomplish the following operations:

a. The cylinder rod moves left when only $V1$ is actuated.

b. The cylinder rod moves right when only $V2$ is actuated.

c. The cylinder rod stops moving when a single actuated valve is unactuated (both valves are unactuated).

d. When both valves are actuated, the cylinder is free (both ends are vented to the atmosphere).

14-30. The pneumatic system in Figure 14-22 is designed to lift and lower two loads using single-acting cylinders connected in parallel. If the load on cylinder 1 is larger than the load on cylinder 2 ($F_1 > F_2$), what will happen when the directional control valve is shifted into the lift mode? Redesign this system so that the cylinders will extend and retract together (synchronized at same speed).

Pneumatic Vacuum Systems

14-31E. How heavy an object can be lifted with a suction cup having a lip with a 7-in outside diameter and a 6-in inside diameter if the vacuum pressure is

    a. −8 psig

    b. A perfect vacuum

    Assume a factor of safety of 3.

14-32E. A pneumatic vacuum lift system has a total volume of 5 ft³ inside the cup and associated pipeline leading to the vacuum pump. The vacuum pump produces a flow rate of 3 scfm when turned on. The desired suction pressure is 5 psia and atmospheric pressure is 30 in Hg abs. Determine the time required to achieve the desired suction pressure.

14-33M. How heavy an object can be lifted with a suction cup having a lip with a 100-mm outside diameter and an 80-mm inside diameter for each suction pressure?

    a. −50 kPa gage

    b. A perfect vacuum

    Assume a factor of safety of 3.

14-34M. A pneumatic vacuum lift system uses six suction cups, each having a lip with a 100-mm outside diameter and an 80-mm inside diameter. The vacuum system is to lift large steel sheets weighing 1500 N. The total volume inside the cup cavities and associated piping up to the vacuum pump is 0.20 m³. If a factor of safety of 2 is used, what flow rate must the vacuum pump deliver if the time required to produce the desired vacuum pressure is 2.0 min?

Sizing Gas-Loaded Accumulators

14-35E. The accumulator of Figure 14-23 is to supply 450 in³ of oil with a maximum pressure of 3000 psig and a minimum pressure of 1800 psig. If the nitrogen precharge pressure is 1200 psig, find the size of the accumulator.

14-36E. For the accumulator in Exercise 14-35, find the load force F_{load} that the cylinder can carry over its entire stroke. What would be the total stroke of the cylinder if the entire output of the accumulator is used?

14-37M. The accumulator of Figure 14-23 is to supply 7370 cm³ of oil with a maximum pressure of 210 bars gage and a minimum pressure of 126 bars gage. If the nitrogen precharge pressure is 84 bars gage, find the size of the accumulator. The hydraulic cylinder piston diameter is 152 mm.

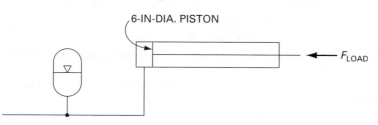

Figure 14-23. Circuit for Exercise 14-35.

14-38M. For the accumulator in Exercise 14-37, find the load force F_{load} that the cylinder can carry over its entire stroke. What would be the total stroke of the cylinder if the entire output of the accumulator were used?

14-39M. An accumulator under a pressure of 10 MPa is reduced in volume from 0.04 m³ to 0.03 m³ while the temperature increases from 40°C to 180°C. Determine the final pressure.

14-40E. When an oscillating load comes to a stop at midstroke, the charge gas in an accumulator expands from 180 in³ to 275 in³ and cools from 200°F to 100°F. If the resulting pressure in the accumulator is 1000 psig, calculate the charge gas pressure.

15 Basic Electrical Controls for Fluid Power Circuits

Learning Objectives

Upon completing this chapter, you should be able to:

1. Understand the operation of the various electrical components used in electromechanical relay control systems.
2. Read and understand the operation of electrical ladder diagrams.
3. Interrelate the operation of electrical ladder logic diagrams with the corresponding fluid power circuits.
4. Troubleshoot basic electrohydraulic and electropneumatic circuits for determining causes of system malfunction.

15.1 INTRODUCTION

Electrical devices have proven to be an important means of improving the overall control flexibility of fluid power systems. In recent years, the trend has been toward electrical control of fluid power systems and away from manual control. One of the reasons for this trend is that more machines are being designed for automatic operation to be controlled with electrical signals from computers.

Basic Electrical Devices

There are seven basic electrical devices commonly used in the control of fluid power systems: manually actuated push-button switches, limit switches, pressure switches, solenoids, relays, timers, and temperature switches. Switches can be wired either normally open (NO) or normally closed (NC). A normally open switch is one in which no electric current can flow through the switching element until the switch is actuated. In a normally closed switch, electric current can flow through the switching element until the switch is actuated. These seven electrical devices are briefly described as follows:

1. Push-button switches. By the use of a simple push-button switch, an operator can cause sophisticated equipment to begin performing complex operations. These push-button switches are used mainly for starting and stopping the operation of machinery as well as providing for manual override when an emergency arises.

2. Limit switches. Limit switches open and close circuits when they are actuated either at the end of the retraction or extension strokes of hydraulic or pneumatic cylinders. Figure 15-1 shows a hydraulic cylinder that incorporates its own limit switches (one at each end of the cylinder). Either switch can be wired normally open or normally closed. The limit switch on the cap end of the cylinder is actuated by an internal cam when the rod is fully retracted. The cam contacts the limit switch about 3/16 in from the end of the stroke. At the end of the cylinder stroke, the cam has moved the plunger and stem of the limit switch about 1/16 in for complete actuation. The limit switch on the head end of the cylinder is similarly actuated. Since these limit switches are built into the cylinder end plates, they are not susceptible to accidental movement, which can cause them to malfunction.

3. Pressure switches. Pressure switches open or close their contacts based on system pressure. They generally have a high-pressure setting and a low-pressure setting. For example, it may be necessary to start or stop a pump to maintain a given pressure. The low-pressure setting would start the pump, and the high-pressure setting would stop it. Figure 15-2 shows a pressure switch that can be wired either normally open (NO) or normally closed (NC), as marked on the screw terminals.

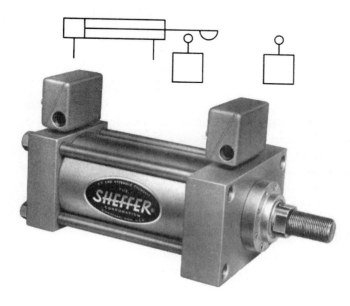

Figure 15-1. Cylinder with built-in limit switches. *(Courtesy of Sheffer Corp., Cincinnati, Ohio.)*

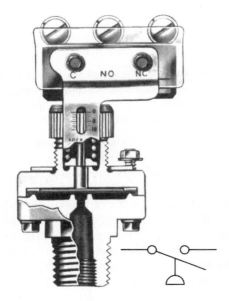
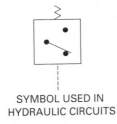

SYMBOL USED IN
HYDRAULIC CIRCUITS

Figure 15-2. Pressure switch.
*(Courtesy of DeLaval Turbine
Inc., Barksdale Control Division,
Los Angeles, California.)*

Pressure switches have three electrical terminals: C (Common), NC (normally closed), and NO (normally open). When wiring in a switch, only two terminals are used. The common terminal is always used, plus either the NC or NO terminal depending on whether the switch is to operate as a normally open or normally closed switch. In Figure 15-2, observe the front scale that is used for visual check of the pressure setting, which is made by the self-locking, adjusting screw located behind the scale. Figure 15-2 also gives the graphic symbol used to represent a pressure switch in hydraulic circuits as well as the graphic symbol used in electrical circuits.

4. Solenoids. Solenoids are electromagnets that provide a push or pull force to operate fluid power valves remotely. When a solenoid (an electric coil wrapped around an armature) is energized, the magnetic force created causes the armature to shift the spool of the valve containing the solenoid.

5. Relays. Relays are switches whose contacts open or close when their corresponding coils are energized. These relays are commonly used for the energizing and de-energizing of solenoids because they operate at a high current level. In this way a manually actuated switch can be operated at low voltage levels to protect the operator. This low-voltage circuit can be used to energize relay coils that control high-voltage contacts used to open and close circuits containing the solenoids. The use of relays also provides interlock capability, which prevents the accidental energizing of two solenoids at the opposite ends of a valve spool. This safety feature can, therefore, prevent the burnout of one or both of these solenoids.

6. Timers. Time delay devices are used to control the time duration of a working cycle. In this way a dwell can be provided where needed. For example, a dwell can be applied to a drilling machine operation, which allows the drill to pause for

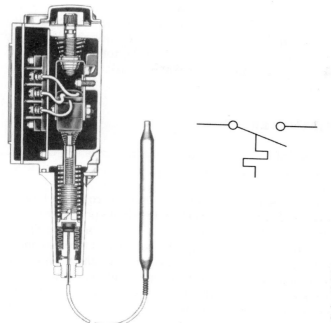

Figure 15-3. Remotely located temperature switch. *(Courtesy of DeLaval Turbine, Inc., Barksdale Control Division, Los Angeles, California.)*

a predetermined time at the end of the stroke to clean out the hole. Most timers can be adjusted to give a specified dwell to accommodate changes in feed rates and other system variables.

7. Temperature switches. Figure 15-3 shows a temperature switch, which is an instrument that automatically senses a change in temperature and opens or closes an electrical switch when a predetermined temperature is reached. This switch can be wired either normally open or normally closed. Note that at its upper end there is an adjustment screw to change the actuation point. The capillary tube (which comes in standard lengths of 6 or 12 ft) and bulb permit remote temperature sensing. Thus, the actual temperature switch can be located at a substantial distance from the oil whose temperature is to be sensed.

Temperature switches can be used to protect a fluid power system from serious damage when a component such as a pump or strainer or cooler begins to malfunction. The resulting excessive buildup in oil temperature is sensed by the temperature switch, which shuts off the entire system. This permits troubleshooting of the system to repair or replace the faulty component.

Circuit Diagrams

When drawing electrohydraulic or electropneumatic circuits, a separate circuit is drawn for the fluid system and a separate circuit is drawn for the electrical system.

Each component is labeled to show exactly how the two systems interface. Electrical circuits use ladder diagrams with the power connected to the left leg and the ground connected to the right leg. It is important to know the symbols used to represent the various electrical components. The operation of the total system can be ascertained by examination of the fluid power circuit and electrical diagram, as they show the interaction of all components.

15.2 ELECTRICAL COMPONENTS

There are five basic types of electric switches used in electrically controlled fluid power circuits: push-button, limit, pressure, temperature, and relay switches.

1. Push-button switches. Figure 15-4 shows the four common types of push-button switches. Figure 15-4(a) and 15-4(b) show the single-pole, single-throw type. These single-circuit switches can be wired either normally open or closed. Figure 15-4(c) depicts the double-pole, single-throw type. This double-contact type has one normally open and one normally closed pair of contacts. Depressing the push button opens the normally closed pair and closes the normally open pair of contacts.

Figure 15-4(d) illustrates the double-pole, double-throw arrangement. This switch has two pairs of normally open and two pairs of normally closed contacts to allow the inverting of two circuits with one input.

2. Limit switches. In Figure 15-5 we see the various types of limit switches. Basically, limit switches perform the same functions as push-button switches. The difference is that they are mechanically actuated rather than manually actuated. Figure 15-5(a) shows a normally open limit switch, which is abbreviated LS-NO. Figure 15-5(b) shows a normally open switch that is held closed. In Figure 15-5(c) we see the normally closed type, whereas Figure 15-5(d) depicts a normally closed type that is held open. There are a large number of different operators available

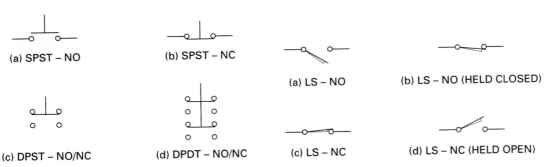

(a) SPST – NO

(b) SPST – NC

(a) LS – NO

(b) LS – NO (HELD CLOSED)

(c) DPST – NO/NC

(d) DPDT – NO/NC

(c) LS – NC

(d) LS – NC (HELD OPEN)

Figure 15-4. Push-button switch symbols.

Figure 15-5. Limit switch symbols.

for limit switches. Among these are cams, levers, rollers, and plungers. However, the symbols used for limit switches do not indicate the type of operator used.

3. Pressure switches. The symbols used for pressure switches are given in Figure 15-6. Figure 15-6(a) gives the normally open type, whereas Figure 15-6(b) depicts the normally closed symbol.

4. Temperature switches. This type of switch is depicted symbolically in Figure 15-7. Figure 15-7(a) gives the symbol for a normally closed type, whereas Figure 15-7(b) provides the normally open symbol.

5. Electrical relays. A relay is an electrically actuated switch. As shown schematically in Figure 15-8(a), when switch 1-SW is closed, the coil (electromagnet) is energized. This pulls on the spring-loaded relay arm to open the upper set of normally closed contacts and close the lower set of normally open contacts. Figure 15-8(b) shows the symbol for the relay coil and the symbols for the normally open and closed contacts.

Timers are used in electrical control circuits when a time delay from the instant of actuation to the closing of contacts is required. Figure 15-9 gives the symbol used for timers. Figure 15-9(a) shows the symbol for the normally open switch that is time closed when energized. This type is one that is normally open but that when energized closes after a predetermined time interval. Figure 15-9(b) gives the normally closed switch that is time opened when energized. Figure 15-9(c) depicts the normally open type that is timed when de-energized. Thus, it is normally open, and when the signal to close is removed (de-energized), it reopens after a predetermined time interval. Figure 15-9(d) gives the symbol for the normally closed type that is time closed when de-energized.

The symbol used to represent a solenoid, which is used to actuate valves, is shown in Figure 15-10(a). Figure 15-10(b) gives the symbol used to represent indicator lamps. An indicator lamp is often used to show the state of a specific circuit component. For example, indicator lamps are used to determine which solenoid operator of a directional control valve is energized. They are also used to indicate whether a hydraulic cylinder is extending or retracting. An indicator lamp wired across each valve solenoid provides the troubleshooter with a quick means of pinpointing trouble in case of an electrical malfunction. If they are mounted on an operator's display panel, they should be mounted in the same order as they are actuated. Since indicator lamps are not a functional part of the electrical system, their inclusion in the ladder diagram is left to the discretion of the designer.

The remaining portion of this chapter is devoted to discussing a number of basic electrically controlled fluid power systems. In these systems, the standard electrical

(a) PS – NO (b) PS – NC

(a) TS – NO (b) TS – NC

Figure 15-6. Pressure switch symbols.

Figure 15-7. Temperature switch symbols.

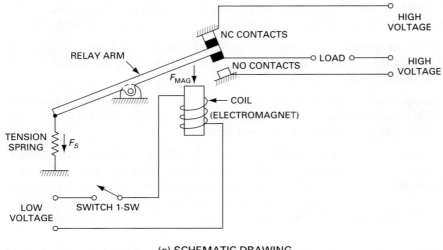

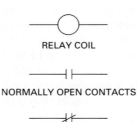

(a) SCHEMATIC DRAWING

RELAY COIL

NORMALLY OPEN CONTACTS

NORMALLY CLOSED CONTACTS

(b) COMPONENT SYMBOLS

Figure 15-8. Electrical relay.

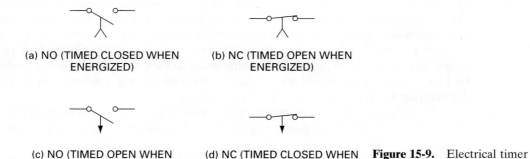

(a) NO (TIMED CLOSED WHEN ENERGIZED)

(b) NC (TIMED OPEN WHEN ENERGIZED)

(c) NO (TIMED OPEN WHEN DE-ENERGIZED)

(d) NC (TIMED CLOSED WHEN DE-ENERGIZED)

Figure 15-9. Electrical timer symbols.

symbols are combined with ANSI fluid power symbols to indicate the operation of the total system. Electrical devices such as pressure switches and limit switches are shown on the fluid power circuits using graphic symbols to correspond to the graphic symbols used in the electrical diagrams.

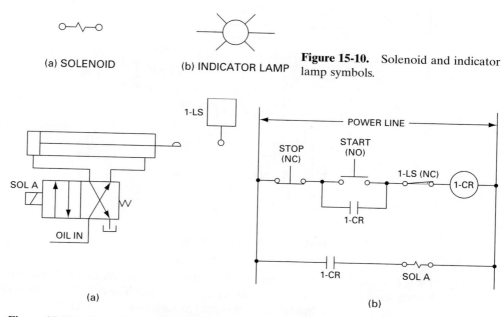

Figure 15-10. Solenoid and indicator lamp symbols.

(a) SOLENOID

(b) INDICATOR LAMP

Figure 15-11. Control of hydraulic cylinder using single limit switch. *(This circuit is simulated on the CD included with this textbook.)*

15.3 CONTROL OF A CYLINDER USING A SINGLE LIMIT SWITCH

Figure 15-11 shows a system that uses a single solenoid valve and a single limit switch to control a double-acting hydraulic cylinder. Figure 15-11(a) gives the hydraulic circuit in which the limit switch is labeled 1-LS and the solenoid is labeled SOL A. This method of labeling is required since many systems require more than one limit switch or solenoid.

In Figure 15-11(b) we see the electrical diagram that shows the use of one relay with a coil designated as 1-CR and two separate, normally open sets of contacts labeled 1-CR (NO). The limit switch is labeled 1-LS (NC), and also included are one normally closed and one normally open push-button switch labeled STOP and START, respectively. This electrical diagram is called a "ladder diagram" because of its resemblance to a ladder. The two vertical electric power supply lines are called "legs" and the horizontal lines containing electrical components are called "rungs."

When the START button is pressed momentarily, the cylinder extends because coil 1-CR is energized, which closes both sets of contacts of 1-CR. Thus, the upper 1-CR set of contacts serves to keep coil 1-CR energized even though the START button is released. The lower set of contacts closes to energize solenoid A to extend the cylinder. When 1-LS is actuated by the piston rod cam, it opens to de-energize coil 1-CR. This reopens the contacts of 1-CR to de-energize solenoid A. Thus, the valve returns to its spring-offset mode and the cylinder retracts. This closes the contacts of

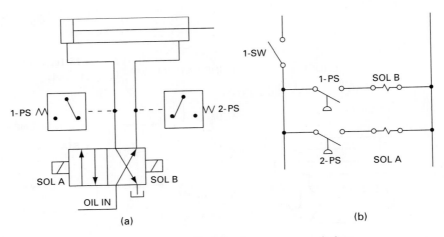

Figure 15-12. Reciprocation of cylinder using pressure switches.

1-LS, but coil 1-CR is not energized because the contacts of 1-CR and the START button have returned to their normally open position. The cylinder stops at the end of the retraction stroke, but the cycle is repeated each time the START button is momentarily pressed. The STOP button is actually a panic button. When it is momentarily pressed, it will immediately stop the extension stroke and fully retract the cylinder.

15.4 RECIPROCATION OF A CYLINDER USING PRESSURE OR LIMIT SWITCHES

In Figure 15-12 we see how pressure switches can be substituted for limit switches to control the operation of a double-acting hydraulic cylinder. Each of the two pressure switches has a set of normally open contacts. When switch 1-SW is closed, the cylinder reciprocates continuously until 1-SW is opened. The sequence of operation is as follows, assuming solenoid A was last energized: The pump is turned on, and oil flows through the valve and into the blank end of the cylinder. When the cylinder has fully extended, the pressure builds up to actuate pressure switch 1-PS. This energizes SOL B to switch the valve. Oil then flows to the rod end of the cylinder. On full retraction, the pressure builds up to actuate 2-PS. In the meantime, 1-PS has been deactuated to de-energize SOL B. The closing of the contacts of 2-PS energizes SOL A to begin once again the extending stroke of the cylinder.

Figure 15-13 gives the exact same control capability except each pressure switch is replaced by a normally open limit switch as illustrated. Observe that switches are always shown in their unactuated mode in the electrical circuits.

15.5 DUAL-CYLINDER SEQUENCE CIRCUITS

Figure 15-14 shows a circuit that provides a cycle sequence of two pneumatic cylinders. When the start button is momentarily pressed, SOL A is momentarily

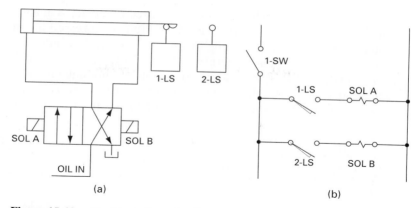

Figure 15-13. Reciprocation of cylinder using limit switches.

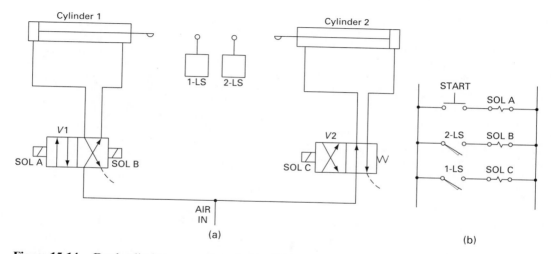

Figure 15-14. Dual-cylinder sequencing circuit. *(This circuit is simulated on the CD included with this textbook.)*

energized to shift valve $V1$, which extends cylinder 1. When 1-LS is actuated, SOL C is energized, which shifts valve $V2$ into its left flow path mode. This extends cylinder 2 until it actuates 2-LS. As a result, SOL B is energized to shift valve $V1$ into its right flow path mode. As cylinder 1 begins to retract, it deactuates 1-LS, which de-energizes SOL C. This puts valve $V2$ into its spring-offset mode, and cylinders 1 and 2 retract together. The complete cycle sequence established by the momentary pressing of the start button is as follows:

1. Cylinder 1 extends.
2. Cylinder 2 extends.
3. Both cylinders retract.
4. Cycle is ended.

A second dual-cylinder sequencing circuit is depicted in Figure 15-15. The operation is as follows: When the START button is depressed momentarily, SOL A is energized to allow flow through valve *V1* to extend cylinder 1. Actuation of 1-LS de-energizes SOL A and energizes SOL B. Note that limit switch 1-LS is a double-pole, single-throw type. Its actuation opens the holding circuit for relay 1-CR and simultaneously closes the holding circuit for relay 2-CR. This returns valve *V1* to its spring-offset mode and switches valve *V2* into its solenoid-actuated mode. As a

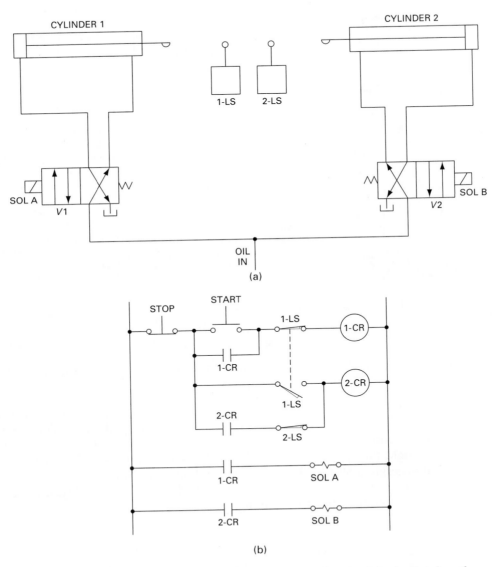

Figure 15-15. Second dual-cylinder sequencing circuit. *(This circuit is simulated on the CD included with this textbook.)*

result, cylinder 1 retracts while cylinder 2 extends. When 2-LS is actuated, SOL B is de-energized to return valve $V2$ back to its spring-offset mode to retract cylinder 2. The STOP button is used to retract both cylinders instantly. The complete cycle initiated by the START button is as follows:

1. Cylinder 1 extends.
2. Cylinder 2 extends while cylinder 1 retracts.
3. Cylinder 2 retracts.
4. Cycle is ended.

15.6 BOX-SORTING SYSTEM

An electropneumatic system for sorting two different-sized boxes moving on a conveyor is presented in Figure 15-16. Low boxes are allowed to continue on the same conveyor, but high boxes are pushed on to a second conveyor by a pneumatic cylinder. The operation is as follows: When the START button is momentarily depressed, coil 2-CR is energized to close its two normally open sets of contacts. This turns on the compressor and conveyor motors. When a high box actuates 1-LS, coil 1-CR is energized. This closes the 1-CR (NO) contacts and opens the 1-CR (NC) contacts. Thus, the conveyor motor stops, and SOL A is energized. Air then flows through the solenoid-actuated valve to extend the sorting cylinder to the right to begin pushing the high box onto the second conveyor. As 1-LS becomes deactuated, it does not de-energize coil 1-CR because contact set 1-CR (NO) is in its closed position. After the high box has been completely positioned onto the second conveyor, 2-LS is actuated. This de-energizes coil 1-CR and SOL A. The valve returns to its spring-offset mode, which retracts the cylinder to the left. It also returns contact set 1-CR (NC) to its normally closed position to turn the conveyor motor back on to continue the flow of boxes.

When the next high box actuates 1-LS, the cycle is repeated. Depressing the STOP button momentarily turns off the compressor and conveyor motors because this causes coil 2-CR to become de-energized. The production line can be put back into operation by the use of the START button.

15.7 ELECTRICAL CONTROL OF REGENERATIVE CIRCUIT

Figure 15-17 shows a circuit that provides for the electrical control of a regenerative cylinder. The operation is as follows: Switch 1-SW is manually placed into the *extend* position. This energizes SOL A, which causes the cylinder to extend. Oil from the rod end passes through check valve $V3$ to join the incoming oil from the pump to provide rapid cylinder extension. When the cylinder starts to pick up its intended load, oil pressure builds up to actuate normally open pressure switch 1-PS. As a result, coil 1-CR and SOL C become energized. This vents rod oil directly back to the oil tank through valve $V2$. Thus, the cylinder extends slowly as it drives its load. Relay contacts 1-CR hold SOL C energized during the slow extension movement of the cylinder to prevent any *fluttering* of the pressure switch. This would occur

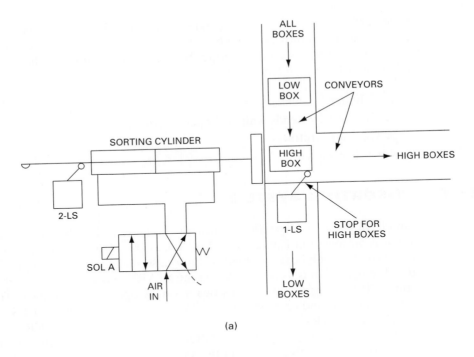

(a)

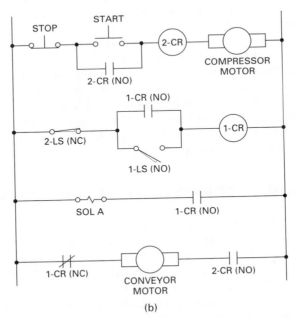

(b)

Figure 15-16. Electropneumatic box-sorting system. *(This circuit is simulated on the CD included with this textbook.)*

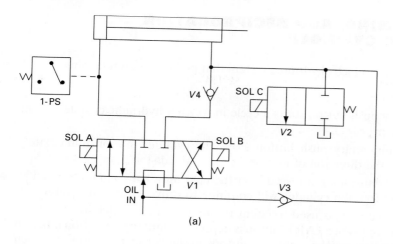

(a)

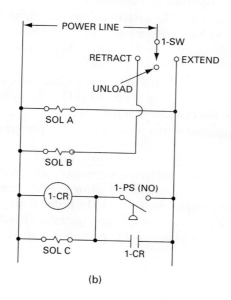

(b)

Figure 15-17. Electrical control of regenerative circuit.

because fluid pressure drops at the blank end of the cylinder when the regeneration cycle is ended. This can cause the pressure switch to oscillate as it energizes and de-energizes SOL C. In this design, the pressure switch is bypassed during the cylinder's slow-speed extension cycle. When switch 1-SW is placed into the *retract* position, SOL B becomes energized while the relay coil and SOL C become de-energized. Therefore, the cylinder retracts in a normal fashion to its fully retracted position. When the operator puts switch 1-SW into the *unload* position, all the solenoids and the relay coil are de-energized. This puts valve V1 in its spring-centered position to unload the pump.

15.8 COUNTING, TIMING, AND RECIPROCATION OF HYDRAULIC CYLINDER

Figure 15-18 shows an electrohydraulic system that possesses the following operating features:

1. A momentary push button starts a cycle in which a hydraulic cylinder is continuously reciprocated.

2. A second momentary push button stops the cylinder motion immediately, regardless of the direction of motion. It also unloads the pump.

3. If the START button is depressed after the operation has been terminated by the STOP button, the cylinder will continue to move in the same direction.

4. An electrical counter is used to count the number of cylinder strokes delivered from the time the START button is depressed until the operation is halted via the STOP button. The counter registers an integer increase in value each time an electrical pulse is received and removed.

5. An electrical timer is included in the system to time how long the system has been operating since the START button was depressed. The timer runs as long as a voltage exists across its terminals. The timer runs only while the cylinder is reciprocating.

6. Two lamps (L1 and L2) are wired into the electric circuit to indicate whether the cylinder is extending or retracting. When L1 is ON, the cylinder is extending, and when L2 is ON, the cylinder is retracting.

7. The cylinder speed is controlled by the pressure- and temperature-compensated flow control valve.

Note that the resistive components (lamps, solenoids, coils, timer, and counter) are connected in parallel in the same branch to allow the full-line voltage to be impressed across each resistive component. It should be noted that switches (including relay contacts) are essentially zero-resistance components. Therefore, a line that contains only switches will result in a short and thus should be avoided.

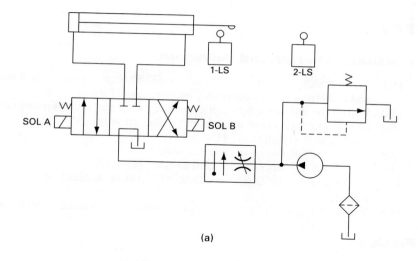

(a)

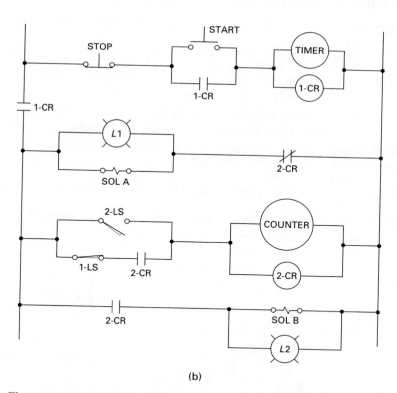

(b)

Figure 15-18. Counting, timing, and reciprocation of a hydraulic cylinder. *(This circuit is simulated on the CD included with this textbook.)*

EXERCISES

Questions, Concepts, and Definitions

15-1. In recent years, the trend has been toward electrical control of fluid power systems and away from manual controls. Give one reason for this trend.

15-2. What is the difference between a pressure switch and a temperature switch?

15-3. How does a limit switch differ from a push-button switch?

15-4. What is an electrical relay? How does it work?

15-5. What is the purpose of an electrical timer?

15-6. How much resistance do electrical switches possess?

15-7. Give one reason for having an indicator lamp in an electrical circuit for a fluid power system.

15-8. What is the difference between a normally open switch and a normally closed switch?

Problems

Electrical Control of Fluid Power (Analysis)

15-9. What happens to the cylinder of Figure 15-19 when the push button is momentarily depressed?

15-10. What happens to cylinders 1 and 2 of Figure 15-20 when switch 1-SW is closed? What happens when 1-SW is opened?

15-11. For the system of Figure 15-21, what happens to the two cylinders in each case?

a. Push-button 1-PB is momentarily depressed.

b. Push-button 2-PB is momentarily depressed.

Note that cylinder 2 does not actuate 1-LS at the end of its extension stroke.

15-12. Explain the complete operation of the system shown in Figure 15-22.

15-13. What happens to cylinders 1 and 2 of Figure 15-23 when switch 1-SW is closed with switch 2-SW open?

15-14. What happens to cylinders 1 and 2 of Figure 15-23 when switch 2-SW is closed with 1-SW open?

15-15. What happens to cylinders 1 and 2 of Figure 15-24 in each case?

a. 1-PB is momentarily depressed.

b. 2-PB is momentarily depressed.

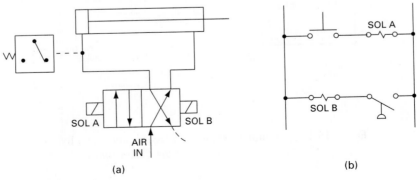

(a)

(b)

Figure 15-19. Circuit for Exercise 15-9.

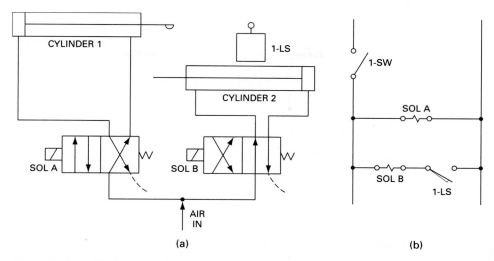

Figure 15-20. Circuit for Exercise 15-10. *(This circuit is simulated on the CD included with this textbook.)*

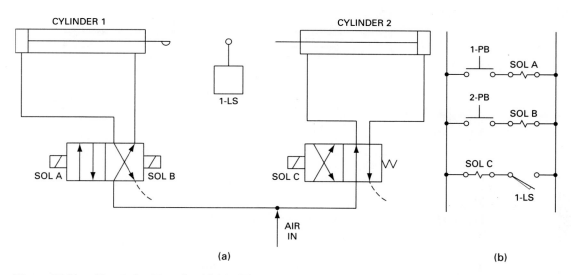

Figure 15-21. Circuit for Exercise 15-11. *(This circuit is simulated on the CD included with this textbook.)*

Electrical Control of Fluid Power (Design)

15-16. For the system of Figure 15-24. show two design changes that can be made to the ladder diagram in which either design change would allow both cylinders to retract fully when 1-PB is momentarily depressed.

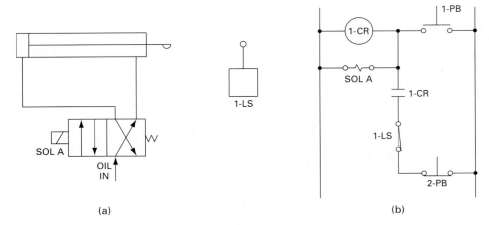

(a)   (b)

Figure 15-22.   Circuit for Exercise 15-12.

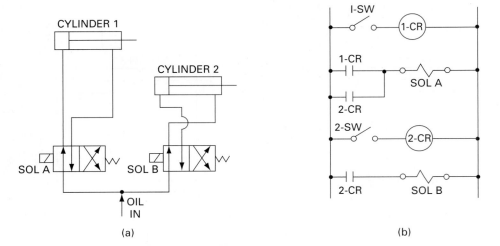

(a)   (b)

Figure 15-23.   Circuit for Exercises 15-13 and 15-14.

15-17. Design a ladder diagram for the electrical control of the regenerative circuit of Figure 15-25 as follows:

1. A manually actuated electric switch is placed into one of its three positions to cause the cylinder to rapidly extend until 1-LS is actuated.
2. Then the cylinder continues to extend at a slower rate until it is fully extended.
3. Then the manually actuated electric switch is placed into a second position to cause the cylinder to fully retract.
4. When the manually actuated electric switch is placed into its third position, the cylinder is hydraulically locked.

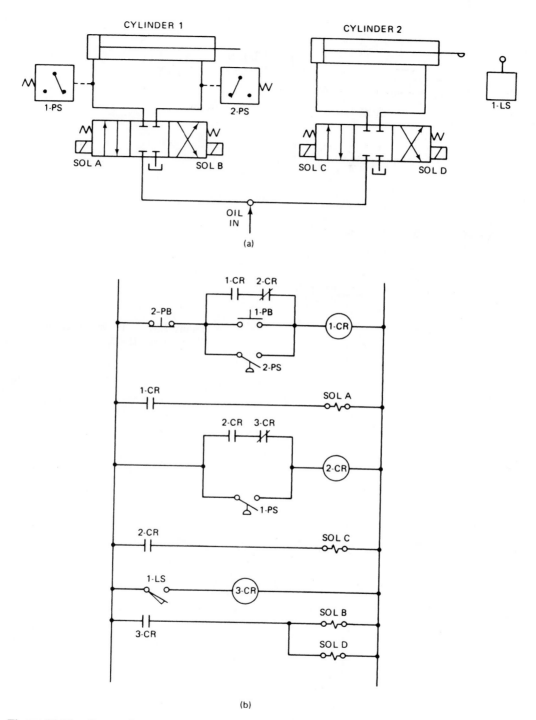

Figure 15-24. System for Exercise 15-15. (*This circuit is simulated on the CD included with this textbook.*)

560

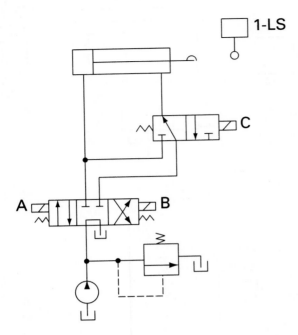

Figure 15-25. Hydraulic circuit for Exercise 15-17.

16 Fluid Logic Control Systems

Learning Objectives

Upon completing this chapter, you should be able to:

1. Explain the operating principles of moving-part logic (MPL) devices.
2. Understand how MPL is used to control fluid power systems.
3. Read MPL-controlled fluid power circuit diagrams and explain the corresponding system operation.
4. Discuss the control functions and establish the truth tables for OR/NOR, AND/NAND, NOT, EXCLUSIVE-OR, and FLIP-FLOP devices.
5. Perform the fundamental operations of Boolean algebra as related to control technology.
6. Apply Boolean algebra techniques to control logic diagrams.
7. Apply Boolean algebra techniques to control fluid power systems.

16.1 INTRODUCTION

Fluid logic control systems use logic devices that switch a fluid, usually air, from one outlet of the device to another outlet. Hence, an output of a fluid logic device is either ON or OFF as it is rapidly switched from one state to the other by the application of a control signal. Fluid logic control systems have several advantages over electrical logic control systems. For example, fluid logic devices are not as adversely affected by temperature extremes, vibration, and mechanical shock. In addition, fluid logic systems are ideally suited for applications where electric arcing or sparks can cause a fire or an explosion. Also, fluid logic devices do not generate electric noise and therefore will not interfere with nearby electric equipment. Devices that use a fluid for control logic purposes are broadly classified as either moving-part logic (MPL) devices or fluidic devices.

Moving-part logic devices are miniature valve-type devices, which—by the action of internal moving parts—perform switching operations in fluid logic control circuits. MPL devices are typically available as spool, poppet, and diaphragm valves, which can be actuated by means of mechanical displacement, electric voltage, or fluid pressure. Moving-part logic circuits provide a variety of logic control functions for controlling the operation of fluid power systems.

Figure 16-1 shows an MPL pneumatic control package with a push button for ON/OFF operation. The subplate and the four valves mounted on it form a single push-button input providing a binary four-way valve output that is pressure and speed regulated by restrictions on the exhaust ports. It is an ideal control for air collet vises, air clamps, assembly devices, indexing positioners, and other air-powered tools and devices.

Fluidic devices use a completely different technique for providing control logic capability as compared to MPL devices. Fluidics is the technology that uses fluid flow phenomena in components and circuits to perform a variety of control functions such as sensing, logic, memory, and timing. The concepts of fluidics are basically simple. They involve the effect of one fluid stream meeting another to change its direction of flow and the effect of a fluid stream sticking to a wall.

Since fluidic components have no moving parts, they virtually do not wear out. However, component malfunction can occur due to clogging of critical flow passageways if contaminants in the air supply are not eliminated by proper filtration. Fluidics is rarely used in practical industrial applications and thus is not covered in this book.

Boolean algebra is a two-valued algebra (0 or 1) that can be used to assist in the development of logic circuits for controlling fluid power systems. Boolean algebra serves two useful functions relative to controlling fluid power systems:

1. It provides a means by which a logic circuit can be reduced to its simplest form so that its operation can be more readily understood.

Figure 16-1. MPL pneumatic control package. *(Courtesy of Clippard Instrument Laboratory, Inc., Cincinnati, Ohio.)*

2. It allows for a quick synthesis of a control circuit that is to perform desired logic operations.

These two useful functions can be accomplished for both MPL control systems and electrical control systems.

16.2 MOVING-PART LOGIC (MPL) CONTROL SYSTEMS

Introduction

Moving-part logic (MPL) control systems use miniature valve-type devices, each small enough to fit in a person's hand. Thus, an entire MPL control system can be placed in a relatively small space due to miniaturization of the logic components. Figure 16-2 shows a miniature three-way limit valve along with its outline dimensions of $1\frac{3}{16}$ in long by $\frac{3}{4}$ in by $\frac{1}{2}$ in. This valve, which is designed to give dependable performance in a small, rugged package, has a stainless steel stem extending $\frac{1}{8}$ in from the top. The valve design is a poppet type with fast opening and high flow 7.0 CFM at 100-psi air (working range is 0 to 150 psi). Mounted on a machine or fixture, the valve is actuated by any moving part that contacts and depresses the stem.

Figure 16-3(a) shows an MPL circuit manifold, which is a self-contained modular subplate with all interconnections needed to provide a "two-hand-no-tie-down" pneumatic circuit. The manifold is designed to be used with three modular plug-in control valves and to eliminate the piping time and materials normally associated with circuitry. The main function of this control system is to require a machine operator to use both hands to actuate the machinery, thus ensuring that the operator's hands are not in a position to be injured by the machine as it is actuated. When used with two guarded palm button valves [see Figure 16-3(b)], which have been

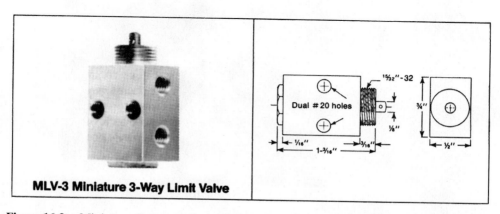

Figure 16-2. Miniature three-way limit valve. *(Courtesy of Clippard Instrument Laboratory, Inc., Cincinnati, Ohio.)*

properly positioned and mounted, the control system provides an output to actuate machinery when inputs indicate the operator's hands are safe.

Moving-part logic circuits use four major logic control functions: AND, OR, NOT, and MEMORY.

(a)

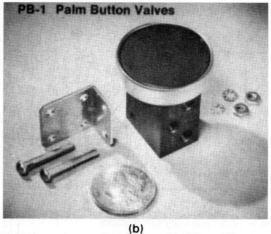

(b)

Figure 16-3. MPL circuit manifold for two-hand, no-tie-down control. *(Courtesy of Clippard Instrument Laboratory, Inc., Cincinnati, Ohio.)*

AND Function

Figure 16-4(a) shows a circuit that provides the AND function, which requires that two or more control signals must exist in order to obtain an output. The circuit consists of three two-way, two-position, pilot-actuated, spring-offset valves connected in series. If control signals exist at all three valves (A, B, and C), then output D will exist. If any one of the pilot signals is removed, output D will disappear.

A second method of implementing an AND function, shown in Figure 16-4(b), uses a single directional control valve and two shuttle valves. Pilot lines A, B, and C must be vented to shut off the output from S to P.

OR Function

An OR circuit is one in which a control signal at any one valve will produce an output. Thus, all control signals must be off in order for the output not to exist. This is accomplished in Figure 16-5(a), in which the three valves are now hooked in parallel. If any one of the valves picks up an air pilot signal, it will produce an output at D. Figure 16-5(b) shows how an OR function can be implemented using one

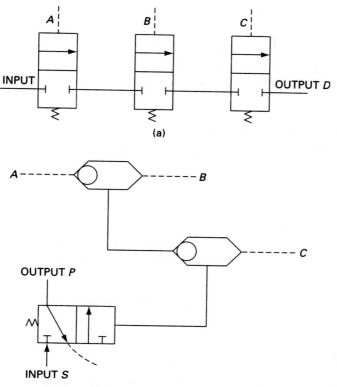

(a)

(b)

Figure 16-4. AND function. (a) Multiple directional control valves. (b) Single directional control valve.

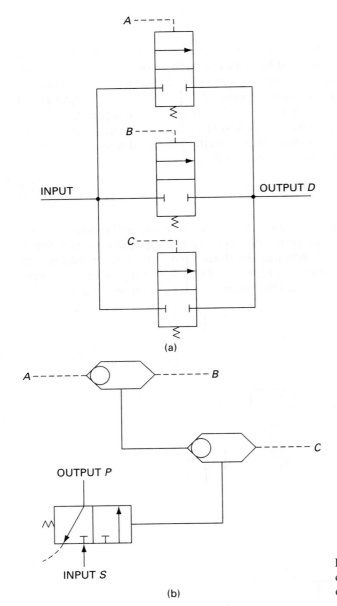

Figure 16-5. OR function. (a) Multiple directional control valves. (b) Single directional control valve.

directional valve and two shuttle valves. In this case, a signal applied at *A, B,* or *C,* will produce an output from *S* to *P.*

NOT Function

In a NOT function, the output is ON only when the single input control signal *A* is OFF, and vice versa. This is illustrated in Figure 16-6(a), which shows that the

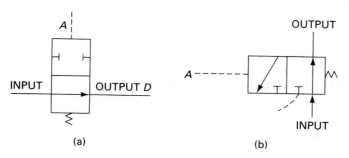

Figure 16-6. NOT function. (a) Two-way valve. (b) Three-way valve.

output will not exist if the control signal A is received. A second way to implement a NOT function is to use a three-way valve, as shown in Figure 16-6(b).

MEMORY Function

MEMORY is the ability of a control system to retain information as to where a signal it has received originated. Figure 16-7(a) shows a MEMORY circuit, which operates as follows: If control signal A is momentarily applied, output C will come on. Conversely, if control signal B is momentarily applied, the output will exist at D. Thus, an output at C means the signal was applied at A, and an output at D means the signal was applied at B. The MEMORY circuit does not function if control signals A and B are applied simultaneously because both ends of the output pilot valve would be piloted at the same time.

A second way to implement a MEMORY function is to use two three-way, double-piloted valves, as shown in Figure 16-7(b).

16.3 MPL CONTROL OF FLUID POWER CIRCUITS

Sequence Control of Two Double-Acting Cylinders

In this section we show the use of MPL control in fluid power circuits. Figure 16-8 shows an MPL circuit, which controls the extension and retraction strokes of two double-acting cylinders. The operation is as follows, assuming that both cylinders are initially fully retracted: When the START valve $V1$ is momentarily depressed, pilot valve $V2$ shifts to extend cylinder 1. At full extension, limit valve $V4$ is actuated to shift valve $V5$ and extend cylinder 2. On full extension, limit valve $V6$ is actuated. This shifts valve $V2$ to retract cylinder 1. On full retraction, limit valve $V3$ is actuated. This shifts valve $V5$ to fully retract cylinder 2. Thus, the cylinder sequence is as follows: Cylinder 1 extends, cylinder 2 extends, cylinder 1 retracts, and finally cylinder 2 retracts. The cycle can be repeated by subsequent momentary actuation of the START push-button valve. The sequence can be made continuous by removing the START valve and adding a limit switch to be actuated at the retraction end of cylinder 2. Upon actuation, this limit switch would pilot-actuate valve $V2$ to initiate the next cycle.

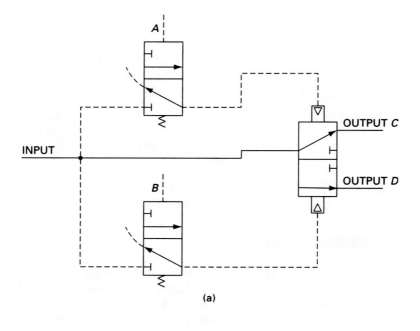

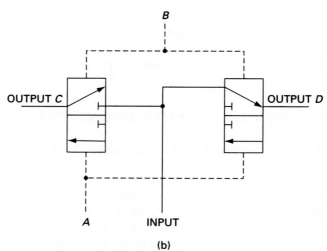

Figure 16-7. MEMORY function. (a) Three-directional control valves. (b) Two-directional control valves.

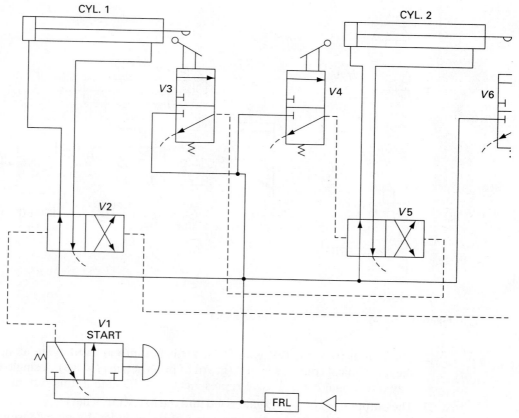

Figure 16-8. MPL cylinder sequencing circuit.

Control of Cylinder with Interlocks

Figure 16-9 shows an MPL circuit that controls the extension of a double-acting cylinder by having the following features:

1. The system provides interlocks and alternative control positions.
2. In order to extend the cylinder, either one of the two manual valves (*A* or *B*) must be actuated and valve *C* (controlled by a protective device such as a guard on a press) must also be actuated.
3. The output signal is memorized while the cylinder is extending.
4. At the end of the stroke, the signal in the MEMORY is canceled.

The circuit operation is described as follows:

1. The input signals *A* and *B* are fed into an OR gate so that either *A* or *B* can be used to extend the cylinder. The OR gate consists of one shuttle valve and two three-way, button-actuated directional control valves.

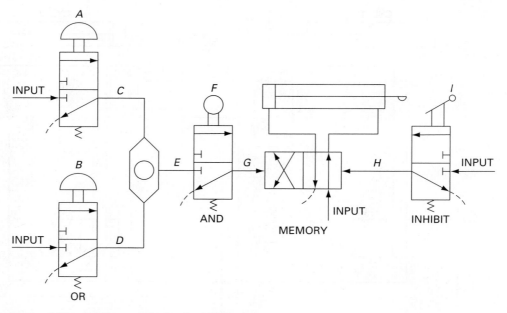

Figure 16-9. MPL control of a single cylinder.

2. The output from the OR gate (*C* or *D*) is fed into an AND gate along with the mechanical control signal *F* (guard of press actuates valve). A single three-way directional control valve represents the AND gate in this system.
3. The output from the AND gate is fed into the MEMORY device, which remembers to keep pressure on the blank end of the cylinder during extension.
4. At the end of the stroke, the inhibit (cancel) limit valve is actuated to cancel the signal in the memory. This stops the extension motion and retracts the cylinder.

It is interesting to note that a single directional control valve (four-way, double-piloted) can function as a MEMORY device. Also note that for the limit valve to provide the inhibit (cancel) function, the operator must release the manual input *A* and *B*.

16.4 INTRODUCTION TO BOOLEAN ALGEBRA

Introduction

The foundations of formal logic were developed by the Greek philosopher Aristotle during the third century B.C. The basic premise of Aristotle's logic is "a statement is either true or false; it cannot be both and it cannot be neither." Many philosophers have tried without success to create a suitable mathematical model of the preceding sentence based on the logical reasoning process of Aristotle. This was finally accomplished in 1854 when George Boole, an English mathematician, developed a two-valued algebra that could be used in the representation of true-false propositions.

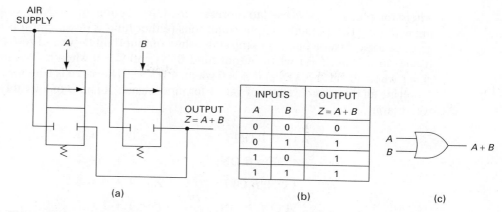

Figure 16-10. The OR function. (a) MPL components. (b) Truth table. (c) Symbol.

In developing this logical algebra (called "Boolean algebra"), Boole let a variable such as A represent whether a statement was true or false. An example of such a statement is "the valve is closed." The variable A would have a value of either zero (0) or one (1). If the statement is true, the value of A would equal one ($A = 1$). Conversely, if the statement is false, then the value of A would equal zero ($A = 0$).

Boolean algebra serves two useful functions relative to controlling fluid power systems:

1. It provides a means by which a logic circuit can be reduced to its simplest form so that its operation can more readily be understood.
2. It allows for the quick synthesis of a control circuit that is to perform desired logic operations.

In Boolean algebra, all variables have only two possible states (0 or 1). Multiplication and addition of variables are permitted. Division and subtraction are not defined and thus cannot be performed.

The following shows how Boolean algebra can be used to represent the basic logic functions (OR, AND, NOT, NOR, NAND, EXCLUSIVE-OR, and MEMORY). The components that perform these functions are called gates.

OR Function

An OR function can be represented in fluid flow systems by the case where an outlet pipe receives flow from two lines containing MPL valves controlled by input signals A and B, as shown in Figure 16-10(a).

Thus, fluid will flow in the outlet pipe (output exists) if input signal A is ON, OR input signal B is ON, OR both input signals are ON. Representing the flow in the outlet pipe by Z, we have

$$Z = A + B \qquad \text{(16-1)}$$

where the plus sign (+) is used to represent the OR function. In this case we are dealing with only two possible output conditions (either fluid is flowing or it is not). We give the logical value one (1) to the state when output fluid flows and zero (0) when it does not. Thus, $Z = 1$ when output fluid flows and $Z = 0$ when it does not. Also, $A = 1$ when signal A is ON and $A = 0$ when A is OFF. The same is true for signal B. Applying all the possible states of values for input signals A and B to logical Eq. (16-1) we obtain:

$$A \text{ OFF, } B \text{ OFF;} \qquad Z = 0 + 0 = 0$$

$$A \text{ OFF, } B \text{ ON;} \qquad Z = 0 + 1 = 1$$

$$A \text{ ON, } B \text{ OFF;} \qquad Z = 1 + 0 = 1$$

$$A \text{ ON, } B \text{ ON;} \qquad Z = 1 + 1 = 1$$

For these two inputs, the logical equation is: $Z = A + B$. For X inputs it is

$$Z = A + B + \cdots + W + X \qquad\qquad \textbf{(16-2)}$$

Since each input variable has two possible states (ON and OFF), for n input variables there are 2^n possible combinations of MPL valve settings. For our two-input-variable case, there are $2^2 = 4$ combinations, as shown in the truth table of Figure 16-10(b). A truth table tells how a particular device behaves. In a truth table, number 0 means OFF and number 1 means ON for all devices. Therefore as shown by the truth table, for an OR gate an output exists if input signal A is ON, OR input signal B is ON, OR both input signals are ON. Figure 16-10(c) shows the graphic symbol used to represent an OR gate. This is a general symbol that is used regardless of the type of component involved (electrical or MPL valve) in the logic system.

AND Function

The AND function can be represented for fluid flow systems by the case in which we have a number of MPL valves connected in series in a pipeline. The simplest case is for two valves with input signals A and B, as shown in Figure 16-11(a).

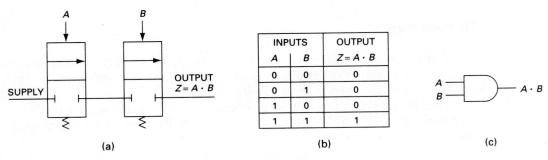

Figure 16-11. The AND function. (a) MPL components. (b) Truth table. (c) Symbol.

As can be seen, the output flow is zero ($Z = 0$) when either valve signal is OFF or both valve signals are OFF. Thus, fluid flows in the outlet pipe (there is an output) only when both A and B are ON. Figure 16-11(b) and (c) show the truth table and general symbol for the AND gate, respectively. The logic AND function for two variables is represented by the equation

$$Z = A \cdot B \qquad \text{(16-3)}$$

and for X variables by

$$Z = A \cdot B \cdots W \cdot X \qquad \text{(16-4)}$$

The dot ($\cdot$) is used to indicate the logic AND connective, and this form of equation is known as the *logic product function*. Inspection of each row of the truth table shows that the numerical value of the logic product function is also equal to the numerical value of the arithmetic product of the variables.

NOT Function

The NOT function is the process of logical inversion. This means that the output signal is NOT equal to the input signal. Since we have only two signal states (0 and 1), then an input of 1 gives an output of 0, and vice versa. Figure 16-12 gives the MPL valve, the truth table, and general symbol for a NOT gate.

A NOT operation is also known as logical complementing or logical negation in addition to logical inversion. It is represented in Boolean algebra by placing a bar over the variable as follows:

$$Z = \text{NOT } A = \overline{A} \qquad \text{(16-5)}$$

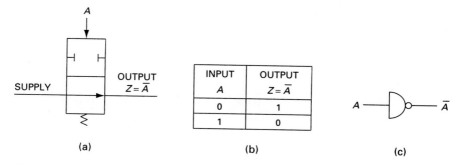

(a)   (b)   (c)

Figure 16-12. The NOT function. (a) MPL component. (b) Truth table. (c) Symbol.

NOR Function

The NOR function has its name derived from the following relationship:

$$NOR = NOT\ OR = \overline{OR} \qquad \textbf{(16-6)}$$

Thus, the NOR function is an inverted OR function whose MPL valve system, truth table, and general symbol are provided in Figure 16-13. Also, as shown in Figure 16-13(d), a NOR function can be created by placing a NOT gate in series with an OR gate. The Boolean relationship is

$$Z = NOT\ (A + B) = \overline{A + B} \qquad \textbf{(16-7)}$$

As shown by the truth table, the output of a NOR gate is ON (1) only when all inputs are OFF (0). One significant feature of a NOR gate is that it is possible to generate any logic function (AND, OR, NOT, and MEMORY) using only NOR gates.

NAND Function

The NAND function has its name derived from the relationship:

$$NAND = NOT\ AND = \overline{AND} \qquad \textbf{(16-8)}$$

Thus, the NAND function is an inverted AND function whose MPL valve system, truth table, and general symbol are provided in Figure 16-14. As can be seen, both signals must be ON to cause a loss of output. Also, as shown in Figure 16-14(d), a

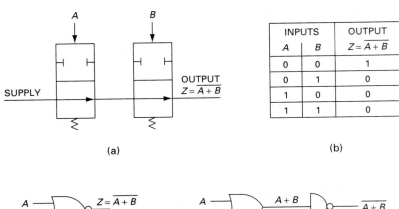

INPUTS		OUTPUT
A	B	$Z = \overline{A + B}$
0	0	1
0	1	0
1	0	0
1	1	0

(a) (b)

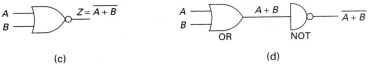

(c) (d)

Figure 16-13. The NOR function. (a) MPL components. (b) Truth table. (c) Symbol. (d) OR/NOT combination = NOR.

NAND function can be created by placing a NOT gate in series with an AND gate. The Boolean relationship is

$$Z = \text{NOT}\,(A \cdot B) = \overline{A \cdot B} \qquad \text{(16-9)}$$

Laws of Boolean Algebra

There are a number of laws of Boolean algebra that can be used in the analysis and design of fluid logic systems. These laws are presented as follows:

1. Commutative law:

$$A + B = B + A$$
$$A \cdot B = B \cdot A$$

2. Associative law:

$$A + B + C = (A + B) + C = A + (B + C) = (A + C) + B$$
$$A \cdot B \cdot C = (A \cdot B) \cdot C = A \cdot (B \cdot C) = (A \cdot C) \cdot B$$

3. Distributive law:

$$A + (B \cdot C) = (A + B) \cdot (A + C)$$
$$A \cdot (B + C) = (A \cdot B) + (A \cdot C)$$

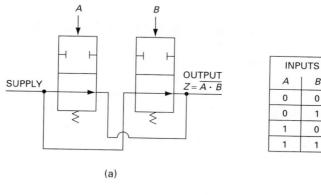

INPUTS		OUTPUT
A	B	$Z = \overline{A \cdot B}$
0	0	1
0	1	1
1	0	1
1	1	0

(a)

(b)

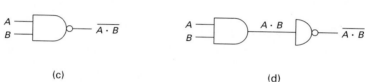

(c)

(d)

Figure 16-14. The NAND function. (a) MPL components. (b) Truth table. (c) Symbol. (d) AND/NOT combination = NAND.

4. DeMorgan's theorem:

$$\overline{A + B + C} = \overline{A} \cdot \overline{B} \cdot \overline{C}$$
$$\overline{A \cdot B \cdot C} = \overline{A} + \overline{B} + \overline{C}$$

Additional theorems that can be used to simplify complex equations and thus minimize the number of components required in a logic system are

5. $A + A = A$
6. $A \cdot A = A$
7. $A + 1 = 1$
8. $A + 0 = A$
9. $A \cdot 0 = 0$
10. $A \cdot 1 = A$
11. $A + (A \cdot B) = A$
12. $A \cdot (A + B) = A$
13. $(\overline{\overline{A}}) = A$
14. $A \cdot \overline{A} = 0$
15. $A + \overline{A} = 1$

It should be noted that all the preceding laws can be proven by the use of truth tables. Within the truth table, all combinations of the variables are listed, and values of both sides of the equation to be proven are computed using the definitions of the operators. If both sides of the equation have exactly the same values for every combination of the inputs, the theorem is proven.

16.5 ILLUSTRATIVE EXAMPLES USING BOOLEAN ALGEBRA

In this section we show how to use Boolean algebra to provide logic control of fluid power.

EXAMPLE 16-1

Prove that $A + (A \cdot B) = A$ using a truth table.

Solution

A	B	$A \cdot B$	$A + (A \cdot B)$
0	0	0	0
1	0	0	1
0	1	0	0
1	1	1	1

The column $A \cdot B$ is obtained based on an AND function, whereas the column $A + (A \cdot B)$ is obtained based on an OR function.

EXAMPLE 16-2

Prove the first DeMorgan theorem using two variables: $\overline{(A + B)} = \overline{A} \cdot \overline{B}$.

Solution

A	B	$\overline{A}$	$\overline{B}$	$\overline{A} \cdot \overline{B}$	$A + B$	$\overline{A + B}$
0	0	1	1	1	0	1
1	0	0	1	0	1	0
0	1	1	0	0	1	0
1	1	0	0	0	1	0

EXAMPLE 16-3

Generate the truth table for the function $Z = A \cdot \overline{B} + \overline{A} \cdot B$. Draw the logic circuit diagram representing the function using OR, AND, and NOT gates.

Solution

Since there are two variables A and B, there are $2^2 = 4$ combinations of input variable values. These are shown in the following truth table along with intermediate values for $A \cdot \overline{B}$ and $\overline{A} \cdot B$ and the values of the output Z.

Inputs				Output
A	B	$A \cdot \overline{B}$	$\overline{A} \cdot B$	$Z = A \cdot \overline{B} + \overline{A} \cdot B$
0	0	0	0	0
0	1	0	1	1
1	0	1	0	1
1	1	0	0	0

Let's examine the first row of the truth table. Since $A = 0$, $A \cdot \overline{B} = 0$. Likewise, since $B = 0$, $\overline{A} \cdot B = 0$, giving an output $Z = 0$. In the second row, $A \cdot \overline{B}$ remains at zero since $A = 0$. However, the product $\overline{A} \cdot B$ equals one since $\overline{A} = 1$ (from Theorem 10). Finally, $Z = 0 + 1 = 1$ from Theorem 8. The rest of the truth table is completed in a similar fashion.

EXCLUSIVE-OR Function

Further examination of the truth table of Example 16-3 reveals that the function is an EXCLUSIVE-OR function. This is the case because an EXCLUSIVE-OR function gives an output only if input A or input B is ON. It differs from an OR function (also called INCLUSIVE-OR), which gives an output when A or B is ON

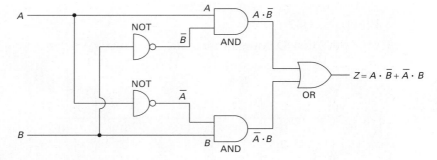

Figure 16-15. EXCLUSIVE-OR logic circuit.

or both A and B are ON. Thus, the Boolean relationship for an EXCLUSIVE-OR gate is

$$Z = A \cdot \overline{B} + \overline{A} \cdot B$$

whereas for an OR gate (INCLUSIVE-OR) it is

$$Z = A + B$$

The logic circuit diagram representing the function Z is given in Figure 16-15. Note that inputs A and $\overline{B}$ are applied to the upper AND gate, so that its output is the function $A \cdot \overline{B}$. These two signals are applied to the OR gate to produce a system output: $Z = A \cdot \overline{B} + \overline{A} \cdot B$.

EXAMPLE 16-4

Determine the logic function generated by the circuit in Figure 16-16. Simplify the function expression developed as much as possible using the theorems and laws of logical algebra.

Solution   First, establish the intermediate outputs:

$$01 = A + B$$
$$02 = \overline{C}$$
$$03 = (01) \cdot (02) = (A + B) \cdot \overline{C}$$
$$04 = A \cdot B$$
$$05 = \overline{04} = \overline{A \cdot B}$$
$$06 = A \cdot (02) = A \cdot \overline{C}$$

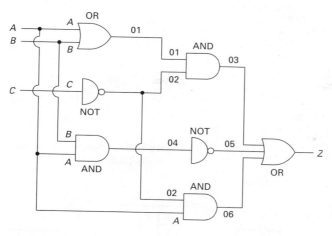

Figure 16-16. Logic circuit for Example 16-4.

Then, the output from the entire circuit is

$$Z = 03 + 05 + 06$$

$$Z = (A + B) \cdot \overline{C} + \overline{A \cdot B} + A \cdot \overline{C}$$

The simplification of the function is accomplished next.

$$
\begin{aligned}
Z &= A \cdot \overline{C} + B \cdot \overline{C} + \overline{A \cdot B} + A \cdot \overline{C} && \text{(distributive law)} \\
&= A \cdot \overline{C} + B \cdot \overline{C} + \overline{A \cdot B} && \text{(Theorem 5)} \\
&= A \cdot \overline{C} + B \cdot \overline{C} + \overline{A} \cdot \overline{B} && \text{(DeMorgan's theorem)} \\
&= (\overline{A} + A \cdot \overline{C}) + (\overline{B} + B \cdot \overline{C}) && \text{(associative law)} \\
&= (\overline{A} + A) \cdot (\overline{A} + \overline{C}) + (\overline{B} + B) \cdot (\overline{B} + \overline{C}) && \text{(distributive law)} \\
&= 1 \cdot (\overline{A} + \overline{C}) + 1 \cdot (\overline{B} + \overline{C}) && \text{(Theorem 15)} \\
&= \overline{A} + \overline{C} + \overline{B} + \overline{C} && \text{(Theorem 10)} \\
&= \overline{A} + \overline{B} + \overline{C} && \text{(Theorem 5)}
\end{aligned}
$$

This example shows how an entire circuit can possess many redundant logic components since $\overline{A} + \overline{B} + \overline{C}$ requires only four gates (three NOTS and one OR), as shown in Figure 16-17.

MEMORY Function

A MEMORY function (SR FLIP-FLOP) can be generated by the use of two NOR gates, as shown in Figure 16-18(a). The inputs are S (SET) and R (RESET) and the outputs are P and $\overline{P}$.

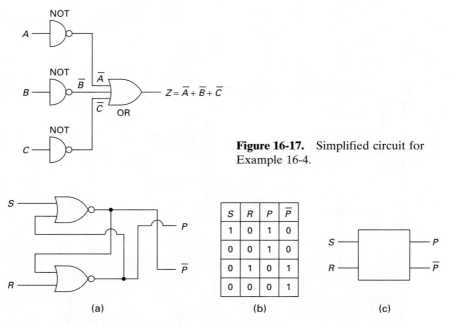

Figure 16-17. Simplified circuit for Example 16-4.

Figure 16-18. A MEMORY function (SR FLIP-FLOP). (a) Two NOR gate combination. (b) Truth table. (c) Symbol.

The truth table for an SR FLIP-FLOP is shown in Figure 16-18(b) and the general symbol in Figure 16-18(c). The truth table shows that when S is ON (R is OFF), P is the output ($P = 1, \overline{P} = 0$). Turning S OFF and then ON repeatedly does not shift the output from P (system has memory). Turning S OFF and R ON causes the output to shift to $\overline{P}(\overline{P} = 1, P = 0)$, which represents the second of two stable states possessed by an SR FLIP-FLOP.

Figure 16-19 shows a two-handed press safety control application of fluid logic. From a safety consideration, it is necessary to ensure that both of an operator's hands are used to initiate the operation of a press. If either hand is removed during the operation of the press cycle, the cylinder retracts. Both hands must be removed from the push-button controls A and B before the next operation can be started. Push-buttons A and B are back-pressure sensors, element C is an interruptible jet sensor, and P_s represents the fluid pressure supply sources.

In Section 17.4 we show how Boolean algebra is used to implement the use of programmable logic controllers (PLCs).

16.6 KEY EQUATIONS

OR function: $Z = A + B + \cdots + W + X$ **(16-2)**

AND function: $Z = A \cdot B \cdots W \cdot X$ **(16-4)**

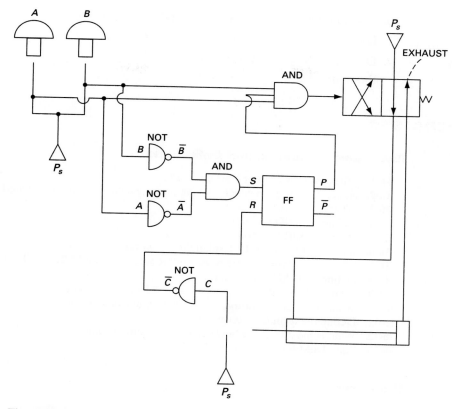

Figure 16-19. Two-handed press control system.

NOT function:

$$Z = \text{NOT } A = \overline{A} \qquad \text{(16-5)}$$

NOR function:

$$\text{NOR} = \text{NOT OR} = \overline{\text{OR}} \qquad \text{(16-6)}$$

NAND function:

$$\text{NAND} = \text{NOT AND} = \overline{\text{AND}} \qquad \text{(16-8)}$$

Laws of Boolean algebra

1. Commutative law:
$$A + B = B + A$$
$$A \cdot B = B \cdot A$$

2. Associative law:
$$A + B + C = (A + B) + C = A + (B + C) = (A + C) + B$$
$$A \cdot B \cdot C = (A \cdot B) \cdot C = A \cdot (B \cdot C) = (A \cdot C) \cdot B$$

3. Distributive law:

$$A + (B \cdot C) = (A + B) \cdot (A + C)$$
$$A \cdot (B + C) = (A \cdot B) + (A \cdot C)$$

4. DeMorgan's theorem:

$$\overline{A + B + C} = \overline{A} \cdot \overline{B} \cdot \overline{C}$$
$$\overline{A \cdot B \cdot C} = \overline{A} + \overline{B} + \overline{C}$$

EXERCISES

Questions, Concepts, and Definitions

16-1. What are moving-part logic devices?
16-2. Name three ways in which moving-part logic devices can be actuated.
16-3. What is fluidics?
16-4. Define an AND function.
16-5. Define an OR function.
16-6. What is a FLIP-FLOP? Explain how it works.
16-7. What is the difference between an OR gate and an EXCLUSIVE-OR gate?
16-8. Define a MEMORY function.
16-9. Name two useful functions provided by Boolean algebra.
16-10. What algebraic operations are permitted in Boolean algebra?
16-11. Define the term *logic inversion*.
16-12. What is the difference between the commutative and associative laws?
16-13. State DeMorgan's theorem.

Problems

Boolean Algebra

16-14. Prove that $A \cdot (A + B) = A$ using a truth table.
16-15. Prove that $A + (B + C) = (A + B) + C$, the first associative law, using a truth table.
16-16. Prove that $\overline{A \cdot B} = \overline{A} + \overline{B}$, DeMorgan's theorem 2, using a truth table.
16-17. Prove that $A \cdot (B + C) = (A \cdot B) + (A \cdot C)$, the second distributive law, using a truth table.
16-18. Prove that $A \cdot (\overline{A} + B) = A \cdot B$, using a truth table.
16-19. Using DeMorgan's theorem, show how the AND function can be developed using NOR gates.
16-20. Using DeMorgan's theorem, design a NOR circuit to generate the function: $Z = \overline{A} + \overline{B} + \overline{C}$.

General Logic Control of Fluid Power

16-21. Describe the operation of the fluid logic system of Figure 16-20.
16-22. Describe the operation of the fluid logic circuit system of Figure 16-21, when the machine guard blocks the back pressure sensor and then unblocks.
16-23. For the logic circuit of Figure 16-22, use Boolean algebra to show that valve 3 is not needed.

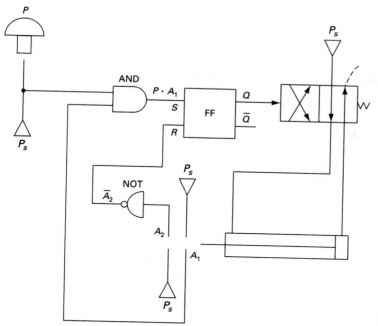

Figure 16-20.   Circuit for Exercise 16-21.

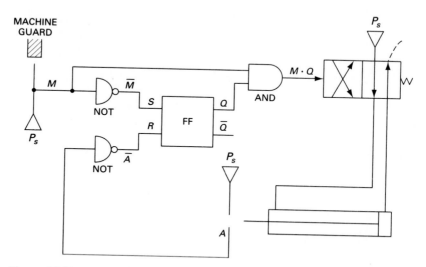

Figure 16-21.   Circuit for Exercise 16-22.

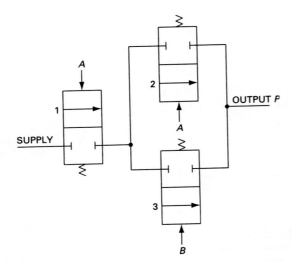

Figure 16-22. Circuit for Exercise 16-23.

17 Advanced Electrical Controls for Fluid Power Systems

Learning Objectives

Upon completing this chapter, you should be able to:

1. Understand the operation of servo valves and transducers.
2. Explain the operation of electrohydraulic servo systems.
3. Discuss the meaning of the block diagrams of electrohydraulic closed-loop systems.
4. Analyze the performance of electrohydraulic servo systems.
5. Explain the operation of programmable logic controllers (PLCs).
6. Appreciate the advantages of PLCs over electromechanical relay control systems.
7. Apply Boolean algebra and ladder logic diagrams to programmable logic control of fluid power systems.

17.1 INTRODUCTION

Electrohydraulic Servo Systems

Electrohydraulic servo systems are widely used in industry because these closed-loop systems provide more precise control of the position and velocity of a load than do the open-loop systems presented in previous chapters. See Figure 17-1 for a block diagram of such a servo system, and notice the closed loop between the electrical input signal from the command module and the mechanical output from the hydraulic actuator that drives the load. The overall operation of an electrohydraulic servo system is described as follows:

An electrical feedback signal from a device called a *feedback transducer* (which is mechanically connected to the hydraulic actuator output shaft) is subtracted

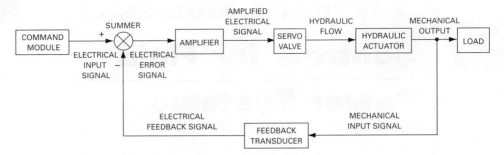

Figure 17-1. Block diagram of an electrohydraulic servo system (closed-loop system).

(negative feedback) from the command module electrical input signal by a device called a *summer* (adds negative feedback signal to input signal). This difference between the two signals (called the *error signal*) is electronically amplified to a higher power level to drive the torque motor of a servo valve. The torque motor shifts the spool of the servo valve, which produces hydraulic flow to drive the load actuator. The velocity or position of the load is fed back in electrical form via the feedback transducer, which is usually either a potentiometer for position or a tachometer for velocity. A feedback device is called a *transducer* because it converts a mechanical signal into a corresponding electrical signal.

The feedback signal (which is an electrical measure of load position or velocity) is continuously compared to the command module input signal. If the load position or velocity is not that called for by the command module, an error signal is generated by the summer to correct the discrepancy. When the desired output is achieved, the feedback and command module signals are equal, producing a zero error signal. Thus, no change in output will occur unless called for by a change in input command module signal. Therefore, the ability of a closed-loop system to precisely control the position or velocity of a load is not affected by internal leakage due to changes in pressure (load changes) and oil viscosity (temperature changes).

Figure 17-2 shows an electrohydraulic, servo-controlled robotic arm that has the strength and dexterity to torque down bolts with its fingers and yet can gingerly pick up an eggshell. This robotic arm is adept at using human tools such as hammers, electric drills, and tweezers and can even bat a baseball. The arm has a hand with a thumb and two fingers, as well as a wrist, elbow, and shoulder. It has 10 degrees of freedom including a 3-degree-of-freedom hand designed to handle human tools and other objects with humanlike dexterity. The servo control system is capable of accepting computer or human operator control inputs. The system can be designed for carrying out hazardous applications in the subsea, utilities, or nuclear environments, and it is also available in a range of sizes from human proportions of 6 ft long.

Figure 17-3 shows an application where a closed-loop electrohydraulic system is used to control the position of the hitch of a tractor while pulling implements such as plows, spreaders, and cultivators through rough terrain. By maintaining the optimum position of the implement-lifting gear and the tractor's forward velocity, the hitch is protected from unbalanced forces and damaging loads.

Figure 17-2. Hydraulically powered dextrous arm. *(Courtesy of Sarcos, Inc., Salt Lake City, Utah.)*

The lifting gear height is sensed with a position transducer and the pulling force is measured by a force sensor. Electronic output signals from the position transducer and force sensor are fed back to the electronic operating console in the driver's cab. This causes a change in electronic signals to the proportional pressure control valve and likewise to the tractor speed-control throttle system. As a result, pressure to the lifting-gear cylinder is varied by the proportional pressure control valve to raise or lower the gear to a new position, and the tractor's velocity is increased or decreased to maintain the optimum pull force.

Programmable Logic Controllers

In recent years, programmable logic controllers (PLCs) have increasingly been used in lieu of electromechanical relays to control fluid power systems. A PLC is a user-friendly electronic computer designed to perform logic functions such as AND, OR, and NOT for controlling the operation of industrial equipment and processes. A PLC consists of solid-state digital logic elements for making logic decisions and providing corresponding outputs. Unlike general-purpose computers, a PLC is designed to oper-ate in industrial environments where high ambient temperature and humidity levels may exist. PLCs offer a number of advantages over electromechanical relay control systems. Unlike electromechanical relays, PLCs are not hard-wired to perform spe-cific functions. Thus, when system operation requirements change, a software program is readily changed instead of having to physically rewire relays. In addition, PLCs are more reliable, faster in operation, smaller in size, and can be readily expanded.

Figure 17-4 shows a PLC-based synchronous lift system used for precise lifting and lowering of high-tonnage objects on construction jobs. Unlike complex and costly electronic lift systems, this hydraulic system has a minimum number of parts and can be run effectively and efficiently by one person. The PLC enables the operator to

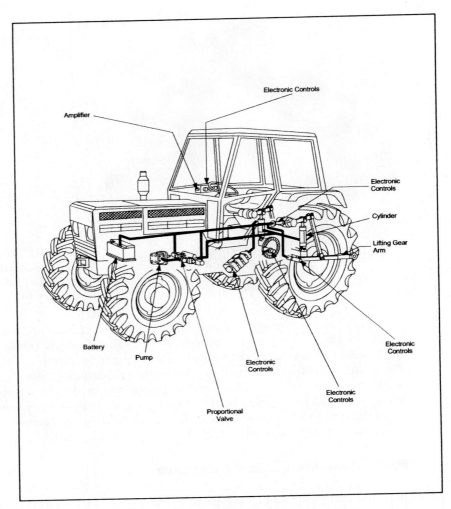

Figure 17-3. Hitch control for a tractor using electrohydraulic servo system. *(Courtesy of National Fluid Power Association, Milwaukee, Wisconsin.)*

quickly and easily set the number of lift points, stroke limit, system accuracy, and other operating parameters from a single location. The PLC receives input signals from electronic sensors located at each lift point, and in turn sends output signals to the solenoid valve that controls fluid flow to each hydraulic cylinder to maintain the relative position and accuracy selected by the operator. Because the sensors are attached directly to the load, they ensure more exact measurement of the load movement. The system accommodates a wide range of loads and is accurate to ±0.040 in (1 mm).

The PLC unit of this system (see Figure 17-5) contains an LCD display that shows the position of the load at each lift point and the status of all system operations so the operator can stay on top of every detail throughout the lift. The PLC unit, which weighs only 37 pounds and has dimensions of only 16 in by 16 in by 5 in, can

Figure 17-4. PLC synchronous lift system. *(Courtesy of Enerpac, Applied Power Inc., Butler, Wisconsin.)*

Figure 17-5. PLC unit of synchronous lift system. *(Courtesy of Enerpac, Applied Power, Inc., Butler, Wisconsin.)*

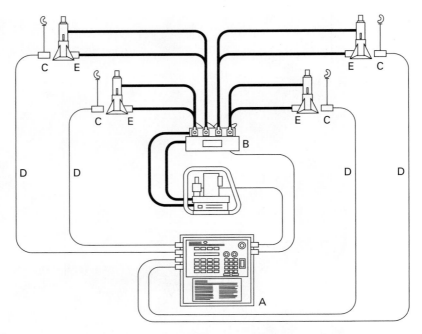

Figure 17-6. Diagram of PLC synchronous lift system. *(Courtesy of Enerpac, Applied Power, Inc., Butler, Wisconsin.)*

control up to eight lifting points. The system diagram is shown in Figure 17-6, in which components are identified using letters as follows:

A: Programmable logic controller

B: Solenoid directional control valve

C: Electronic load displacement sensors

D: Sensor cables

E: Hydraulic cylinders with flow control valves to regulate movement

17.2 COMPONENTS OF AN ELECTROHYDRAULIC SERVO SYSTEM

The primary purpose of a fluid power circuit is to control the position or velocity of an actuator. All the circuits discussed in previous chapters accomplished this objective using open-loop controls. In these systems, precise control of speed is not possible. The speed will decrease when the load increases. This is due to the higher pressures that increase internal leakage inside pumps, actuators, and valves. Temperature changes that affect fluid viscosity and thus leakage also affect the accuracy of open-loop systems.

In Figure 17-7 we see a closed-loop (servo) electrohydraulic control system. It is similar to the open-loop electrohydraulic systems of Chapter 15 except the solenoid-actuated directional control valves and flow control valves have been replaced by electrohydraulic servo valves such as the one shown in Figure 17-8. An electrohydraulic servo valve contains an electrical torque motor that positions a

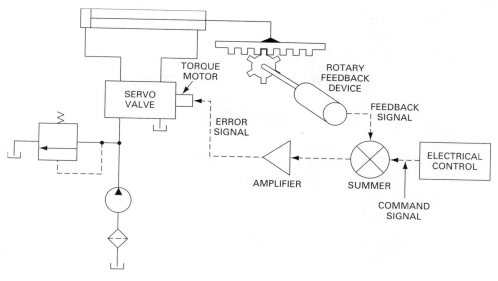

Figure 17-7. Electrohydraulic servo system.

sliding spool inside the valve to produce the desired flow rates. The spool position is proportional to the electrical signal applied to the torque motor coils (see Figure 17-9 for a schematic cross-sectional drawing of an electrohydraulic servo valve).

The system of Figure 17-7 is a servo system because the loop is closed by a feedback device attached to the actuator. This feedback device, either linear or rotary, as shown in Figure 17-7, senses the actuator position or speed and transmits a corresponding electrical feedback signal. This signal is compared electronically with the electrical input signal. If the actuator position or speed is not that intended, an error signal is generated by the electronic summer. This error signal is amplified and fed to the torque motor to correct the actuator position or speed. Therefore, the accuracy of a closed-loop system is not affected by internal leakages due to pressure and temperature changes.

Feedback devices are called *transducers* because they perform the function of converting one source of energy into another, such as mechanical to electrical. A velocity transducer is one that senses the linear or angular velocity of the system output and generates a signal proportional to the measured velocity. The type most commonly used is the tachometer/generator, which produces a voltage (AC or DC) that is proportional to its rotational speed. Figure 17-10 shows a DC tachometer/generator. A positional transducer senses the linear or angular position of system output and generates a signal proportional to the measured position. The most commonly used type of electrical positional transducer is the potentiometer, which can be of the linear or rotary motion type. Figure 17-11 shows a rotary potentiometer in which the wiper is attached to the moving member of the machine and the body to the stationary member. The positional signal is taken from the wiper and one end.

The electrical control box, which generates the command signal (see Figure 17-7), can be a manual control unit such as that shown in Figure 17-12. The scale graduations can represent either the desired position or velocity of the hydraulic actuator. An operator would position the hand lever to the position location or velocity level desired.

Figure 17-8. Electrohydraulic servo valve. *(Courtesy of Moog Inc., Industrial Division, East Aurora, New York.)*

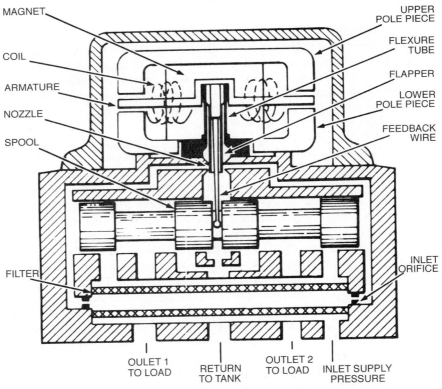

Figure 17-9. Schematic cross section of electrohydraulic servo valve. *(Courtesy of Moog Inc., Industrial Division, East Aurora, New York.)*

Figure 17-10. DC tachometer/ generator. *(Courtesy of Oil Gear Co., Milwaukee, Wisconsin.)*

Figure 17-11. Rotary feedback poten- tiometer. *(Courtesy of Bourns, Inc., Riverside, California.)*

Figure 17-12. Remote manual control unit. *(Courtesy of Oil Gear Co., Milwaukee, Wisconsin.)*

17.3 ANALYSIS OF ELECTROHYDRAULIC SERVO SYSTEMS

Block Diagram of Single Component

The analysis of servo system performance is accomplished using block diagrams in which each component is represented by a rectangle (block). The block diagram representation of a single component is a single rectangle shown as follows:

$$\text{INPUT} \longrightarrow \boxed{G} \longrightarrow \text{OUTPUT}$$

The gain, G, or transfer function of the component, is defined as the output divided by the input.

$$G = \text{gain} = \text{transfer function} = \frac{\text{output}}{\text{input}} \qquad \textbf{(17-1)}$$

For example, the block diagram for a pump is as follows, where the input to the pump is shaft speed N (rpm) and the pump output is fluid flow Q (gpm). If a pump delivers 10 gpm when running at 2000 rpm, the pump gain is

$$N\text{ (rpm)} \longrightarrow \boxed{G_p} \longrightarrow Q\text{ (gpm)}$$

G_P = pump gain = pump transfer function

$$= \frac{Q\text{ (gpm)}}{N\text{ (rpm)}} = \frac{10}{2000} = 0.005 \text{ gpm/rpm}$$

General Block Diagram of Complete System

A block diagram of a complete closed-loop servo system (or simply servo system) is shown in Figure 17-13. Block G represents the total gain of all the system components between the error signal and the output. This total gain (transfer function)

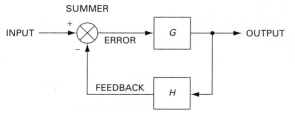

Figure 17-13. Block diagram of a closed-loop system.

is commonly called the "forward transfer function" because it is in the forward path. Block H represents the gain of the feedback component between the output and the summer (error detector). This transfer function is commonly called the "feedback transfer function" because it is in the feedback path. The sum of the input signal and negative feedback signal represents the error signal.

Another important parameter of a servo system is the open-loop gain (transfer function), which is defined as the gain from the error signal to the feedback signal. Thus, it is the product of the forward path gain and feedback path gain.

$$\text{open-loop gain} = GH \tag{17-2}$$

The closed-loop gain (transfer function) is defined as the system output divided by the system input.

$$\text{closed-loop transfer function} = \frac{\text{system output}}{\text{system input}} \tag{17-3}$$

It can be shown that the closed-loop transfer function (which shows how the system output compares to the system input) equals the forward gain divided by the expression 1 plus the product of the forward gain and feedback gain.

$$\text{closed-loop transfer function} = \frac{G}{1 + GH} \tag{17-4}$$

Thus, we have

$$\text{closed-loop transfer function} = \frac{\text{forward gain}}{1 + \text{open-loop gain}} \tag{17-5}$$

Frequency and Transient Response

Each component of a servo system generates a phase lag as well as an amplitude gain. For a sinusoidal input command (which can be used to determine the frequency response of a servo system), if the open-loop gain is greater than unity when the phase shift from input signal to feedback signal is 180°, the system will be unstable. This is because the feedback sine wave is in phase with the input sine wave, causing the output oscillations to become larger and larger until something breaks or corrective action is taken.

Figure 17-14(b) and (c) shows the transient response of a stable and an unstable, respectively, servo system responding to a step input command. Figure 17-14(a) shows the step input command, which is a constant input voltage applied at time $t = 0$ to produce a desired output. Observe that even though the stable system exhibits overshoot, the output oscillation quickly dampens as it approaches the steady-state desired value. The unstable system is one that depicts repeated undamped oscillations, and thus the desired steady-state output value is not achieved.

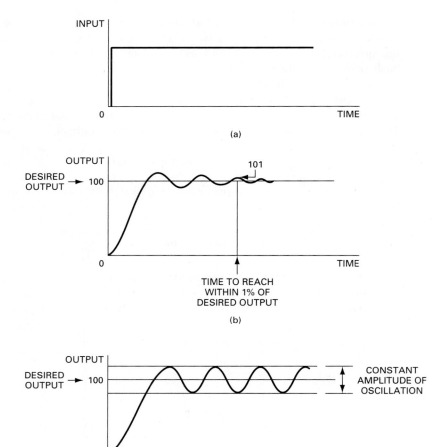

Figure 17-14. Stable and unstable transient responses to a step input command. (a) Step input command. (b) Stable (output dampens out). (c) Unstable (output oscillations repeat).

Detailed Block Diagram of Complete System

Figure 17-15 shows a more detailed block diagram of an electrohydraulic positional closed-loop system. The components in the forward path include the amplifier, servo valve, and hydraulic cylinder. The gain of each component is specified as follows:

$$G_A = \text{amplifier gain (milliamps per volt, or mA/V)}$$

$$G_{SV} = \text{servo valve gain}\left(\text{in}^3/\text{s per milliamp, or }\frac{\text{in}^3/\text{s}}{\text{mA}}\right)$$

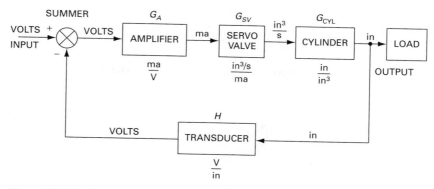

Figure 17-15. Detailed block diagram of electrohydraulic positional closed-loop system.

$$G_{cyl} = \text{cylinder gain } (\text{in}/\text{in}^3)$$

$$H = \text{feedback transducer gain (volts per in, or V/in)}$$

The open-loop gain can be determined as a product of the individual component gains.

$$\text{open-loop gain} = GH = G_A \times G_{SV} \times G_{cyl} \times H \qquad \textbf{(17-6)}$$

$$= \frac{\text{mA}}{\text{V}} \times \frac{\text{in}^3/\text{s}}{\text{mA}} \times \frac{\text{in}}{\text{in}^3} \times \frac{\text{V}}{\text{in}} = \frac{1}{\text{s}}$$

Thus, for a positional system, the units of open-loop gain are the reciprocal of time.

Deadband and Hysteresis

The accuracy of an electrohydraulic servo system depends on the system deadband and open-loop gain. System deadband is the composite deadband of all the components in the system. Deadband is defined as that region or band of no response where an input signal will not cause an output. For example, with respect to a servo valve, no flow will occur until the electric current to the torque motor reaches a threshold minimum value. This is shown in Figure 17-16, which is obtained by cycling a servo valve through its rated input current range and recording the output flow for one cycle of input current. The resulting flow curve shows the deadband region and hysteresis of the servo valve. Hysteresis is defined as the difference between the response of the valve to an increasing signal and the response to a decreasing signal.

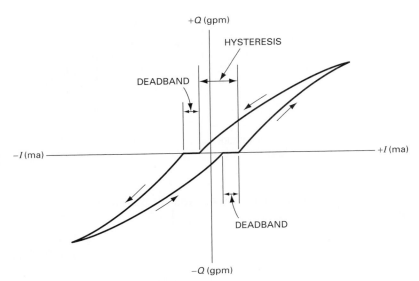

Figure 17-16. Flow curve showing deadband and hysteresis for a servo valve.

System Accuracy (Repeatable Error)

The repeatable error (discrepancy from the programmed output position) can be determined from Eq.(17-7):

$$RE = \text{repeatable error (in)} = \frac{\text{system deadband (mA)}}{G_A \text{ (mA/V)} \times H \text{ (V/in)}} \tag{17-7}$$

If the transducer gain (H) has units of V/cm, the units for repeatable error (RE) become cm.

To regulate system output accurately requires that the open-loop gain be large enough. However, the open-loop gain is limited by the component of the loop having the lowest natural frequency. Since hydraulic oil (although highly incompressible) is more compressible than steel, the oil under compression is the lowest natural frequency component. The following equation allows for the calculation of the natural frequency of the oil behaving as a spring-mass system.

$$\omega_H = A\sqrt{\frac{2\beta}{VM}} \tag{17-8}$$

where ω_H = natural frequency (rad/s),
        A = area of cylinder (in^2, m^2),
        β = bulk modulus of oil (lb/in^2, Pa),
        V = volume of oil under compression (in^3, m^3),
        M = mass of load (lb · s^2/in, kg).

The volume of oil under compression equals the volume of oil between the servo valve and cylinder in the line that is pressurized. For a typical servo system, the approximate value of the open-loop gain is one-third of the natural frequency of the oil under compression.

$$\text{open-loop gain} = \frac{v_H(\text{rad}/\text{s})}{3} \tag{17-9}$$

It should be noted that the amount of open-loop gain that can be used is limited. This is because the system would be unstable if the feedback signal has too large a value and adds directly to the input command signal.

Examples 17-1 and 17-2 show how to determine the accuracy of an electrohydraulic servo system.

EXAMPLE 17-1

An electrohydraulic servo system contains the following characteristics:

a. $G_{SV} = (0.15 \text{ in}^3/\text{s})/\text{mA}$. The servo valve saturates at 300 mA to provide a maximum fluid flow of 45 in³/s.

b. $G_{cyl} = 0.20 \text{ in/in}^3$. The cylinder piston area equals 5 in², allowing for a maximum velocity of 9 in/s. Cylinder stroke is 6 in.

c. $H = 4 \text{ V/in}$. Thus, the maximum feedback voltage is 24.

d. Volume of oil under compression = 50 in³.

e. Weight of load = 1000 lb.

$$\text{mass} = \frac{1000 \text{ lb}}{386 \text{ (in/s}^2)} = 2.59 \text{ lb} \cdot \text{s}^2/\text{in}$$

f. System deadband = 4 mA.

g. Bulk modulus of oil = 175,000 lb/in².

Determine the system accuracy.

Solution First, calculate the natural frequency of the oil. This is the resonant frequency, which is the frequency at which oil would freely vibrate as would a spring-mass system.

$$\omega_H = 5\sqrt{\frac{(2)(175,000)}{(50)(2.59)}} = 260 \text{ rad/s}$$

The value of the open-loop gain is

$$v_H/3 = 86.7/s$$

Solving for the amplifier gain from Eq. (17-6), we have

$$G_A = \frac{\text{open-loop gain}}{G_{SV} \times G_{cyl} \times H} = \frac{86.7}{0.15 \times 0.20 \times 4} = 723 \text{ mA/V}$$

The repeatable error (RE) can now be calculated using Eq. (17-7):

$$RE = \frac{4}{723 \times 4} = 0.00138 \text{ in}$$

EXAMPLE 17-2

Determine the system accuracy for a servo system containing the following characteristics (note that this system is identical to that given in Example 17-1 except that the units are metric rather than English).

a. $G_{SV} = (2.46 \text{ cm}^3/\text{s})/\text{mA}$.
b. $G_{cyl} = 0.031 \text{ cm/cm}^3$, cylinder area = 32.3 cm².
c. $H = 1.57 \text{ V/cm}$.
d. $V_{oil} = 819 \text{ cm}^3$.
e. Mass of load = 450 kg.
f. System deadband = 4 mA.
g. Bulk modulus of oil = 1200 MPa.

Solution

$$\omega_H = 32.3 \times 10^{-4} \sqrt{\frac{(2)(1200 \times 10^6)}{(819 \times 10^{-6})(450)}} = 260 \text{ rad/s}$$

$$\text{open-loop gain} = \frac{260}{3} = 86.7/\text{s}$$

$$G_A = \frac{86.7}{2.46 \times 0.031 \times 1.57} = 724 \text{ mA/V}$$

$$RE = \frac{4}{(724)(1.57)} = 0.00352 \text{ cm} = 0.00138 \text{ in}$$

Tracking Error

Another parameter that identifies the performance of a servo system is called *tracking error*, which is the distance by which the output lags the input command signal while the load is moving. The maximum tracking error (TE) is mathematically defined as

$$\text{TE (in)} = \frac{\text{servo valve maximum current (mA)}}{G_A \text{ (mA/V)} \times H \text{ (V/in)}} \tag{17-10}$$

If the transducer gain (H) has units of V/cm, the units for the maximum tracking error (TE) become cm.

Positional Versus Velocity Systems

Although the analysis presented here is applicable for a cylinder positional system, the technique can be applied to hydraulic motor drive positional systems. In addition, if velocity (angular or linear) is to be controlled, then velocity units would be used instead of distance units.

Since the block diagram of Figure 17-15 represents a positional system rather than a velocity system, the position of the cylinder is being controlled rather than its velocity. Thus, the transducer senses the cylinder's position rather than its velocity. However, the output of the servo valve is a volume flow rate that is the input to the cylinder. Only when the servo valve output flow rate goes to zero is a given cylinder position achieved. For a given servo valve output flow rate, the cylinder output is a velocity rather than a fixed position, as can be seen from the following equation:

$$y_{cyl} = \frac{Q_{SV}}{A_{cyl}} = \frac{\text{in}^3/\text{s}}{\text{in}^2} = \text{in/s}$$

Note, however, that in order to represent a positional system in Figure 17-15, the cylinder output is designated as a position with units of in rather than a velocity with units of in/s. This is why the transducer has a gain with units of V/in rather than V/in/s. The transducer senses cylinder position in inches and delivers a corresponding voltage to the summer.

EXAMPLE 17-3

For Examples 17-1 and 17-2, find the maximum tracking error.

Solution

a. For English units:

$$\text{TE} = \frac{300 \text{ mA}}{724 \text{ mA/V} \times 4 \text{ V/in}} = 0.104 \text{ in}$$

Thus, the hydraulic cylinder position will lag its desired position by 0.104 in based on the command signal when the cylinder is moving at the maximum velocity of 9 in/s.

b. For metric units:

$$TE = \frac{300 \text{ mA}}{724 \text{ mA/V} \times 1.57 \text{ V/cm}} = 0.264 \text{ cm} = 0.104 \text{ in}$$

17.4 PROGRAMMABLE LOGIC CONTROLLERS (PLCs)

Introduction

A programmable logic controller (PLC) is a user-friendly electronic computer designed to perform logic functions such as AND, OR, and NOT for controlling the operation of industrial equipment and processes. PLCs, which are used in lieu of electromechanical relays (which are described in Chapter 15), consist of solid-state digital logic elements for making logic decisions and providing corresponding outputs. Unlike general-purpose computers, a PLC is designed to operate in industrial environments where high ambient temperature and humidity levels may exist. In addition, PLCs are designed not to be affected by electrical noise commonly found in industrial plants.

Figure 17-17 shows a PLC designed for a wide variety of automation tasks. This PLC provides user-friendly service, from installation to troubleshooting and maintenance. Its compact size (3.1 in by 5.1 in by 2.4 in) permits the unit to be mounted directly onto an installation panel.

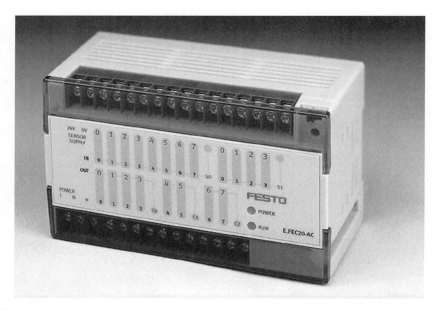

Figure 17-17.   Programmable logic controller (PLC). *(Courtesy of Festo Corp., Hauppauge, New York.)*

PLCs provide the following advantages over electromechanical relay control systems:

1. Electromechanical relays (as shown in Chapter 15) have to be hard-wired to perform specific functions. Thus, when system operation requirements change, the relays have to be rewired.
2. They are more reliable and faster in operation.
3. They are smaller in size and can be more readily expanded.
4. They require less electrical power and are less expensive for the same number of control functions.

Major Units of a PLC

As shown in Figure 17-18, a PLC consists of the following three major units:

1. Central processing unit (CPU). This unit represents the "brains" of the PLC. It contains a microprocessor with a fixed memory and an alterable memory. The fixed memory contains the program set by the manufacturer. It is set into integrated circuit (IC) chips called *read only memory* (ROM) and this memory cannot be changed during operation or lost when electrical power to the CPU is turned off. The alterable memory is stored on IC chips that can be programmed and altered by the user. This memory is stored on random access memory (RAM) chips and information stored on RAM chips is lost (volatile memory) when electrical power is removed. In general, the CPU receives input data from various sensing devices such as switches, executes the stored program, and delivers corresponding output signals to various load control devices such as relay coils and solenoids.

The PLC of Figure 17-17 stores user programs in nonvolatile flash memory for safety. Thus, no battery is needed to prevent loss of user programs if electrical power is lost.

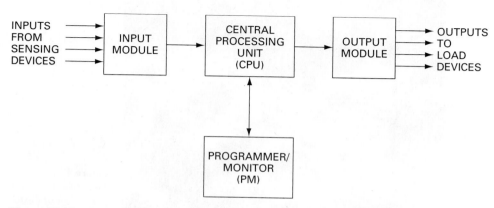

Figure 17-18. Block diagram of a PLC.

2. Programmer/monitor (PM). This unit allows the user to enter the desired program into the RAM memory of the CPU. The program, which is entered in relay ladder logic (similar to the relay ladder logic diagrams in Chapter 15), determines the sequence of operation of the fluid power system being controlled. The PM needs to be connected to the CPU only when entering or monitoring the program. Programming is accomplished by pressing keys on the PM's keypad. The programmer/monitor may be either a handheld device with a light-emitting diode (LED) or a desktop device with a cathode-ray tube (CRT) display.

Figure 17-19 shows the remote keypad interface panel used for the PLC of Figure 17-17. This remote keypad panel provides the operator with the ability to interact with the PLC during machine setup and operation. It features a high-visibility display with a wide viewing angle for displaying stored messages. The keypad allows the operator to run programs continuously or in single-step mode, monitor all real-time functions, edit or monitor programs, and print programs and output code. Figure 17-20 shows an application of this PLC where doors are being assembled onto automobiles using a pneumatic clamping system.

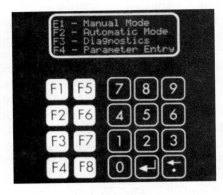

Figure 17-19.   Remote keypad interface panel for PLC. *(Courtesy of Festo Corp., Hauppauge, New York.)*

Figure 17-20.   Automobile assembly application of PLC. *(Courtesy of Festo Corp., Hauppauge, New York.)*

3. Input/output module (I/O). This module is the interface between the fluid power system input sensing and output load devices and the CPU. The purpose of the I/O module is to transform the various signals received from or sent to the fluid power interface devices such as push-button switches, pressure switches, limit switches, motor relay coils, solenoid coils, and indicator lights. These interface devices are hard-wired to terminals of the I/O module.

The PLC of Figure 17-17 contains a powerful 16-bit, 20-MHz processor. Multi-tasking of up to 64 programs allows the user to divide the control task into manageable objects. As shown in Figure 17-17, this PLC contains 12 inputs and 8 outputs, each of which are monitored by light-emitting diodes. This PLC can be programmed off-line in a ladder logic diagram using a microcomputer. Capabilities include project creation, program editing, loading, and documentation. The system allows the user to monitor the controlled process during operation and provides immediate information regarding the status of timers, counters, inputs, and outputs.

PLC Control of a Hydraulic Cylinder

To show how a PLC operates, let's look at the system of Figure 15-11 (repeated in Figure 17-21), which shows the control of a hydraulic cylinder using a single limit switch.

For this system, the wiring connections for the input and output modules are shown in Figure 17-22(a) and (b), respectively. Note that there are three sensing input devices to be connected to the input module and one output control/load device to be connected to the output module. The electrical relay is not included in the I/O connection diagram since its function is replaced by an internal PLC control relay.

The PLC ladder logic diagram that would be constructed and programmed into the memory of the CPU is shown in Figure 17-23(a). Note that the layout of the PLC ladder diagram [Figure 17-23(a)] is similar to the layout of the hard-wired relay ladder diagram (Figure 17-21). The two rungs of the relay ladder diagram are converted to two rungs of the PLC ladder logic diagram. The terminal numbers used on the I/O connection diagram are the same numbers used to identify the electrical devices on the PLC ladder logic diagram. The symbol —$\mathcal{V}$— represents a normally closed set of contacts and the symbol —||— represents a normally open set of contacts. The symbol —( )— with number 030 represents the relay coil that controls the two sets of contacts with number 030, which is the address in memory for this internal relay. The relay coil and its two sets of contacts are programmed as internal relay equivalents. The symbol —( )— with number 010 represents the solenoid.

A PLC is a digital solid-state device and thus performs operations based on the three fundamental logic functions: AND, OR, and NOT. Each rung of a ladder diagram can be represented by a Boolean equation. The hard-wired ladder diagram logic is fixed and can be changed only by modifying the way the electrical components are wired. However, the PLC ladder diagram contains logic functions that are programmable and thus easily changed. Figure 17-23(b) shows the PLC ladder logic diagram

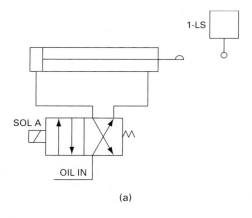

(a)

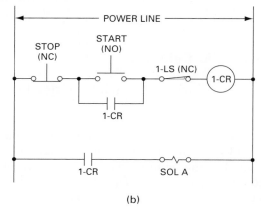

(b)

Figure 17-21. Control of a hydraulic cylinder using a single limit switch.

of Figure 17-23(a) with capital letters used to represent each electrical component. With the letter designation, Boolean equations can be written for each rung as follows:

Top rung:

$$\overline{A} \cdot (B + C) \cdot \overline{D} = E$$

This equation means: NOT A AND (B OR C) AND NOT D EQUALS E. It can also be stated as follows, noting that 0 is the OFF state and 1 is the ON state: E is energized when A is NOT actuated AND B OR C is actuated AND D is NOT actuated.

Bottom rung:

$$F = G$$

G is energized when F is actuated.

Since PLCs use ladder logic diagrams, the conversion from existing electrical relay logic to programmed logic is easy to accomplish. Each rung contains devices

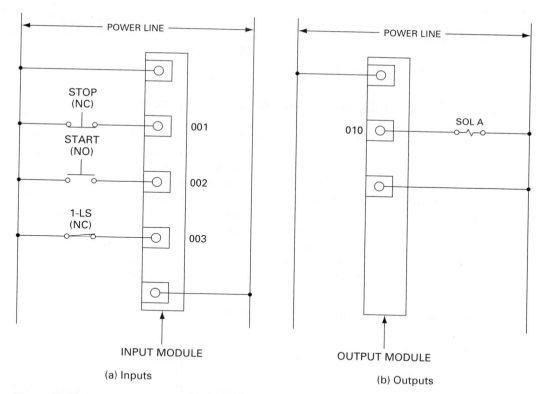

Figure 17-22. I/O connection diagram.

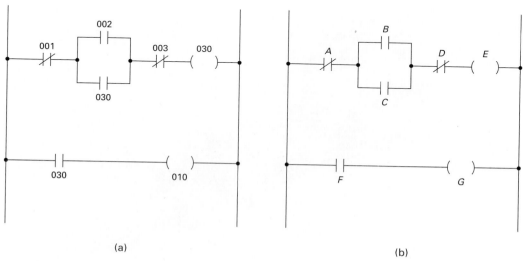

Figure 17-23. PLC ladder logic diagrams.

connected from left to right, with the device to the far right being the output. The devices are connected in series or parallel to produce the desired logical result. PLC programming can thus be readily implemented to provide the desired control capability for a particular fluid power system.

PLC Control of Electrohydraulic Servo System

Figure 17-24 shows a PLC-controlled electrohydraulic linear servo actuator designed to move and position loads quickly, smoothly, and accurately. As shown in the cutaway view of Figure 17-25, the linear actuator includes, in its basic configuration, a hydraulic servo cylinder (with built-in electronic transducer) complete with electrohydraulic valves and control electronics so that it forms one integrated machine element. As shown in the block diagram of Figure 17-26, the control electronics is of the closed-loop type, and programmable cycles are capable of generating the reference command signal. Typical applications for this system include sheet metal punching and bending, lifting and mechanical handling, radar and communications control systems, blow molding machines, machine tools, and mobile machinery.

PLC Control of Jackknife Carloader

Figure 17-27 shows an application in which an electrohydraulic robotic system is used for loading cars onto a trailer. Conventional car-transport trailers are difficult to load and keep secure, and the cars they carry are at the mercy of the environment. This system overcomes all these problems. It uses a closed trailer with a built-in electrohydraulic system that picks up a car and inserts it into the trailer. The work is done with articulating arms, flush to the inside walls of the trailer.

Figure 17-24.   PLC-controlled electrohydraulic linear servo actuator. *(Courtesy of Atos Systems, Inc., East Brunswick, New Jersey.)*

A robot carriage rides along horizontal tracks just under the trailer's roof. The robot carriage travels to the designated position and deposits its load on heavy pins that swing out automatically.

Eight hydraulic cylinders power the linkages, commanded by an integral micro-computer that is programmed to perform the desired task. The keyboard is on a pendant, connected to the electrohydraulic system via a plug-in umbilical. The

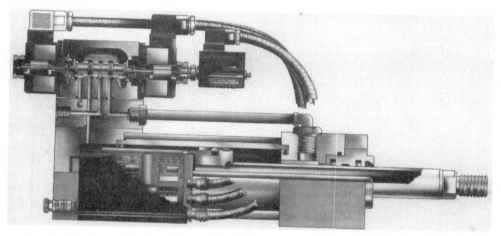

Figure 17-25. Cutaway view of PLC-controlled electrohydraulic linear servo actuator. *(Courtesy of Atos Systems, Inc., East Brunswick, New Jersey.)*

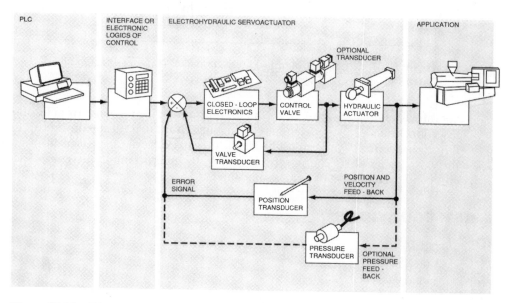

Figure 17-26. Block diagram of PLC control for electrohydraulic linear servo actuator. *(Courtesy of Atos Systems, Inc., East Brunswick, New Jersey.)*

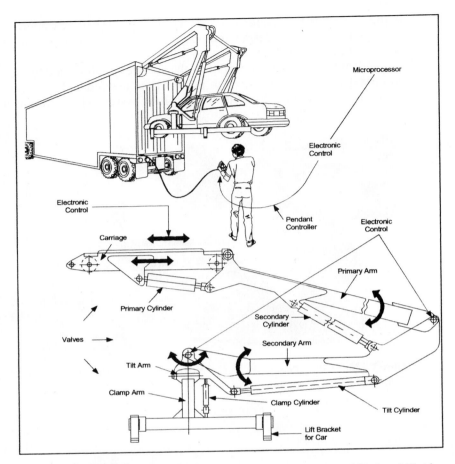

Figure 17-27. PLC control of jackknife carloader. *(Courtesy of National Fluid Power Association, Milwaukee, Wisconsin.)*

operator, standing alongside, may call for fully automatic response from liftoff to full insertion, or choose step-by-step motions such as lift and tilt forward. Proportional directional control valves adjust hydraulic flow to each cylinder. Electronic signals to the valve solenoids are purposely ramped by the microcomputer to control acceleration and deceleration of linkage movements.

17.5 KEY EQUATIONS

Electrohydraulic servo systems

$$\text{Open-loop gain} = GH \tag{17-2}$$

$$\text{Closed-loop transfer function} = \frac{\text{system output}}{\text{system input}} \tag{17-3}$$

$$\text{Closed-loop transfer function} = \frac{\text{forward gain}}{1 + \text{open loop gain}} = \frac{G}{1 + GH} \qquad \textbf{(17-4) and}$$
$$\textbf{(17-5)}$$

$$\text{Open-loop gain} = GH = G_A \times G_{SV} \times G_{cyl} \times H \qquad \textbf{(17-6)}$$

$$RE = \text{repeatable error (in)} = \frac{\text{system deadband (mA)}}{G_A(\text{mA/V}) \times H(\text{V/in})} \qquad \textbf{(17-7)}$$

$$\text{Natural frequency of oil} = \omega_H = A\sqrt{\frac{2\beta}{VM}} \qquad \textbf{(17-8)}$$

$$\text{Open-loop gain} = \frac{v_H(\text{rad/s})}{3} \qquad \textbf{(17-9)}$$

$$\text{Tracking error} = TE \text{ (in)} = \frac{\text{servo valve maximum current (mA)}}{G_A(\text{mA/V}) \times H(\text{V/in})} \qquad \textbf{(17-10)}$$

EXERCISES

Questions, Concepts, and Definitions

17-1. Name two reasons an open-loop system does not provide a perfectly accurate control of its output actuator.

17-2. Name two types of feedback devices used in closed-loop systems. What does each device accomplish?

17-3. What is the definition of a *feedback transducer?*

17-4. What two basic components of an open-loop system does a servo valve replace?

17-5. What is a transfer function?

17-6. What is the difference between deadband and hysteresis?

17-7. Define the term *open-loop gain.*

17-8. Define the term *closed-loop transfer function.*

17-9. What is meant by the term *repeatable error?*

17-10. What is meant by the term *tracking error?*

17-11. What is the difference between the forward and feedback paths of a closed-loop system?

17-12. What is a programmable logic controller?

17-13. How does a PLC differ from a general-purpose computer?

17-14. What is the difference between a programmable logic controller and an electromechanical relay control?

17-15. Name three advantages that PLCs provide over electromechanical relay control systems.

17-16. State the main function of each of the following elements of a PLC:

   a. CPU

   b. Programmer/monitor

   c. I/O module

17-17. What is the difference between read only memory (ROM) and random access memory (RAM)?

Problems

Note: The letter *E* following an exercise number means that English units are used. Similarly, the letter *M* indicates metric units.

Performance of Electrohydraulic Servo Systems

17-18E. An electrohydraulic servo system contains the following characteristics:

$$G_{cyl} = 0.15 \text{ in/in}^3, \text{ cylinder pistion area} = 6.67 \text{ in}^2$$
$$H = 3.5 \text{ V/in}$$
$$V_{oil} = 40 \text{ in}^3$$
$$\text{weight of load} = 750 \text{ lb}$$
$$\text{system deadband} = 3.5 \text{ mA}$$
$$\text{bulk modulus of oil} = 200{,}000 \text{ lb/in}^2$$
$$\text{system accuracy} = 0.002 \text{ in}$$

Determine the gain of the servo valve in units of $(\text{in}^3/\text{s})/\text{mA}$.

17-19E. What is the closed-loop transfer function for the system of Exercise 17-18?

17-20M. An electrohydraulic servo system contains the following characteristics:

$$G_{cyl} = 0.04 \text{ cm/cm}^3, \text{ cylinder pistion area} = 25 \text{ cm}^2$$
$$H = 1.75 \text{ V/cm}$$
$$V_{oil} = 750 \text{ cm}^3$$
$$\text{mass of load} = 300 \text{ kg}$$
$$\text{system deadband} = 3.5 \text{ mA}$$
$$\text{bulk modulus of oil} = 1400 \text{ MPa}$$
$$\text{system accuracy} = 0.004 \text{ cm}$$

Determine the gain of the servo valve in units of $(\text{cm}^3/\text{s})/\text{mA}$.

17-21M. What is the closed-loop transfer function for the system of Exercise 17-20?

17-22E. For the system of Exercise 17-18, if the maximum amplifier current is 250 mA, find the maximum tracking error.

17-23M. For the system of Exercise 17-20, if the maximum amplifier current is 250 mA, find the maximum tracking error.

Programmable Logic Control of Fluid Power

17-24. Draw the PLC ladder logic diagram and write the Boolean statements and equations for the hard-wired relay ladder diagram shown in Figure 15-12.

17-25. Draw the PLC ladder logic diagram and write the Boolean statements and equations for the hard-wired relay ladder diagram shown in Figure 15-14.

17-26. Draw the PLC ladder logic diagram and write the Boolean statements and equations for the hard-wired relay ladder diagram shown in Figure 15-20.

17-27. Draw the I/O connection diagram and PLC ladder logic diagram that will replace the hard-wired relay logic diagram shown in Figure 15-21.

17-28. Draw the I/O connection diagram and PLC ladder logic diagram that will replace the hard-wired relay logic diagram shown in Figure 15-24.

17-29. Draw the PLC ladder logic diagram rung for each of the following Boolean algebra equations:

   a. $Z = A + B$

   b. $Z = A \cdot B$

   c. $Z = A \cdot (B + C)$

   d. $Z = (A + B) \cdot C \cdot D$

   e. $Z = A \cdot \overline{B} \cdot C + \overline{D} + E$

Automation Studio™ Computer Software

Learning Objectives

Upon completing this chapter and playing the CD included with this textbook, you should be able to:

1. Identify the salient features of Automation Studio computer software.
2. Obtain a dynamic and visual presentation of the creation, simulation, analysis, and animation of many of the fluid power circuits studied in class or assigned as homework exercises.
3. Understand how the design and creation of fluid power circuits for a given application are accomplished using Automation Studio.
4. Explain the difference between the simulation of circuits and the animation of components.
5. Appreciate the capability of Automation Studio when connected to a fluid power trainer or another automated system.
6. Describe the use of Automation Studio in the design and operation of virtual systems.

18.1 INTRODUCTION

Definition of Automation Studio

Automation Studio is a computer software package that allows users to design, simulate, and animate circuits consisting of various automation technologies including hydraulics, pneumatics, PLCs, electrical controls, and digital electronics. Automation Studio is readily useable by instructors and students (education version) as well as by technicians and engineers (professional version). The professional version allows technicians and engineers to design, mathematically analyze,

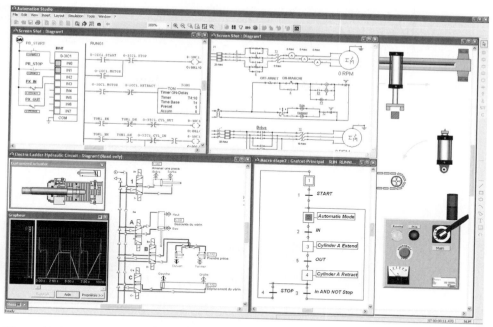

Figure 18-1. Automation Studio allows for the use of multiple technologies that are linked during simulation. *(Courtesy of Famic Technologies, Inc., St.-Laurent, QC, Canada.)*

test, and troubleshoot projects in industry to improve product quality, reduce costs, and increase productivity.

The education version of Automation Studio readily integrates into course content. For classroom presentations, instructors can customize their circuits and insert faulty components or electrical sequences to enhance student troubleshooting abilities. For course projects students can design/create virtual circuits, simulate circuit behavior, animate component operation, and perform a mathematical analysis to optimize system performance. As shown in the computer screen photograph of Figure 18-1, Automation Studio allows for the use of multiple technologies in projects that are linked during simulation.

Automation Studio CD

Included with this textbook is a CD (produced by Famic Technologies, Inc.) that illustrates how Automation Studio is used to create, simulate, and animate the following 16 fluid power circuits presented throughout the book:

- Four hydraulic circuits: Figures 9-3, 9-5, 9-9, and 9-16.
- Four pneumatic circuits: Figures 14-7, 14-11, 14-18, and 14-19.
- Four electrohydraulic circuits: Figures 15-11, 15-15, 15-18, and 15-24.
- Four electropneumatic circuits: Figures 15-14, 15-16, 15-20, and 15-21.

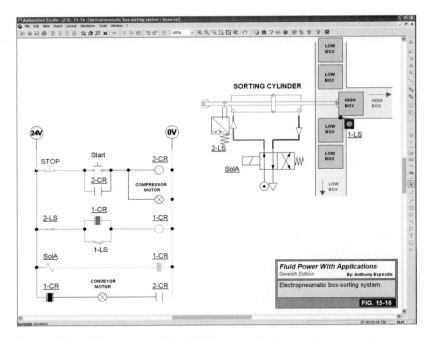

Figure 18-2. Simulation of the electropneumatic box sorting system in Figure 15-16. *(Courtesy of Famic Technologies, Inc., St.-Laurent, QC, Canada.)*

A notation is made on each of the above 16 figures within the textbook, that the corresponding circuit is represented on the CD. By playing this CD on a personal computer, the student obtains a dynamic and visual presentation of the creation, simulation, analysis, and animation of many of the fluid power circuits studied in class or assigned as homework exercises. Figure 18-2 is a computer screen photograph that shows the electropneumatic box sorting system in Figure 15-16 as it is being simulated on the CD. Sections 18.2 through 18.6 describe the salient features of the education version of Automation Studio.

18.2 DESIGN/CREATION OF CIRCUITS

The design and creation of circuits for a given application is readily accomplished using Automation Studio. Using a personal computer, the user simply clicks on the selected component (shown in graphical symbol form), drags the component, and drops it onto a workspace on the computer screen. The components desired are found either from a main library provided or from a library customized by the user. The library of components is displayed in drop-down menus on the left side of the computer screen.

Complying with ISO standards, the Hydraulics Library contains all the component symbols required to create hydraulic systems. The library includes hundreds of symbols of hydraulic components such as directional control/flow control/pressure control valves, pumps, reservoirs, cylinders, motors, and accumulators to create all types of systems, from simple to complex. Components are preconfigured but can be sized to realistically reproduce the system behavior by considering parameters such as pressure, flow, and pressure drops. Simulation parameters such as external loads, fluid leaks, thermal phenomena, fluid viscosity, and flow characteristics can also be configured.

The Pneumatic Library contains all the symbols necessary to create pneumatic and moving part logic systems. As in the Hydraulic Module, the parameters of pneumatic components can be configured to show a realistic behavior. The Electrical Controls Library interacts with all the components from other libraries to create electrically controlled systems. It includes switches, relays, solenoids, push buttons, and many other electrical components.

Automation Studio also contains three libraries of PLC Ladder Logic, including all ladder logic functions such as contacts, input/output, timers, counters, logic test, and mathematical functions. This allows the user to create and simulate the control part of an automated system. Combined with the other libraries, the Programmable Logic Controller Libraries allow for the implementation of a complete virtual factory. Figure 18-3 is a photograph that displays the complete library of components in a drop-down menu located on the left side of the computer screen.

18.3 SIMULATION OF CIRCUITS

Circuit simulation begins as soon as a valve, electric switch, or some other component is actuated on the computer screen. In this phase the operation of the entire circuit is deployed in a dynamic and visual way showing how it actually works. During simulation, components are animated and pipelines and electric wires are color-coded according to their state. For example, high-pressure hydraulic pipelines are colored in red and low-pressure pipelines are colored in blue. Simulation can then help to explain system operation, from the component up to the system level, and allow the student to more readily assimilate theories and concepts studied in class. The simulation paces "Normal," "Slow Motion," "Step by Step," and "Pause" allow the user to control the simulation speed of selected diagrams.

During simulation a mathematical analysis can be performed to determine whether all selected component characteristics are adequate for proper system operation. For example, in a hydraulic system the pump flow rate, pump discharge pressure, and cylinder external load can be numerically specified. Automation Studio will then calculate the remaining system parameters such as cylinder piston diameter and velocity. System parameters can be readily changed and new calculations performed until optimum components are determined. With a simple drag-and-drop operation, the user can also plot simulated parameters and variables such as cylinder velocity and pump discharge pressure. The results can be exported into a text file or spreadsheet for further analysis. Figure 18-4 is a computer screen photograph showing an analysis example dealing with a double-acting hydraulic cylinder.

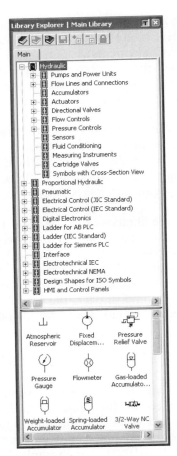

Figure 18-3. Complete display of components in main library. *(Courtesy of Famic Technologies, Inc., St.-Laurent, QC, Canada.)*

18.4 ANIMATION OF COMPONENTS

While circuit simulation is occurring, any component can be animated. Automation Studio does this by displaying in color, a cross-sectional view of the component such as a pump, cylinder, or pressure relief valve with its internal parts in motion. These animations thus illustrate the internal operating features of components. An example is shown for a single-acting, spring return cylinder in the computer screen photograph of Figure 18-5. The animations are synchronized with the circuit simulation.

18.5 INTERFACES TO PLCs AND EQUIPMENT

Automation Studio can also be connected to a hydraulic or pneumatic trainer or another automated system. This can be done by using an I/O interface kit or OPC module. The I/O interface kit is a hardware solution that allows connecting 8 inputs

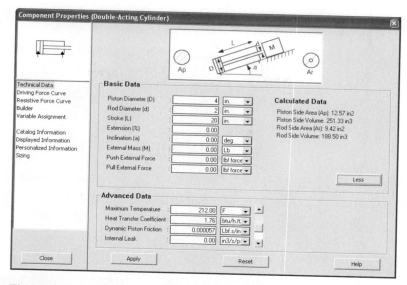

Figure 18-4. Analysis of a double-acting hydraulic cylinder. *(Courtesy of Famic Technologies, Inc., St.-Laurent, QC, Canada.)*

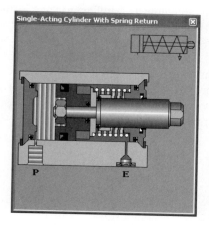

Figure 18-5. Cross-section animations illustrate the internal functioning of components. *(Courtesy of Famic Technologies, Inc., St.-Laurent, QC, Canada.)*

and 8 outputs directly to a PLC or to real equipment such as relays, contacts, valves, and sensors. The OPC module is a standard software interface that allows Automation Studio to exchange data with any PLC or other control devices for which a manufacturer supplies OPC server software.

18.6 VIRTUAL SYSTEMS

Virtual System gives the user access to inputs and outputs that can be controlled by the Electrical Control library, PLC libraries, and the SFC (Grafcet) module. The

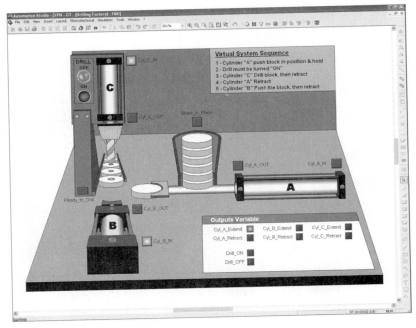

Figure 18-6. Simulation of an automated drilling machine Virtual System using Electrical, PLC, or SFC modules. *(Courtesy of Famic Technologies, Inc., St.-Laurent, QC, Canada.)*

user needs to correctly link sensors, switches, lights, conveyors, etc., in order to make the Virtual System function properly. This is accomplished in a safe, virtual environment. Figure 18-6 is a computer screen photograph that illustrates the simulation of an automated drilling machine virtual system.

Appendixes

Sizes of Steel Pipe (English Units)

Nominal Pipe Size (in)	Outside Diameter (in)	Inside Diameter (in)	Wall Thickness (in)	Internal Area (in^2)
Schedule 40				
⅛	0.405	0.269	0.068	0.0568
¼	0.540	0.364	0.088	0.1041
⅜	0.675	0.493	0.091	0.1910
½	0.840	0.622	0.109	0.304
¾	1.050	0.824	0.113	0.533
1	1.315	1.049	0.133	0.864
1¼	1.660	1.380	0.140	1.496
1½	1.900	1.610	0.145	2.036
2	2.375	2.067	0.154	3.36
2½	2.875	2.469	0.203	4.79
3	3.500	3.068	0.216	7.39
3½	4.000	3.548	0.226	9.89
4	4.500	4.026	0.237	12.73
Schedule 80				
⅛	0.405	0.215	0.095	0.0364
¼	0.540	0.302	0.119	0.0716
⅜	0.675	0.423	0.126	0.1405
½	0.840	0.546	0.147	0.2340
¾	1.050	0.742	0.154	0.432
1	1.315	0.957	0.179	0.719
1¼	1.660	1.278	0.191	1.283

Nominal Pipe Size (in)	Outside Diameter (in)	Inside Diameter (in)	Wall Thickness (in)	Internal Area (in²)
		Schedule 80		
1½	1.900	1.500	0.200	1.767
2	2.375	1.939	0.218	2.953
2½	2.875	2.323	0.276	4.24
3	3.500	2.900	0.300	6.61
3½	4.000	3.364	0.318	8.89
4	4.500	3.826	0.337	11.50

Sizes of Steel Pipe (Metric Units)

Nominal Pipe Size (in)	Outside Diameter (mm)	Inside Diameter (mm)	Wall Thickness (mm)	Internal Area (mm²)
Schedule 40				
⅛	10.3	6.8	1.7	36.6
¼	13.7	9.2	2.2	67.1
⅜	17.1	12.5	2.3	123.1
½	21.3	15.8	2.8	195.9
¾	26.7	20.9	2.9	343.9
1	33.4	26.6	3.4	557.3
1¼	42.2	35.1	3.6	964.5
1½	48.3	40.9	3.7	1312.8
2	60.3	52.5	3.9	2163.8
2½	73.0	62.7	5.2	3087.3
3	88.9	77.9	5.5	4767.0
3½	101.6	90.1	5.7	6375.4
4	114.3	102.3	6.0	8208.9
Schedule 80				
⅛	10.3	5.5	2.4	23.4
¼	13.7	7.7	3.0	46.2
⅜	17.1	10.7	3.2	90.6
½	21.3	13.9	3.7	151.0

Nominal Pipe Size (in)	Outside Diameter (mm)	Inside Diameter (mm)	Wall Thickness (mm)	Internal Area (mm²)
		Schedule 80		
¾	26.7	18.8	3.9	278.8
1	33.4	24.3	4.5	463.8
1¼	42.2	32.5	4.9	827.2
1½	48.3	38.1	5.1	1139.5
2	60.3	49.3	5.5	1904.1
2½	73.0	59.0	7.0	2733.0
3	88.9	73.7	7.6	4259.2
3½	101.6	85.4	8.1	5731.2
4	114.3	97.2	8.6	7413.6

Sizes of
Steel Tubing
(English Units)

Outside Diameter (in)	Wall Thickness (in)	Inside Diameter (in)	Inside Area (in²)	Outside Diameter (in)	Wall Thickness (in)	Inside Diameter (in)	Inside Area (in²)
⅛	0.028	0.069	0.00374	½	0.035	0.430	0.1452
	0.032	0.061	0.00292		0.049	0.402	0.1269
	0.035	0.055	0.00238		0.065	0.370	0.1075
					0.083	0.334	0.0876
³⁄₁₆	0.032	0.1235	0.01198				
	0.035	0.1175	0.01084	⅝	0.035	0.555	0.2419
					0.049	0.527	0.2181
¼	0.035	0.180	0.02544		0.065	0.495	0.1924
	0.049	0.152	0.01815		0.083	0.459	0.1655
	0.065	0.120	0.01131				
				¾	0.049	0.652	0.3339
⁵⁄₁₆	0.035	0.2425	0.04619		0.065	0.620	0.3019
	0.049	0.2145	0.03614		0.083	0.584	0.2679
	0.065	0.1825	0.02616		0.109	0.532	0.2223
				⅞	0.049	0.777	0.4742
⅜	0.035	0.305	0.07306		0.065	0.709	0.4359
	0.049	0.277	0.06026		0.083	0.709	0.3948
	0.065	0.245	0.04714		0.109	0.657	0.3390

Outside Diameter (in)	Wall Thickness (in)	Inside Diameter (in)	Inside Area (in²)	Outside Diameter (in)	Wall Thickness (in)	Inside Diameter (in)	Inside Area (in²)
1	0.049	0.902	0.6390	1¾	0.065	1.620	2.061
	0.065	0.870	0.5945		0.083	1.584	1.971
	0.083	0.834	0.5463		0.109	1.532	1.843
	0.109	0.782	0.4803		0.134	1.482	1.725
1¼	0.049	1.152	1.0423	2	0.065	1.870	2.746
	0.065	1.120	0.9852		0.083	1.834	2.642
	0.085	1.084	0.9229		0.109	1.782	2.494
	0.109	1.032	0.8365		0.134	1.732	2.356
1½	0.065	1.370	1.474				
	0.083	1.334	1.398				
	0.109	1.282	1.291				

Sizes of Steel Tubing (Metric Units)

Outside Diameter (mm)	Wall Thickness (mm)	Inside Diameter (mm)	Inside Area (mm²)	Outside Diameter (mm)	Wall Thickness (mm)	Inside Diameter (mm)	Inside Area (mm²)
4	0.5	3	7.1	20	2.0	16	201.0
6	1.0	4	12.6	20	2.5	15	176.6
6	1.5	3	7.1	22	3.0	14	153.9
8	1.0	6	28.3	22	1.0	20	314.0
8	1.5	5	19.6	22	1.5	19	283.4
8	2.0	4	12.6	22	2.0	18	254.3
10	1.0	8	50.2	25	3.0	19	283.4
10	1.5	7	38.5	25	4.0	17	226.9
10	2.0	6	28.3	28	2.0	24	452.2
12	1.0	10	78.5	28	2.5	23	415.3
12	1.5	9	63.4	30	3.0	24	452.2
12	2.0	8	50.2	30	4.0	22	380.0
14	2.0	10	78.5	35	2.0	31	754.4
15	1.5	12	113.0	35	3.0	29	660.2
15	2.0	11	95.0	38	4.0	30	706.5
16	2.0	12	113.0	38	5.0	28	615.4
16	3.0	10	78.5	42	2.0	38	1133.5
18	1.5	15	176.6	42	3.0	36	1017.4

Unit Conversion Factors

English to Metric Conversions

Parameter	English Unit	Metric Unit
Area	1 in^2	6.452 cm^2
	1 ft^2	0.0929 m^2
Density	1 slug/ft^3	515 kg/m^3
Energy	1 ft · lb	1.356 J
Flow rate	1 ft^3/s	0.0284 m^3/s
	1 gpm	0.06309 Lps
Force	1 lb	4.448 N
Length	1 ft	0.3048 m
	1 in	2.540 cm
Mass	1 slug	14.59 kg
Power	1 ft · lb/s	1.356 W
	1 hp	745.7 W
	1 Btu/min	0.0176 kW
Pressure	1 psi	6895 Pa
	1 psi	0.06895 bars
	1 standard atmosphere (14.7 psia)	101.3 kPa abs
Specific weight	1 lb/ft^3	157 N/m^3
Velocity	1 ft/s	0.3048 m/s
Viscosity (absolute)	1 lb · s/ft^2	47.88 Pa · s
Viscosity (kinematic)	1 ft^2/s	0.0929 m^2/s
Volume	1 in^3	16.39 cm^3
	1 gal	3.785 L

English to English Conversions

Parameter	First English Unit	Second English Unit
Energy	1 Btu	778 ft $\cdot$ lb
Flow rate	1 ft^3/s	449 gpm
Power	1 hp	550 ft $\cdot$ lb/s
	1 hp	42.4 Btu/min
Velocity (angular)	1 rpm	0.1047 rad/s
Volume	1 gal	231 in^3
	1 gal	0.135 ft^3

Metric to Metric Conversions

Parameter	First Metric Unit	Second Metric Unit
Force	1 N	10^5 dyn
Pressure	1 bar	10^5 Pa
Velocity (ang.)	1 rpm	0.1047 rad/s
Viscosity (abs.)	1 N $\cdot$ s/m^2	10 poise
	1 poise	1 dyn $\cdot$ s/cm^2
Viscosity (kin.)	1 m^2/s	10,000 stokes
	1 stoke	1 cm^2/s
Volume	1 m^3	1000 L
	1 cm^3	0.001 L

Temperature Conversions

$$T(°F) = 1.8\, T(°C) + 32 \qquad T(°C) = \frac{T(°F) - 32}{1.8}$$

$$T(°R) = 1.8\, T(K) \qquad T(K) = \frac{T(°R)}{1.8}$$

$$T(°R) = T(°F) + 460 \qquad T(K) = T(°C) + 273$$

Nomenclature

Roman Alphabetic Symbols

Symbol	Definition	Units
a	Acceleration	in/s^2, ft/s^2, cm/s^2, m/s^2
A	Area	in^2, ft^2, cm^2, m^2
BP	Burst pressure	psi (lb/in^2), MPa
C	Flow coefficient	None
CF	Coefficient of friction	None
CR	Compression ratio	None
C_v	Air flow capacity constant	None
C_v	Liquid capacity coefficient	$gpm/\sqrt{psi}$, $Lpm/\sqrt{kPa}$
D	Diameter	in, ft, cm, m
D_i	Inside diameter	in, ft, cm, m
D_o	Outside diameter	in, ft, cm, m
E	Energy	$lb \cdot in$, $lb \cdot ft$, $N \cdot cm$, $N \cdot m$
F	Force	lb, N
FS	Factor of safety	None
f	Friction factor	None
g	Acceleration of gravity	in/s^2, ft/s^2, cm/s^2, m/s^2
G	Gain (transfer function)	in/V, cm/V, mm/V
G_A	Amplifier gain	mA/V
G_{SV}	Servo valve gain	$(in^3/s)/mA$, $(cm^3/s)/mA$
G_{cyl}	Cylinder gain	in/in^3, cm/cm^3
H	Feedback transducer gain	V/in, V/cm, V/mm
H	Head	in, ft, cm, m
HP	Horsepower	hp
H_L	Head loss	in, ft, cm, m
H_m	Motor head	in, ft, cm, m

Symbol	Definition	Units
H_p	Pump head	in, ft, cm, m
ID	Inside diameter	in, ft, cm, m
J	Joule	$N \cdot m$
K	K factor	None
KE	Kinetic energy	$lb \cdot in, lb \cdot ft, N \cdot cm, N \cdot m$
L	Length	in, ft, cm, m
L_e	Equivalent length	in, ft, cm, m
m	Mass	Slugs, kg
n	Rotational (speed)	rpm (rev/min), rad/s
N	Force in newtons	$kg \cdot m/s^2$
N_R	Reynolds number	None
OD	Outside diameter	in, ft, cm, m
p	Pressure	$lb/in^2, lb/ft^2, N/m^2$, bars
Pa	Pascal	N/m^2
Q	Volume flow rate	$in^3/s, ft^3/s, cm^3/s, m^3/s$, gpm, Lps
Q_A	Actual flow rate	$in^3/s, ft^3/s, cm^3/s, m^3/s$, gpm, Lps
Q_T	Theoretical flow rate	$in^3/s, ft^3/s, cm^3/s, m^3/s$, gpm, Lps
RE	Repeatable error	in, cm, mm
s	Tensile strength	lb/in^2 (psi), N/m^2 (Pa)
S	Distance	in, ft, cm, m
SG	Specific gravity	None
t	Time	s
T	Temperature	°F, °R, °C, K
T	Torque	$lb \cdot in, lb \cdot ft, N \cdot cm, N \cdot m$
T_A	Actual torque	$lb \cdot in, lb \cdot ft, N \cdot cm, N \cdot m$
T_T	Theoretical torque	$lb \cdot in, lb \cdot ft, N \cdot cm, N \cdot m$
TE	Tracking error	in, cm, mm
υ	Velocity	in/s, ft/s, cm/s, m/s
V	Volume	in^3, ft^3, cm^3, m^3, L
V_D	Volumetric displacement	in^3, ft^3, cm^3, m^3, L
VI	Viscosity index	None
w	Weight flow rate	lb/s, lb/min, N/s, N/min
W	Watt	J/s
WP	Working pressure	lb/in^2 (psi), N/m^2 (Pa)
Z	Elevation	in, ft, cm, m
Δp	Pressure drop	$lb/in^2, lb/ft^2, N/cm^2, N/m^2$

Greek Alphabetic Symbols

Symbol	Phonetic	Definition	Units
β	Beta	Bulk modulus	lb/in^2 (psi), Pa
γ	Gamma	Specific weight	lb/in^3, lb/ft^3, N/cm^3, N/m^3
η	Eta	Efficiency	None
η_m	—	Mechanical efficiency	None
η_o	—	Overall efficiency	None
η_v	—	Volumetric efficiency	None
θ	Theta	Angle	Degrees, radians
ω	Omega	Natural frequency	rads/s
μ	Mu	Absolute viscosity	lb · s/ft^2, poise, cP, N · s/m^2
ν	Nu	Kinematic viscosity	ft^2/s, cm^2/s, stoke, cS, m^2/s
π	Pi	A constant	None
ρ	Rho	Density	slugs/ft^3, kg/m^3
τ	Tau	Shear stress	lb/in^2, lb/ft^2, N/m^2

Metric System Prefixes

Symbol	Name	Multiplication Factor
T	tera	10^{12}
G	giga	10^{9}
M	mega	10^{6}
k	kilo	10^{3}
c	centi	10^{-2}
m	milli	10^{-3}
μ	micro	10^{-6}
n	nano	10^{-9}
p	pico	10^{-12}

Fluid Power Symbols

G

THE SYMBOLS SHOWN CONFORM TO THE AMERICAN NATIONAL STANDARDS INSTITUTE (ANSI) SPECIFICATIONS. BASIC SYMBOLS CAN BE COMBINED IN ANY COMBINATION. NO ATTEMPT IS MADE TO SHOW ALL COMBINATIONS.			
LINES AND LINE FUNCTIONS		**LINES AND LINE FUNCTIONS (CONT.)**	
LINE, WORKING	————	DIRECTION OF FLOW, HYDRAULIC PNEUMATIC	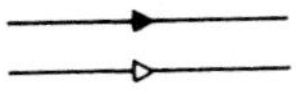
LINE, PILOT (L>20W)	– – – –		
LINE, DRAIN (L<5W)	– – – – –	LINE TO RESERVOIR ABOVE FLUID LEVEL BELOW FLUID LEVEL	
CONNECTOR	•		
LINE, FLEXIBLE		LINE TO VENTED MANIFOLD	
LINE, JOINING		PLUG OR PLUGGED CONNECTION	✕
LINE, PASSING		RESTRICTION, FIXED	

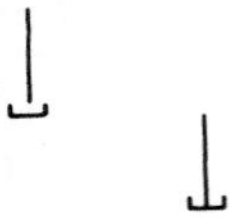

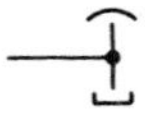

634

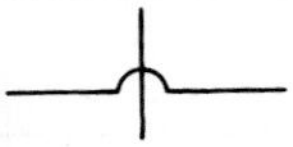

PUMPS	
PUMP, SINGLE FIXED DISPLACEMENT	
PUMP, SINGLE VARIABLE DISPLACEMENT	
MOTOR AND CYLINDERS	
MOTOR, ROTARY FIXED DISPLACEMENT	
MOTOR, ROTARY VARIABLE DISPLACEMENT	
MOTOR, OSCILLATING	
CYLINDER, SINGLE-ACTING	
CYLINDER, DOUBLE-ACTING	
CYLINDER, DIFFERENTIAL ROD	
CYLINDER, DOUBLE-END ROD	
CYLINDER, CUSHIONS BOTH ENDS	
MISCELLANEOUS UNITS	
DIRECTION OF ROTATION (ARROW IN FRONT OF SHAFT)	
COMPONENT ENCLOSURE	
RESERVOIR, VENTED	
RESERVOIR, PRESSURIZED	
PRESSURE GAGE	

MISCELLANEOUS UNITS (CONT.)	
TEMPERATURE GAGE	
FLOW METER (FLOW RATE)	
ELECTRIC MOTOR	
ACCUMULATOR, SPRING-LOADED	
ACCUMULATOR, GAS-CHARGED	
FILTER OR STRAINER	
HEATER	
COOLER	
TEMPERATURE CONTROLLER	
INTENSIFIER	
PRESSURE SWITCH	
BASIC VALVE SYMBOLS	
CHECK VALVE	
MANUAL SHUT-OFF VALVE	
BASIC VALVE ENVELOPE	
VALVE, SINGLE-FLOW PATH, NORMALLY CLOSED	
VALVE, SINGLE-FLOW PATH, NORMALLY OPEN	
VALVE, MAXIMUM PRESSURE (RELIEF)	
BASIC VALVE SYMBOL, MULTIPLE FLOW PATHS	
FLOW PATHS BLOCKED IN CENTER POSITION	
MULTIPLE FLOW PATHS (ARROW SHOWS FLOW DIRECTION)	

VALVE EXAMPLES		METHODS OF OPERATION	
UNLOADING VALVE, INTERNAL DRAIN, REMOTELY OPERATED		PRESSURE COMPENSATOR	
DECELERATION VALVE, NORMALLY OPEN		DETENT	
SEQUENCE VALVE, DIRECTLY OPERATED, EXTERNALLY DRAINED		MANUAL	
PRESSURE-REDUCING VALVE		MECHANICAL	
COUNTERBALANCE VALVE WITH INTEGRAL CHECK		PEDAL OR TREADLE	
TEMPERATURE- AND PRESSURE-COMPENSATED FLOW CONTROL WITH INTEGRAL CHECK		PUSH BUTTON	
DIRECTIONAL VALVE, TWO-POSITION, THREE-CONNECTION		LEVER	
DIRECTIONAL VALVE, THREE-POSITION, FOUR-CONNECTION		PILOT PRESSURE	
PROPORTIONAL DIRECTIONAL CONTROL VALVE, INFINITE POSITIONING (INDICATED BY HORIZONTAL BARS)		SOLENOID	
		SOLENOID-CONTROLLED, PILOT-PRESSURE-OPERATED	
		SPRING	
		SERVO	

Derivation of Key Equations

1. Pressure at the Bottom of a Column of Liquid (Equation 2-9)

The derivation of Eq. (2-9) is provided as follows:

Refer to Figure I-1 that shows an open cylindrical container filled with a liquid to a height H. The container has a bottom surface whose area A supports the weight W of the entire column of liquid. The liquid has a specific weight γ and volume V. We simply use Eq. (2-8) to determine the pressure p at the bottom of the container. The result is Eq. (2-9).

$$p = \frac{F}{A} = \frac{W}{A} = \frac{\gamma V}{A} = \frac{\gamma(AH)}{A} = \gamma H \tag{2-9}$$

Note in the preceding equation that the liquid column cross-sectional area parameter A cancels out since it is a term in both the numerator and denominator. The result shows that the pressure at the bottom of a liquid column does not depend on the cross-sectional area of the column.

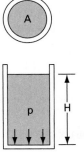

Figure I-1. Pressure p developed at bottom of liquid column of height H.

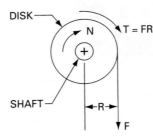

Figure I-2. Force F applied to periphery of disk creates torque T in shaft to drive load such as a pump, at rotational speed N.

2. Brake Horsepower (Equation 3-7)

Figure I-2 shows a force F applied to a string wrapped around the periphery of a circular disk mounted on a shaft. The shaft is mounted in bearings so that it can rotate at an rpm speed N as it drives a connecting shaft of a load such as a pump. The force F creates a torque T whose magnitude equals the product of the force F and its moment arm R that equals the radius of disk. The amount of horsepower (called *brake horsepower*) transmitted to the load via the shaft is determined as follows:

Per Eq.(3-3) we can determine the work W done by the force F moving through a distance S.

$$W(\text{in} \cdot \text{lb}) = F(\text{lb}) \times S(\text{in})$$

When the disk makes one complete revolution, the force F moves through a distance S equal to the circumference of the disk.

$$S(\text{in}) = 2\pi R \ (\text{in})$$

The time t for the disk to rotate one revolution is

$$t(\text{min}) = \frac{1}{N(\text{rev/min})} = \frac{1}{N(\text{rpm})}$$

Since power equals work per unit time we have

$$\text{Power}\left(\frac{\text{in} \cdot \text{lb}}{\text{min}}\right) = \frac{F(\text{lb}) \times S(\text{in})}{t(\text{min})} = F(\text{lb}) \times 2\pi R \ (\text{in}) \times N \ (\text{rpm})$$

$$= 2\pi T(\text{in} \cdot \text{lb}) \times N(\text{rpm})$$

But
$$1 \text{ HP} = 33{,}000 \, \frac{\text{ft} \cdot \text{lb}}{\text{min}} \times \frac{12 \text{ in}}{1 \text{ ft}} = 396{,}000 \, \frac{\text{in} \cdot \text{lb}}{\text{min}}$$

Thus, we have the desired result.

$$\text{HP} = 2\pi T \ (\text{in} \cdot \text{lb}) \times N \ (\text{rpm}) \times \frac{1 \text{ HP}}{396{,}000 \text{ in} \cdot \text{lb/min}} = \frac{T \ (\text{in} \cdot \text{lb}) \times N \ (\text{rpm})}{63{,}000} \qquad \textbf{(3-7)}$$

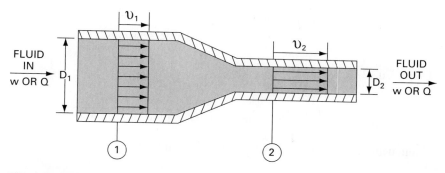

Figure I-3. Continuity of flow.

3. Continuity of Flow (Equation 3-17)

Figure I-3 shows a pipe in which fluid is flowing at a weight flow rate w that has units of weight per unit time. The pipe has two different-sized cross-sectional areas identified by stations 1 and 2. If no fluid is added or withdrawn from the pipeline between stations 1 and 2, then the weight flow rate at stations 1 and 2 must be equal:

$$w_1 = w_2$$

This means that during a time t the weight W_1 of the cylindrical chunk of fluid passing station 1 equals the weight W_2 of the cylindrical chunk of fluid passing station 2:

$$W_1 = W_2$$

But the weight of a cylindrical chunk of fluid equals the product of its specific weight γ and volume V. Thus, we have

$$\gamma_1 V_1 = \gamma_2 V_2$$

Since the volume of a cylinder equals the product of its cross-sectional area A and length L, we have

$$\gamma_1 A_1 L_1 = \gamma_2 A_2 L_2$$

Dividing both sides of the preceding equation by the time t yields

$$\gamma_1 A_1 (L_1/t) = \gamma_2 A_2 (L_2/t)$$

Since L/t equals velocity v, we have the final result:

$$\gamma_1 A_1 v_1 = \gamma_2 A_2 v_2 \qquad \textbf{(3-17)}$$

4. Pump Head (Equation 3-29)

The equation for determining the pump head H_p can be derived by combining Eqs. (2-9) and (3-25) as follows:
    Equation (2-9) is

$$p = \gamma H$$

Equation (3-25) is

$$\text{HHP} = \frac{p(\text{psi}) \times Q(\text{gpm})}{1714}$$

Using Eq. (2-9), we have

$$H_p(\text{ft}) = \frac{p(\text{lb/in}^2)}{\gamma(\text{lb/in}^3)} \times \frac{1\,\text{ft}}{12\,\text{in}} = \frac{p(\text{psi})}{12\gamma_{H_2O}(\text{lb/in}^3) \times (SG)}$$

Then, using Eq. (3-25), we obtain the final result:

$$H_p(\text{ft}) = \frac{1714 \times (\text{HHP})/Q(\text{gpm})}{12 \times 0.0361 \times (SG)} = \frac{3950 \times (\text{HHP})}{Q(\text{gpm}) \times (SG)} \qquad \textbf{(3-29)}$$

5. Brake Power (Equation 3-37)

Figure I-4 shows a force F applied to a string wrapped around the periphery of a circular disk mounted on a shaft. The shaft is mounted in bearings so that it can rotate at an rpm speed N as it drives a connecting shaft of a load such as a pump. The force F creates a torque T whose magnitude equals the product of the force F and its moment arm R that equals the radius of the disk. The amount of power (called *brake power*) transmitted to the load via the shaft is determined as follows:

The work W done by the force F moving through a distance S is

$$W(\text{N} \cdot \text{m}) = F(\text{N}) \times S(\text{m})$$

where distance S for one revolution of shaft rotation equals $2\pi R(\text{m})$.

$$\text{Also time } t(\text{min}) \text{ for one revolution} = \frac{1}{N(\text{rev/min})} = \frac{1}{N(\text{rpm})}$$

Thus, the power is

$$\text{power}\left(\frac{\text{N} \cdot \text{m}}{\text{min}}\right) = \frac{F(\text{N}) \times S(\text{m})}{t(\text{min})} = F(\text{N}) \times 2\pi R(\text{m}) \times N(\text{rpm})$$

$$= 2\pi T(\text{N} \cdot \text{m}) \times N(\text{rpm})$$

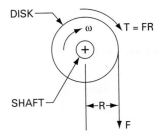

Figure I-4. Force F applied to periphery of disk creates torque T in shaft to drive a load such as a pump, at rotational speed N.

or

$$\text{power} \left(\frac{N \cdot m}{s} \right) = \text{power (W)}$$

$$= \frac{2\pi T \ (N \cdot m) \times N \ (\text{rpm})}{60} = \frac{T \ (N \cdot m) \times N(\text{rpm})}{9.55}$$

Also, since 2π radians = 1 revolution we have

$$N \ (\text{rpm}) = \omega \ (\text{rad/s}) \times \frac{1 \ \text{rev}}{2\pi \text{rad}} \times \frac{60 \ \text{s}}{1 \ \text{min}} = \frac{60}{2\pi} \ \omega (\text{rad/s}) = 9.55 \ \omega (\text{rad/s})$$

Thus,

$$\text{power (kW)} = \frac{T \ (N \cdot m) \times \omega \ (\text{rad/s})}{1000} = \frac{T(N \cdot m) \times N(\text{rpm})}{9550} \qquad \textbf{(3-37)}$$

Answers to Selected Odd-Numbered Exercises

Chapter 2

2-21. 0.881, 1.71 slugs/ft^3
2-23. 896 kg/m^3
2-25. (a) 0.00123, (b) 813
2-27. (a) 8700 N/m^3, (b) 888 kg/m^3,
(c) 0.889
2-29. 11.4 psi
2-31. 99 kPa abs

2-33. 19.5 psi
2-35. −0.0633 in^3
2-37. 507,000 psi
2-39. 43.3 cS, 39.0 cP
2-41. 1 lb · s/ft^2 = 47.88 N · s/m^2
2-43. 0.0145 ft^2/s
2-45. 0.00115 lb · s/ft^2

Chapter 3

3-15. 1000 N
3-17. 15,700 N
3-19. 900 lb
3-21. 26,565 N, 1.63%
3-23. 320.4 lb, 0.0437 hp
3-25. 30 in^2
3-27. 0.015 m^2
3-29. −1.56 psig
3-31. 0.0881 cm
3-33. 24.5 gpm
3-35. 0.0188 m
3-37. 1.273
3-39. 3.27 s
3-41. 20.4 gpm
3-43. v_2 = 5.12 m/s, v_3 = 6.97 m/s
3-45. 8.57 gpm

3-47. Derivation
3-49. 0.5 m/s
3-51. (a) 0.004 m^2, (b) 0.0015 m^3/s,
(c) 15 kW, (d) 15 kW, (e) 0.00404 m^2,
0.00153 m^3/s, 15.3 kW, 15 kW, 98.0%
3-53. (a) 33,200 N · m, (b) 481 kPa,
(c) 3.32 kW, (d) 0.0259 m/s,
(e) 0.00518 m^3/s
3-55. 3590 Pa gage
3-57. 51.8 ft · lb
3-59. 0.00618 m^3/s
3-61. (a) 13.7 psig, (b) 3.94 psig
3-63. 27.4 kPa
3-65. Q_{in} = 0.00190 m^3/s, (a) $p_B - p_A$ = 0,
(b) $p_B - p_A$ = 781 kPa

Chapter 4

4-9. 3096
4-11. Increase
4-13. **(a)** 0.0735, **(b)** 0.0245
4-15. **(a)** 0.16, **(b)** 0.032
4-17. Square
4-19. 0.0274 bars
4-21. 0.473, 4.43
4-23. 0.313 ft

4-25. 231.7 psi
4-27. 1600 kPa
4-29. −65.1 psi
4-31. 91.4 psi
4-33. 38,400 lb
4-35. 0.340 ft/s
4-37. 4.15 kW

Chapter 5

5-37. 0.371 in
5-39. 0.00106 m³/s
5-41. 9.9°
5-43. $Q(m^3/min) = V_D(m^3) \times N(rpm)$
5-45. 86.1%
5-47. $\eta_m = 96\%$, frictional HP = 0.96
5-49. 10 dB

5-51. **(a)** $\eta_o = 80.2\%$, **(b)** $T_T = 956$ in · lb
5-53. **(a)** 81.8%, **(b)** 109.6 N · m
5-55. 504.2 in · lb
5-57. **(a)** 5.47 in³, **(b)** 25.2 hp,
(c) 882 in · lb, **(d)** 72.2%
5-59. **(a)** $11,820/yr, **(b)** 0.30 years

Chapter 6

6-15. **(a)** Force does not change. Time
increases by a factor of 2, **(b)** Force
increases by a factor of 4. Time
increases by a factor of 4, **(c)** Force
increases by a factor of 4. Time
increases by a factor of 8.
6-17. 6.00 gpm
6-19. **(a)** 3.98 MPa, **(b)** 1.27 m/s,
(c) 6.37 kW, **(d)** 5.31 MPa,
(e) 1.70 m/s, **(f)** 8.50 kW

6-21. $1/4 \times p \times A_{piston}$
6-23. 354 psi
6-25. 2.19 in
6-27. **(a)** 4444 N, 2222 N, 8888 N,
(b) 4444 N, 2222 N, 8888 N
6-29. 8480 lb
6-31. 816 psi

Chapter 7

7-19. 1423 psi
7-21. 1654 in · lb
7-23. **(a)** 577.5 rpm, **(b)** 1911 in · lb,
(c) 17.5 hp
7-25. 8.16 gpm, 4.76 hp
7-27. 93.3 hp
7-29. Friction

7-31. 144 in · lb
7-33. **(a)** 1339 in · lb, **(b)** 80.9%
7-35. **(a)** 92.4%, **(b)** 94.2%, **(c)** 87.0%,
(d) 57.1 hp
7-37. 82.4%
7-39. **(a)** 8.01 in³, **(b)** 1756 in · lb

Chapter 8

8-43. **(a)** 462 psi, **(b)** 769 psi
8-45. **(a)** 3.81 MPa, **(b)** 6.10 MPa
8-47. 29.2 hp
8-49. 22.4 kW
8-51. 713 gpm
8-53. Quicker to use but not as accurate.

8-55. Valve no. 1
8-57. 60.8 Lpm
8-59. **(a)** $v_p = 37.2$ in/s, **(b)** $v_p = 27.4$ in/s
8-61. **(a)** $v_p = 0.938$ m/s, **(b)** $v_p = 0.691$ m/s

Chapter 9

9-13. 785 lb, 9.81 in/s

9-15. (a) 9.63 in/s, 15,000 lb,
(b) 9.63 in/s, 15,000 lb

9-19. As cylinder 1 extends, cylinder 2 does not move.

9-21. Cylinder 1 extends, cylinder 2 extends. Cylinder 1 retracts, cylinder 2 retracts. Cycle repeats.

9-23. Both cylinder strokes would be synchronized.

9-25. 1000 psi

9-27. 1030 psi

9-29. 12.57 MPa

9-31. Unloading Valve (1480 kPa)
Pressure Relief Valve (10,860 kPa)

9-33. 47,900 lb

9-35. 2750 BTU/hr

9-37. 0.835 kW

9-39. 11,600 N

9-41. 4.20 hp, 10,700 BTU/hr

9-43. Upper position of DCV (2.96 in/s) Spring centered position of DCV (11.8 in/s) Lower position of DCV (3.95 in/s)

9-45. $1.15 \dfrac{\text{gpm}}{\sqrt{\text{psi}}}$

9-47. (a) $p_1 = 1600$ psi, $p_2 = 1560$ psi, $p_3 = 0$
(b) $p_1 = 1600$ psi, $p_2 = 3470$ psi, $p_3 = 0$

Chapter 10

10-27. 0.639-in ID

10-29. 22-mm ID

10-31. Square

10-33. $C_1 = 0.321$, $C_2 = 1$

10-35. Inlet: need larger size than maximum given in Figure 10-7, Outlet: 28-mm OD, 23-mm ID

10-37. 1680 psi

10-39. (a) 0.75-in OD, 0.049-in wall thickness, (b) 0.75-in OD, 0.049-in wall thickness

10-41. 40 MPa

10-43. 1.145 in

Chapter 11

11-27. 0.18 m³

11-29. 0.000333 m³/s, 210 bars

11-31. 68.7°C

11-33. 12.31 kW

11-35. 19,000 BTU

11-37. 48 kJ/min

Chapter 12

12-49. None

12-51. 96.5%

12-53. Identifies a particle size of 10 μm and a Beta ratio of 75

12-55. Per milliliter of fluid there are 640,000 particles of size greater than 5 μm and 5 particles of size greater than 15 μm.

Chapter 13

13-19. 193.7 psig

13-21. 36.6 psig

13-23. 19 bars gage

13-25. 2.45 bars gage

13-27. 71.1°C, 620°R, 344.1°K

13-29. 221 ft³, 176 ft³

13-31. (a) 6.22 m³, (b) 4.98 m³

13-33. 566 kPa abs

13-35. 80 psig

13-37. Valve is choked.

13-39. 0.203 std m³/min

13-41. 68.0 scfm

13-43. 1.11 std m³/min, 1.63 kW

13-45. (a) 0.0257 L, (b) 0.0257 L, (c) 22.5 mm, (d) 1.64 L, (e) 0.0257 L/s, (f) 0.811 std m³/min

Chapter 14

14-15. 12.0 psi
14-17. 97.6 kPa
14-19. 23.7 mm
14-21. $9600/yr
14-23. $6280/yr
14-25. **(a)** Nothing if fully extended. Extends and stops if fully retracted. **(b)** Cylinder extends and retracts continuously.

14-27. Insert a pilot check valve in the line connected to the blank end of cylinder 1.
14-31. **(a)** 126 lb, **(b)** 189 lb
14-33. **(a)** 179 N, **(b)** 264 N
14-35. 7.27 gal
14-37. 27.5 L
14-39. 19.3 MPa

Chapter 15

15-9. The cylinder extends, retracts, and stops.
15-11. **(a)** Cylinder 1 extends, cylinder 2 extends. **(b)** Cylinder 1 and 2 retract together.
15-13. Initially cylinder 1 is fully retracted and cylinder 2 is fully extended. Cylinder 1 fully extends while cylinder 2 fully retracts. End of cycle.
15-15. **(a)** Cylinder 1 extends, cylinder 2 extends. **(b)** If 2-PB is depressed while cylinder 1 is extending, cylinder 1 stops. If 2-PB is depressed while cylinder 2 is extending, cylinder 2 continues to extend.

Chapter 16

16-15. 3 variables produce $2^3 = 8$ possible combinations.

A	B	C	$A + B$	$B + C$	$(A + B) + C$	$A + (B + C)$
0	0	0	0	0	0	0
1	0	0	1	0	1	1
1	1	0	1	1	1	1
1	0	1	1	1	1	1
0	1	0	1	1	1	1
0	0	1	0	1	1	1
0	1	1	1	1	1	1
1	1	1	1	1	1	1

16-17.

A	B	C	$B + C$	$A \cdot (B + C)$	$A \cdot B$	$A \cdot C$	$(A \cdot B) + (A \cdot C)$
0	0	0	0	0	0	0	0
1	0	0	0	0	0	0	0
1	1	0	1	1	1	0	1
1	0	1	1	1	0	1	1
0	1	0	1	0	0	0	0
0	0	1	1	0	0	0	0
0	1	1	1	0	0	0	0
1	1	1	1	1	1	1	1

16-19. From DeMorgan's theorem we have

$$\overline{A \cdot B \cdot C} = \overline{A} + \overline{B} + \overline{C}$$

hence

$$A \cdot B \cdot C = \overline{\overline{(A \cdot B \cdot C)}}$$
$$= \overline{(\overline{A} + \overline{B} + \overline{C})}$$
$$= \text{NOT} \, (\overline{A} + \overline{B} + \overline{C})$$

16-21. When the cylinder is fully retracted, the signals from A_1 and A_2 are both ON. The extension stroke begins when push-button A is pressed, since the output $P \cdot A_1$ of the AND gate produces an output Q from the flip-flop. The push button can be released because the flip-flop maintains its Q output even though

$P \cdot A_1$ is OFF. When the cylinder is fully extended, A_2 is OFF, causing $\overline{A}_2$ to go ON switching the flip-flop to output $\overline{Q}$. This removes the signal to the DCV, which retracts the cylinder. The push button must be pressed again to produce another cycle. If the push button is held depressed, the cycle repeats continuously.

16-23. $P = A \cdot (A + B)$
$P = A \cdot A + A \cdot B$
$P = A + A \cdot B$
(using Theorem 6)
Thus, output P is ON when A is ON, or A and B are ON. Therefore, control signal B (applied to valve 3) is not needed.

Chapter 17

17-19. 0.283 in/V
17-21. 0.565 in/V
17-23. 0.286 cm

17-25.

Index

Absolute humidity, 482
Absolute pressure, 36–37, 459
Absolute temperature, 459–460
Absolute viscosity, 43–45
Absorbent filters, 435
Accumulators
   application, 390–394
   auxiliary power source, use as, 391
   bladder accumulator, 389–390
   classification, 384
   definition, 383–384
   diaphragm accumulator, 387–388
   emergency power source, as, 392–393
   gas-loaded accumulators, 385–390
   hydraulic shock absorber, 393–394
   hydropneumatic accumulators, 385–390
   leakage compensator, as, 392
   non-separator-type accumulator, 386
   piston accumulator, 386–387
   separator-type accumulator, 386
   sizing of, 527–530
   spring-loaded accumulator, 384–385
   water hammer, 393
   weight-loaded accumulator, 384
Actual torque, 179
Actuators, pneumatic
   air requirements, 500–502
   application, 457
   definition, 2, 455–456
   pneumatic cylinders, 495
   pneumatic rotary actuator, 496–497
   problems, 456
   rotary air motors, 497–499
Adsorbent filters. 435
Aerospace industry, application of
   fluid power in, 5–7

Aftercooler, 480–481
Air capacity ratings of compressors, 471–472
Air control valves
   check valve, 487–488
   flow control valves, 493–494
   four-way directional control valves, 491–492
   pressure regulator, 487–488
   shuttle valve, 488–489
   three-way directional control
      valves, 491–492
   two-way directional control valves, 490
Air coolers, 480–481
Air dryers, 481–482
Air filter, 474–475
Air lubricator, 476–478
Air-over-oil circuits, 337–338, 398–399
Air pressure regulator, 475–476
Air tank, 15, 18–19
Air-to-hydraulic pressure booster, 75–77
AIT. See Autogenous ignition temperature
American National Standards Institute
      (ANSI), 308
Ancillary hydraulic devices. See specific
      devices
AND function, 565
Angular motion, 63–65
ANSI. See American National Standards
      Institute
Associative law, 575
Atmospheric pressure, 35
Autogenous ignition temperature, 427
Automation Studio, 310, 614–620
Automatic cylinder reciprocating system, 322
Automotive shock absorber, 51–53
Aviation industry, application of
   fluid power in, 5–7

Axial piston motor (bent-axis design), 240–244

Axial piston pump (bent-axis design), 170–172

Axial (propeller) pumps, 153–154

Balanced vane pump, 167–168

Ball check valve, 131

Barometers, 36

Bernoulli's equation, 91–93

Beta ratios, 438–439

BHP. See Brake horsepower

Bladder accumulator, 389–390

Block diagrams, 585–586, 594–597

Boolean algebra

  associative law, 575

  commutative law, 575

  DeMorgan's theorem, 576

  distributive law, 575

  EXCLUSIVE-OR function, 577–578

  AND function, 572–573

  INCLUSIVE-OR function, 578

  laws of, 575–576

  MEMORY function, 579–580

  NAND function, 574–575

  NOR function, 574

  NOT function, 573

  OR function, 571–572

Bourdon gage, 413–414

Box-sorting system, 551–552

Boyle's Law, 460–461

BP. See burst pressure

Brake horsepower, 65

Brake power, 65

Bulk modulus, 42–43

Burst pressure, 357–358

Capillary tube viscometer, 47–48

Cartridge valves, 295–300

Cavitation, pump, 188–189, 444

Celsius comparisons, 40, 495–460

Centrifugal (impeller) pumps, 153–155

CF. See Coefficient of friction

CGS metric system. See Metric system, CGS

Charles' Law, 461–462

Check valve, 263–265

Circuit design and analysis

  air-over-oil circuit, 337–338

  American National Standards Institute

    specifications, 308

  analysis of hydraulic system with frictional

    losses, 338–342

  automatic cylinder reciprocating system, 322

  Automation Studio, 310

  closed loop system, 342–343

  computer analysis and simulation, 308–310

  considerations, 308

  counterbalance valve application, 320–321

  cylinder synchronizing circuits, 320–322

  double-acting hydraulic cylinder,

    control of, 311

  double-pump hydraulic system, 317–318

  extending velocity of cylinder with meter-in

    flow control valve, 328–330

  fail-safe circuits, 325–327

  hydraulic cylinder sequencing

    circuits, 320–322

  hydraulic motor braking system, 334–335

  hydrostatic transmission system, 334–337

  locked cylinder using pilot check

    valves, 322–323

  mechanical-hydraulic servo system, 342–343

  meter-in versus meter-out flow control valve

    systems, 330

  pump-unloading circuit, 316–317

  regenerative cylinder circuit, 312–314

  regenerative cylinder extending speed, 314

  single-acting hydraulic cylinder, control of, 310

  speed control of hydraulic cylinder, 328–330

  speed control of hydraulic motor, 333–334

Circuit simulation, 308–310, 617–618

Closed loop system, 292–293, 585–586, 596–597

Closed loop transfer function, 595

Coefficient of friction, 209, 428–429

Commutative law, 575

Compound pressure relief valves, 278–280

Compression packings, 402–403

Compressor, 18–19, 465–471

Computer simulation, 308–310, 617–618

Computer software, 308–310, 614–620

Conductors and fittings

  flexible hoses, 369–371

  fluid velocity, 355

  implications, 353–354

  metric steel tubing, 374–375

  pipe burst pressure, 357–358

  pipe tensile stress, 356–357

  pipe working pressure, 357–359

  plastic tubing, 368–369

  pressure ratings, 356–360

  quick disconnect couplings, 374

  selection process, 358

  sizing, 354–355

  steel pipes, 360–362

  steel tubing, 362–367

Conservation of energy. See Energy

Constant force or torque, 8–9

Contamination, evaluation of, 441–442

Continuity equation, 84–86

Continuous rotation hydraulic motor, 234–249

Coolers, use of, 408–410
Corrosion, 425–426
Counterbalance valves, 284–285
Cushion, 217–219

Darcy's equation, 123–124
DB. See Decibels
DCV. See Directional control valves
Deadband, 597–598
Decibels, 184–187
DeMorgan's theorem, 576
Density, 30–31
Density-specific weight relationship, 31
Diaphragm accumulator, 387–388
Directional control valves, 262–275
Dissolved air, 444–445
Distributive law, 575
Double-acting cylinder, 201–202
Double-pump hydraulic system, 317–318
Double-rod cylinder, 210–211
Durometer hardness tester, 407–408
Dynamic (nonpositive displacement) pumps,
    149, 153–155
Dynamic seals, 400

Efficiency, 66
Electrical controls
    application, 540
    box-sorting system, 551–552
    circuit diagrams, 543–544
    components, 544–546
    counting, timing, and reciprocation, 554–555
    devices, basic, 540–543
    dual-cylinder sequence circuits, 548–551
    electrical relays, 545–546
    limit switches, 541
    manually actuated push-button switches, 541
    normally closed switches, 540–541
    normally open switches, 540–541
    pressure switch, 541–542
    push-button switches, 541
    regenerative circuit, electrical
        control of, 551–553
    relays, 542
    solenoids, 542–543
    temperature switches, 543
    timers, 542
Electrical relays, 542
Electric motor, 16
Electrohydraulic servo system. See also Servo
    analysis, 594–601
    application, 585–587
    block diagrams, 585–586, 594–597

closed-loop transfer function, 595
components, 590–593
deadband, 597–598
electrohydraulic servo valve, 590–592
frequency and transient response, 595–596
hysteresis, 597–598
natural frequency of oil, 598
open-loop gain, 595
positional versus velocity systems, 601
repeatable error, 598–599
system accuracy, 598–600
tracking error, 600–601
transducers, 591–593
Electrohydraulic servo valves, 590–592
Electronic digital readout, 418
Elevation head, 93
Energy
    calculation of, 62, 82–84
    conservation of, 82–84
    definition, 62
    hydraulic system role, 59–61
    losses, 66
    sources, 59–60
    versatility, 61
Energy equation, 93
Engineer, fluid power, 20
Entrained gas, 442–443
Environmental rules and regulations,
    compliance with, 425, 449–450
Equivalent length method, 132–136
Error signal, 586
EXCLUSIVE-OR function, 577–578
External gear pumps, 156–160

Fahrenheit relationship, 40, 459–460
Fail-safe circuits, 325–327
Feedback transducer, 585–586, 591–593
Filters and strainers, 431–437
Fire point, 427
Fire resistance fluids, 428
First-class lever system, 212–213
Flash point, 427
Flexible hoses, 369–371
Flow control valves, 284–292
Flow meters, 415–418
Flow rate, 84–86
Fluid cleanliness levels, 439–442
Fluid conditioners
    aftercooler, 480–481
    air dryers, 481–482
    air filter, 474–475
    air lubricator, 476–477
    air pressure regulator, 475–476
    application, 457

650

Fluid conditioners (Continued)
  pneumatic pressure indicator, 478–479
  pneumatic silencer, 480
Fluidic devices, 562
Fluid logic control systems, 561–581
Fluid lubricating ability, 428–429
Fluid neutralization number, 429
Fluid power
  advantages, 8–10
  application, 11–15
  definition, 1–2
  disadvantages 11,
  engineer, fluid power, 20
  future, 20–21
  history, 4–7
  mechanics, fluid power, 20
  National Fluid Power Association, 20
  personnel, 20
  sales figures, annual, 20
  size and scope, 20
  technicians, fluid power, 20
Fluid transport systems, 2
Fluid velocity, 84–86, 119–120, 355
Foam-resistant fluids, 428
Force, 32
Force multiplication ratio, 67–71
Four-way valves, 267–270
Free air, 511
Frequency and transient response, 595–596
Frictional losses, 124–128

Gage, 36–37
Gas-loaded accumulators, 385–390
Gas problems
  dissolved air, 444–445
  entrained gas, 442–443
  free air, 442
Gate valve, 130
Gay-Lussac's Law, 463
Gear motors, 234–237
Gear pumps,
  external gear pumps, 156–160
  Gerotor pump, 161–162
  internal gear pump, 160–161
  lobe pump, 160–162
  pump slippage, 158
  screw pump, 162–164
  theoretical flow rate, 156–158, 172–173
  volumetric displacement, 156–157, 164–165
  volumetric efficiency, 158
General gas law, 464–465
Gerotor pump, 161–162
Global warming, 11–12, 14–16
Globe valve, 129–130

Hagen-Poiseuille equation, 124
Hand-operated hydraulic jack, 72–73
Harris formula, 512
Head, 33–34, 93
Head loss, 93
Heat exchangers, 408–413
HHP. See Hydraulic horsepower
High-water-content fluids, 427–428
Horsepower, 63–65
Humidity, 482
Hydraulic circuit analysis, 136–141, 338–342
Hydraulic cylinders
  application, 199–200
  barrel, 201
  cushion, 217–219
  cylinder extension force, 205–206
  cylinder extension velocity, 205–206
  cylinder retraction force, 205–206
  cylinder retraction velocity, 205–206
  designs, special, 210–211
  double-acting cylinder, 201–202
  double-rod cylinder, 210–211
  extension stroke, 201, 205
  first-class lever system, 212–213
  force, 205–206
  inertial force, 208–209
  mechanical linkages, 203
  mountings, 203–204
  operating features, 201–202
  output force, 205–206
  power, 205–207
  retraction stroke, 201, 205
  second-class lever system, 214
  single-acting design, 201
  telescopic cylinder, 211
  third-class lever system, 214–215
  universal mounting accessory, 204–205
  velocity, piston, 205–206
Hydraulic cylinder sequencing circuits, 320–322
Hydraulic fluid, 23–24
Hydraulic horsepower, 86–89
Hydraulic jack, 72–73
Hydraulic motor braking system, 334–335
Hydraulic motors
  application, 231
  axial piston motor (bent-axis design), 240–244
  continuous rotation hydraulic motor, 234–249
  efficiencies, 246–248
  gear motors, 234–237
  hydrostatic transmissions (See Hydrostatic transmissions)
  in-line-piston motor (swash plate design), 238–242
  limited rotation hydraulic motors, 231–235
  mechanical efficiency, 247

metric unit performance, 253–255
oscillating motor, 231–235
overall efficiency, 247–248
performance, 246
piston motors, 238–244
rotary actuator, 231–235
theoretical flow rate, 242–245
theoretical power, 242–245
theoretical torque, 233, 242–245
vane motors, 237–240
volumetric efficiency, 246
Hydraulic power, 86–90, 102
Hydraulic pump performance
actual torque, 178–180
comparison factors, 183–184
definition, 176–177
efficiencies, 177–180
mechanical efficiency, 178–179
metric unit ratings, 190–191
overall efficiency, 180
performance curves, 181–183
theoretical torque, 178–180
volumetric efficiency, 158
Hydraulic pumps
application, 149
axial (propeller) pumps, 153–154
centrifugal (impeller) pumps, 153–155
classification, 149
dynamic (nonpositive displacement)
pumps, 149
gear pumps, 156–160
piston pumps, 170–176
positive displacement pumps, 149
pumping theory, 151–153
shutoff head, 154–155
vane pumps, 164–168
variable displacement pumps, 149
Hydraulic robot, 5–16, 13–15
Hydraulic shock absorber, 220–223
Hydraulic system, 15–17
Hydroforming, 78–82
Hydropneumatic accumulators, 385–390
Hydrostatic transmissions, 250–252
Hysteresis, 597–598

INCLUSIVE-OR function, 578
Incompressibility, degree of, 42
Industrial Revolution, role of the, 4
Inertial force, 208–209
In-line piston motor (swash plate design),
238–242
In-line piston pump (swash plate design),
174–176
Input/output module (I/O), 605

Integrated hydraulics technology, 295–300
Internal gear pumps, 160–161
ISO standards, 439–441, 616

KE. See Kinetic energy
K factor, 129
Kinematic viscosity, 45–48
Kinetic energy, 84

Laminar flow, 119–120, 124
Laminar friction factor, 124
Laws of motion, 61
Le Systeme International d'Unites, 38
Limited rotation motors, 231–235
Limit switches, 541
Linear motion, 62–63
Linear power, 62–63
Linear velocity, 62
Lobe pumps, 160–162
Locked cylinder using pilot check valves,
322–323
Loss coefficient, 129
Losses in valves and fittings
common valves and fittings, 129–131
K factor, 129–131
loss coefficient, 129–131
pressure drop versus flow rate curves, 132

Maintenance
breakdown, common causes of, 423
contamination, evaluation of, 441–442
corrosion, 425–426
environmental considerations, 425, 449–450
filters and strainers, 431–437
fire-resistant fluids, 426–428, 430
fluid cleanliness levels, 439–442
fluid lubricating ability, 428–429
fluid neutralization number, 429
foam-resistant fluids, 428
gas problems (See Gas problems)
ISO standards, 439–441
maintaining and disposing of fluids, 430–431
oxidation, 425–426
preventive maintenance program,
importance of, 423–425
report and records system, importance of, 424
rust, 425
safety considerations, 425, 449
sampling and testing of fluid, 423
solid-particle contamination, 439–442
troubleshooting, 447–448
water content, determination of, 423

Manually actuated valves, 267–269
Mechanical efficiency, 178–179, 247
Mechanical filters, 433
Mechanical-hydraulic servo system, 342–343
Mechanical linkages, 203
Mechanically actuated valves, 270
Mechanical power, 106
Mechanical-type servo valves, 292–293
Mechanics
    angular motion, 63–65
    definition, 61–62
    horsepower, 63–65
    laws of motion, 61
    linear motion, 62–63
    linear velocity, 62
    power, 63–65
    torque horsepower, 64–65
    velocity, 62
MEMORY function, 579–580
Meter-in versus meter-out flow control
    systems, 330
Metric steel tubing, 374–375
Metric system, CGS, 38, 45
Metric system, SI, 38–41, 101–103
Moisture removal from air, analysis of, 482–485
Moody Diagram, 126–128
Motion, law of, 61
Moving-part logic control systems, 561–580
MPL. See Moving-part logic control systems
Multiplication of force (Pascal's Law), 67–71

NAND function, 574–575
National Fluid Power Association, 20
Natural frequency of oil, 598
Needle valves, 286–288
Newton's law of motion, 61
Non-pressure-compensated valves, 290–291
Non-separator-type accumulator, 386
NOR function, 574
Normally closed (NC) switches, 540–541
Normally open (NO) switches, 540–541
NOT function, 573

Occupation Safety and Health Administration
    (OSHA), 449
OHP. See Output horsepower
Open-loop gain, 595
OR function, 571–572
O-rings, 400–402
Oscillating motor, 231–235
OSHA. See Occupational Safety and Health
    Administration
Output horsepower, 87

Overall efficiency, 180–181, 247–248
Overheating of hydraulic fluid, 408–410
Oxidation, 425–426

Pa. See Pascals
Pascals, 39
Pascal's law, 67–71
Perfect gas laws
    Boyle's Law, 460–461
    Charles' Law, 461–462
    Gay-Lussac's Law, 463
    General gas law, 464
Petroleum-base versus fire-resistant fluids, 430
Pilot-actuated valves, 270–271
Pilot-operated check valve, 265
Pipe burst pressure, 357–358
Pipe roughness, effect of, 125–126
Pipe tensile stress, 356–357
Pipe working pressure, 357–359
Piping, 16–17, 360–362
Piston accumulator, 386–387
Piston compressor, 465–468
Piston cup packings, 403–405
Piston motors, 238–244
Piston pumps
    analysis, 172–173
    application, 184
    axial piston pump, 170–173
    in-line piston pump, 174–176
    radial piston pump, 174–177
    theoretical flow rate, 172–173
    volumetric displacement, 172–173
Plastic tubing, 368–369
Pneumatic circuits and applications
    air pressure loss in pipelines, 512–514
    analysis using metric units, 530–533
    application, 509–510
    circuits, 517–522
Pneumatic cylinders, 495–496
Pneumatic pressure indicator, 478–479
Pneumatic rotary actuator, 496–499
Pneumatics
    absolute pressure, 459
    absolute temperature, 459–460
    actuators (See Actuators)
    air, properties of, 457–459
    air capacity ratings of compressors, 471–472
    air control valves, 487–494
    air flow rate, 485–487
    air receiver size, 472–473
    application, 18–19
    Boyle's Law, 460–461
    Charles' Law, 461–462
    compressors, 465–471

definition, 2, 455–456
dew point, 482
flow rate of air through an orifice, 485–487
fluid conditioners, 474–478
Gay-Lussac's Law, 463
general gas law, 464
humidity, 482
moisture removal from air, analysis of, 482–485
motors, 497–499
power requirement, 473–474
pressure regulators, 475–476
saturation, 482
standard air, 458
Pneumatic silencer, 480
Pneumatic suction lift force, 522–525
Poise, 45
Poppet design, 264
Position versus velocity systems, 601
Positive displacement pumps, 149, 155–156
Potential energy, 83–84
Pour point, 51
Power, 86–90
Pressure, 32
Pressure-compensated valves, 290–292
Pressure-compensated vane pump, 166–167
Pressure control valves, 275–284
Pressure-force relationship, 32
Pressure gages, 413–415
Pressure head, 93
Pressure intensifiers, 394–397
Pressure ratings, 158, 184, 357–359
Pressure-reducing valves, 280–281
Pressure regulator, 475–476
Pressure switches, 541–542
Preventive maintenance, 423–425
Programmable logic controllers (PLCs)
    advantages, 602–603
    application, 587–590
    capabilities, 587–590
    central processing unit (CPU), 603
    electrohydraulic servo system, control of, 608–609
    hydraulic cylinder, control of, 605–608
    input/output module (I/O), 605
    programmer/monitor (PM), 604
    random access memory (RAM), 603
    read only memory (ROM), 603
Programmer/monitor (PM), 604
Pump
    cavitation, 188–189
    efficiency, 158, 177–180
    noise, 184–188
    selection, 189–190
    slippage, 158

volumetric efficiency, 178
Pumping theory, 151–153
Pump-unloading circuit, 316–317
Push-button switches, 541

Quick disconnect couplings, 374

Radial piston pump, 174–177
Random access memory (RAM), 603
Read Only Memory (ROM), 603
Regenerative cylinder circuit, 312–314
Relative roughness, 125–126
Relays, 542
Repeatable error, 598–599
Reservoirs, 381–383
Return bend, 131
Reynolds number, 120–122
Robots, 5–6, 13–15
Rotameter, 415–416
Rotary actuators, 231–235
Rotary air motors, 497–499
Rust, 425

Saturation, 482
Saybolt Universal Second (SUS), 46
Saybolt viscometer, 45–46
Schrader gage, 413–414
Screw compressor, 468–470
Screw pumps, 162–164
Sealing devices
    application, 400
    compression packings, 402–403
    durometer hardness tester, 407–408
    dynamic seals, 400
    O-rings, 400–402
    piston cup packings, 403–405
    piston rings, 404–406
    static seals, 400
    wiper rings, 406–407
Second-class lever system, 214
Separator-type accumulator, 386
Sequence valves, 283–284
Servo valves, 292–295, 590–592
SG. See Specific gravity
Shutoff head, 154–155
Shuttle valves, 275
Sight flow indicator, 415–416
SI metric system. See Metric system, SI
Simple pressure relief valves, 275–277
Single-acting cylinder, 73
Siphon, 99–101
Solenoid-actuated valves, 270–273

Solenoids, 270–272
Solid-particle contamination, 439–442
Sound intensity, 185–187
Specific gravity, 30
Specific weight, 28–29
Spring-loaded accumulator, 384–385
Standard air, 458
Standard atmospheric pressure, 35, 458
Steel pipes, 360–362
Steel tubing, 362–367
Stoke, 45
Straight synthetics, 427
Suction, 37
Suction lift force analysis, 524–527
Summer, 586
SUS. See Saybolt Universal Second
System accuracy (repeatable error), 598–600

Tank (reservoir), 15–16, 381–383
Technicians, fluid power, 20
Tee, 131
Telescopic cylinder, 211
Temperature comparisons, 40, 459–460
Temperature switches, 543
Tensile stress, 356–357
Theoretical flow rate, 158
Theoretical power, 179–180
Theoretical torque, 179
Thick-walled conductors, 359–360
Third-class lever system, 214–215
Three-way valves, 265–267
Timers, 542
Torque, 63–65
Torque horsepower, 65
Torricelli's theorem, 97–98
Tracking error, 600–601
Transducers, 591–593
Troubleshooting, 445–449
Tubing, 362–366
Turbine flowmeter, 417
Turbulent flow, 119–120, 125–128
Two-handed safety system, 326–327, 580–581

Unbalanced vane pumps, 167
Unloading valves, 281–282

Vacuum, 37
Vacuum systems, 522–527
Valves, hydraulic
    application, 262
    cartridge valves, 295–300
    classification, 262
    directional control valves, 262–275
    electrohydraulic proportional valves,
        292–295
    flow control valves, 284–292
    hydraulic fuses, 300, 302
    orifice flow rate, 284–286
    pressure control valves, 275–284
    proportional control valves, 295–297
    servo valves, 292–295
Vane compressor, 470–471
Vane motors, 237–240
Vane pumps
    balanced vane pump, 167–168
    design and operation, 164
    pressure-compensated vane pump, 166–167
    variable displacement pump, 165
    volumetric displacement, 164–165
Vapor pressure, 443–444
Variable displacement pumps, 165
Velocity, 62
Velocity head, 93
Venturi application, 94–95
VI. See Viscosity index
Viscosity, 42–48
Viscosity index, 48–51
Volumetric displacement, 156–157, 164–165,
    172–173
Volumetric efficiency, 158, 246

Water content, determination of, 423, 482–485
Water-glycol solutions, 427
Water hammer, 393
Water hydraulics, 450–452
Water-in-oil emulsions, 427
Waterjet cutting, 150–152
Weight-loaded accumulator, 384
Wind turbines, 14–16
Wiper rings, 406–407
Work, 62
Working pressure, 357–359